The Vegetation of Egypt

PLANT AND VEGETATION

Volume 2

Series Editor: M.J.A. Werger

For other titles published in this series, go to www.springer.com/series/7549

The Vegetation of Egypt

2nd edition

by

M.A. Zahran

Emeritus Professor of Plant Ecology,
Department of Botany,
Faculty of Science,
Mansoura University, Egypt

In association with

A.J. Willis†

Springer

Prof. M.A. Zahran
Mansoura University
Faculty of Sciences
Dept. Botany
Mansoura 35516
Egypt
Email, personal: zahrancabi2001@yahoo.com

Prof. A.J. Willis†
Emeritus Prof. of Plant Ecology,
Dept. of Animal & Plant Sciences,
The University of Sheffield,
S10 2TN, UK.

ISBN: 978-94-017-8144-2 ISBN: 978-1-4020-8756-1 (eBook)

Softcover re-print of the Hardcover 2nd edition 2009

Printed on acid-free paper

9 8 7 6 5 4 3 2 1

springer.com

- *To the Egyptian-British Scientific Cooperation*
- *To the Soul of the Late Professor A.J. Willis*
- *To Professor M. Kassas, Cairo University*
- *To my Colleagues and Students*
- *To my family: Ekbal, Ahmed, Amal & Eman*

(Prof. Dr. M.A. Zahran)

Contents

Foreword: First Edition

Egypt is a cross-road territory with its Mediterranean front connecting it with Europe with which it has had biotic exchanges during the Glacials and the Interglacials, and today we know that routes of migratory birds converge through Egypt. Two highway corridors join Egypt with tropical Africa and beyond: the Nile Valley and the basin of the Red Sea. The Sinai Peninsula is the bridge between Africa and Asia. Its cultural and ethnic history bears testimony to complexities of this position, as does its natural history. Attempts to unravel the mysteries of its cultural history have involved scholars from all over the world, and collections of its legendary heritage abound in museums of the capitals of the world. The natural history of Egypt was not less fortunate, contributions of international scientists to biological, geological and geographical surveys of Egypt include a wealth of research, and this book, compiled by two scholars from Mansoura and Sheffield, is a most welcome example of international collaboration.

The history of vegetation antedates that of human culture, but plant life as we see it today has been influenced in every way by human action, exploitation, destruction, husbandry, introductions, etc. An attempt to compile a comprehensive inventory of various aspects of plant growth and ecological relationships in plant communities requires indefatigable enthusiasm and stamina. The authors have both given of their time, energy and toil with infinite generosity, and achieved a formidable objective.

The plan of the book is set in a sequence that makes it readable and that facilitates access to detailed description of sample areas. Introductory parts are brief and the main space (Chapters 3–6) is devoted to addressing available information on plant life in the chief eco-geographic sections of the country: Western Desert, Eastern Desert, Sinai Peninsula and the Nile region. A final chapter refers to the history of the vegetation and to topics on which further investigation is required. With this structure the text will be most useful for students and for research workers interested in pursuing studies on the ecology and the geography of plant life in Egypt. It is hoped that it will interest school teachers and encourage them to take their pupils out to the nearby fields and adjoining deserts.

For me, it is a very special pleasure, having now completed 50 years of studying plant life in Egypt, to welcome this book and to congratulate Professor Zahran and Professor Willis for their remarkable achievement and to thank them for the unremitting effort that they have both invested in this worthwhile work.

Cairo M. Kassas
October 1989

Foreword: Second Edition

For the 2nd edition of the book, apart from updating the ecological information of the different vegetation types of the four main regions of Egypt (Western Desert, Eastern Desert, Sinai Peninsula, and River Nile), two more chapters have been added. Chapter 8 contains basic knowledge on the Remote Sensing Technology and its use in the vegetation mapping; three case studies from Egypt are described. Chapter 9 is concerned with the sustainable development of the Egyptian deserts using their renewable natural resources with particular references to the naturally growing xerophytes and halophytes. Selected species proved to have agro-industrial potentialities have been demonstrated. Both chapters contain valuable information necessary for the undergraduate, postgraduate students as well as for scientists interested in the vegetation of the arid-land areas.

Cairo
(April 2008)

M. Kassas

Preface

This book is an attempt to compile and integrate the information documented by many botanists, both Egyptians and others, about the vegetation of Egypt. The first treatise on the flora of Egypt, by Petrus Forsskal, was published in 1775. Records of the Egyptian flora made during the Napoleonic expedition to Egypt (1778–1801) were provided by A.R. Delile from 1809 to 1812 (Kassas, 1981).

The early beginning of ecological studies of the vegetation of Egypt extended to the mid-nineteenth century. Two traditions may be recognized. The first was general exploration and survey, for which one name is symbolic: Georges-Auguste Schweinfurth (1836–1925), a German scientist and explorer who lived in Egypt from 1863 to 1914. The second tradition was ecophysiological to explain the plant life in the dry desert. The work of G. Volkens (1887) remains a classic on xerophytism. These two traditions were maintained and expanded in further phases of ecological development associated with the establishment of the Egyptian University in 1925 (now the University of Cairo). The first professor of botany was the Swedish Gunnar Tackholm (1925–1929). He died young, and his wife Vivi Tackholm devoted her life to studying the flora of Egypt and gave leadership and inspiration to plant taxonomists and plant ecologists in Egypt for some 50 years. She died in 1978.

The second professor of botany in Egypt was F.W. Oliver (1929–1932) followed by the British ecologist F.J. Lewis (1935–1947). This episode marked the beginning of plant ecological studies by Egyptian scientists in two principal traditions: ecophysiological and synecological studies of the vegetation. The pioneers were A.M. Migahid, A.H. Montasir and M. Hassib who started their scientific work in 1931. About 1950, two schools of research emerged. These were mainly concerned with a survey of natural vegetation and the phytosociological analysis of plant communities. One was centred in the University of Alexandria led by T.M. Tadros (1910–1972) who followed the Zurich-Montpellier School. The second is centred in the University of Cairo and led by M. Kassas who followed the Anglo-American school of phytosociology. During the last 30 years, researches in plant ecology continue with refined methodologies and creation of new research units in the several provincial universities opened in Egypt.

We warmly thank Professor Dr M. Kassas, Faculty of Science, University of Cairo, for his great encouragement and assistance in the production of this book, and for supplying many references. We are also much indebted to Dr Sekina M. Ayyad for her help with the section on the history of the vegetation, to Dr P.D. Moore for his useful comments on this section, and to Professor L. Boulos, Dr T.A. Cope and Professor M.N. El-Hadidi for their kind assistance with nomenclature. The valued sponsorship the first edition of this book by UNEP and UNESCO is highly appreciated and has much facilitated its production. The valuable contributions of Prof. Dr. Boshra B. Salem, Alexandria University, Egypt and Dr. Gidska, L. Andersen, Bergen University, Norway, are the backbone of the new chapter (No. 8) of Remote Sensing. We are deeply thankful to them.

My thanks to Dr. H. Kashaba and my son Ahmed for their sincere help.

Egypt M.A. Zahran
U.K. A.J. Willis†

About the Authors

Professor Mahmoud Abdel Kawy Zahran was born in Samalut (Minya Province, Upper Egypt) on 15 January 1938. He graduated (BSc 1959) from the Faculty of Science, Cairo University where he got his MSc (1962) and PhD (1965) degrees in the field of plant ecology.

Professor Zahran worked as research assistant and researcher in the National Research Centre (1959–1963) and Desert Research Institute (1963–1972) of Cairo. In October 1972 he was appointed Assistant Professor in the Faculty of Science, Mansoura University and promoted to the professorship of plant ecology in November 1976. He joined the Faculty of Meteorology and Environmental Studies of King Abdul Aziz University, Jeddah, Saudi Arabia from November 1977 to March 1983.

For his scientific achievements in plant ecology, Professor Zahran received the State Prize of Egypt from the Academy of Scientific Researches and Technology (1983), the First Class Gold Medal of the Egyptian President (1983), the Diploma of the International Cultural Council of Mexico (1987) and the major Prize of Mansoura University in Basic Sciences (1991).

Apart from this book, Prof. Zahran is the author and co-author of 12 books, contributor of 12 books (13 chapters), compiler of one book and translator of one book. These books have been/to be published in Egypt, USA, UK, UAE, Netherlands, Germany, KSA, Belgium and Pakistan.

Emeritus Professor Arthur J. Willis, Ph.D., D.Sc, F.I. Biol, F.L.S., graduated in Botany at the University of Bristol, England, and joined the staff there in 1947 as Demonstrator. He subsequently became Junior Fellow in Physiological Ecology, Lecturer and Reader in Botany, but left Bristol in 1969 to become the Head of the Department of Botany of the University of Sheffield. Here he remained Head and also Honorary Director of the Natural Environment Research Council Unit of Comparative Plant Ecology until retirement in 1987.

Professor Willis is the author of An Introduction to Plant Ecology (1971) and a contributor to a number of books, most recently (1990) the last edition of the Weed Control Handbook: Principles. He has written about a hundred of papers in scientific journals, spanning the fields of plant ecology, the British flora, bryophytes, coastal systems, particularly sand dunes, plant physiology, especially nitrogen metabolism and water relations, and palaeobotany. He was a general editor of the extensive series of books titled *Contemporary Biology*, an editor of the *Journal of Ecology* and the *Biological Flora of the British Isles*.

Unfortunately, Prof. Willis died during summer 2006 leaving behind a wealth of knowledge in the fields of plant ecology, ecophysiology, etc…

Introduction

Six zones of vegetation have been recognized by phytogeographers on a global scale. Each zone is occupied by similar types of vegetation, with the same periods of growth and the same general adaptations to environment. The divisions are exclusively climatic and ecological; the systematic relations of the plants are not taken into consideration. These zones of vegetation are: the northern glacial zone, with a very short growth period (in the arctic and high altitudes); the northern zone of cold winters, with a growth period of 4–7 months; the northern zone of hot summers, comprising regions of the subtropics; the tropical zone, with no significant seasonal interruption of growth; and in the southern hemisphere the zone of the hot summers; and the cold zone. In the northern zone of hot summers there is no real winter, but there may be some interruption of growth in January. Xerophytism is well marked, although some regions are wet. Forest, maquis, chaparral, steppe and prairie are common in this zone. As indicated by Hassib (1951), the vegetation of Egypt belongs to this northern zone of hot summers.

According to Eig's system (1931–1932), Egypt comprises four floral provinces:

1. *Mediterranean Province*: This comprises the region around the Mediterranean Sea. It has mild winters with plentiful rain and dry summers. It is the region of evergreen maquis (except in Egypt) and forest associations. The northern Mediterranean coast of Egypt belongs here.
2. *North African-Indian Desert Province*: This is also known as the Saharo-Sindian Province. It encompasses the great desert from the Atlantic coast of Morocco to the deserts of Sind, Punjab and South Afghanistan. The air is extremely dry, temperatures are high, rainfall is low, salty ground is abundant, there are few species and individual plants and the vegetation is uniform. The greater part of Egypt belongs here.
3. *Central Asiatic Province*: This is also known as the Irano-Turanian Province. It comprises a large region stretching east towards China west to the Mediterranean, north to the Northern extratropical deserts and south to the North African-Indian deserts. There is little rain, rather long dry periods, great temperature differences, an almost complete absence of forest growth, and a rich occurrence of species and endemics. The mountain region of Sinai and certain enclave areas in the Eastern Desert, e.g. Galala mountains of Egypt, belong here.

4. *African Forest and Steppe Province*: This is also known as the Sudano-Deccanian Province. It comprises a belt of broad steppes and savannas from the Atlantic Ocean south of Sahara and north of the Equatorial Forest region, through Sonegambia to Eritrea and Ethiopia and through tropical Arabia and India, including the Deccan. There are tropical summer rains and dry and warm winters. The vegetation is dominated by, for example, tropical *Acacias* and the grasses *Panicum* and *Andropogon*. This is the region of steppes and savannas and the park forests which lose their leaves during the dry period. As an enclave the Gebel Elba mountainous region in the southeast of Egypt belongs here.

For its unique position midway between Africa and Asia, with its long coasts of both the Mediterranean Sea in the north (c. 970 km) and the Red Sea in the east (c. 1100 km), Egypt has attracted the attention of explorers and botanists for very many years. Hundreds of studies on the vegetation of Egypt have been published which when assembled together and integrated, as attemped here, would form a valuable scientific base for further studies.

In the numerous descriptions of vegetation and plant communities given in this book it has inevitably been necessary to rely heavily on accounts compiled by many authors. The majority of these accounts follow the Anglo-American School of phytosociology, referring to dominant and associated species, and characterizing communities by their dominants or co-dominants. Some accounts, however, of types of vegetation are in accordance with continental phytosociology and original descriptions are necessarily followed here.

In general, types of communities are distinguished mainly on the basis of features of the plants, including their structure, the floristic composition of the vegetation and its overall appearance (physiognomy). Characteristics of the habitat are, however, also taken into account, including, for example, the geomorphology.

Among important structural features of the vegetation are the number of layers which may be recognized: often a tree layer, shrub layer, subshrub layer or suffrutescent layer and a ground layer, but one or more of these may be lacking The layer containing the dominant, which usually constitutes the major part of the perennial plant cover, normally has the greatest effect on the physiognomy. The habit of the plant may also be distinctive e.g. succulents, grasses and woody species. Important characters of the habitat concern the nature of the substratum and geomorphological features such as the situation of the community or the stand (a visually fairly homogeneous unit of vegetation, often of a single species) in relation to drainage systems and the nature and depth of surface material or deposits forming the soil. The texture and depth of soil control the capacity for the storage of water; a shallow soil soon dries after the rainy season whereas a deep soil may provide a subsurface reserve of moisture.

This book is divided into nine chapters. The first presents Egypt as a part of the arid region of the world and describes how far the River Nile is important to its life and fertility.

In Chapter 2 the physiography of Egypt including its main geological and geographical characteristics, climatic features and soil-vegetation relationships are described.

The main subject of the book, the description of the vegetation types of Egypt, is covered in four chapters entitled the Western Desert, the Eastern Desert, the Sinai Peninsula and the Nile region.
The chapter on the Western Desert is in four parts:

1. The western section of the Mediterranean coastal land, i.e. the coast of the Western Desert;
2. The Inland Oases and Depressions;
3. Gebel Uweinat (Uweinat Mountain);
4. The Gelf Kebir.

There are two parts in the chapter on the Eastern Desert:

1. The Red Sea coastal land;
2. The inland desert.

The chapter on the Sinai Peninsula is in two parts:

1. The coastal belts:
    (a) The eastern section of the Mediterranean coastal land of Egypt;
    (b) The west coast of the Gulf of Aqaba and the east coast of the Gulf of Suez;
2. The inland desert and mountains.

The chapter on the Nile region is a short account of the plant life of the River Nile and its banks from Aswan northwards to its mouth in the Mediterranean Sea. It describes also the vegetation of the northern lakes and that of the middle (deltaic) section of the Mediterranean coastal land of Egypt.

In each of the four main chapters, before descriptions of the vegetation types, the local geomorphology, climate and habitat types of that particular region are described.

A concluding chapter provides an account on the history of the vegetation, indicates fields in plant ecology on which further research is needed and give a summary of the main types of vegetation in Egypt.

For the second edition of the book, apart from the above mentioned chapters of the 1st edition of the book, two more chapters are added. Chapter 8 entitled "Remote Sensing and Vegetation Map of Egypt" and Chapter 9 entitled: "Sustainable Development of Egypt's Deserts". Also, updating of the ecological information has been considered.

Chapter 1
Egypt: The Gift of the Nile

The land of Egypt occupies the northeastern part of the African continent. It is roughly quadrangular, extending about 1073 km from north to south and about 1229 km from east to west. Thus, the total area of Egypt is a little more than one million square kilometers (1019 600 km^2) occupying nearly 3% of the total area of Africa (Ball, 1939; Said, 1962; Abu Al-Izz, 1971). Egypt is bordered on the north by the Mediterranean Sea, on the south by the Republic of Sudan, on the west by the Republic of Libya and on the east by the Gulf of Aqaba and the Red Sea (Fig. 1.1).

Egypt extends over about 10 degrees of latitude, being bounded by Lat. 22 °N and 32 °N, i.e. lies mostly within the temperate zone, less than a quarter being south of the Tropic of Cancer. The whole country forms part of the great desert belt that stretches from the Atlantic across the whole of North Africa through Arabia.

Egypt is characterized by a hot and almost rainless climate. The average annual rainfall over the whole country is only about 10 mm. Even along the narrow northern strip of the Mediterranean coastal land where most of the rain occurs, the average annual rainfall is usually less than 200 mm and the amount decreases very rapidly inland (southwards). The scanty rainfall accounts for the fact that the greater part of Egypt is barren and desolate desert. Only through the River Nile is a regular and voluminous supply of water secured, coming from the highlands hundreds of kilometres to the south. This is channelled by artificial canals over the narrow strip of alluvial land on both sides of the river, the Fayium Depression and the delta expanse. These tracts of fertile land, covering less than 3% of the total area of Egypt, support a dense population. According to Said (1962), the average density of population in the agricultural lands of Egypt is more than 600 persons/km^2, whereas in the vast desert areas, which represent more than 97% of the total area, there is only one inhabitant/7 km^2.

The River Nile, therefore, is a salient geographical feature that has shaped not only the physical tracts of Egypt but also its history and the nature of its human settlements. Herodotus (484–425 BC) states that "Egypt is the Gift of the Nile". This is very true as the Nile gave Egypt, out of all regions of the great North African Sahara, a fertility that made possible not only the development of the famed ancient agricultural civilization, but also the growth of this civilization in peace and stability.

M.A. Zahran, A.J. Willis, *The Vegetation of Egypt*,

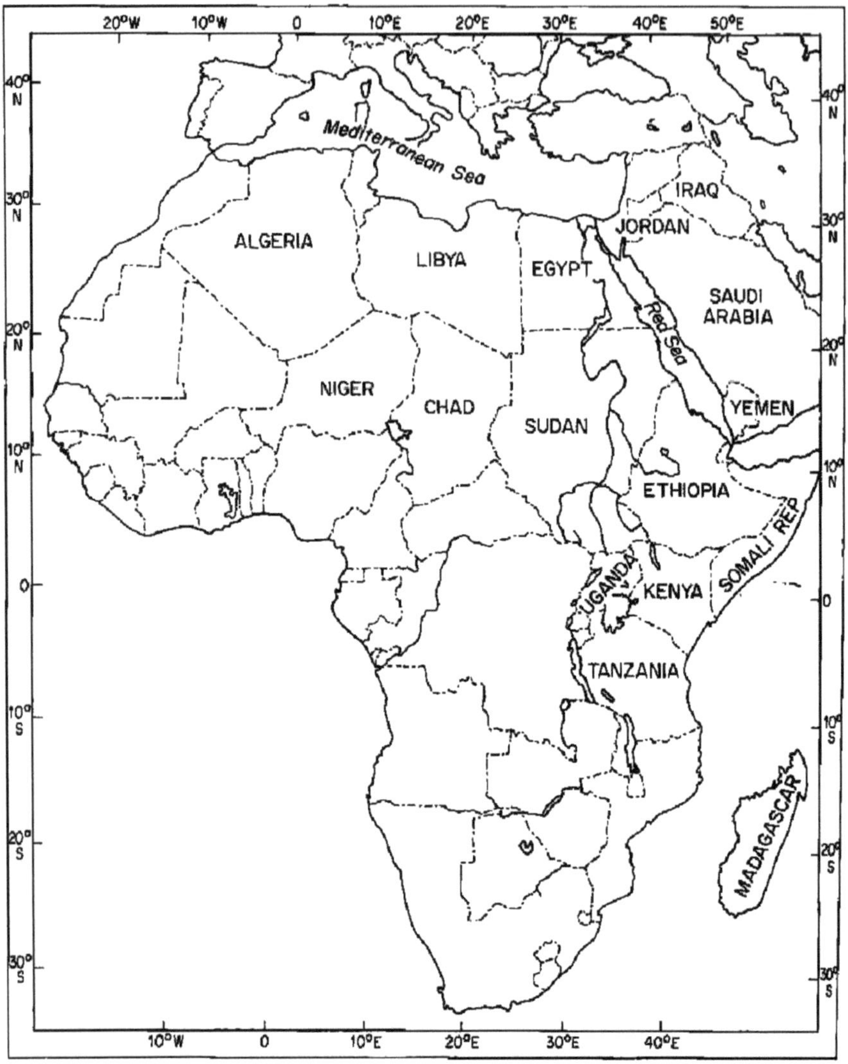

Fig. 1.1 The position of Egypt in Africa

Chapter 2
Physiography, Climate and Soil-Vegetation Relationships

2.1 Geological Characteristics

In early geological time, Egypt (as well as other countries of North Africa) was invaded on several occasions by the Sea of Tethys. This old geologic sea, probably of the Precambrian Era, is the antecedent of the Mediterranean Sea and has always encroached on Egypt from the north. This means that Egypt's past land-sea distribution has not always been the same as that of today. During late geological periods, the land of Egypt was uplifted, such uplift leading to a retreat of the Sea of Tethys. The retreating sea must have left behind sediments and remnants of the living organisms which it contained. Proof of this is the great quantity of sea shells spread over the surface of the Egyptian inland deserts, in places far from the sea and at elevations much higher than present sea level (Abu Al-lzz, 1971).

The oldest rocks in Egypt are Archaean, covering at present about 10% (about 93 000 km^2) of the area of Egypt (Fig. 2.1). They constitute the most rugged section of the country, including the highest peaks in the Red Sea mountains, mountains of south Sinai and mountains of Uweinat in the southwestern corner of the Western Desert. Archaean rocks also occur scattered along the Nile Valley, e.g. Kalabsha Gorge.

The formations of the Carboniferous period are found in three areas of Egypt: western Sinai, Wadi Araba in the North Eastern Desert and Uweinat mountain. These formations cover about 1200 km^2. Some Triassic deposits are present in a small area of northeastern Sinai and the Khashm El-Galala area along the Gulf of Suez. The Triassic deposits cover only some 50 km^2 of Egypt.

Jurassic formations are limited to small patches of total area about 450 km^2. These include Gebel El-Maghara and the El-Tih Plateau in Sinai as well as the Galala El-Bahariya Plateau along the Gulf of Suez.

The Cretaceous exposures cover about two-fifths of Egypt. They also extend under more recent formations in about half of the country. This means that Cretaceous

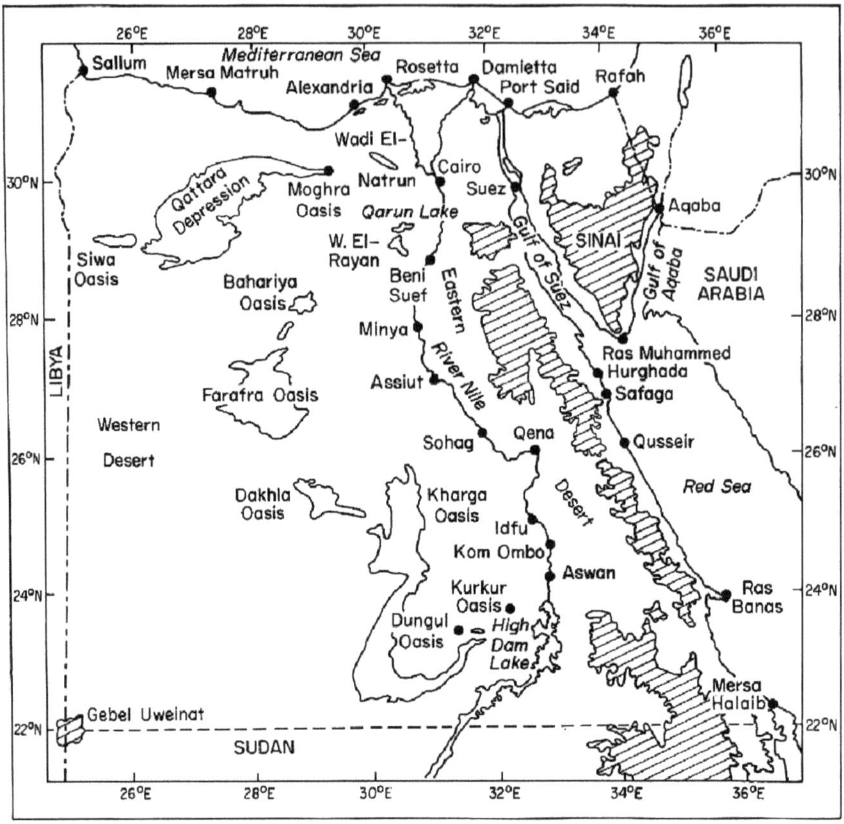

Fig. 2.1 The major topographical features of Egypt (mountainous areas cross-shaded)

rocks, whether exposed or buried, cover about nine-tenths of Egypt's area. The Cretaceous formations are of two main groups:

1. A lower group, of a massive layer of sandstone (Nubian), is about 500 m thick and accounts for about 29% of Egypt's area in the Western, Eastern and Sinai deserts.
2. The formations of the Upper Cretaceous group also have an average depth of 500 m but are formed of chalks, clay and mud plates. The chalk covers about 12% of the area of Egypt in the Western Desert and the El-Tih Plateau of Sinai.

The Eocene formations cover about 20% of the total area of Egypt distributed in the Eastern and Western Deserts between latitudes of Isna and Cairo having elevation of several hundreds of meters above the Nile Valley. In Sinai desert, the Eocene rocks cover vast areas including the central and northern parts of the Al-Ugma and Tih Plateaus, some high plateaus in the west (such as Al-Raha and Um Khushayb Plateaus) and some in the east (such as the Gebel Al Ayn amd Al-Qusayame Plateaus).

The Eocene strata has a thickness of 700 m and reach 1000 m in Sinai. The strata consists of limestone, marl and clay and contain in most parts marine fossil with round shapes resembling coins and called, therefore: "nummulities". Egyptian Eocene limestone are frequently termed "nummulitic limestone" because of the presence of these fossils.

The limestone in Egypt can be divided into three categories:

1. The lower group, belonging to the lower Eocene, is called "Thebes Limestone" which is widely distributed in: (a) southern Egypt particularly in the eastern and western fringes of the Farafra Oasis and (b) Sinai Desert.
2. The middle group which belongs to the middle Eocene and is known as the "Lower Muqattam". It consists primarily of limestone interbedded with layers of marl and mud shales. This group appears clearly in Minya Governorate (Upper Egypt) and in the Al-Ugma Plateau of Sinai but never seen south of Lat. 27°10'.
3. The upper group dating to Upper Eocene is identified as the Upper Muqattam. Its brownish color differentiates it from the lower layers which are bright white.

The Oligocene Formations cover about 1.5% of Egypt in isolated patches of the Western Desert between the Fayium and Bahariya Depressions, the Cairo-Suez desert road and Gulf of Suez. Miocene formations occupy less than one-eighth of Egypt's area, extending west of Cairo in a triangular shape with the apex to the northwest of Cairo and base along the Egyptian-Libyan border (Said, 1960). Pliocene formations occupy a limited area of not more than 700 km^2 (c. 0.68% of Egypt) and are present in several isolated spots in the Western Desert, the Gulf of Suez, Sinai and wadis of the Eastern Desert.

The Pleistocene and Holocene formations cover a large area amounting to 165000 km^2 (16% of the total). The Pleistocene deposits may be divided into three main classes:

1. Marine deposits exemplified by oolitic limestone along the Mediterranean coast south of Alexandria and along the shore-lines and coral reef of the Red Sea and the Gulf of Suez;
2. Fluvio-marine deposits, mostly in the deltas of old wadis at their points of entry to the Red Sea or the Mediterranean Sea;
3. Continental deposits accumulated by the agents of erosion; they may be aeolian (e.g. sand dunes) or lacustrine.

The Holocene deposits include both marine and fluvio-marine deposits.

2.2 Geographical Characteristics

Egypt comprises four main geographical units (Ball, 1939; Said, 1962):

2.2.1 The Nile Valley and the Delta;

2.2.2 The Western Desert; The Eastern Desert;

2.2.3 The Sinai Peninsula.

The Nile is more than 6650 km long from its source near Lake Tanganyika to its mouth in the Mediterranean Sea, but only about 1530 km lies within Egypt and in the whole of this part there is not one tributary. On entering Egypt from the Sudan a little north of Wadi Haifa (Fig. 2.1), the Nile flows more than 300 km in a narrow valley, with cliffs of sandstone and granite on both its eastern and western sides before reaching the First Cataract, about 7 km upstream of Aswan. North of Aswan the Nile Valley broadens and the flat strips of cultivated land, extending between the river and the cliffs bounding the valley on both sides, gradually increase in width northwards. The average width of the flat alluvial floor of the Nile Valley between Aswan and Cairo is about 10 km and that of the river itself about 0.75 km.

After passing Cairo, the Nile takes a north-westerly direction for some 20 km, then divides into two branches, each of which meanders separately through the delta to the sea: the western branch (c.239 km long) reaches the Mediterranean Sea at Rosetta and the eastern one (c.245 km long) at Damietta.

Closely connected with the River Nile is the Fayium Depression (c.1700 km^2) which lies a little to the west of the Nile Valley and to which it is connected by a narrow channel through the distant hills. The lower part of the depression (25 m below sea level, 200 km^2) is occupied by a shallow saline lake called Birket Qarun. The depression floor slopes downward to the lake in a northwesterly direction from about 23 m above sea level. It is a rich alluvial land irrigated by the Bahr Yusuf canal that enters it from the Nile.

The Western Desert stretches westward from the Nile Valley to the border of Libya with an area (exclusive of Fayium) of some 681 000 km^2, more than two-thirds of that of the whole of Egypt. Its surface is for the most part composed of bare rocky plateaux and high-lying stony and sandy plains with few distant drainage lines. True mountains are to be seen only in the extreme southwestern part where the highest peak of Gebel Uweinat is 1907 m. In northern and central parts of the Western Desert the plateau surface is broken at intervals by great depressions and oases.

The Eastern Desert (c. 223 000 km^2) extends eastwards from the Nile Valley to the Red Sea. It consists essentially of a great backbone of high mountains more or less parallel to the Red Sea. It is dissected by deeply incised valleys (wadis), some of which drain westward to the Nile and others eastward to the Red Sea.

The Sinai Peninsula (c.61 000 km^2) is separated from the Eastern Desert by the Gulf of Suez. It is a complex of high mountains intensely dissected by deep canyon-like wadis draining to the Gulf of Suez, the Gulf of Aqaba and to the Mediterranean Sea.

2.3 The Climate of Egypt

Although Egypt is an arid country, its climate was wet in geological times. The history of the climate in Egypt has been subject to many speculations based on inference from geomorphological and archaeological studies: see for example, Sandford (1934), Murray (1951) and Butzer (1959). Murray (1951) concludes that regular rainfall ceased over Egypt below the 500 m contour some time about the close of the Plio-Pleistocene period, three-quarters of a million years ago and, though torrents from the Red Sea Hills have been able to maintain their courses to the Nile through the foothills of the Eastern Desert, the Western Desert has ever since been exposed to erosion by wind alone. The earlier European glaciations seem to have left the Egyptian desert dry, but the long span of drought was broken by at least two rainy interludes; the first when the desert, both east and west of the Nile, were habitable in Middle Palaeolithic times, the second, with light rainfall, from about 8000–4000 BC. An occurrence of subsoil water near the surface in the southern part of the Western Desert permitted people to live there in oases till about 3000 BC when a drop of the water-table rendered these places uninhabitable.

The source of surface water all over the Eastern Desert is the rainfall on the chains of the Red Sea mountains. These, mountains seem to intercept some orographic rain from the continental northerlies which absorb their moisture through passage over the warm water of the Red Sea. The mountain rains may feed the wadis of the Eastern Desert with considerable torrential flows (Hassib, 1951).

According to Ayyad and Ghabbour (1986), Egypt can be divided into two hyperarid and two arid provinces as follows:

1. Hyperarid provinces
   (a) Hyperarid with a mild winter (mean temperature of the coldest month between 10° and 20 °C) and a very hot summer (mean temperature of the hottest month more than 30 °C), including the southwestern part of the Western Desert.
   (b) Hyperarid with a mild winter and a hot summer (mean temperature of the hottest month 20–30 °C) covering the Eastern Desert and the northeastern part of the Western Desert and Gebel Uweinat area.
2. Arid provinces
   (a) The northern section with winter rainfall which extends along the Mediterranean coast and the Gulf of Suez. This section is divided into two provinces by the UNESCO/FAO map of 1963: the coastal belt province under the maritime influence of the Mediterranean, with a shorter dry period (attenuated), and a more inland province with a longer dry period (accentuated) and an annual rainfall of 20–100 mm. Both provinces are characterized by a mild winter and a hot summer.
   (b) A southern section with winter rainfall which includes one province – the Gebel Elba area of the Red Sea coast of Egypt.

2.4 Soil-Vegetation Relationships

Soils of the hot arid regions are estimated (Dregne, 1976) to occupy 31.5% of the land area of the world (excluding the polar deserts). Africa has the largest area (17.7 million km^2) while Australia has the greatest percentage (82.1%) followed by Africa (59.2%), with Asia 33.0% and Europe only 6.6%.

In Egypt, the soils fall into two main categories or orders as recognized by the US Comprehensive System of soil classification (Dregne, 1976): Aridisols (essentially desert soils) and Entisols (alluvial soils and soils of sandy and stony deserts). Aridisols, which are confined to arid regions, are mineral soils distinguished by the presence of horizons showing accumulations, e.g. carbonates, soluble salts, in the profile typical of development in dry regions. These horizons have been formed under recent conditions of climate or those of earlier pluvial periods. The horizon at the surface (the epipedon) of this soil is light-coloured and there may be a salic (salty) horizon near to the surface or an argillic (clayey) horizon. These saline soils are well represented in the coastal plain of the Red Sea.

Most of the time when temperatures are favourable for plant growth aridisols are dry or salty, with consequent restrictions on growth. Entisols, the most common type in arid regions, are mineral soils with little or no development of horizons. This lack of pedogenic horizons is because the soils are young as a result of recent deposition of material, or the former surface has been lost by erosion. Saline soils of this type, with a water-table sufficiently near to the surface for salts to move upwards and be deposited at or near the surface at some time in the year, are represented in the lower Nile and in the Qattara Depression of the Western Desert (Dregne, 1976).

The general characteristics of the soils of arid regions and their relationship with climate and vegetation are described by many authors, e.g. Shreve (1942), Kassas (1953a), Kassas and Imam (1954), Chapman (1960), Zohary (1962), Zahran (1972, 1977), Ayyad and Ammar (1974), Dregne (1976), Ayyad (1981) and Younes et al. (1983). Such studies show that the soils of Egypt (and of other arid regions) have many features distinguishing them from their better-known counterparts of humid regions. Usually, soils of arid lands have a low level of organic matter, are slightly acid to alkaline in reaction (pH) at the surface, show an accumulation of calcium carbonate within the topmost 1.5 m (5 ft), have weak to moderate profile development, are of coarse to medium texture and have a low biological activity (Dregne, 1976). Frequently, and especially in upland areas, aridisols show a thin surface layer of stones, pebbles and gravels that constitutes a desert pavement, from which the fine particles have been lost by the action of wind or water. In some soils soluble salts may be present in sufficient quantities to influence the growth of plants significantly, particularly in poorly drained depressions, along coastal deserts and where there are appreciable amounts of gypsum. The lack of organic matter and the large particle size of some soils result in a low water-holding capacity, with also relatively low levels of micro-organisms.

Variability of ecosystem structure and function is generally a product of interactions between its different components. In the extreme environmental conditions of arid lands these interactions are of high significance, so that slight irregularities in one component of the ecosystem are likely to lead to substantial variations in others, so creating distinct microhabitats. In arid lands, the interrelationships between soils, vegetation and atmosphere are so interconnected that, in an ecological perspective, they can hardly be considered as separate entities.

Climatically induced processes of weathering, erosion and deposition are continuously at work, dissecting the desert landscape into a variety of landforms and fragmenting the physical environment into a complex mosaic of microenvironments. The impact of rainfall is unmistake-able. A decade or even a century may pass before a desert ecosystem experiences a heavy precipitation, but when rain does fall, it results in a great deal of erosion and deposition owing to the sparseness of vegetation which offers little or no protection to the soil. Major erosional forms now present in deserts generally result from fluvial action (Hills et al., 1966). Some, such as wadis and their affluents, are undoubtedly relict features derived from past periods of heavier rainfall, but many are attributed to occasional heavy rainfall at the present time. Because of the scarcity of rainfall, the high evaporation rate and the sparseness of vegetation in arid lands, salt accumulation close to the soil surface is a common phenomenon. This is obvious in the coastal belts affected mainly by maritime factors and in the inland depressions where water-table is very shallow or exposed.

The role played by vegetation in the development of desert soils varies with the degree of aridity. In extremely arid regions with very scanty vegetation, as in most areas of the Egyptian deserts, the role of vegetation is insignificant and soil development is essentially a geomechanical process where calcareous, siliceous and gypseous crusts or subsurface pans are formed. As rainfall increases, as in the northern Mediterranean coastal belt in Egypt, the vegetation becomes more dense and plants assume an important role in modifying edaphic conditions. Batanouny and Batanouny (1969) show that desert plants may also play an active role in stabilizing surface deposits. Some are capable of building mounds and hillocks which form suitable micro-habitats for certain annuals, whereas others are instrumental in arresting the movement of large dunes, rendering them less mobile and more suitable for colonization by other plants.

Kassas (1953a) described the relationships between the landforms and the plant cover in the Egyptian deserts. He attributed the importance of landform to two factors: first, its controlling influence on water resources; and second, the landform may make the area accessible for grazing and human interference or make it far from such destructive agencies. Water resources and human interference are among the most important factors controlling the plant life in the deserts.

Apart from the wadis, the Egyptian deserts are characterized by rock surfaces, erosion pavements, gravel deserts, slopes and cliffs. Each has its vegetation type. The rocky substratum of a desert plateau represents a habitat of extreme aridity and provides little opportunity for plant growth. Chasmophytes, which can send their roots into the rock crevices, may be present in this habitat, e.g. *Erodium glaucophyllum, Fagonia mollis, Helianthemum kahiricum, Iphiona mucronata, Reaumuria hirtella* and *Stachys aegyptiaca*. On the rocky surface rainfall produces shallow depressions, holes, or cavities where some water and perhaps some soil may collect during the rainy season. In these, an ephemeral plant cover may appear in the late winter and early spring.

Erosion pavement is an erosion surface overlain by a layer of soft rock waste and a surface of boulders. The surface of the erosion pavement may be flat or undulating. Run-off water collects in water-ways which form drainage systems each with a main channel and numerous affluents. The beds of these water-ways are covered with layers of soft material. In these water-ways plants find favourable conditions. There is a clear distinction between the vegetation of the affluents with shallow and limited water resources and that of the main channels. In the affluents vegetation is either of ephemerals, e.g. *Anastatica hierochuntica, Diplotaxis acris* and *Pteranthus dichotomus,* or of perennials which acquire a summer-deciduous growth-form, e.g. *Asteriscus graveolens, Farsetia aegyptia* and *Iphiona mucronata.* In the main channels, there are greater amounts of water and soil. The vegetation is evergreen and richer in both number of species and individual plants, e.g. *Zilla spinosa* and *Zygophyllum coccineum.*

In the gravel deserts, the surface deposits are mainly transported material (not waste material produced *in situ* as in the erosion pavement). The surface flint gravels are usually globose rather than angular. The deposits of the gravel desert are essentially siliceous whereas those of the erosion pavements are calcareous as their parent rock is limestone. Thus, chemical difference subjects the deposits of limestone origin to the surface accumulation of salts derived from the gypsum and rock-salt veins that fill the limestone joints. The sandy materials of the gravel desert are usually poor in salt content. The gravel cover (desert "armour") is barren except for the growth of lichens in certain localities. However, in the gaps between the gravels some plants appear, especially in the rainy season, e.g. *Aizoon canariense, Centaurea aegyptiaca, Fagonia glutinosa, Mesembryanthemum crystallinum* and *Polycarpaea repens.* The undulating surface of the gravel desert forms networks of furrows which guide the run-off water. These furrows are lined with water-borne silt and provide a favourable habitat for certain species, e.g. *Farsetia aegyptia, Heliotropium arbainense* and *Pancratium sickenbergeri* and many ephemerals.

The flat parts of the gravel deserts are subject to the deposition and accumulation of wind-borne materials which produce sand sheets where ephemeral plants grow. As the sheet becomes deeper more species find the habitat favourable and the vegetation acquires a more permanent appearance. The gradual building up of the surface sandy deposits coincides with the progressive modification of the plant cover. Among the common species of this type of habitat are *Hammada elegans, Panicum turgidum* and *Zilla spinosa.* Associate species include *Artemisia monosperma, Astragalus spinosus, Convolvulus lanatus, Lasiurus hirsutus, Moltkiopsis ciliata* and *Pituranthos tortuosus.*

The slopes, which are well represented on the plateau edges, wadi sides and mountain and hillsides, are usually covered with rock detritus of favourable texture. There are always little pockets among the surface fragments where some soil accumulates and where conditions permit the growth of plants. The effect of exposure is especially marked on the vegetation of the slopes. In contrast to the north-facing slopes, the south-facing ones are nearly always barren. Among the species characteristic of the slopes on the wadi sides of the Egyptian deserts are *Diplotaxis harra, Fagonia kahirina, Gymnocarpos decander, Limonium pruinosum, Reaumuria hirtella* and *Salsola volkensii.* On high mountains, where the slopes are gentle, the plant cover may show zonation in relation to altitude. The lower levels of the slopes receive more water and are less exposed than higher levels. This can be seen on the slopes of the mountains of the Red Sea coastal land and those of the Sinai Peninsula.

The cliffs represent an exceptionally dry habitat for the growth of plants; these are essentially chasmophytes inhabiting the rock joints. There is no possibility of surface accumulation of soil. The plants are usually confined to high levels. Some water soaks into the surface layers of the rock and through the crevices. The cliff-side habitat is a type inaccessible to grazing and human interference which is an advantage. The plant cover of these cliffs includes very characteristic species but these are only few. The most common species is *Capparis spinosa;* others include *Cocculus pendulus, Fagonia mollis, Iphiona mucronata, Limonium pruinosum* and *Zygophyllum coccineum*. The cliffs of steep waterfalls where rainfall water accumulates in deep pot-holes (about 4 m deep) support the growth of species such as *Ficus pseudosycomorus (F. palmata)*. On the sides of the pot-holes ferns, e.g *Adiantum capillus-veneris,* may occur.

Wadis represent one of the main ecosystems of the Egyptian deserts. A wadi has the great merit of being a drainage system collecting water from an extensive catchment area, so that the water supply in the immediate vicinity of a wadi is relatively higher than that of the slopes between which it runs.

As wadis contain vegetation richer than that of other types of desert habitats, and are accessible to bedouins and their domestic animals, they are subject to serious grazing. The most common species are the least grazed. The cutting and lumbering effect is specially marked on woody plants that are valuable for fuel. These destructive agencies deprive the soil of its plant cover and hinder the natural development of the habitat and its vegetation.

The soils of the wadi beds are usually composed of rock waste varying in texture from fine particles to gravel and boulders (Kassas and Imam, 1954). Wadi beds are often seen to be covered with layers of fine materials alternating with layers of coarse gravels. Alternation of layers of different texture has a substantial influence on the water available to plants.

The soil depth is by far the most important factor restricting the type of vegetation in the Egyptian desert wadis. A thin soil will be moistened during the rainy season but will be dried by the approach of the dry season, here ephemeral vegetation appears. A deep soil allows for the storage of some water in the subsoil. This will provide a continuous supply of moisture for the deeply seated roots of herbaceous perennials, undershrubs, shrubs and trees.

The plants of the sand drifts and dunes of the Egyptian deserts are of much ecological interest. These plants, psammophytes, when growing in the path of air currents, usually form mounds of accumulated wind-borne material around them but they may be overwhelmed by extensive deposition of sand. They are, however, often saved by their ability to produce adventitious roots on stems that are covered by sand, and new shoots replace the buried ones. By this ability, plant growth keeps ahead of the influx of sand. These plants are good sand collectors and binders, producing phytogenic mounds, hillocks or dunes. Such species include *Ammophila arenaria, Anabasis articulata, Atriplex farinosa, Cornulaca monacantha, Halopyrum mucronatum, Hammada elegans, Nitraria retusa, Panicum turgidum. Stipagrostis scoparia* and *Tamarix* spp.

In Egypt, the salt marshes are littoral and inland. The littoral salt marshes occur along the coasts of the Mediterranean Sea, Red Sea, Gulfs of Aqaba and Suez and also around the northern lakes: Mariut, Idku, Burullus, Manzala and Bardawil. The inland salt marshes, which are far from the reach of maritime influences, are represented by the sabkhas and playas of the oases and depressions of the inland deserts. Being lower in level than the surrounding territories, the inland salt marshes are characterized by a shallow underground water-table. In certain localities the water is exposed, forming lakes of brackish or saline nature.

The climatic conditions of Egypt have pronounced effects on the edaphic characteristics of the salt marshes. Aridity of climate leads to high rates of evaporation and as rainfall is low, particularly in the inland and Red Sea salt marshes, there is insufficient leaching to prevent the accumulation of salts in the form of surface crusts. Thus, the total amounts of soluble salts are generally high in the salt marshes of Egypt but the amounts are greater in the inland and Red Sea salt marshes than in the Mediterranean ones (Zahran, 1982a). The surface layers usually contain the highest proportion of soluble salts – up to 60.5% in a stand dominated by *Arthrocnemum macrostachyum* in the Red Sea coast, but the amount of soluble salts drops abruptly in the subsurface and bottom layers of this site to 2.9% and 2.7% respectively. In the swampy habitats, e.g. that dominated by *Typha elephantina* in the Wadi El-Natrun Depression, the amount of soluble salts in the mud is low (0.4%). The soil here is permanently under water.

The tidal mud of the mangrove vegetation of the Egyptian Red Sea coastal belt is usually grey or black and is foul smelling. Its total soluble salts range from 1.2 to 4.3%. However, there is a notable difference between the muddy substratum of *Avicennia marina* mangrove and that of *Rhizophora mucronata* mangrove. The soil of *A. marina* mangrove contains 4.5–19.5% calcium carbonate whereas that of *R. mucronata* is highly calcareous, containing up to 80% of its weight of calcium carbonate (Kassas and Zahran, 1967).

Chapter 3
The Western Desert

3.1 General Features

Until the beginning of the 20th century it was customary to refer to the whole North African desert as the Sahara (Sahra is the Arabic name for a desert), but recently it has become usual to divide the entire desert region of North Africa into Libyan (on the east) and Saharan (on the west) sections (Mitwally, 1953). The Libyan portion of the Sahara is now called the Western Desert of Egypt as it occurs west of the River Nile (Fig. 2.1). It extends from the Mediterranean coast in the north to the Egyptian-Sudanese border in the south (c.1073 km) and from the Nile Valley in the east to the Egyptian-Libyan border in the west (width ranges between 600 and 850 km), i.e. it covers about two-thirds of Egypt (c.681,000 km^2).

Except for the narrow Mediterranean coastal belt, which is the wettest region of Egypt, the whole Western Desert is one of the extremely arid parts of the world. Its very great aridity results from its distant position from seas, coupled with the absence of high altitudes which may attract orographic rain.

The drainage lines which define the courses of former streams of the occasional torrents, which follow the rainfall in certain desert regions, are almost entirely absent from the Western Desert. There are a few gullies draining into the Mediterranean Sea from the northern edge of the plateau and a few others along the eastern border draining into the Nile Valley but none extends far back into the rocky platform. The vast interior of this desert is flat and devoid of any sign of drainage lines belonging to a comparatively recent age.

In this respect the Western Desert contrasts with its neighbour, the Eastern Desert, where landscape is characterized by several wadis.

Another salient feature in the physiography of the Western Desert, resulting from arid conditions, is the uniformity of the surface. The interior of the plateau is flat; as far as the eye can see, there is nothing but plains or rocks either bare or covered with sand and detrital material. This surface is seldom broken by any conspicuous relief feature (Hume, 1925), except along the northern margins and the Nile Valley. The Western Desert thus appears as a huge rocky plateau of moderate altitude. The mean elevation is 500 m above sea level.

M.A. Zahran, A.J. Willis, *The Vegetation of Egypt*,

Another distinctive feature of the Western Desert is the nature and distribution of its water sources (Said, 1962). Along the narrow belt of the Mediterranean Sea, there are wells and cisterns fed by local rainfall. At the foot of Gebel Uweinat are springs fed by the occasional rains which fall on the mountain mass; but the land between is almost rainless. The oases of, for example, Siwa, Bahariya, Kharga and Dakhla are in great depressions where the ground water supplies can rise to the surface, but the vast intervening areas of high plateau are waterless.

Arid conditions in the Western Desert allow the free interplay of sand and wind. Sand driven by wind accumulates to build up sand dunes which become a dominant feature of the landscape. Most of this sand originated from the Miocene rocks forming the northern parts of the Western Desert (Abu Al-Izz, 1971). These sand dunes are always moving in the direction of the prevailing wind. They are of various forms and are scattered over large areas of the desert surface. As a result of their mobility the details of the desert landscape are constantly changing. Ball (1927) estimated that these dunes move at a rate of 10 m/year.

The Western Desert as a whole, though considered barren, supports certain types of plant which occur in areas with enough water resources (rainfall and/or underground).

Ecologically, the Western Desert comprises four main regions, namely:

1. The Western Mediterranean Coastal Belt
2. The Inland Oases and Depressions
3. The Gebel Uweinat
4. The Gelf Kebir

3.2 The Western Mediterranean Coastal Belt

3.2.1 Physiography

The Mediterranean coastal land of Egypt (the northern coast) extends from Sallum eastward to Rafah for about 970 km. It is the narrow, less arid belt of Egypt which is divided, ecologically, into three sections (Zahran et al., 1985a, 1990): western, middle and eastern. The western section (Mariut coast) extends from Sallum to Abu Qir for about 550 km, the middle section (Deltaic coast) runs from Abu Qir to Port Said for about 180 km and the eastern section (Sinai Northern coast) stretches from Port Said to Rafah for about 240 km (Fig. 2.1). The western section is the northern coast of the Western Desert. It is a thin belt of land parallel to the Mediterranean Sea that narrows or widens according to the position of its southern boundary – the Western Desert Plateau. Its average north-south width, from sea landward, is about 20 km and it is bordered by Lake Mariut on the east (Fig. 3.1).

The most remarkable feature of the Mediterranean coast west of Alexandria is the prevalence of ridges of soft oolitic limestone, often 20 m or more high, extending parallel to the shore for long distances (Ball, 1939). Commonly one line of ridges skirts the coast closely, while another runs parallel with it a few kilometres inland, and there is sometimes a third ridge between the second and the edge of the Western Desert Plateau.

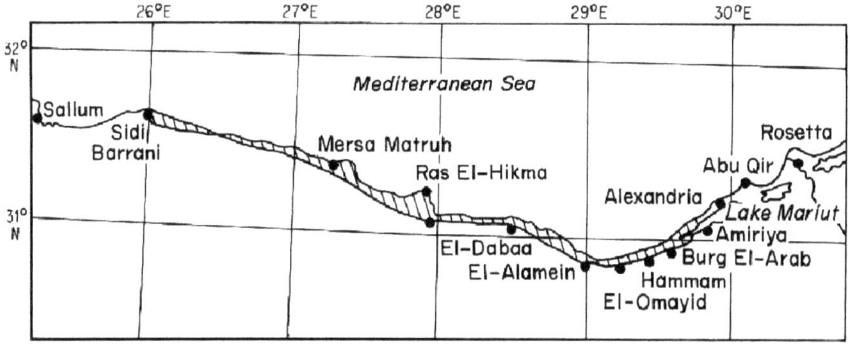

Fig. 3.1 The Western Mediterranean coastal belt of Egypt

In some places between the coastal ridge and the one next inland are salt-lagoons and marshes and in others a tract of loamy ground. A second strip of loamy ground usually separates the second ridge from the third or from the Western Desert Plateau.

In the western province of the Mariut coast, the plain is narrow or lacking. The southern tableland extends southwards to the Qattara Depression. It increases gradually in level westward and attains a maximum elevation of 200 m above sea level at Sallum, sloping gently northwards. Eastward it decreases gradually in level until it loses its line of demarcation with the coastal plain (Ayyad and Hilmy, 1974).

3.2.2 Climate

The Mediterranean coastal land of Egypt, in general, belongs to the dry arid climatic zone of Koppen's (1931) classification system (as quoted by Trewartha, 1954), and the Mediterranean bioclimatic zone of Emberger (1955). The bioclimatic map of UNESCO/FAO (1963) indicates that it is of a subdesertic warm climate. Climate data for five meteorological stations along the Western Mediterranean coast of Egypt, namely: Sallum, Sidi Barrani, Mersa Matruh, El-Dabaa and Alexandria, are given in Table 3.1 (see also Fig. 3.1). The annual mean maximum temperatures range between 25.3 °C and 23.8'C and the annual mean minimum between 13.3 °C and 15.1 °C. The mean relative humidities are: 67–74%, 60–70%, 59–71% and 59–67% in summer, autumn, winter and spring respectively (Ayyad and Hilmy, 1974). The mean annual evaporation ranges between 5.2 and 8.7 mm Piche/day with a maximum of 11.1 mm Piche/day in Sallum and a minimum of 3.8 mm Piche/day in Alexandria. Rainfall occurs mainly during the October–March period (60% or more); summer is virtually dry. The maximum amount falls during either January or December and varies appreciably between the different stations. Rainfall of torrential nature may be expected—"values [in one day] up to 120.8 mm were recorded" (Shaltout, 1983).

Dew in arid and semi-arid regions is a valuable source of moisture to plants. It has been repeatedly observed that some perennials, especially on sand dunes, produce ephemeral rootlets during the dry season which may absorb dew as it moistens the

Table 3.1 Climatic data of five meteorological stations along the Western Mediterranean coast of Egypt (*Climatic Normals of Egypt,* 1960). AMMx = annual mean maximum, AMMn = annual mean minimum, HAbs Mx = highest absolute maximum, LAbs Mn = lowest absolute minimum, MA = mean annual, Mx = maximum, Mn = minimum, M = mean, TA = total annual

Stations	*Temperature (°C)*				*Relative Humidity (%)*			*Evaporation (mm Piche/day)*			*Rainfall (mm)*
	AMMx	*AMMn*	*HAbs Mx*	*LAbs Mn*	*MA (6 a.m.)*	*MA (12 noon)*	*MA (6 p.m.)*	*MA*	*Mx*	*Mn*	*TA*
Sallum	25.3	13.3	44.1	0.0	65	50	65	8.7	11.1	7.2	119.7
Sidi Barrani	23.8	15.1	42.6 (June)	2.6 (Feb.)	67	51	66	6.5	7.4 (Apr.)	5.2 (Dec.)	138.5
Mersa Matruh	24.3	14.3	43.2 (June)	2.6 (Feb.)	67	51	67	8.3	9.7 (Sep.)	6.5 (Dec.)	144.0
El-Dabaa	24.4	14.4	43.8 (June)	0.0 (Jan.)	69	54	70	6.0	7.0 (Sep.)	3.9 (Dec.)	142.6
Alexandria	24.9	14.9	42.1 (June)	24.0 (Jan.)	70	55	72	5.2	5.8 (May)	3.8 (Dec.)	192.1

surface layer of the soil (Kassas, 1955). Roots of xerophytes, characteristic of dry conditions, typically have two features not seen in roots of other plants: the ability of old roots, although lignified and corky, to form young rootlets with great rapidity in moist conditions, and root hairs (active in water absorption) are not restricted to a narrow zone behind the growing point (Evenari et al., 1971). The gain in moisture content due to water vapour condensation on sand dunes of of the Western Coast was estimated by Migahid and Ayyad (1959) as ranging between 2.4 and 4.7% at Ras El-Hikma and a total amount of dewfall of 11.5 mm was recorded during 1955. At Burg El-Arab (50 km west of Alexandria) Abdel Rahman et al. (1965a) recorded gains in soil moisture content from water vapour condensation of between 0.47 and 1.4%.

Winds in the Western Mediterranean coast of Egypt are generally strong, and violent dust storms and pillars are not rare. Dry hot dust-laden winds from the south known as Khamsin blow occasionally for about 50 days during spring and early summer. During winter and early spring winds blow strongly with an average velocity of about 20–23 km/h. Wind speed decreases in May and June, but July is windy. The end of summer records many calm days and the average wind speed drops to 15 km/h (Shaltout, 1983). The mean annual potential evapotranspiration, as estimated by Thornthwaite's formula is about 995 mm in Burg El-Arab (Ayyad, 1973).

According to Murray (1951) the climate of the Western Mediterranean coast of Egypt has not changed since Roman times (200C years ago). Kassas (1972a) reports that Sutton (1947) quotes records of annual rainfall made by Thurnburn (1847–1849) and brought up to 1970 as follows: 1847–1849 = 191 mm, 1881–1886 = 209 mm 1901–1906 = 217 mm, 1921–1926 = 178 mm, 1939–1941 = 161 mm 1951–1956 = 187 mm, 1966–1970 = 207 mm.

3.2.3 Land Use

The Western Mediterranean coastal land of Egypt is called the Mareoti District, being related to Mariut Lake (Fig. 3.1). In the past this lake was a fresh-water one. Kassas (1972a) states "Strabo (66–24 BC) records that Lake Marea is filled by many canals from the Nile through which a greater quantity of merchandise is imported". De Cosson (1935) notes that the lake was rather deep fresh water and adds: "There seems to be little doubt that 2000 years ago it was of greater extent than in modern times. The Canopic Nile Branch and the other canals that fed the lake gradually silted and its water receded. Thus, Lake Mariut was in Graeco-Roman times a fresh-water lake, the water of which was used for irrigating the fields. This source of freshwater gradually diminished and by the end of the twelfth century the lake became saline."

Kassas (1970) infers that, in the Western Mediterranean coast c Egypt, agriculture and horticulture have become established under a resident population of cultivators. The farms depended partly on irrigation from an ancient branch of the Nile (the Canopic) that extended for some distance west of the present site of Mariut, but the location of farms far beyond the reach of this branch indicates that effective methods of dryland farming were used. According to Kasss (1972a), the Mareotis district was an area of prosperous cultivation particularly vineyards, and was well inhabited. Good wine was produced in such quantity that Mareotis wine was racked in order that it might be kept to be old. By the tenth century, the district gradually declined and the vineyards were replaced by desert. It is unlikely that there have been major climatic changes during the last 2000 years that could have caused the deterioration of the area. Also, there is evidence that the fresh water of Lake Mariut and its arm that extended westward for 79 km was used for irrigating farms and orchards fringing shores of the lakes and banks of its western arm. These strips of irrigated agriculture must have been limited in extent because of the topography.

Earlier this century some attention was given to the Mareotis region. The extension of a railway westward of Alexandria to Mersa Matruh, and the plantation of vine, olive and date palm at Ikingi (20–25 km west of Alexandria) were "early steps towards regeneration" (De Cosson, 1935). Several attempts have been made to reintroduce a variety of orchard crops in Mareotis: vine *(Vitis vinifera),* fig *(Ficus carica),* date palm *(Phoenix dactylifera),* olives *(Olea europaea),* carob *(Ceratonia siliqua),* almond *(Prunus amygdalus)* and pistachio *(Pistacia vera)* (Kassas, 1972b).

At present the main land uses of Mareotis are grazing and rain-fed farming (or irrigated by underground and run-off water). The main annual crop is barley *(Hordeum vulgare).* Figs are successful on calcareous coastal dunes and olives, almonds and pistachio in inland alluvial depressions. Irrigated agriculture of pasture and grain crops and fruit trees (mainly vine) is spreading after the extension of irrigation canals from the Nile up to 60 km west of Alexandria (Ayyad, 1983).

3.2.4 Plant Cover

(a) Floristic Analysis

The Western Mediterranean coastal belt is by far the richest part of Egypt in its floristic composition owing to its relatively high rainfall. The number of species in this belt makes up about 50% of the total of the Egyptian flora which is estimated to be about 2000 (Oliver, 1938) about 2080 species (Täckholm, 1974), 2094 species by Boulos (1995). However, Boulos (1999, 2000, 2002, 2005) recorded a total of 2125 species of which 50 species are cultivated. Most of these species are therophytes that flourish during the rainy season, giving the coastal belt a temporary showy grassland desert. During the longer dry period, only the characteristic woody shrubs and perennial herbs are evident; these constitute the scrub vegetation of the area, scattered sparsely in parts and grouped in denser more distinct patches in others (Tadros, 1956).

Hassib (1951) describes the percentage distribution of both annual and perennial species among the life-forms in this coastal belt as follows. Neither mega- and mesophanerophytes nor epiphytes are represented. The micro- and nanophanerophytes are represented by 3.2%, stem succulents by 0.1%, chamaephytes by 9.2%, hemicryptophytes by 11.7%, geophytes by 11.9%, hydrophytes and helophytes by 4.0%, therophytes by 58.7% and parasites by 1.1%. Maquis vegetation that characterizes the other Mediterranean countries is not represented in Egypt. The prevailing life-form of the perennials is chamaephytes; nanophanerophytes are less abundant.

The floristic elements of the Western Mediterranean coastal belt enjoy better climatic conditions than those of the other parts of Egypt. There are more species and great numbers of individual plants and the vegetation is more or less continuous, not like that in the inland desert areas where the plant communities are separated by large stretches of barren ground. In the autumn numerous geophytes make an attractive show of flowers and in late spring grasses and members of the Leguminosae, Compositae and Cruciferae are particularly abundant.

Xerophytes make up about 90% of the total number of species in this coastal belt; most are therophytes (67%), followed by geophytes (11%), halophytes and helophytes (11%), chamaephytes (6.6%), micro-and nanophanerophytes (3%), parasites (1.2%) and stem succulents (0.1%). The common xerophytes include: *Achillea santolina, Ammophila arenaria, Anabasis articulata, Euphorbia paralias, Gymnocarpos decander, Hammada scoparia, Helianthemum lippii, Lygos raetam, Ononis vaginalis, Pancratium maritimum, Plantago albicans, Thymelaea hirsuta* and *Thymus capitatus*.

The halophytes (submerged including macroscopic algae and terrestrial) include about 45 species. Algae are well developed in the rock coastal areas but apparently absent from the loose soil. The submerged phanerophytes include: *Cymodocea major, Posidonia oceanica* and *Zostera noltei*. Terrestrial halophytes include *Arthrocnemum macrostachyum* (*A. glaucum*, Meikle, 1985), *Atriplex* spp., *Halimione portulacoides, Halocnemum strobilaceum, Inula crithmoides, Juncus acutus, J. rigidus, Limoniastrum monopetalum, Nitraria retusa, Saiicornia fruticosa, Suaeda fruticosa, S. pruinosa, Tamarix nilotica* and *Zygophyllum album*.

The helophytes and fresh-water hydrophytes represent about 4% of the total number of the flora of this coastal belt. They include: submerged species, e.g. *Ceratophyllum demersum* and *Potamogeton crispus*, floating species, e.g. *Eichhornia crassipes* and *Lemna* spp., reeds, e.g. *Phragmites australis* and *Typha domingensis*, and sedges, e.g. *Cyperus* spp. and *Scirpus* spp.

(b) The Vegetation of Mariut Lake

Mariut Lake is a closed lake covering about 23,960 feddans (feddan = 4200 m^2). The northern coast of the lake is 9 km long and the southern coast 13 km. It is broadest in the middle, and has no bays or bogs. A western arm of the lake extends to the south, along a hollow between the Al-Maks-Abu Sir range in the north and the Mariut range. The width of the 35 km long arm is from 2 to 5 km. This is the depression of Mallahet Mariut which has become a group of shallow saline lagoons. The water level of these lagoons is high in winter and low in summer, at which time a layer of white salt is present. The western end of Mallahet Mariut is no longer covered by water: some halophytic shrubs and grasses grow in it. The central section of the arm is nearly always dry, but covered with a layer of salt. The lower eastern end is always covered with salt water (Abu Al-Izz, 1971).

Lake Mariut was fed by the Canopic branch of the Nile, but in the 12th century that branch was filled with silt and the connection of the lake with the Nile was thus cut. Thereafter Lake Mariut formed a number of insignificant stagnant pools whose level was related only to local winter rain. The lake has been fed since 1892 by drainage canals.

The vegetation associated with Mariut Lake comprises communities of both aquatic and terrestrial habitats (Tadros and Atta, 1958a). In the aquatic habitat *Phragmites australis* grows luxuriantly and densely in the shallow water (30–50 cm depth). Inwards, in deeper water, an almost pure population of *Eichhornia crassipes* is present and in still deeper parts there are submerged communities of *Potamogeton pectinatus* associated with *Ceratophyllum demersum* and *Lemna gibba*.

Towards the shore of the lake, the soil is saline and halophytic vegetation prevails. The vegetation of this terrestrial habitat can be distinguished into distinct zones. In the submerged soil is a community dominated by *Scirpus tuberosus* associated with *S. litoralis* and *Typha domingensis. T. domingensis* dominates a zone close to that of *Phragmites australis* and passes gradually into a *S. tuberosus* community which merges, as the level of the ground increases so that it become less liable to flooding, into a community dominated by either *Salicornia herbacea* or by *Juncus rigidus. S. herbacea* gradually diminishes and is replaced by *Salicornia fruticosa* which passes gradually to a typical *Salicornia fruticosa-Limoniastrum monopetalum* zone. The *Juncus rigidus* community, on the other hand, is replaced by a community co-dominated by *Salicornia fruticosa-Suaeda salsa* which passes gradually to a typical *S. fruticosa-Limoniastrum monopetalum* type. In both situations the ground becomes very dry and saline and a *Halocnemum strobilaceum* community replaces that of *Salicornia-Limoniastrum*. On the elevated border of the dry saline beds of the western extension of Lake Mariut is a community dominated by *Salsola tetrandra* associated with *Atriplex halimus, Frankenia revolute, Limoniastrum monopetalum, Limonium pruinosum* and *Sphenopus divaricatus*. In the less saline stands of this community *Pituranthos tortuosus, Thymelaea hirsuta, Trigonella maritima* and other non-halophytic species may grow. This community has also certain affinities with the non-halophytic communities. The *Salsola tetrandra* zone gradually gives way to a community whose principal constituents are *Limoniastrum monopetalum* and *Lycium europaeum*. Associate species include *Asphodelus microcarpus, Bassia muricata, Carthamus glaucus, Cutandia dichotoma, Echinops spinosissimus, Filago spathulata, Helianthemum lippii, Ifloga spicata, Launaea nudicaulis, Noaea mucronata, Picris radicata, Plantago albicans, Reaumuria hirtella, Salvia lanigera* and *Suaeda pruinosa*.

(c) The Vegetation of the Coastal Land

The vegetation of the western coastal land associated with the Mediterranean Sea of Egypt may be considered as follows:

i. Main habitats and their communities;
ii. Vegetation of Ras El-Hikma coastal area;

iii. Vegetation of Sallum coastal area
iv. Vegetation of a transect along the Western Desert.

(i) Main Habitats

In spite of the relative simplicity of the relief and the apparent uniformity of the climate, the plant habitats in the region present some diversity. For the casual observer, however, the physiognomy of the vegetation seems monotonous over large tracts of land, owing to the prevailing life-form of the perennial plants, being mostly chamaephytes and to a less extent nanophanerophytes with scattered distribution. The only variation in the physiognomy is the change from the short vernal (spring) aspect of the vegetation to the longer aestival (summer) aspect (Tadros, 1956).

The distribution of plant communities in the Western Mediterranean coastal land is controlled by topography, the origin and nature of the parent material and the degree of degradation influenced by human manipulation (Ayyad and El-Ghareeb, 1984). Generally, the vegetation of this coastal belt belongs to the *Thymelaeion hirsutae* alliance with two associations:

1. *Thymelaea hirsuta, Noaea mucronata* association with two variants dominated by *Achillea santolina* and
2. *Anabasis articulata, Suaeda pruinosa* association (El-Ghonemy and Tadros, 1970).

The local distribution of communities in different habitats is linked primarily to physiographic variations. According to these variations two main sets of habitats may be distinguished – one on ridges and plateaux and the other in depressions. Ridge and plateau habitats may be further differentiated into two main types. The coastal ridge is composed mainly of snow-white oolitic (calcareous) sand grains and is overlain by dunes in most places and inland are less calcareous ridges and the southern tableland. The southern tableland is characterized by the dissection of the landscape into an extensive system of wadis which drain into the Mediterranean Sea and form a distinct type of habitat. Inland siliceous dunes are sporadically distributed on the southern tableland and support a community different from that of calcareous dunes on the coastal ridge. Habitats of depressions differ according to the relative proximity of the water-table to the surface and consequently to the level of salinity and extent of waterlogging. Therefore, five main types of ecosystems may be recognized (Ayyad and El-Ghareeb, 1984):

1. Sand dunes (coastal calcareous and inland siliceous);
2. Rocky ridges and plateaux with skeletal shallow soils;
3. Saline depressions;
4. Non-saline depressions;
5. Wadis.

Besides these five habitats, two others are described by Tadros (1956) and Batanouny (1973):

6. The uncultivated desert areas;
7. The sand plains.

1. Sand Dunes

Along the Western Mediterranean coast lies a chain of intensely white calcareous granular sand dunes. They are formed of loose oval pseudo-oolitic grains, each composed of a series of successive coats of calcium carbonate. These dunes form a fairly continuous ridge with an undulating surface and present a type of habitat notable for its monotony. However, such monotony does not invariably mean that either the soil or the vegetation lacks variety. Owing to proximity to the sea, the dunes are more humid and exposed to the immediate effect of the northerly winds. They are also reached by sea spray (Ahmed and Mounir, 1982). Certain sections of the coast are devoid of dunes.

A short distance from the beach, fresh water is frequently obtained by digging carefully in the sand to a depth of 3–4 m. This fresh water is undoubtedly rain water, which, having a lower specific gravity than saline water below, may form a layer above it; there may be a hard pan of limestone rock underlying the sand which prevents percolation of rain water, the sand acting as a reservoir for fresh water.

Plants growing in sand dunes are highly specialized and many have the ability to elongate vertically on burial with sand (Girgis, 1973). They are also subject to partial exposure of their underground organs, often without being seriously affected. The coarse grain and loose texture of the sand result in poor water-retention because of rapid percolation. Many psammophytes develop extensive superficial roots that make use of dew.

The vegetation of these sand dunes has been studied by Oliver (1945), Tadros (1953, 1956), Tadros and Atta (1958a), Tadros and El-Sharkawi (1960), El-Sharkawi (1961), El-Ghonemy (1973), Girgis (1973), Ayyad (1973), Ayyad and El-Bayyoumi (1979), Ayyad and El-Ghareeb (1984) etc. Bordering the sea, a community of *Ammophila arenaria* and *Euphorbia paralias* can be usually distinguished on the mobile young calcareous sand dunes. Associates include *Lotus polyphyllos* and *Sporobolus virginicus*.

The vigorous growth made by *Ammophila* when sand covered enables it to dominate the mobile dunes. It is a pioneer species in invading mobile coastal dunes and is consequently extensively used for sand-dune fixation. On the older, advanced and higher dunes, where the sand may be consolidated in parts, *Crucianella maritima* and *Ononis vaginalis* predominate. Associate species include *Ammophila arenaria, Cakile maritima, Centaurea pumila, Echinops spinosissimus, Echium sericeum, Elymus farctus, Euphorbia paralias, Hyoseris lucida, Launaea tenuiloba, Lotus polyphyllos, Lygos raetam, Pancratium maritimum, Plantago albicans, Reseda alba, Salvia lanigera,* and *Silene succulenta*. In the more advanced stages of dune stabilization, communities of *Crucianella maritima, Echinops spinosissimus, Elymus farctus, Euphorbia paralias, Pancratium maritimum* and *Thymelaea hirsuta* become successively more common. When the coastal ridge is fairly exposed a community of *Globularia arabica, Gymnocarpos decander, Helichrysum conglobatum* and *Thymus capitatus* predominates.

The inland siliceous dunes are dominated by communities of *Plantago albicans, P. squarrosa* and *Urginea maritima.*

2. Rocky Ridges

Two (or sometimes three) ridges run south of the sand dune zone extending parallel to the Western Mediterranean coast of Egypt and are separated from the sea by the sand dunes. These ridges are composed of oolitic sand and shell debris, often 20 m or more high with smooth rounded summits. The outer ridge closely skirts the coast while the second one runs parallel with it at a distance of a few kilometres inland. The third ridge, when present, is between the second one and the edge of the Western Desert.

The vegetation of the rocky ridges is an association of *Thymelaea hirsuta* and *Gymnocarpos decander* (Tadros and Atta, 1958a). However, local variation in the nature of the position and degree of slope lead to parallel variations in the distribution of the vegetation. The characteristic species of this community include *Aegilops kotschyi, Arisarum vulgare, Bupleurum nodiflorum, Carduus getulus, Chenolea arabica, Erodium cicutarium, Limonium tubiflorum, Lotus corniculatus, L. creticus, Lygeum spartum, Malva aegyptia, Medicago minima, Moricandia suffruticosa, Orlaya maritima, Plantago notata, Reaumuria hirtella, Reichardia orientalis, Scorzonera alexandrina, Stipa capensis, S. parviflora* and *Teucrium polium.*

Rocky sites with low moisture availability are dominated by communities of *Globularia arabica* and *Thymus capitatus* while sites with fairly deep soils and high moisture availability are dominated by communities of *Asphodelus microcarpus, Herniaria hemistemon, Plantago albicans* and *Thymelaea hirsuta*. In sites of intermediate rockiness and moisture availability, *Echinops spinosissimus, Helianthemum stipulatum, Noaea mucronata, Pituranthos tortuosus* and *Scorzonera alexandrina* are abundant (Ayyad and Ammar, 1974).

These communities extend to the plateau of the south tableland. Two other communities dominated by *Hammada scoparia* and *Anabasis articulata* are found on degraded shallow skeletal soils subjected to active erosion. Associate species of this community include *Asphodelus microcarpus, Atriplex halimus, Carthamus mareoticus, Noaea mucronata, Pituranthos tortuosus, Verbascum letourneuxii* and *Zilla spinosa. Salsola tetrandra, Suaeda fruticosa* and *Suaeda pruinosa* are poorly represented. Bushes of *Capparis spinosa* and *Ephedra alata* often grow in vertical rock.

3. Saline Depressions

The saline depressions (littoral salt marshes) are a common habitat of the Western Mediterranean coastal belt. Tadros (1956) recognized two series of salt marshes in this coastal belt. One is formed from depressions directly adjacent to the dune strips. Salinity of this series results from the evaporation of seepage water, where the water-table is exposed or near the surface and where there is poor drainage. The soil is mostly calcareous-sandy due to the encroachment of sand from the neighbouring dunes. In certain places in these salt marshes, low bushes of *Arthrocnemum macrostachyum* and *Halocnemum strobilaceum* and other species eventually become buried under moist conditions, forming dense black rotten material from which frequently the smell of hydrogen sulphide can be detected. The second series of salt marshes is formed from the dried bed of Lake Mariut lying between the two ridges. The causes of salinity are essentially as in the first series, but the soil texture is different, having a considerable proportion of silt, regarded as derived from the Nile during its previous connection with the lake.

The littoral salt marsh vegetation of the Western Mediterranean coast of Egypt has been described by several authors: Oliver (1938); Hassib (1951); Tadros (1953, 1956); Migahid et al. (1955); Tadros and Atta (1958a, b); Tadros and El-Sharkawi (1960); Ayyad and El-Ghareeb (1982, 1984); Ahmed and Mounir (1982), etc.

Apart from the communities of the swamp vegetation dominated by *Phragmites australis, Scirpus tuberosus* and *Typha domingensis*, the halophytic vegetation is characterized by some 11 communities:

1. *Salicornia fruticosa-Suaeda salsa* community. This usually occupies the zone on the more elevated banks with less submerged saline soil. Associate species are *Phragmites australis* and *Salicornia herbacea.*
2. *Juncus rigidus* community. This occupies lower parts of the marsh with high moisture content where the calcareous sand fraction dominates the soil texture. Associate plants include *Halimione portulacoides. Inula crithmoides, Juncus acutus, Limonium pruinosum* and *Sporobolus pungens*. In certain patches of this community, there are societies dominated by *Schoenus nigricans*.
3. *Sporobolus pungens* community. This occupies higher parts of the marsh, especially where calcareous sand is plentiful. The associate species are *Juncus rigidus* and *Limonium pruinosum.*
4. *Halocnemum strobilaceum* community. This community occurs over a wide range of fluctuations of salt concentration between the wet and dry seasons where there is a high proportion of fine fractions affecting soil texture. Associate species are *Arthrocnemum macrostachyum, Juncus rigidus* and *Salicornia fruticosa.*
5. *Salicornia fruticosa—Limonium pruinosum* community. This is present in somewhat more elevated and less saline parts than that of the *H. strobilaceum* community. Common associate species include *Inula erithmoides, Juncus rigidus, Parapholis marginata, Plantago crassifolia* and *Sphenopus divaricatus. Halimione portulacoides* and *Phragmites australis* dominate in some patches, the latter species being associated with depressed areas with high water content.
6. *Arthrocnemum macrostachyum-Limoniastrum monopetalum* community. This occurs on even more elevated substrates than the *S. fruticosa-L. pruinosum* community. Characteristic species are *Cressa cretica, Frankenia revoluta, Mesembryanthemum nodiflorum* and *Parapholis marginata.*
7. *Zygopkyllum album* community. *Z. album* frequently forms an almost pure community on saline patches recently covered by drifted sand in shallow layers. It is also found in communities with other species in similar habitats.

8. *Lygeum spartum* community. This occurs in less saline parts with high organic matter content. Associate species are *Frankenia revoluta, Halimione portulacoides, Limoniastrum monopetalum* and *Limonium pruinosum.*
9. *Salsola tetrandra* community. This community is usually present on the elevated border of the dry saline beds of the marshy valleys. *S. tetrandra* is a very efficient soil conserver against wind blowing as well as being a soil builder. The associate species include *Anthemis cotula, Coris monspeliensis, Frankenia revoluta, Haplophyllum tuberculatum, Limoniastrum monopetalum, Salicornia fruticosa, Sphenopus divaricatus, Suaeda fruticosa, S. pruinosa* and *Traganum nudatum.*
10. *Limoniastrum monopetalum-Lycium europaeum* community. This is another community rich in floristic composition. It may follow in succession the community dominated by *Salsola tetrandra.* Associate species include *Asphodelus microcarpus, Bassia muricata, Carthamus glaucus, Cutandia dichotoma, Echinops spinosissimus, Ifloga spicata, Lotus villosus, Noaea mucronata, Orlaya maritima, Plantago albicans, Reaumuria hirtella* and *Suaeda pruinosa.*
11. *Atriplex halimus-Picris radicata* community. This is the richest of all the communities of the salt-affected land. It occurs on deep sandy loam at the edges and upper parts of valleys where the vegetation covers the soil almost completely. Associate species include *Anthemis microsperma, Chenolea arabica. Chrysanthemum coranarium, Koeleria phleoides, Lolium rigidum, Lycium europaeum, Medicago minima, Picris radicata, Salvia lanigera, Schismus barbatus* and *Stipa capensis.*

4. Non-Saline Depressions

The non-saline depressions (the barley fields) are the most fertile areas of the Western Mediterranean coastal belt of Egypt. These depressions are mainly limited to the plains south of the second ridge in the eastern section of the coast, but are widespread in the valley and plains of the western section.

The soils of these depressions, e.g. Abu Sir depression, are variable (Ayyad, 1976). In some parts, highly calcareous soils are derived from drifted oolitic grains of the coastal ridge; in other parts alluvial, less calcareous, loamy soils are derived from the Abu Sir ridge.

The non-saline depressions provide favourable conditions for cultivation; extensive areas are occupied by barley, figs and olives. Fanning operations promote the growth of a considerable number of species, mostly therophytes. Weeds of barley fields are recognized as the *Achilleetum santolinae mareoticum* association, with subassociation of *Chrysanthemetosum coronariae* and *Arisaretosum vulgare* (Tadros and Atta, 1958b), composed of the following characteristic species: *Achillea santolina, Anagallis arvensis, Calendula aegyptiaca, Carthamus glaucus, Convolvulus althaeoides, Echinops spinosissimus, Echium sericeum, Eryngium creticum, Hordeum murinum, Koeleria phleoides, Lathyrus cicera, Muscari comosum* and *Vicia cinerea.*

According to Ahmed and Mounir (1982), there are still other species of different communities occasionally present in the barley fields, e.g. *Atriplex halimus, Trifolium tomentosum* and *Suaeda fruticosa*. These species may indicate possible affinities with other associations. The "accidental" species recorded include *Anchusa hispida, Anthemis cotula, Asteriscus graveolens, Avena sterilis, Beta vulgaris, Bupleurum subovatum, Crucianella maritima, Echiochilon fruticosum, Emex spinosus, Filago spathulata, Francoeuria crispa, Gagea fibrosa, Helianthemum stipulation, Hippocrepis bicontorta, Hymenocarpus nummularius, Hyoseris lucida, Ifloga spicata, Koniga arabica, Limonium tubiflorum, Lotus creticus, Malva parviflora, Moricandia nitens, Ononis vaginalis, Orlaya maritima, Ornithogalum trichophyllum, Papaver hybridum, Reseda alba. Salvia aegyptiaca, Scorzonera alexandrina, Silene villosa, Thesium humile* and *Verbascum letourneuxii*.

The vegetation of the non-saline depressions belongs to the *Plantagineto-Asphodeletum microcarpae* associations (Tadros and Atta, 1958b). The *Anabasis articulata* community is found on more or less sandy soils with low contents of calcium carbonate, a *Zygophyllum album* community where the soil content of calcium carbonate and salinity are higher, a *Plantago albicans* community where salinity is lower and an *Asphodelus microcarpus-Thymelaea hirsuta* community on fine-textured soils (Ayyad, 1976). The characteristic species include *Alkanna tinctoria, Brachypodium distachyum, Brassica tournefortii, Bupleurum subovatum, Carthamus glaucus, Centaurea glomerata, Linaria haelava, Lolium perenne, Malva parviflora, Medicago littoralis, Onopordum alexandrinum, Orobanche ramosa, Papaver rhoeas, Polygonum equisetiforme, Raphanus raphanistrum, Reseda alba, R. decursiva* and *Zygophyllum album*.

5. The Wadis

The landscape of the Western Mediterranean coastal land of Egypt is dissected by a drainage system (wadis) which originates from a southern limestone plateau which lies parallel to the Mediterranean Sea. The plateau reaches a maximum elevation of about 200 m above sea level at Sallum and slopes gently to a coastal plain west of Mersa Matruh which varies from 10 to 20 m above sea level. These wadis drain northwards into the Mediterranean Sea. An ecological account of four of these wadis: Wadi Habis, Wadi Hashem, Wadi Zeitouna and Wadi Shabbat is given.

a. Wadi Habis

Wadi Habis (31°24'N, 27°03'E) is of ecological and historical interest. In this wadi there are archaeological remains of apparently Graeco-Roman age (about 300 BC–600 AD). The Graeco-Roman occupation of the wadi was restricted to its mouth and its immediate vicinity.

According to El-Hadidi and Ayyad (1975), Wadi Habis is characterized by nine habitats: fallow saline areas, fallow non-saline areas, barley fields, olive orchards, wadi bed, lower position of slopes, middle slopes, upper slopes and plateau.

The saline fallow areas are co-dominated by *Reseda decursiva* and *Asphodelus tenuifolius*. The abundant associates are *Carthamus glaucus* and *Onopordum alexandrinum* while other associates include *Centaurea glomerata, Chrysanthemum coronarium, Echium sericeum, Glaucium corniculatum, Malva parviflora, Papaver rhoeas, Paronychia argentea, Plantago albicans. Salvia lanigera, Senecio desfontainei* and *Trigonella maritima.*

In the non-saline fallow areas, the most abundant species are: *Chrysanthemum coronarium, Picris sprengeriana* and *Trigonella maritima.* Other associates include *Asphodelus tenuifolius, Chenopodium murale* v. *microphyllum, Emex spinosus, Eragrostis pilosa, Erucaria pinnata, Lolium rigidum, Matthiola longipetala, Schismus barbatus, Silene apetala* and *Trifolium tomentosum.*

The barley fields support about 40 species which are co-dominated by *Chrysanthemum coronarium, Convolvulus althaeoides, Launaea nudicaulis* and *Plantago albicans*. Other associates include *Achillea santolina, Adonis dentata, Anagallis arvensis, Arisarum vulgare, Avena sterilis* ssp. *ludoviciana, Beta vulgaris, Brassica tournefortii, Echinops spinosissimus, Echium setosum, Erodium laciniatum, Lamarckia aurea, Lathyrus aphaca, Linaria haelava, Lotus creticus, Medicago littoralis, Noaea mucronata, Papaver hybridum* and *Senecio desfontainei.*

The olive orchards of the frontal section are characterized by a dense cover of weeds which may be distinguished into two main synusiae. The upper is co-dominated by grasses such as *Hordeum leporinum, Lolium rigidum* and *Lophochloa cristata* and the lower by *Achillea santolina, Astragalus boeticus* and *Matthiola longipetala* ssp. *aspera*. Other associates include *Anchusa milleri, Emex spinosus, Euphorbia parvula, Filago desertorum, Fumaria bracteosa, Glaucium corniculatum, Hippocrepis cyclocarpa, Reichardia orientalis, Roemeria hybrida, Schismus barbatus, Scorpiurus muricatus* v. *subvillosus* and *Spergularia diandra.*

The vegetation of the wadi bed is sparse, but the number of species is high. In this habitat, fine soil material has little chance to settle owing to the high velocity of the water stream during the rainy season. The wadi bed is filled mainly with large boulders, the sparse vegetation being largely restricted to shallow soil accumulation between rock fragments. Common perennials in the wadi bed are *Allium erdelii, Echium sericeum, Euphorbia terracina* and *Salvia lanigera.* Less common ones include *Allium aschersonianum, A. barthianum, Arisarum vulgare* v. *veslingii, Cynara sibthorpiana, Lygos raetam, Scorzonera alexandrina, Silybum marianum* and *Suaeda pruinosa*. Common annuals include *Astragalus boeticus, Erodium gruinum* and *E. hirtum*. Less common annuals include *Aizoon hispanicum, Chenopodium murale* v. *microphyllum, Emex spinosus, Fumaria bracteosa, Mesembryanthemum nodiflorum, Minuartia geniculata* v. *communis, Polycarpon succulentum, Polygonum equisetiforme, Rumex vesicarius, Spergula fallax, Spergularia diandra* and *Trifolium formosum*. More than two-thirds of the taxa recorded in the wadi bed are Mediterranean. The lower gentle slopes support meadow-like vegetation of annual species; the most common are *Astragalus hamosus, Hippocrepis bicontorta, Medicago littoralis, M. truncatula* and *Spergula fallax*. Perennial associates include *Allium barthianum, Asphodelus micro-carpus, Cynara sibthorpiana, Salsola longifolia, Salvia lanigera, Scorzonera alexandrina, Silybum marianum* and *Traganum nudatum.*

On the middle slopes the vegetation is dominated by shrubby species including *Artemisia inculta, Gymnocarpos decander, Limonium sinuatum* and *L. tubiflorum* and grasses such as *Hyparrhenia hirta* and *Stipa capensis*. Other associates include *Allium erdelii, Asparagus stipularis* v. *tenuispinus, Avena sterilis* ssp. *ludouiciana, Brassica tournefortii, Bromus rubens, Carduus getulus, Erucaria pinnata, Hammada scoparia,* (= *Hyloxylon scoparium*) *Limonium thouini, Lycium europaeum, Mesembryanthemum nodiflorum, Noaea mucronata, Phalaris minor, Picris sprengeriana, Pituranthos tortuosus, Plantago albicans, P. squarrosa, Reichardia orientalis, Salvia verbenaca, Spergula fallax, Spergularia diandra, Suaeda pruinosa, Traganum nudatum, Trifolium scabrum, T. stellatum* and *Umbilicus horizontalis*.

The upper slopes are usually steep and almost completely devoid of soil cover. They support a typical cliff vegetation dominated by *Asparagus stipularis, Capparis orientalis, Ephedra aphylla, Lycium europaeum, Periploca angustifolia, Phlomis floccosa* and *Umbilicus horizontalis*. Common perennials include *Allium barthianum, Asphodelus microcarpus, Echinops spinosissimus, Gymnocarpos decander, Hammada scoparia, Hyparrhenia hirta, Micromeria nervosa, Noaea mucronata, Scorzonera alexandrina* and *Thymus capitatus*. Common annuals include *Echium setosum, Mesembryanthemum forsskaolii* (= *Opophyllum forsskaolii*), *Picris sprengerana, Reichardia orientalis* and *Thesium humile* v. *maritima*. Less common are *Anagallis arvensis, Arisarum vulgare* v. *veslingii, Astragalus asterias, Carthamus glaucus, Convolvulus althaeoides, Cutandia dichotoma, Echium sericeum, Fagonia cretica, Globularia arabica, Helianthemum ciliatum, Hippocrepis cyclocarpa, Leontodon hispidulus* (= *Crepis bulbosa*), *Limonium thouini, Lotus creticus, Malva parviflora, Medicago aschersoniana, Pallenis spinosa, Plantago crypsoides, Pteranthus dichotomus, Ranunculus asiaticus, Salvia lanigera, S. verbenaca* and *Valantia hispida*.

In the plateau of the wadi, the vegetation is co-dominated by *Gymnocarpos decander, Hammada scoparia* and *Phagnalon rupestre*. In this habitat the fewest associate species have been recorded, including *Artemisia inculta. Asparagus stipularis* v. *tenuispinus, Atractylis prolifera, Echinops spinosissimus, Ephedra aphylla, Filago desertorutn, Globularia arabica, Helianthernum ciliatum, Lycium europaeum, Micromeria nervosa, Noaea mucronata, Periploca angustifolia, Reichardia orientalis, Reseda decursiva, Rumex vesicarius. Salvia lanigera* and *Thymus capitatus*.

b + c. Wadi Hashem and Wadi Zeitouna

Tadros (1956) states that there are certain patches of vegetation in obscure and relatively isolated and protected ends of valleys. An example is the community present at Wadi Hashem, 45 km east of Mersa Matruh and about 3 km south of the sea coast, where the following species grow: *Artemisia inculta, Asparagus aphyllus, Atriplex halimus, Capparis spinosa, Lycium europaeum, Phlomis floccosa, Rhamnus oleoides* v. *libyca* and *Varthemia candicans*. At the uppermost end of Wadi Hashem is a single tree of *Ceratonia siliqua* (semi-wild). Another patch of rich vegetation is sheltered in Wadi Zeitouna, about 45 km west of Wadi Hashem. The species recorded include *Asphodelus microcarpus, Asparagus aphyllus, Atractylis flava, Atriplex halimus, Phlomis floccosa, Pituranthos tortuosus* and *Rhamnus oleoides* v. *libyca*. Single trees of *Olea europaea* and *Ceratonia siliqua* have been recorded in the upper part of the wadi.

d. Wadi Shabbat

This is a deep well-defined wadi that runs in the dissected tableland in the Fuka area about 88 km east of Mersa Matruh. The water course of this wadi cuts across the tableland forming successive hollows by progressively narrower strips. The breadth of the water channel increases concomitantly with decreased height of the bounding rocky banks until it fans out through the piedmont plain (Ahmed, 1983).

In the upstream part of Wadi Shabbat, where the soil is formed mainly from sand, silt and shales, covered and intermixed with rock detritus, is a flourishing dense community dominated by the fodder salt-tolerant shrub *Atriplex halimus*. Associate species include *Anabasis articulata, Atractylis prolifera, Gymnocarpos decander, Halogeton alopecuroides, Salsola tetrandra, S. vermiculata, Suaeda pruinosa* and *Thymelaea hirsuta*.

Nearer to the outlet of the wadi the bed becomes patchy; on deep soil the community of *Atriplex halimus* merges gradually into shallow scrubland of *Astragalus spinosus, Gymnocarpos decander* and *Lycium europaeum*.

6. Uncultivated Desert Areas

These are the rocky and gravelly areas that form a distinct plant habitat with characteristic communities whose scattered individuals grow in cracks and concavities filled with transported debris. Local weathering forming residual soil is also commonly detectable where soil depth may be up to 40 cm. In addition, there are large areas covered with stones and gravels more or less cemented together by soil and forming a hard surface, which though not cultivated is quite favourable to some desert shrubs and small herbs (Ahmed and Mounir, 1982).

Three communities have been recognized in this habitat dominated by:

1. *Thymelaea hirsuta-Gymnocarpos decander;*
2. *Asphodelus microcarpus-Plantago albicans;*
3. *Anabasis articulata-Hammada scoparia.*

The *Thymelaea hirsuta-Gymnocarpos decander* community occupies mainly the rocky ridges. The plants grow either in the cracks filled with soil or on residual and accumulated soil. Owing to grazing, this community rarely attains full growth. *Thymelaea* and *Gymnocarpos* are palatable shrubs and usually suffer heavy grazing (Heneidy, 1986). Characteristic species of this community include *Aegilops kotschyi, Arisarum vulgare, Bupleurum nodiflorum, Carduus pycnocephalus, Carrichtera annua, Chenolea arabica, Dactylis hispanica, Erodium cicutarium, Helianthemum stipulatum, Helichrysum conglobatum, Herniaria hemistemon, Iris sisyrinchium, Limonium tubiflorum, Lotus corniculatus, L. creticus, Lygeum spartum, Medicago minima, Moricandia nitens, Nonea viviani, Onobrychis crista-galli, Orlaya maritima, Plantago notata, Reaumuria hirtella, Reichardia tingitana, Scorzonera alexandrina, Stipa capensis*, S. *parviflora* and *Teucrium polium*.

According to Migahid et al. (1955), El-Ghonemy and Tadros (1970), Zahran and Boulos (1973), El-Ghonemy et al. (1977), Shaltout (1983) and Bornkamm and Kehl (1990) *Thymelaea hirsuta*, a perennial evergreen shrub 40–200 cm tall, is circum-Mediterranean in distribution but of minor importance along the European coastal belt. It is one of the most common and widespread species in the western Mediterranean coastal land of Egypt. The gradual landward (southward) change of climatic conditions (decrease in rainfall and humidity and rise in temperature and evaporation rate as well as reduction in soil moisture content) is associated with progressive decline in the number, abundance and vigour of *T. hirsuta*. This shrub is intolerant of dry conditions. Its most southern limit is 70–75 km from the coast.

Along the western Mediterranean coast, the *T. hirsuta* community is one of the commonest features of the xerophytic vegetation. Dominance by this shrub has been observed in three habitat types (Zahran and Boulos, 1973):

1. Downstream parts of water runnels and wadis with alluvial silty soil that is compact and consolidated, containing a high proportion of fine material (>75%);
2. Sandy sheets formed of aeolian material which are loose and contain almost equal proportions of fine and coarse sand and silt;
3. Rocky ridges with a high percentage of coarse material (>60%).

The growth form, abundance, cover and flora of the *T. hirsuta* community vary in these three habitats. The highest cover (50–60%) is recorded in the silty habitat where *T. hirsuta* is vigorous. In the sandy habitat, individual bushes are also healthy but the total plant cover is low (20–30%). In the rocky ridges the shrubs are stunted and pale green, and total cover of the stands is usually less than 10%. In these sites the water-table is deeper and the growth of plants seems to depend mainly on rainfall. Associate species in the three habitats are *Anabasis articulata, Artemisia inculta, Linaria aegyptiaca, Lycium europaeum, Noaea mucronata, Pituranthos tortuosus* and *Salsola tetrandra. Cynodon dactylon* and *Hammada scoparia* are recorded only in the silty habitat, *Halogeton alopecuroides* only in the sandy habitat whereas *Artemisia monosperma* and *Convolvulus lanatus* are only in the rocky habitat. *Gymnocarpos decander* and *Atractylis flava* are present in the sandy and rocky habitats.

The *Asphodelus microcarpus-Plantago albicans* community occurs on shallow pebbly soil as well as in vast desert areas left fallow for a long time. Characteristic species include *Alkanna tinctoria, Astragalus forsskaolii, Brachypodium distachyum, Brassica tournefortii, Bupleurum semicompositum, B. subovatum, Carthamus glaucus, Centaurea glomerata, Chrysanthemum coronarium, Convolvulus althaeoides, Enarthrocarpus strangulatus, Linaria haelava, Lolium perenne, Malva parviflora, Medicago littoralis, Onopordum alexandrinum, Orobanche ramosa, Polygonum equisetiforme, Raphanus raphanistrum, Reseda alba, R. decursiva* and *Zygophyllum album.*

Asphodelus microcarpus is circum-Mediterranean in its distribution. In Egypt, it is restricted to the Mediterranean coastal land, particularly in the western region where it co-dominates a community with *Plantago albicans* in sites with deep sandy or sandy loam soils and more rarely in those with shallow pebbly soils overlying rocks of limestone hills; it is also co-dominant in another community with *Artemisia inculta* on the slopes and plains with shallow or deep loamy soils (Long, 1955; Ayyad and Hilmy, 1974). Features favouring *Asphodelus microcarpus* are moist conditions, high levels of nitrogen and moderate levels of calcium carbonate (Ayyad and Hilmy,

1974). The *Anabasis articulata-Hammada scoparia* community is fairly common in the southern plateau of the coastal land. The soil here is gravelly, compact and heavy, and in many stands it has a reddish tinge implying the presence of iron compounds; it is probably also not as rich in calcareous matter as that of the other habitat. The flora of this community is less rich than that in the uncultivated desert areas. This might be due to climate. Ahmed and Mounir (1982) state that it is very likely that the amount of rainfall in the southern plateau where the *A. articulata-H. scoparia* community dominates is less than in the more northern parts nearer to the coast where the principal species – A. *articulata, H. scoparia* and *Zilla spinosa* – are rarely found. The average atmospheric humidity may be lower in the conditions under which this community exists. *Asphodelus microearpus, Atriplex halimus, Carthamus mareoticus, Noaea mucronata, Pituranthos tortuosus, Salsola tetrandra, Suaeda fruticosa, S. pruinosa* and *Verbascum letourneuxii* are commonly present.

7. The Sand Plains

According to Batanouny (1973), the sand plain habitat of the Western Mediterranean coastal belt may be represented by narrow strips south of the coastal sand dunes, depressions between the rocky ridges or vast areas about 2 km or more south of the shore. These plains are more or less flat with slight undulation and some rock outcrops. The land rises southwardly to the Libyan plateau. The widest part of the sand plain habitat is that between Mersa Matruh and Sallum. These plains receive run-off water from the high plateau in the south. Run-off water may be several times as great as rain water. Its amount depends on soil depth and slope.

The sand plains could be classified provisionally according to soil depth into plains with deep loose sandy soil, plains with shallow soil, and plains with very shallow soil.

Plains with Deep Loose Sandy Soil

Soil of this habitat is deep, loose, easily penetrable and with relatively high moisture content in the deep layers. This habitat is represented by wide areas in the coastal zone with dense vegetation and cultivated patches. Numerous species grow here with a cover of 80% during winter-spring (rainy season) though in summer it is only some 25% or even less. Cover of the xerophytes is high in winter and spring. An association of *Plantago albicans-Echiochilon fruticosum* is present in these plains. The common associate species include *Asphodelus microcarpus, Cutandia dichotoma, Cyperus conglomeratus, Echinops spinosissimus, Helianthemum lippii* and *Thymelaea hirsuta. Urginea maritima* v. *pancratium* dominates localized patches.

Plains with Shallow Soil

Soils in this habitat are more compact and shallow, with higher contents of fine sand and calcium carbonate than in the habitat mentioned above. Because of these characteristics, they are relatively dry, with low water content in the deep layers. *Artemisia inculta* dominates this habitat. Plant cover ranges between 20% in summer and 50% in winter. Associate species include *Anthemis microsperma, Arisarum vulgare, Asphodelus microcarpus, Chrysanthemum coronarium, Daucus syrticus, Erodium hirtum, Koeleria phleoides, Papauer rhoeas, Salvia lanigera* and *Thymelaea hirsuta.*

Plains with Very Shallow Soil

This habitat has degraded soils with stones and boulders in the profile and on the surface. The soil is compact with a hard surface crust and its depth does not exceed 100cm. This habitat is subject to wind and water erosion, characters which collectively lead to diminished water resources. Moreover, except for salt marsh soils, the soil of this habitat has the highest salt content. The plant cover is very low, being 40% in winter and only 15% in summer. Common species in these plains are *Anabasis articulata, Carthamus mareoticus* and *Hammada scoparia.*

(ii) Vegetation of Ras El-Hikma Coastal Area

Ras El-Hikma is a little village on the Mediterranean coast lying about 230km west of Alexandria at latitude 31°15'N, longitude 27°51'E (Fig. 3.1) where a cape projects a long distance into the sea. The beach is rocky in some parts, particularly on the west of the cape, and sandy in other parts.

Ras El-Hikma is approximately 5km long, and is characterized by alternating elevations and depressions; a few elevations rise to 60m and some depressions are less than 2m above sea level. To the south the area becomes less variable (Migahid et al., 1955; Ayyad, 1969).

The climate at Ras El-Hikma is comparable to that of Mersa Matruh (the nearest city with a meteorological station, Table 3.1). It is maritime with moderate temperature, high relative humidities and high wind velocity. At Ras El-Hikma, wind velocity may be as high as 90km/h and as low as 10km/h (Ayyad, 1957). Mean annual rainfall is 158mm, most of which falls between November and February. The monthly mean air temperature varies between 12.4°C in Januar y and 25.5°C in August.

According to Migahid et al. (1955), a variety of distinct habitats can be recognized at Ras El-Hikma, namely rocky ridges, slopes, sand plains, sand dunes and salt marshes. These habitats differ in exposure, microclimate, water supply, soil depth and other soil conditions. Each habitat supports a particular type of vegetation with its characteristic flora. Although some species are common to several habitats, their abundance, frequency, cover and vigour differ in the different habitats.

Rocky Ridges (Rocky Plateaux)

In these habitats strong winds and run-off contribute to the extreme aridity. The plant cover is, consequently, sparse and occurs only where the microrelief allows for soil and moisture to accumulate.

Only very few species, e.g. *Globularia arabica* and *Thymus capitatus*, can grow on solid rocks. These two species are narrowly distributed, being found on the western ridges only. The plant cover in the rocky ridges does not exceed 2–5%, and plants are stunted.

Two plant communities associated with two types of habitats are recognized in the rocky plateaux (Ayyad, 1969):

1. The inland rocky plateaux support *Gymnocarpos decander* and *Thymelaea hirsuta* as dominants. Other common species are *Dactylis hispanica, Globularia arabica, Helianthemum kahiricum, Herniaria hemistemon, Thymus capitatus* and *Varthemia candicans.*
2. The coastal rocky plateaux are characterized by *Herniaria hemistemon, Inula crithmoides, Limonium pruinosum* and *Reaumuria hirtella.*

Migahid et al. (1955) listed *Crucianella maritima, Echium sericeum, Helichrysum conglobatum, Hippocrepis bicontorta, Ononis vaginalis, Phagnalon rupestre* and *Zygophyllum album.*

Slopes

Slopes support a denser vegetation than that of the rocky ridges. They are more sheltered. Different slopes, however, differ in the condition of the vegetation according to their steepness, exposure, distance from the sea and direction in relation to the prevailing wind. If they are steep, only a small proportion of the run-off is retained, but if gentle or nearly flat a considerable fraction of the run-off water may be retained and absorbed. Gentle slopes, therefore, are more favourable for plant growth than steep slopes.

The plant cover gradually increases with distance down the slope. The vegetation is a mixture of rock ridge and sand formation types, with a preponderance of the former towards the top and the latter towards the base. Near the top and in the middle zone *Launaea tenuiloba* is dominant and associated species include *Centaurea pumila, Crucianella maritima, Echinops spinosissimus, Echiochilon fruticosum, Echium sericeum, Helianthemum stipulatum, Lotus creticus, Pituranthos tortuosus, Plantago albicans, Polycarpon arabicum, Thymelaea hirsuta* and *Zygophyllum album.*

Near the base of the gentle slopes, *Plantago albicans* dominates. Associate species include *Echiochilon fruticosum* (abundant), *Centaurea pumila, Crucianella maritima, Echinops spinosissimus, Launaea tenuiloba, Lotus creticus, Ononis vaginalis, Pancratium maritimum, Salvia lanigera* and *Thymelaea hirsuta.*

Where leeward slopes are steep, loose sand deposits heavily and forms a mobile substratum that supports a thin vegetation, especially near the top of the slopes. Near the base of the slopes the substratum is more stable and the vegetation denser. At the high zones of these slopes *Thymelaea hirsuta* dominates and *Launaea tenuiloba* is commonly present. Other associates include *Echiochilon fruticosum, Lotus creticus, Ononis vaginalis, Pituranthos tortuosus, Reaumuria hirtella, Suaeda pruinosa* and *Teucrium polium.* At the base, *Suaeda pruinosa* dominates and associates include *Aeluropus repens, Frankenia revoluta, Reaumuria hirtella, Thymelaea hirsuta* and *Zygophyllum album.*

Sand Plains

Sand plains of Ras El-Hikma cape occupy depressions between ridges, where shelter is greatest and soil deepest. Vegetation is most dense and vigorous in this type of habitat. The ground is generally flat, with only slight irregularities in the microrelief. Water is plentiful and soil water content is high; sand plains receive run-off water in addition to the normal rainfall. The soil is partly sand and partly alluvium.

Deep soil, rich water supply, and protection from wind and desiccation make sand plains a favourable habitat for plants. Three communities have been recognized on these plains (Ayyad, 1969).

1. The first is co-dominated by *Plantago albicans-Asphodelus microcarpus*. Common associates include *Echiochilon fruticosum, Echinops spinosissimus* and *Stipa lagascae*.
2. The second is co-dominated by *Artemisia inculta-Asphodelus microcarpus*. Among common associates are *Matthiola humilis, Medicago minima, Noaea mucronata, Picris radicata* and *Salvia lanigera*.
3. The cultivated barley fields form a third type of plain habitat. Farming operations result in the appearance of a weed flora dominated by *Achillea santolina. Arisarum vulgare. Calendula aegyptiaca, Chrysanthemum coronarium, Eryngium creticum, Hordeum murinum* and *Muscari comosum* are common species.

However, wherever soils are shallow or when rocky slopes adjoin the plain habitats, transitional types occur; *Gymnocarpos decander* and *Thymelaea hirsuta* are common.

The floor of the sand plain also includes *Allium roseum, Ammophila arenaria. Asparagus stipularis, Centaurea alexandrina. Convolvulus althaeoides, Echium sericeum, Elymus farctus, Herniaria hemistemon, Limonium pruinosum, Lotus polyphyllos, Lycium europaeum, Lygeum spartum, Matthiola livida, Ononis vaginalis, Pancratium maritimum, Phlomis floccosa, Polygonum equisetiforme, Reaumuria hirtella, Salvia lanigera* and *Zygophyllum album*.

Sand Dunes

These are little hillocks of loose white coarse sand, emerging slightly above the general level of the surrounding land. They have an irregular undulating surface and may cover extensive areas. Certain plants, mainly grasses, act as wind breaks, wind-borne sand being deposited around their bases.

Sand dunes occur all over the cape on elevated parts of sand plains bordering salt marshes, on gentle slopes and ridges and on exposed areas near the sea-shore.

The flora of the sand-dune vegetation is essentially the same on all dunes. *Ammophila arenaria* and *Elymus farctus* exchange dominance, being associated with *Pancratium maritimum* in the coastal dunes together with *Centaurea pumila* and *Echinops spinosissimus*. The last-named species is rarely found on the dunes themselves but often occurs at their base together with *Crucianella maritima, Echiochilon fruticosum* and *Ononis vaginalis. Zygophyllum album* may occasionally be present on the dune itself, mixed with the clumps of the dominant grass. *Silene succulenta* is a characteristic dune plant, but is very limited in its distribution and shows local abundance. Other associates recorded in this habitat are *Aeluropus repens, Lotus polyphyllos, Lycium europaeum, Suaeda maritima* and *Thymelaea hirsuta.*

Salt Marshes

The main factor affecting plant growth in this habitat is obviously the high soil salt content. The ground is only slightly above sea level, so the water-table is usually near the soil surface.

The salt marshes of Ras El-Hikma may be distinguished into northern and southern areas. In the northern marshes (near the sea) in winter, sea water occasionally overflows the beach and collects in the depressed central part where it stands for a long time. In these marshes, especially in the lowest part, a hard pan is found a little below the surface, preventing rapid percolation and drainage of sea water and maintaining the surface soil saturated. The water-table, again derived from the sea water, is at only a small depth below the surface. As surface evaporation proceeds, salt accumulates in the upper layer and the soil solution becomes more and more concentrated. The southern marshes are some distance from the sea. They are surrounded on nearly all sides by high ridges that give shelter from wind.

In the middle of both the north and south salt marshes, where the ground level is lowest, the soil is wet and salinity is relatively high; there are a few scattered plants of *Salicornia fruticosa* forming small green bushes with very thin cover (<2%). In the other parts of these salt marshes four plant communities have been distinguished:

1. *Halocnemum strobilaceum* community associated with *Arthrocnemum macrostachyum, Limoniastrum monopetalum* and *Salicornia fruticosa;*
2. *Suaeda fruticosa* community with *Mesembryanthemum crystallinum* and *M. nodiflorum* as common associates;
3. *Salsola tetrandra* and *Suaeda pruinosa* community;
4. *Lygeum spartum* community.

In the southern marshes the middle depressed areas are surrounded by a belt of slightly higher level, the difference being about 0.5 m. This belt supports *Frankenia revoluta, Mesembryanthemum nodiflorum, Salicornia fruticosa* (= *Sarcocornia fruticosa*, Boulos 1995) and *Suaeda pruinosa* with plant cover up to 40%. Another slightly higher belt formed of dry, saline sand may be present in certain patches. On these elevated areas *Aeluropus repens* and *Sphenopus divaricatus* occur in addition to the species mentioned above.

External to these belts the ground level continues to rise, though imperceptibly. The content of water and salts decreases and the habitat becomes progressively more favourable to plant growth. In consequence other species appear, increasing gradually in number from the centre of the salt marsh towards the edge. The density of the vegetation increases and the plant cover rises to about 50%. Towards the edge of the southern salt marsh, progressive preponderance of less halophytic species occurs. These include *Achillea santolina, Centaurea pumila, Echiochilon fruticosum, Helianthemum stipulatum, Lotus creticus, L. polyphyllos, Pancratium maritimum* and *Thymelaea hirsuta.*

(iii) Vegetation of Sallum Coastal Area

Sallum coastal area is located in the most western section of Egypt's Mediterranean coastal desert in the border with Libya: Lat. 31°30'–31°15'N, Long. 25°09'–25°35'E (Fig. 3.1). Its flora comprises 111 species: (75 perennials and 36 annuals) belong to 92 genera and 34 families (Salama et al., 2005).

The vegetation of Sallum coastal desert is classified under five main communities dominated by: *Haloxylon salicornicum, H. salicornicum-Thymelaea hirsuta, Thymelaea hirsuta-Anabasis articulata, Haloxylon salicornicum- Atriplex portulacoides and Salsola tetrandra- Limoniastrum monopetalum.*

1. *H. salicornicum* community
   The habitat of this community is the sand sheets on the elevated calcareous soil with moderate moisture content (3.16%), very low organic matter (0.08%) and low soil salinity (EC=0.36 mScm-^{1}). The common perennial associates are: *Retama raetam, Lycium europaeum, Farsetia aegyptia, Periploca angustifolia, Euphorbia retusa, Citrullus colocynthis, Marrubiuim alysson,* and *Anabasis articulata.* The annual associates are many (23 species) which usually make dense ground layer after rainfall and may be considered a winter rangeland. These include: *Anthemis microsperma, Astragalus hamosus, Brassica tournifortii, Cutandia memphetica, Erodium pulverulantum, Malva parviflora, Medicago laciniata, Reichardia tingitana,* and *Schismus barbatus.*
2. *H. salicornicum-T. hirsuta community*
   This community inhabits the sand plains with deep-loose soil having low moisture content (2.12%), organic matter (0.19%) and salinity (EC=0.45 mScm-^{1}). It represents a transitional zone between the non saline and saline habitats. Eleven associates are recorded including 7 annuals and 4 perennials, e.g. *Periploca angustifolia, Deverra tortusa, Globularia arabia* and *Zilla spinosa* (Perennials), *Asphodelus tenuifolius, —Astragalus peregrinus, Bupleurum lacifolium* and *Hordeum murinum. subsp. leporinum* (annuals).
3. *T. hirsuta-Anabasis articulata community.*
   The proper habitat of this community is the non-saline depression (EC=41 mScm-^{1}) with low moisture content (2.2%) amd organic matter (0.2%), the common perennial associates include: *Gymnocarpos decander, Asphodelus ramosus, Astragalus siebri, Atriplex portulacoides, Echinops spinosus* and *Halocnemum strobilaceum.* The rare perennial associates include: *Cardunicellus mareoticus, Farsetia aegyptia, Hyoscyamus muticus,* and *Lycium europium.* Annual associates included: *Asphodelus tenuifloius, Centaurea glomerata, Hordeum murinum* subsp. *leporinum* and *Lotus angustissimus.*
4. *H. salicornicum-Atriplex portulacoides communities*
   This community inhabits the saline depression with soil having EC=0.6 mScm-^{1}, low moisture content (2.5%) and organic matter (0.3%). The associated perennials are mixture of halophytes e.g. *Nitraria retusa*, and *Salsola tetrandra* as well as xerophytes e.g. *Anabasis atriculata, Carduncellus mareoticus, Carthamus glaucus, Deverra tortusa, Noaea mucronata,* and *Thymelaea hirsuta etc...* and annuals e.g. *Asphodelus tenuifolius, Astragalus hamosus, Bassia muricata* and *Centaurea glomerata.*

5. *S. tetrandra-L. monopetalum*
   This community inhabits the salt marsh habitat with soil containing relatively higher amounts of soluble salts (EC = 3.3 $mScm^{-1}$) and moisture content (6.7%) but with low organic matter (0.2%). Most of the perennial associates are halophytes and includes: *Arthrocnemum macrostachyum, Atriplex portulacoides, Halocnemum strobilaceum, Limonuim pruinosum,Sporobolus spicatus, Suaeda maritima,* and *Zygophyllum album*. The other perennial associates are xerophytes e.g. *Astragalus sieberi, Deverra tortusa, Lygeuim spartum* and *Thymeleae hirsuta* and annuals: *Astragalus hispidulus, Bassia tournefortii, Centaurea glomerata* and *Hordeum murinum* subsp. *leporinum.*

(iv) Vegetation Types of a Sector Along the Western Desert

In sections (i) and (ii) of Section 3.2.4(c) an ecological account was given of the plant cover of the narrow western Mediterranean coastal belt that extends landwards for about 30 km. The present account includes a description of the different communities of the xerophytic (desert) vegetation of a sector which extends for about 200 km in a northwest-southeast direction crossing the Western Mediterranean coastal belt southward to Cairo (Giza) along the Cairo-Alexandria desert road (Fig. 3.2). The ecological studies carried out on this part of the Western Desert of Egypt include those by Girgis (1970), El-Ghonemy and Tadros (1970), Batanouny and Abou El-Souod (1972) and Ayyad and El-Ghonemy (1976).

Geology and Geomorphology

Depending on interpretation from aerial photographs (FAO/SE: 16/UAR), Batanouny and Abu El-Souod (1972) describe the geological formations of this part of the Egyptian desert. To the south of the study sector is the mountainous area of Abu Rawash. This area has escarpments separated by intervening wadis. The part of the road traversing the Abu Rawash hills to 26 km from Cairo (Giza) crosses a mosaic of Pliocene, Eocene, Cenomanian and Turanian Formations.

From 26 km to 55 km north of Cairo the tableland is almost level, with hills covered with gravels. The terrain is of rather strongly denuded soft formations, presumably of middle Miocene age. North of 55 km north of Cairo are low sandy stretches. The area is rather undulating with gravel deposits predominating on the higher parts.

From north of 90 km to 110 km from Cairo, the terrain is characterized by many small, shallow deposits with short drainage runnels but without outlets. In the higher parts, the gravels are so closely strewn as to form a protective pebble armour.

Further northwards to 135 km from Cairo, the terrain becomes more or less level, with sandy soils covered with pebbles and small gravels. In shallow depressions, coarse sand with fine pebbles predominates on the surface. The elevated parts have gentle slopes and lack vegetation.

From 135 km to 163 km north of Cairo the Cairo-Alexandria desert road traverses an area of wind-blown sand with undulating topography. Dunes of varying heights occur here which decrease northwards and dwindle to a thin blanket of sand

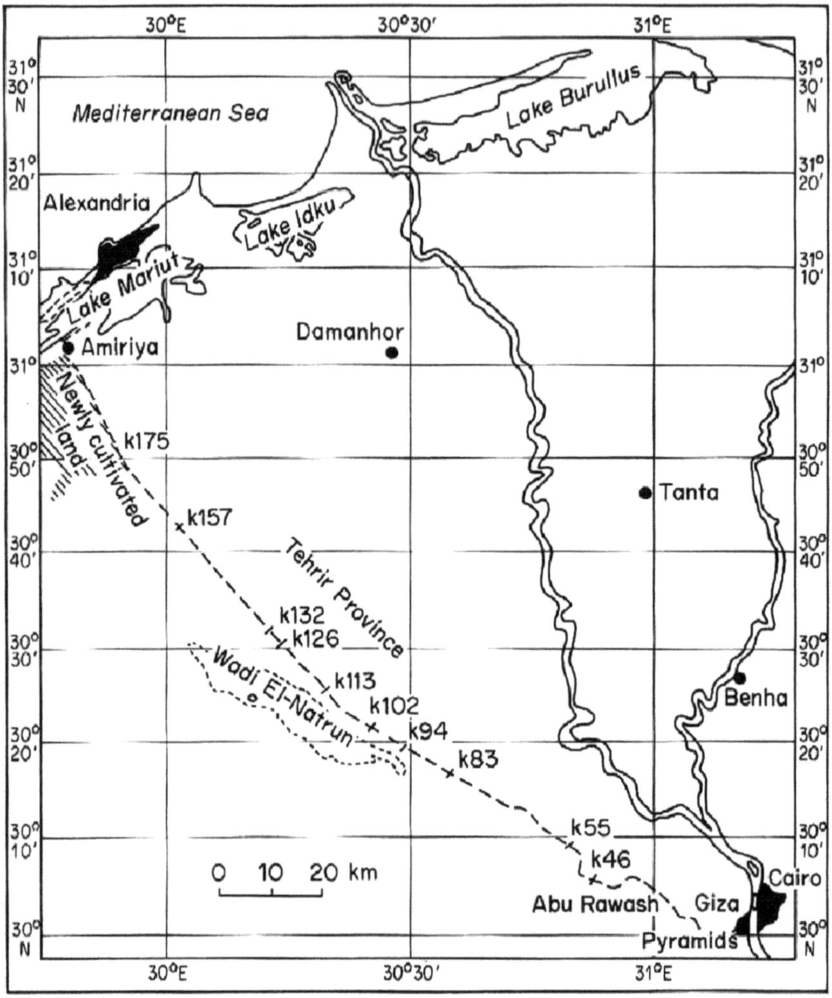

Fig. 3.2 Position of the transect studied in the Western Desert of Egypt

covering the compact soils of the northern part. The soil is very deep and coarse-textured, becoming progressively finer towards the north. The coarse wind-deposited sand is mixed with fragments of land-snails (desert *Helix* spp.) which are very much more common in the northern part of the study section. The landscape is yellowish-brown and becomes increasingly darker towards the north.

In the part bordering the section from 163 km to 170 km north of Cairo, the terrain becomes more or less even with very shallow windblown sand of varying thickness at the surface which is covered with hard lime concretions. Most of the perennial plants in this part form mounds.

From about 170 km northwards to Alexandria, the area is a part of the coastal belt. Here, there is a high ridge on which Amiriya town is situated (about 34 km SW of Alexandria, Fig. 3.2). At the northern foot of this ridge is Lake Mariut, bordered on its northern side by ridges leading to the coastal plain.

Climate

In spite of the relatively short distance from north to south, the studied sector has more than one type of climate. It includes an extremely arid part in the south and an arid one in the north. The records of three meteorological stations, namely Giza to the extreme south, Tehrir near to the middle and Alexandria to the extreme north, represent the climate, but data from Amiriya Station are also relevant.

In the extreme north the average yearly precipitation is less than 200 mm, but it decreases very rapidly southwards. Thus, whereas Alexandria on the Mediterranean coast has an average rainfall of 182 mm, Amiriya has an average of 138 mm, Tehrir which lies at about 85 km north of Cairo has only 38 mm and Giza (Cairo) only 24 mm rainfall. In all stations most of the rainfall is recorded during the November-February period. Summer is usually rainless.

Air temperature shows very similar trends at all stations, being low in winter and high in summer. The range of mean temperature in January and July is 12.1 °C at Alexandria and increases landwards, being 14.8 °C at Tehrir and 15 °C at Giza. The annual range between the mean minima in January and the mean maxima in July are very wide and tend to increase inland, being about 19.5 °C at Alexandria, 27.2 °C at Tehrir and 28.3 °C at Giza.

The relative humidity is high along the Mediterranean coast, being 70% or even more but it decreases landward. Generally, the months with low humidity are those during the blowing of the unfavourable Khamsin winds, mainly in April and May. The mean monthly relative humidity in April is about 57% at Alexandria and 34% at both Tehrir and Giza.

The wind force decreases from high values along the coast to lower ones inland. The range of the mean monthly wind velocity also decreases landward. The rate of air movement varies from completely still air to storms of high velocity: sandstorms of the violent Khamsin winds (= fifty windy days).

Applying Emberger's (1955) formula to the stations considered, the Pluviothermic Quotient[1] is 10.5 for Alexandria, 4.8 for Tehrir and 2.7 for Giza.

Plant Cover

In the extreme southern part of this sector and from Abu Rawash hills to 46 km north of Cairo plant cover is very thin. A few scattered plants of *Stipagrostis plumosa* and *Eremobium diffusum* are found in patches (of 4–25 m^2) separated by extensive areas of barren ground at distances of one or more kilometres. Only isolated specimens of

---

[1] Emberger's Pluviothermic Quotient (rain temperature coefficient) is calculated as follows: 1000 P/(M + m/2) (M – m), where *P* is annual rainfall (mm), *M* mean maximum temperature of the hottest month (°K) and *m* mean minimum temperature of the coldest month (°K). The lower the value of the quotient the more desertic the conditions.

Calligonum comosum are present close to the paved Cairo-Alexandria desert road. The hilly area of Abu Rawash has scanty vegetation, mainly restricted to depressions. *Zygophyllum coccineum* is the abundant succulent xerophyte of these depressions.

From 46 km north of Cairo northwards to the southern border of the Mediterranean coastal belt (i.e. to about 30 km south of Alexandria), nine main communities have been recognized. These are dominated by *Stipagrostis plumosa, Pituranthos tortuosus, Artemisia monosperma* and *Helianthemum lippii* (in the southern section) and *Anabasis articulata, Helianthemum stipulatum, Echiochilon fruticosum, Noaea mucronata* and *Thymelaea hirsuta* (in the northern section).

Communities of the Southern Section

These include the communities present in the area between 46 and 157 km north of Cairo.

1. *Stipagrostis plumosa* community. The landscape supporting the *Stipagrostis* community (from 46 to 94 km from Cairo) is gravelly with sandy areas covered with a thin sheet of wind-blown sand. The vegetation is restricted to low sandy parts. It is notable that wherever there is a fall, even a slight one, in ground level below that of the surrounding area, a sand sheet covers the ground. The scanty vegetation grows on such a thin sheet. The density of the vegetation is affected by the depth of the accumulated sand, being denser on deep soil. The dominant grass collects sand, forming small accumulations 5–15 cm high.

   The soil of the *S. plumosa* community is shallow, hardly penetrable and contains a high percentage of coarse sand (more than 60%). The silt and clay contents increase slightly with depth. The carbonate content is low in the upper layer (2.2–6.0%) and increases abruptly at a depth of 50 cm (17.3%). The total soluble salts are low and show only a slight increase at deep layers (35–67 ppm). Organic carbon content is low (0.012–0.025%) and pH values are higher than 7.

   The *S. plumosa (Aristida plumosa,* Täckholm, 1974) community is very open, with low plant cover, not usually exceeding 5%, and rarely reaching 15% during winter in rainy years. The number of species recorded is 14 perennials and 9 ephemerals. The co-dominant species with considerable presence value are the sand dwellers *Asthenatherum forsskaolii* and *Polycarpaea repens*. The most common associates are *Convolvulus lanatus, Moltkiopsis ciliata* and *Panicum turgidum*. The growth of *P. turgidum* is determined by the soil depth; it is more common and with better growth on deep soils than on shallow ones. It is a sand-binding grass forming phytogenic hillocks of considerable height (50 cm or more). Other associate species include *Artemisia monosperma, Calligonum comosum, Fagonia glutinosa, Helianthemum lippii, Monsonia nivea* and *Zygophyllum album* (perennials) and *Astragalus gyzensis* (= *A. haurensis*), *Eremobium diffusum, Filago spathulata, Ifloga spicata, Launaea cassiniana, Neurada procumbens, Schismus barbatus* and *Senecio desfontainei* (annuals).
2. *Pituranthos tortuosus community*. This community is found in patches scattered in the area from 83 to 113 km from Cairo along the desert road. This area merges

with those bearing communities of *S. plumosa* and *Artemisia monosperma* to the south and north, respectively. A slight variation in the local topography which affects the soil characters, particularly soil texture, may lead to the appearance or disappearance of such communities.

The habitat supporting the *P. tortuosus* community is represented by either runnels crossing the gravelly area or depressions in it. The density of the vegetation and the vigour of the plants are affected by the soil characters, which are closely related to the width and length of the runnels. Wide, long runnels support denser vegetation than short runnels.

Soil is coarse-textured with a hard pan at a depth of about 50 cm. The carbonate content (67–107 ppm) is low showing an increase with depth. The pH values are in the range 8.35–8.8.

The plant cover of this community is low, ranging from <5% to 20%. The plants are stunted and widely spaced except in stands with deep soils. The number of species recorded is 40, including 23 perennials and 17 biennials and annuals. Common associates are *Fagonia arabica, Moltkiopsis ciliata, Polycarpaea repens* and *Stipagrostis plumosa*. Besides the associate species of the *S. plumosa* community, the following are present: *Atractylis flava, Echiochilon fruticosum, Euphorbia kahirensis, Gymnocarpos decander, Launaea nudicaulis* and *Salvia aegyptiaca* (perennials); *Launaea capitata* (biennial); *Erodium laciniatum, Gastrocotyle hispida, Matthiola livida. Ononis serrata, Stipa capensis* and *Zygophyllum simplex* (annuals).

3. *Artemisia monosperma community*. This community is present in areas extending along the Cairo-Alexandria desert road between 102 and 126 km from Cairo. The habitat is sandy with elevated sand dunes increasing northwards. The soils supporting this community are deeper than those of the above-mentioned communities. The soil is more or less loose, with very low salinity (all soluble salts: chlorides, sulphates, bicarbonates etc.) in the upper layers (27–90 ppm) increasing considerably at 50–100 cm (260–273 ppm). The organic carbon content is negligible at the surface and below (0.01–0.02%) and pH ranges between 7.5 and 9.3.

   The vegetation of this community is open with low plant cover, ranging from 5 to 20%. Its flora comprises 45 species (26 perennials and 19 annuals). The most common associates are *Convolvulus lanatus, Moltkiopsis ciliata, Pituranthos tortuosus, Polycarpaea repens* and *Stipagrostis plumosa*. Other associates include some species of the previously described communities as well as *Atractylis flava, Carduncellus eriocephalus, Echinops spinosissimus, Halogeton alopecuroides* and *Plantago cylindrica* (perennials) and *Astragalus bombycinus, Carthamus lanatus, Mesembryanthemum forsskaolii, Orobanche ramosa* and *Polycarpon succulentum* (annuals).

4. *Helianthemum lippii* community. This community is found in sandy areas with slight undulations bordering the road from 132 to 157 km from Cairo. The soil is loose without gravels. Shells *of Helix* spp. are scattered on the soil surface. Wind plays a great role in the plant life, leading to stunted growth, especially in the southern area. The plant cover of this community is low in the southern stands, not

exceeding 15%. Northwards the cover increases and may rise to 40% in the winter months with a minimum of 15% in summer. The number of species recorded is 26 perennials and 13 annuals. *Convolvulus lanatus* and *Cyperus capitatus* are the frequent associates. Common perennials present include *Anabasis articulata, Aristida ciliata, A. plumosa, Artemisia monosperma, Asthenatherum forsskaolii, Cornulaca monacantha, Dipcadi erythraeum, Echiochilon fruticosum, Gymnocarpos decander, Helianthemum lippii, Moltkiopsis ciliata, Plantago cylindrica, Polycarpaea repens* and *Thymelaea hirsuta. Anabasis articulata* and *Thymelaea hirsuta* are recorded only in the northern stands. Of the annuals the most common are *Cutandia dichotoma, Erodium laciniatum, Lotus pusillus* and *Silene villosa*; all are sand dwellers. Other therophytes include *Centaurea pallescens, Eremobium diffusum, Ifloga spicata, Medicago hispida, Neurada procumbens, Ononis serrata, Scabiosa arenaria* and *Schismus barbatus*.

Communities of the Northern Section

The northern section of this sector occupies the area from 157 km north of Cairo and extends northwards to the southern border of the coastal belt. This is the Mariut Tableland (Mariut Plateau). In this plateau, five communities dominated by *Anabasis articulata, Helianthemum stipulatum, Echiochilon fruticosum, Noaea mucronata* and *Thymelaea hirsuta* have been recognized.

1. *Anabasis articulata* community. The habitat of this community is characterized by a slightly undulating landscape with elevated mounds. The dominant plant as well as some associated perennials, e.g. *Astragalus spinosus, Lycium europaeum, Noaea mucronata, Salsola tetrandra* and *Suaeda pruinosa* (not recorded in the southern communities of this sector) can form mounds. Other species occupy the ground between these mounds.

   Soil supporting the *Anabasis articulata* community is shallow with a hard crust on the surface and becomes harder with increasing depth. The layers below 20 cm are compact with lime concretion. The accumulated soil around the plants forming the mounds is looser and darker than that between the mounds. The mounds 20–50 cm high are of accumulated wind-blown and/or water-borne material. Soil characters are widely different from those supporting other communities. In this community, soil has the highest moisture content (4.9–15.7%) whereas in the soils of the other communities this value never exceeds 9%. Also, the soils of the *A. articulata* community contain the highest salinity (51–1275 ppm), carbonate (11.2–19.5%), and organic carbon (0.06–0.08%) content. The carbonate, salinity, moisture content and organic carbon content of this sector's soils increase northwards.

   The plant cover of the *A. articulata* community is higher than in the communities southwards, being 40% on average and reaching 70% in the rainy season. The number of species in the southern stands is smaller than that in the northern ones. This community has the highest number of associates, 72, half of which are perennials. The most common associate perennials are: *Astragalus spinosus, Atractylis flava, Helianthemum lippii, Launaea nudicaulis, Noaea rnucronata,*

Salsola tetrandra, Stipagrostis plumosa and *Thymelaea hirsuta*. The last species shows higher abundance in this community than in that dominated by *Helianthemum lippii*. Common perennial associates are *Argyrolobium uniflorum, Artemisia inculta, Echiochilon fruticosum, Gymnocarpos decander, Helianthemum kahiricum, Plantago albicans, P. cylindrica, Salvia aegyptiaca*, and *S. lanigera*. The less common perennial associates include *Astragalus trigonus, Atriplex stylosa, Convolvulus althaeoides, C. lanatus, Cynodon dactylon, Echinops spinosissimus, Haplophyllum tuberculatum, Herniaria hemistemon. Pancratium sickenbergeri, Scolymus hispanicus, Scorzonera alexandrina* and *Traganum nudatum*. Although *Artemisia monosperma* dominates a community in the southern part of this sector, it is absent when *Anabasis articulata* dominates.

The associate annuals include most of the species recorded in the other communities to the south as well as: *Anthemis microsperma, Asteriscus pygmaeus, Avena alba, Bupleurum semicompositum, Carduus getulus, Carrichtera annua, Crucianella herbacea, Daucus syrticus, Diplotaxis simplex, Hippocrepis bicontorta, Koeleria phleoides, Launaea resedifolia* (biennial), *Mesembryanthemum nodiflorum, Medicago minima, Onobrychis crista-galli, Parapholis incurva, Plantago coronopus, Silene oliveriana, Sonchus oleraceus* and *Trigonella maritima*.

2. *Helianthemum stipulatum* community. *H. stipulatum* occupies the highest level area with the lowest water resources and the deepest water-table. The surface deposits are friable coarse-medium sand with very little of the finer fractions. These deposits are poor in moisture storage. Plant cover is very sparse, not exceeding 10%, contributed mainly by the dominant. *Thymelaea hirsuta* and *Echiochilon fruticosum* are the abundant associates. The commonly present species are *Anabasis articulata, Artemisia monosperma, Asthenatherum forsskaolii, Convolvulus lanatus, Cyperus conglomeratus, Gymnocarpos decander, Pituranthos tortuosus* and *Stipagrostis plumosa*. Other associates include *Astragalus spinosus, Atractylis flava, Carduncellus eriocephalus, Centaurea pallescens, Moltkiopsis ciliata, Noaea mucronata, Polycarpaea repens, Salsola tetragona* and *Traganum nudatum*.
3. *Echiochilon fruticosum* community. According to Girgis (1970), this community is closely related to that dominated by *H. stipulatum*; both occupy comparable areas with respect to relief and surface deposits. The plant cover of the *E. fruticosum* community is thin (5–10%), mostly contributed by the dominant shrublets. The abundant species recorded in all of the stands are *Anabasis articulata, Convolvulus lanatus, Helianthemum stipulatum* and *Thymelaea hirsuta* making up much of the plant cover. The common associates are *Atractylis flava, Centaurea pallescens, Gymnocarpos decander, Pituranthos tortuosus* and *Stipagrostis plumosa*. Other less common species include *Artemisia monosperma, Asthenatherum forsskaolii, Astragalus spinosus, Cyperus conglomeratus, Launaea nudicaulis, Lycium europaeum, Moltkiopsis ciliata, Noaea mucronata, Salsola tetragona, Salvia lanigera* and *Traganum nudatum*.
4. *Noaea mucronata* community. The community dominated by *N. mucronata* is less common in the Mariut Plateau. It is represented by a few patches, mainly confined to the northern boundaries. The surface deposits are sandy with appreciable amounts of fine fractions. The catchment areas of the patches

dominated by *N. mucronata* are greater than those of the other communities of the Mariut Plateau, and hence the water resources are higher. The soil is heavier, with greater moisture storage and availability of moisture to plant growth.

N. mucronata, a thorny shrub, gives the character of this phytocoenosis. The total plant cover ranges between 15 and 25%, contributed mainly by the dominant and partly by the most abundant associates *Anabasis articulata, Gymnocarpos decander* and *Thymelaea hirsuta*. Other common associates include *Aristida plumosa, Atractylis flava, Centaurea pallescens, Echiochilon fruticosum, Helianthemum stipulatum, Pituranthos tortuosus* and *Salsola tetragona*. Less common are, e.g. *Artemisia inculta, Convolvulus lanatus, Cynodon dactylon, Cyperus conglomeratus, Halogeton alopecuroides, Launaea nudicaulis, Linaria aegyptiaca, Lycium europaeum* and *Salvia lanigera*.

5. *Thymelaea hirsuta* community. *T. hirsuta* is the most widespread species in the Mariut Desert. Its community is also the commonest and occupies about 40% of the whole area (Girgis, 1970). The catchment area where *T. hirsuta* dominates is rather larger than those of the previous communities. The surface deposits are sandy and overlie the calcareous consolidated loamy soil. The presence of this consolidated layer seems instrumental in the establishment of the permanently wet underground layer favourable for this community.

   The cover of the *T. hirsuta* community ranges between 15 and 30%, contributed mainly by the dominant shrub and partly by the abundantly present associates, namely: *Anabasis articulata, Centaurea pallescens, Convolvulus lanatus, Echiochilon fruticosum, Gymnocarpos decander, Helianthemum stipulatum* and *Noaea mucronata*. Common associates are *Aristida plumosa, Atractylis flava* and *Pituranthos tortuosus*. Less common are *Artemisia monosperma, Asthenatherum forsskaolii. Astragalus spinosus, Carduncellus eriocephalus, Cyperus conglomeratus, Halogeton alopecuroides, Linaria aegyptiaca, Lycium europaeum, Moltkiopsis ciliata, Panicum turgidum* and *Traganum nudatum*.

3.3 The Oases and Depressions

3.3.1 Origin and Formation

The Western Desert of Egypt is characterized by a number of oases and depressions, namely: Siwa Oasis, Moghra Oasis, Bahariya Oasis. Farafra Oasis, Dakhla Oasis, Kharga Oasis, Kurkur Oasis, Dungul Oasis, Qattara Depression, Wadi El-Natrun Depression and other small oases and depressions (Fig. 2.1).

The origin of these oases and depressions has long been the subject of controversy. According to Said (1962) because these depressions are closed inland basins with no access to the sea, they demonstrate clearly the effect of deflation in arid regions; their formation is the result of arid action (as mentioned by Ball, 1927), the depth of their floor being governed by ground water level. Large quantities of the sandy constituents are removed by the wind from a great chain of sand dunes.

Beadnell (1909) believed that the greater part of the floor of the Kharga Oasis was at one time the site of an immense lake which originated in early prehistoric or late Pleistocene times when the climate was much more humid than at present. He postulated that there are no definite grounds for considering that the erosion of the depressions of the Western Desert result from the action of previously existing rivers and no evidence for assuming that they were formed by local subsidence of portions of the earth's crust. Hume (1925) contended that there must have been a vast primary marine denudation as the anticlinal area now occupied by the oases was formed beneath the sea. The agencies of sand abrasion and wind transport have subsequently carved the depressions in the Western Desert. Ball (1939) stated that enormous amounts of subaerial denundation took place during Pleistocene and Recent periods. The final sculpturing of the depressions has been accomplished by the erosive action of sand-laden winds. However, M.M. Ibrahim (1952, unpublished) supported the view of wind action only as a secondary factor and emphasized the effect of static electrical charges on wind erosion.

El-Shazly and Shata (1960) believed that tectonic and stratigraphic conditions are the essential causal mechanisms in the formation of the depressions. They considered the role of wind as comparatively recent and as essentially depositional rather than eroding. On the other hand, Said (1960) excluded the possibility of a tectonic origin for these depressions. He believed that the uplifting was not accompanied by significant tensional stresses.

Ezzat et al. (1968) concluded that there were steps of land rises before the formation of the depressions of the Western Desert. The rises caused the substantial regression of the shore-line since the Lower Eocene. Vast marine denudation action on the elevated parts gave characteristic deposits such as the conglomeratic bed underlying the Middle Eocene formations in some localities. Thus, these depressions resulted from cycles of evolution which involved rising, surface erosion, thinning of the cover and ended with subsidence movements that formed the depressions which often became basins collecting drainage water.

3.3.2 Historical Relationships

The history of the relations between the oases of the Western Desert of Egypt and the Nile Valley has been discussed by Mitwally (1952) who divided that history into five periods: the Prehistoric Period, the Dynastic Period, the Classical Period, the Mediaeval Period and the Modern Period.

(a) The Prehistoric Period

The Western Desert of Egypt at present is one of the most arid and desolate parts of the world. However, during the Prehistoric period climatic conditions were favourable, supporting the growth of grasses, trees and ferns, particularly on the edges of the oases. Recent archaeological discoveries show that man roamed all over the desert in this period. The discoveries are mainly confined to rock pictography and stone implements, mostly incised on exposed boulders, hillsides and faces of escarpments. Subjects represented are animals, e.g. ibex, oryx, gazelle, cattle and occasionally ostrich and human activities.

(b) The Dynastic Period

The interrelationship between the oases and the Nile Valley in the Dynastic period is shown not only by hieroglyphics depicted on the walls of the tombs and temples of the Ancient Egyptians in the Nile Valley, but also by archaeological finds in the oases themselves. All major oases known at present were also known to the Dynastic Egyptians as follows:

1. Siwa was known to them as "Sekhet Amit" or the field of date palms;
2. Bahariya was known as "Ouhat Meht" or the oasis of the north (Bahariya is the Arabic word for northern);
3. Farafra was known as "Ta-ahed" or the land of the cattle;
4. Dakhla was known as "Desedes" (Aset Ahed as capital) or the seat of the Moon God;
5. Kharga was known as "Kenmit" or "Uahat Rist", the oasis of the south;
6. Wadi El-Natrun Depression was known as "Sekhet Heman" or the salt field.

During the rule of the Libyan Dynasty (22nd Dynasty) there were the routes of donkeys and horses between the oases and the Nile Valley.

(c) The Classical Period

This period began with the Persian invasion of Egypt in 525 BC and ended with the Arab conquest in 640 AD. Many wells were dug and aqueducts constructed in Kharga, Dakhla, Farafra, Bahariya and Siwa and extensive areas of land were accordingly put under cultivation. Many of these old wells are still discharging large volumes of water.

From the economic stand-point the oases gave great prosperity. The land was thoroughly cultivated and there was trade in wine and many other commodities produced in the oases. At the beginning of this period the use of the camel for desert transport was first introduced into Egypt and numerous caravan routes connected the oases with the Nile Valley.

(d) The Mediaeval Period

This period started in 640 AD, when Egypt was invaded by the Islamic Army, and extended to the beginning of the 19th century. That commercial relations existed between the Oasis Parva (Bahariya) and the Oasis Magna (Kharga and Dakhla) on one hand and the Nile Valley on the other during the Roman period, coupled with the proximity of the oases to the valley and their accessibility to the Arabs who were desert travellers by nature, suggests that the Arabs dominated the oasis region from the very beginning of their invasion of Egypt. The oasis dwellers accepted the Arab rule and adopted Islam without any bloodshed. The fact is well manifested in the Friday ceremonial mass meeting of the Moslems of the oases.

(e) The Modern Period

Early in the 19th century Mohamed Ali Pasha came to the throne of Egypt. He initiated an era of improvement, involving nearly all aspects of life. The influence of this movement reached as far as the oases, e.g. mineral exploitation and drilling of wells. By the beginning of the 20th century a railway connected the Kharga Oasis with the Nile Valley.

In recent years, the Egyptian government has been giving more attention to the oases of the Western Desert. The limited area of the land of the Nile Valley and Nile Delta (<4% of Egypt) has become so narrow that the density of people here is among the highest in the world (>800 persons/km^2). The population of Egypt increased from less than 6,000,000 at the start of the 20th century to more than 55,000,000 in 1990 and expected to be 90 000 000 on 2020. However, the population of the oases is very small (1 person/7 km^2 in the desert including the oases). Thus, the oases are the hope of the future. The many developments to use the oases and their resources include paved road constructions between the big cities of the Nile Valley and the Mediterranean coast and the oases, e.g. Matruh-Siwa road, Giza-Bahariya road, Assiut-Kharga road; land reclamation; well drilling; house building; plantations and introduction of new varieties; development of olive production; mining. Such developments will attract Egyptians to the oases.

3.3.3 Ecological Characteristics

Ecologically, the oases and depressions of the Western Desert of Egypt may be categorized into two groups: the northern and southern (Fig. 2.1). The northern group includes oases and depressions located north of Latitude 26 °N and the southern group include those located south of that latitude.

(a) The Northern Oases and Depressions

These include (from north to south) Wadi El-Natrun Depression, the Qattara Depression, Moghra Oasis, Siwa Oasis, Wadi El-Rayan Depression, Bahariya Oasis and Farafra Oasis. An ecological account on 6 of these is given.

I. Wadi El-Natrun Depression, Bir El-Shab

General Remarks

The Wadi El-Natrun Depression is situated west of the Nile Delta. It is a NW-SE depression between Lat. 30'17' and 30°38'N and between Long. 30°2' and 30'30'E (Fig. 3.2). Its northern extremity is 180 km from Alexandria and its southern limit is 70 km from Cairo.

Wadi El-Natrun, which is mentioned in the writing of such scientists as Lucas (1912), Hume (1925), Tousson (1932), Sandford and Arkell (1939), Pavlov (1962) and Abu Al-Izz (1971), is about 3 km long. It is narrow at both ends (2.6 km in the north and 1.24 km in the south) and wider in the middle (8 km). It lies 23 m below sea level and is characterized by the lakes in the bottom of the wadi, aligned with its general axis. Except for Lake Al-Gaar, the lakes are closer to the northwest side of the depression. The principal lakes (Fig. 3.3) are as follows:

1. Lake Fasida, which is oval and about 8 km from the southern end of the depression. Its area is 1.5 km^2. It dries up completely during the summer and is 21 m below sea level in the north. On the bottom of the lake thick deposits of salts have accumulated. The water of the lake is reddish and the amount of natron (sodium carbonate) which it contains is very low but the lake is surrounded by crusts of natron.
2. Lake Um Risha (2.9 km^2) is 21.9 m below sea level. About two-thirds of this lake is dry in the summer and there are thick deposits in its bottom. The amount of natron is limited and its water is reddish.
3. Lake Al-Razoniya (1.05 km^2) dries up in summer and contains little natron.

4,5. Lakes Abu Gubara and Hamra form one lake during the summer (as a result of water seepage from the summer floods of the Nile) and separate during the rest

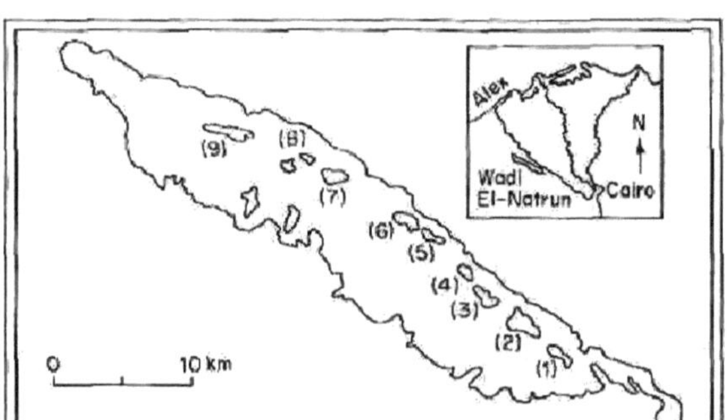

Fig. 3.3 Wadi El-Natrun Depression, Western Desert. 1, Lake Fasida; 2, Lake Um Risha; 3, Lake Al-Razoniya; 4, Lake Abu Gubara; 5, Lake Hamra; 6, Lake El-Zugm; 7, Lake Al-Bida; 8, Lake Khadra; 9, Lake Al-Gaar

of the year. Their combined area is 2.1 km^2. However, as no floods now occur (with the construction of the Aswan High Dam) the water level of the lakes no longer increases.

6. Lake El-Zugm (1.9 km^2) is in the centre of the wadi; it also dries up in summer and has deep natron deposits.
7. Lake Al-Bida is the largest lake in the wadi (3.5 km^2); it dries up in summer and has the highest salinity, with little natron on the lake bed.
8. Lake Khadra (0.77 km^2) is greenish and dries up during the summer.
9. Lake Al-Gaar (1.9 km^2) is in the extreme north of the depression; it never dries up.

El-Fayoumi (1964) distinguished three groups of recent lacustrine deposits of these lakes:

1. Fresh-water marshes. These occupy especially the northern portion of the lakes and are formed where the fresh-water table cuts the bottom surface of the depression. The surface deposits comprise a complex of dark clayey soil resulting from weathering of the gypseous materials, aeolian sand and decaying organic matter.
2. Salt deposits. These are mostly beneath the shallow water of the salt lake. Such deposits are especially rich in natural soda or natron.
3. Wet salt marshes. These occupy the areas immediately east of the lakes, i.e. the areas affected by seasonal fluctuations of the lake water. Surface deposits of these salt marshes are essentially aeolian grains of quartz cemented by salts resulting from the evaporation of the lake water.

The second geomorphological feature of the Wadi El-Natrun Depression is the formations of aeolian sand deposits (loose quartz) on the west of the lakes. These formations are represented by hummocks bordering the lakes and undulating sheets of sand further westwards. This part is also below sea level, the ground level showing a gradual rise westward, associated with a gradual decrease in the salt water.

To the south and west of the depression is the sand-and-gravel country which includes fossil wood. These gravel deposits are associated with a network of water runnels lined with alluvial sand and silt.

The Wadi El-Natrun Depression gets its water from two sources – the springs in the bottom (e.g. in Lake Hamra), and seepage into the lakes. The direction of the lateral seepage is generally from the northeast. There is a hydrostatic connection between the Nile (Rosetta Branch) and the depression. This connection is confirmed by Shata et al. (1962a, b) and Abu Al-Izz (1971).

The climate of Wadi El-Natrun is arid: low and very variable rainfall, a long dry summer, a high rate of evaporation, low humidity etc. The annual rainfall is about 55 mm with November and December having the greatest amount. The mean annual temperature varies from 22.8 °C in January to 28.8 °C in August; mean monthly relative humidity ranges from 52% in May to 70% in November–December. Wind velocities range between 11 and 20 km/h in winter and summer respectively, winds being usually from the north, northeast and northwest.

The Plant Cover

From an ecological point of view Wadi El-Natrun includes two principal ecosystems, namely: the salt marsh ecosystem of the depression proper and the gravel desert ecosystem of the surrounding highlands (Zahran and Girgis, 1970).

Salt Marsh Ecosystem

The vegetation of this system is much influenced by soil salinity and level of the ground water. Both factors are subject to seasonal fluctuations. The level of the ground water-table is nearer the surface in autumn and winter and is at a greater depth in spring and summer. The areas adjacent to the lakes are, consequently, affected by these changes. The extent of this effect is primarily dependent on relief. The localities with lowest relief have continuous underground water feed, swamp conditions predominating. These localities are always continuous with the lakes and represent the typical reed habitat. The successively higher ground to the east of the lakes is rarely inundated but is saturated and slippery in the wet season and compact in the dry season. Where the water-table is shallow, the soil is darkish-brown and very rich in organic matter; evaporation and lack of rainfall with no leaching increase the salinity and create the wet salt marsh habitat and its communities. On the west side of the lakes the ground rises and the underground water is deeper. Also, the extent of the salt-water seepage from the lakes is very little. The sandy soil is relatively dry but still saline, with a low content of organic matter. This is the habitat of dry salt marsh communities that extend westward to the approach of the gravel desert.

A third type of habitat affected by the water of the lakes and the shallow underground water is the sand terraces (sand embankments) to the east of the lakes. This is the habitat of the halfa grassland type. Thus, three vegetation types are associated with these types of habitat, namely:

1. Reed swamp vegetation
2. Salt marsh vegetation
    a) Wet salt marsh communities
    b) Dry salt marsh communities
3. Halfa grassland

Reed Swamp Vegetation

This vegetation grows in the swamp lands bordering the lakes where there is a rich and continuous feed of fresh and brackish waters. The soil is muddy and rich in organic matter. Two species of *Typha* dominate: *T. elephantina* and *T. domingensis*. The former is widespread in Wadi El-Natrun but its presence elsewhere in Egypt is uncertain (Boulos, 1962). *T. domingensis* is recorded wherever marshy conditions prevail in Egypt. In Wadi El-Natrun limited patches of *T. domingensis* are recorded in some of the swamps. *T. elephantina* is abundant along most of the lakes. It forms extensive reed thickets in the swamps and also on sandy terraces bordering the lakes from the west. The distribution of *T. elephantina* within Wadi El-Natrun deserves special note. Along the western side of the four intermediate lakes (Hamra, El-Zugm, Al-Bida and Abu Gubara) are extensive stretches of sand covered with thickets of *T. elephantina*. These extend landward for about 3 km to the west of Hamra lake. Dense thickets of *T. elephantina* are present in the swamps of the northern lake (Al-Gaar, 15 km north of Wadi El-Natrun Village) whereas it has been almost totally exterminated from Al-Razoniya lake very near to the village, *Typha* is clearly thinning at a higher rate in the lakes nearer to the village but is still dense in the lakes distant from it. However, it may be noted that *Typha* thickets are gradually thinning everywhere and that the process of regeneration is rather poor. Very dense thickets of *Typha* were recorded by Stocker (1927). Information provided by local inhabitants indicates that in the past few decades, *T. elephantina* formed very dense thickets which were almost continuous along all the lakes. Among the obvious causes for the decline of *Typha* are cutting for fuel and for making mats and huts, grazing and lowering of the water-table due to gradual thickening of the sand deposits and to land reclamation operations. As the prevailing wind is northwest, the growth of *Typha* on the dunes west of the lake acts as windbreaks, preventing the encroachment of sand on the lake and the villages lying to the east. The destruction *oiTypha* will leave the dunes bare and cause their movement towards the lake and villages.

The swamps of *Typha* provide a suitable habitat for several water-loving plants. These include *Berula erecta, Cyperus articulatus, C. mundtii, C. papyrus, Lemna gibba* (El-Hadidi, 1971) and numerous algae. On the sand dunes the growth of *T. elephantina* is associated with *Desmostachya bipinnata, Nitraria retusa, Sporobolus spicatus* and *Zygophyllum album.*

Salt Marsh Vegetation

The outstanding feature of this habitat is the high salt content of the soil. According to the relief, degree of saturation and concentration of the salt in the soil, communities of the salt marsh vegetation of Wadi El-Natrun may be classified into wet and dry salt marsh communities.

Wet Salt Marsh Communities

These are closely associated with the eastern side of the lakes where the soil is saturated with water, especially in the wet season. The water is shallow or exposed. During the dry season evaporation is high and the effect of water washing is limited so salinity increases. Thus, the degree of salinization of the soil is changeable throughout the year. This is the habitat where the *Cyperus-Juncus* complex is composed of two strata: an upper stratum (restricted) dominated by *Juncus acutus* and a lower stratum (extensive) dominated by *Cyperus laevigatus*. The plant cover is usually high (up to 100%). *J. acutus* is often cut for making mats. *C. laevigatus,* on the other hand, is severely grazed; the height of plants does not exceed a few centimetres and inflorescences are rare.

Differences in land level have a profound effect on plant growth. Generally the level is gradually sloping towards the lakes. The elevated areas are relatively dry and have salt crusts on the surface. In these areas *C. laevigatus* is dominant with an abundance of *Paspalidium geminatum* and *Sporobolus spicatus. P. geminatum* is locally dominant (cover of 90%) in patches which are covered with powdery salt (total soluble salts 7.6%). As the level decreases (towards the lakes) the water content of the soil increases and the soil becomes less saline (total soluble salts 0.95%) due to the leaching effect of water drainage from the higher ground. In these areas adjacent to the lakes, *J. acutus* and *C. laevigatus* co-dominate with an almost complete cover of *Cyperus* lawns. Associates are *Berula erecta, Panicum repens* and *Samolus valerandi*. Limited patches of *Juncus rigidus* are recorded with *J. acutus* in certain localities.

Dry Salt Marsh Communities

These are the communities of the sand formation that occupy ground higher than that of the wet marshes. The sandy deposits are saline, moist in the subsurface layers and poor in organic matter. Evaporation often leaves a mantle of salt on the surface. These dry marshes extend further from the lakes (east and west) where the water-table is relatively deeper. The dominant species of these communities can build sand mounds (*Sporobolus spicatus* and *Zygophyllum album*) or sand hillocks (*Nitraria retusa* and *Tamarix* spp.).

1. *Sporobolus spicatus* community. S. *spicatus* is a halophytic grass widely distributed in the Egyptian inland and littoral salt marshes (Kassas and Zahran, 1967; Zahran, 1982a). It is a good sand binder and may build mounds of moderate size. It is heavily grazed by goats and camels up to a certain height where its tussocks become rigid and spinescent.

   In Wadi El-Natrun, the *S. spicatus* community covers wide areas, particularly west of the lakes. On the east side of the lakes the relatively high salt content of the soil (up to 4.4%) seems to limit the number of species. S. *spicatus* in this part usually forms pure stands or is associated with a few halophytes, e.g. *Cyperus laevigatus, Juncus acutus, Nitraria retusa* and *Zygophyllum album*. West of the lakes this community is widespread especially in the low parts where the water-table is shallow and the soil contains less soluble materials (0.35–0.15%). The number of the associate species is higher. In addition to those mentioned above *Artemisia monosperma, Panicum turgidum, Phoenix dactylifera* and *Saccharum spontaneum* v. *aegyptiacum* are also present.
2. *Zygophyllum album* community. *Z. album* is of wide ecological amplitude. In the littoral salt marshes its community may be present within any of the salt marsh zones (Kassas and Zahran, 1967). In the inland deserts, the *Z. album* community is recorded in the wadis of the limestone country (Kassas and Girgis, 1964). In Wadi El-Natrun this community is common in the sand formations west of the lakes more distant than areas occupied by the *S. spicatus* community. The deep deposits are mainly sands with a relatively high content of soluble material. S. *spicatus* is also recorded in water runnels lined with sand deposits and dissecting the gravel deposits where the salinity is lower. In both situations the deposits are mainly loose sand medium-coarse in texture with low contents of silt and clay.

   The growth of *Z. album* varies according to habitat conditions. In water runnels, plants are stunted and build no sand mounds. In some of the localities where the catchment areas are wide, *Z. album* grows luxuriantly and builds hummocks.

Z. album is a succulent undershrub rarely grazed. The flora of its community includes 18 associate species which indicate that conditions are more favourable for growth than in the community previously described. The majority are characteristic of sandy habitats and there are some salt marsh species. Variability of species assemblage is dependent on the localities: stands nearer to the lakes include halophytes whereas those nearer to the desert plain include xerophytes. The species of the first category include *Cyperus laevigatus, Nitraria retusa, Sporobolus spicatus* and *Tamarix nilotica*. Species of the second category include *Alhagi graecorum, Artemisia monosperma, Asthenatherum forsskaolii, Calligonum comosum, Convolvulus lanatus, Cornulaca monacantha, Helianthemum stipulatum, Monsonia nivea, Neurada procumbens, Panicum turgidum, Polycarpaea repens* and *Stipagrostis plumosa*. There are a few trees of the semi-wild *Phoenix dactylifera*.

The phytocoenosis of the *Z. album* community often consists of three layers:

(b) (a) Frutescent layer including *Calligonum comosum, Nitraria retusa, Phoenix dactylifera* and *Tamarix nilotica*; suffrutescent layer including *Artemisia monosperma, Panicum turgidum* and *Z. album;*
(c) Ground layer including *Cyperus laevigatus, Monsonia niuea* and *Neurada procumbens*.

3. *Nitraria retusa* community. *N. retusa* in Wadi El-Natrun is recorded in some of the water runnels to the west of Hamra Lake and on terraces east of Khadra and Al-Gaar Lakes.

   N. retusa builds huge hummocks and hillocks. Most of these, because of prolonged drought, are barren with recognizable dead remains of *Nitraria*. The hillocks are widely spaced and may be mixed with *Tamarix* hillocks. *Z. album* and S. *spicatus* may grow between these hillocks.
4. *Tamarix* spp. communities. *Tamarix* shrubs are subject to destructive cutting for fuel and other household purposes. In Wadi El-Natrun two species of *Tamarix* are present – *T. nilotica* and *T. passerinoides* v. *macrocarpa*. The first is a common shrub in the littoral and inland salt marshes of Egypt whereas the second is very common in Wadi El-Natrun though rare elsewhere in Egypt. It is recorded by Zahran (1962) in El-Mallaha marsh near Ras Gharib along the Gulf of Suez.

   In Wadi El-Natrun *Tamarix* scrub is confined to localities with extensive catchment areas (e.g. west of Hamra Lake) or high sand dunes with fresh underground water supplies (e.g. the sand dunes west of the lakes). In localities with catchment areas, *Tamarix* forms huge hillocks of fine sand that contain considerable amounts of soluble material (3.4%). The ground deposits include much alluvium which is a mixture of coarse and fine sand and silt. Many of the Tamarix-built hillocks lack living cover. The deposits of these hillocks are aeolian. Associate species include *N. retusa, S. spicatus* and *Z. album*.

   Tamarix spp. are excretive halophytes (Walter, 1961). They excrete salt from their leaves and branches; these crystals are hygroscopic and can absorb atmospheric moisture. In the early morning drops of water are seen falling from the leaves and branches.

The Halfa Grassland

The habitats of this type of vegetation are sand terraces or sand dunes associated with the eastern side of the lakes. The grassland forms a continuous zone eastward of that of the S. *spicatus* community. The dominant grass is *Desmostachya bipinnata.* West of the lakes this grass is confined to some oasis-like depressions with limited patches of soft sand. In both sites the water-table is not deep. *D. bipinnata* is a rigid grass which may reach 150 cm high but as it is extensively grazed and often burned its height rarely exceeds 50 cm. In a protected area east of Al-Razoniya Lake, *Desmostachya* is an exceptionally abundant associate. *D. bipinnata* is an effective sand binder and protects the soil against erosion. Associate species of this community are *Artemisia monosperma, Juncus acutus, Panicum turgidum, Phoenix dactylifera, Sporobolus spicatus* and *Tamarix* spp. The growth of date-palms is an indicator of the presence of a freshwater layer (Abdel Rahman et al., 1965b). Economically, *D. bipinnata* could be considered as a fiber producing plant. It contains 45% crude fiber, 1.96% crude protein, 0.31% total nitrogen & 13.6% moisture (Eisa, 2007).

The Gravel Desert Ecosystem

The desert surrounding the Wadi El-Natrun Depression is a gravel part of the Western Desert dissected by drainage runnels which vary in size. Therophytes and a few perennials with short root systems grow in the smaller runnels that are lined with fine sand. Larger runnels have deeper deposits that favour the growth of perennials with deep roots.

Plants in this sand-and-gravel desert depend mainly on the amount of rain, since it is far from the extent of seepage of lake water. The growth of ephemerals is subject to notable seasonal variations. The noticeable feature of this ecosystem in the Wadi El-Natrun area is the mosaic pattern of the vegetation suggesting that the plants are affected by several interacting factors rather than a single dominant. Most of the perennials are capable of producing new shoots and branches when buried with sand; this advantageous feature enables the plants to survive burial.

Two communities have been recognized in this ecosystem dominated by *Artemisia monosperma* and *Panicum turgidum.*

1. *Artemisia monosperma* community. *A. monosperma* occurs in the main runnels cutting across the gravel and slopes of the Wadi El-Natrun area. The surface deposits are loose mixed sand. The water-table is deep and the salt content is very low.

   The dominant is an undershrub that builds mounds. It is highly palatable and extensively grazed. The plant cover is variable: in the small runnels it is very low (<5%) whereas in the large runnels it is greater (20–30%). The flora of this community includes 13 associate perennials and four ephemerals *(Cotula cinerea, Euphorbia granulata, Ifloga spicata* and *Neurada procumbens),* all characteristic of sandy habitats. The perennials include *Asthenatherum forsskaolii, Convolvulus lanatus, Monsonia nivea, Polycarpaea repens* and *Stipagrostis plumosa* as common associates. Less common associates are *Calligonum comosum, Cornulaca monacantha, Echinops spinosissimus, Eremobium aegyptiacum, Helianthemum stipulatum, Moltkiopsis ciliata* and *Zygophyllum album.*

2. *Panicurn turgidum* community. The presence of the *P. turgidum* community in Wadi El-Natrun gravel desert is restricted to the channels of the main runnels west of the lakes where the bed deposits are deep fine sand, e.g. the wadis at the foot of the old Suriania and Anba Bishoi Monasteries. East of the lakes *P. turgidum* is recorded on fine sand terraces of Al-Gaar Lake.

The dominant is a tussock-forming plant that builds mounds of moderate size, being an active sand binder. This advantage is offset by its being a favourite fodder grass that is extensively grazed. The plant cover of this community ranges from 15 to 20% and its flora includes associates characteristic of sand habitats: *Artemisia monosperma, Centaurea glomerata, Convolvulus lanatus, Cornulaca monacantha, Helianthemum vesicarium, Heliotropium luteum, Moltkiopsis ciliata, Phlomis floccosa, Stipagrostis plumosa* and S. *scoparia.*

(ii) The Qattara Depression

The Qattara Depression is the largest depression in the Western Desert (Fig. 2.1) and is one of the greatest depressions in the world. It is bordered by high scarps on the north and west, while it is open to the east and south. This makes it difficult to calculate the area of the depression accurately, unless the zero contour line is taken as its boundary. On this basis, its length from NE to SW is 289 km, its width is 145 km and its area is 19,500 km^2.

There are two oases within the depression: Moghra and Qara. The first is at the north eastern end (about 56 km from the Mediterranean coast); the other is at the western edge, about 80 km from the nearest settlement in Siwa Oasis.

The average elevation of the floor of the depression is about 60 m below sea level. The lowest spot is −134 m, south of Qara Oasis. More than two-thirds of the depression lies below −50 m. About 26.3% of the area is covered by playas, occasionally filled with water. In most places the playas have a thin hard crust over a sticky layer of mud. The remainder of the depression is covered by sand, gravel and limestone formation (Abu Al-Izz, 1971).

The Mediterranean Sea and the Qattara Depression were never joined; the water of the playa lakes has no relation to the water of the sea. Also the water cannot be derived from local rain. Ball (1927) considers, probably rightly, that the playas and the salty water were produced by great and continuous amounts of groundwater coming to the depression from the same supply that feeds the other depressions in the Western Desert. The playas are almost vegetationless.

The Qattara Depression area is characterized by two main vegetation types, namely (Ahmed, 2002):

A. Halophytic vegetation inhabiting the saline floor of the depression, and
B. Xerophytic vegetation inhabiting the surrounding desert plateau with non-saline soil.

A. Halophytic Vegetation

This vegetation abounds in three habitats: reed swamps, wet salt marshes, and dry salt marshes.

a. Reed Swamps

The reed swamps of the Qattara Depression are inhabited by one community dominated by *Phragmites australis* where soil salinity ranges between 11 305 and 106 000 ppm, high enough to prohibit the growth of other reed flora. However, few associated halophytes, namely: *Arthrocnemum macrostachyum, Juncus rigidus, Nitraria retusa* and *Tamarix mannifera* (*T. nilotica*) grow.

b. Wet Salt Marshes

This vegetation type is represented by four communities dominated by: *Cyperus laevigatus, Juncus rigidus, Arthrocnemum macrostachyum* and *Halocnemum strobilaceum.*

(i) ***Cyperus laevigatus* Community**
This community abounds in the periphery of the reed swamps where soil is subjected to periodic flooding. Lawns of the dominant sedge cover completely (100%) certain areas around the Qifar well which is continuously flowing. Associated species are: *Centaurium spicatum, Juncus acutus, Phragmites australis* and *Tamarix nilotica.*

(ii) ***Juncus rigidus* Community**
This community is widespread in the depression and its cover may reach 100% in areas with shallow or exposed water table. Total soil salinity ranges between 8,893 and 31,131 ppm. The floristic assemblage of this community is formed only of halophytes, namely: *A. macrostachyum, Alhagi graecorum, Cressa cretica, H. stobilaceum, Inula crithmoides, Nitraria retusa* and *Salsola tetrandra.* Parts of this community are subjected to the drifting of sands where *Tamarix nilotica, Tragaunm nudatum* and *Zygophyllum album* grow. Groves of *Phoenix dactylifera* are also growing indicating the presence of fresh water in the deeper layer.

(iii) ***Arthrocnemum macrostachyum* Community**
This community is confined to the slightly elevated wet land associated with the lakes of the depression. The stands of *A. macrostachyum* are usually pure, soil salinity is up to 61,720 ppm which seems to prohibit the growth of associates.

(iv) ***Halocnemum strobilaceum* Community**
H. strobilaceum community abounds in wide areas of dthe depression where water table is close to the soil surface or shallow (70 cm deep). Soil salinity ranges between 42,245 and 52,859 ppm and total plant cover between 5–50%.

Six associates occur; namely: *A. macrostachyum, J. rigidus, P. australis, N. retusa, T. nilotica* and *Z. album.*

c. Dry Salt Marshes

This vegetation type comprises five communities dominated by *Desmostachya bipinnata, Nitraria retusa, Salsola tetrandra, Tamarix nilotica* and *Zygophyllum album.*

(i) ***D. bipinnata* Community**
D. bipinnata is rarely recorded in the Qattara Depression and its growth and domination are confined to the sand bars fringing water bodies. The associated species include: *Sporobolus spicatus, Tragnum nudatum* and *Zygophyllum album.* Depth of water table is 70 cm and soil salinity is up to 9,316 ppm.

(ii) *Z. album* Community

Z. album and its community is widespread in the Qattara Depression being recorded as associated species in all communities of the dry and wet salt marshes as well as those of the non-saline habitat. Its community abounds in the sand embankments fringing the wet salt marshes where soil salinity is up to 8,220 ppm. The total plant cover of this community ranges between 20 and 35% contributed mainly by the dominant species.

Z. album community comprises the relatively highest number of associate species (30 species) comprising perennial halophytes and xerophytes and annuals as well. These include e.g. *Desmostachya bipinnata,, Nitraria retusa, Sporobolus spicatus*, and *T. nilotica* (halophytes), and *Calligonum comosum, Cornulaca monacantha, Convolvulus lanatus, Fagonia arabica, Artemisia monosperma, Anabasis articulata, Hyosyamus muticus, Plantago albicans, Stipagrostis scoparia,* and *Traganum nudatum* (xerophytes). The annual associates include: *Anthemis melapodina, Bassia muricata, Cotula cinerae, Emex spinosa, Mesembryanthemum forsskaolii* (= *Opophytum forsskaolii, Pteranthus dichotomus, Ifloga spicata, Reseda pruinosa* and *Trigonella stellata*. Individuals of semi-wild palm (*Phoenix dactylifera*) occasionally present.

(iii) *Nitraria retusa* Community

Like *Z. album, N. retusa* has been recorded as an associate species in the communities of all habitats of the Qattara Depression. However, its community abounds in two different habitats:

1. dry peripheral zone of the saline playa where soil salinity is up to 19,370 ppm. Here, *N. retusa* indviduals are stunted and scattered with plant cover < 5% forming pure stands. Depth of water table is about 65–75 cm
2. sand embankments with less saline soil (8,893 ppm) and deep water table (110–170 cm). The plant cover of *N. retusa* community in this habitat is up to 60% and the associated species are almost comparable to those of *Z. album*.

(iv) *Salsola tetrandra* community

This community has been observed in two habitats of the salt affected lands of the Qattara Depression. These are:

(1) short runnels dissecting the front slopes of the northern escarpment near the salt springs. In this habitat, ground water occurs at about 65–95 cm below soil surface and soil salinity ranges between 12,672 and 98,500 ppm. Three common associate halophytes recorded are: *Halocnemum strobilacium, Nitraria retusa* and *Tamrix nilotica. Anabasis articulata* (xerophyte) grows in few stands.

(2) premature undulated gravelly plain covered with sand formation ranging between few centimeters to sand dunes. The water table of this habitat is very deep (> 2 m) while soil salinity is low (< 5,000 ppm). The associate species are mostly xerophytes, e.g. *Cornulaca monacantha, Paronychia desertorum, Stipagrostis ciliata, S. plumosa* and *Traganum nudatum, Z. album* is the only halophyte associate species.

(v) ***Tamarix nilotica* Community**
T. nilotica is a widespread tree-shrub in the Qattara Depression being associate species with all communities of the salt affected and non-saline habitats. Its community abounds in the sand dune habitat where it is associated with four species, namely: *Alhagi graecorum, Inula crithmoides, Stipagrostis scoparia* and *Zygophylllum album*. Domination of *T. nilotica* occurs also in wide areas of the dry salt marshes within the sterile sabkhas with thick salt crusts (>15 cm), here the stands are pure and its individuals are stunted.

B. Xerophytic Vegetation

The plant life of the Qattara Depression inhabiting the non-saline habitats is organized into 6 communities dominated by *Stipagrostis scoparia, S. plumosa, Anabasis articulata, Traganum nudatum, Zygophyllum coccineum* and *Acacia raddiana*.

(i) ***Stipagrostis scoparia* Community**
In the Qattara Depression area, *S. scoparia* is of a limited occurrence. Its community is confined to the low sand dune chains where water table is deep (>2 m). The floristic composition comprises: *Artemisia monosperma, Astenatherum forsskaolii, Calligonum comosum, Polycarpaea repens* and *Zygophyllum album*.

(ii) ***Stipagrostis plumosa* Community**
This community occupies vast areas of the gentle slopes of the premature undulated gravelly plains covered with thin sheets of drifted sands. Being sand binding grass, *S. plumosa* is capable of building small sand mounds. The assemblages of associate species include: *Astenatherum forsskaolli, Cornulaca monocantha, Monsonia nivea, Stipagostis ciliata, Tragaunm nudatum* and *Zygophyllum album*.

(iii) ***Anabasis articulata Community***
In the Qattara Depression area, the community dominated by *A. articulata* occupies the runnels traversing the gravelly plain fringing the salt marsh habitat. The associated species include: *Cornulaca monocantha, Monsonia nivea, Neurada procumbens, S. plumosa* and *Zygophyllum album*.

(iv) ***Traganum nudatum Community***
T. nudatum is a leaf succulent xerophyte with halophytic nature that may grow in the xeric and halic habitats. This is a quite obvious in the Qattara Depression where *T. nudatum* has been recorded as associate species with the communities dominated by *Desmestachya bipinnata, Juncus rigidus, Nitraria retusa, Salsola tetrandra* and *Zygophyllum album*. Also, it is a common associate in the community dominated by the psammophytic grass *Stipagrostis scoparia*. Its community is confined to areas of drifted sands covering the gravelly plains with thin plant cover (<5%). The floristic assemblage comprises xerophytes, namely: *Calligonum comosum, Cornulaca monocantha, Asteratherum forsskaolii* and *Paronychia desertorum* and halophytes, namely: *Nitraria retusa, Tamarix nitlotica* and *Zygophyllum album*.

(v) *Zygophyllum coccineum* Community

In the Qattara Depression area the shallow runnels dissecting the limestone plateau are the proper habitat for *Z. coccineum* community. Here, the depth of surface deposits is shallow (50–70 cm) and plant cover is relatively high (40–80 %) contributed mainly by the dominant xerophyte. The number of the associate species of this community varies between the different stands. In some stands *Cleome droserifolia* is the only associate whereas in others *Acacia raddiana* and *Nitraria retusa* commonly occur. In a third group of stands the associates are many. The perennial associates are: *Atriplex leucoclada, Fagonia cretica, Francoeuria crispa, Heliotropium bacciferum, Hyoscyamus muticus, Pergularia tomentosa, Pulicaria undulata, Salsola tetrandra, Salvia aegyptiaca, Stipagrostis plumosa* and *Zygophyllum album. Anastatica hierochuntica* is the most common annual associate in addition to rare presence of *Euphorbia granulata.*

(vi) *Acacia raddiana* Community

A. raddiana (= *A. tortilis* spp. *raddiana*, Boulos, 1999) represents the climax stage of xerophtic vegetation. In the Qattara Depression, area, the community dominated by *A. raddiana* scrub could be considered as a desert open-forest with tree layer formed solely from the dominant spiny tree. The habitats of this community are the rocky upstream part of the wadis dissecting the desert plateau. The associate species are mainly perennial xerophytes with few halophytes and annuals, e.g.: *Anabasis articulata, Asthenatherum forsskaolii, Anastatica hierochuntica* (annual), *Astragalus siebri, Brassica tournefortii* (annual), *Calligonum comosum, Cleome droserifolia, Convolvulus lanatus, Fagonia arabica, F. indica, Francoeuria crispa* (*Pulicaria crispa*, Boulos, 1995), *Hyoscyamus muticus, Moltkiopsis ciliata, Monsonia nivea, Plantago albicans, Polycarpaea repens, Pulicaria undulata, Salvia aegyptiaca, Stipagrostis ciliata, Trigonella stellata* (annual), *Paronychia desertorum* (annual) and *Zygophyllum album.* Scattered individuals of the two halophytes: *Tamarix nilotica* and *Zygophyllum album* are growing on the phytogenic sandy hillocks.

(iii) Moghra Oasis

Geomorphology and Climate

Moghra Oasis is a small uninhabited oasis situated on the northeastern edge of the Qattara Depression and bordered by a brackish water lake (about 4 km^2). The lake represents the lowest part (−38 m) of the oasis. The shallow water-table and outward seepage of the lake's water accompanied by very high evaporation create the wet salt marshes that skirt the lake. Thick surface crusts of salt form, which may prohibit the growth of several species. The surface deposits were originally aeolian but run-off water has deposited brownish silt on the surface which is slippery when wet and cracked when dry.

Sand formations are dominant in the western and southern edges of Moghra lake. The deposits are in the form of dunes in areas adjacent to the lake and deep sheets of sand elsewhere.

The Wadi El-Natrun Depression and Moghra Oasis are located on almost the same latitude (Fig. 2.1); their climatic features are very comparable. Thus, water supply of the oasis is mainly the underground artesian water derived from the Nubian Sandstone series (Attiah, 1954).

Plant Cover

The plant cover of the Moghra Oasis has certain common characteristics and may combine all or some of the features of reed swamp vegetation, salt marsh vegetation and sand formation vegetation (Girgis et al., 1971).

Relicts of *Phragmites australis* reed are present in one locality of Moghra lake. The salt-marsh vegetation comprises two groups: communities of saline flats and communities of piled-up sand. The ecological conditions in these two groups are different. In the saline flats the underground water is shallow (0–30 cm), the soil is wet, dark in colour and contains a high proportion of fine sand and silt. In the sand formations, the underground water is relatively deep and the soil is yellowish sand poor in organic matter.

Saline Flats

The vegetation of the saline flats is mostly dominated by extensive growth of *Juncus rigidus* covering the area between the lake and sand formation. Differences in the microhabitats result in notable variations in composition, structure and density of the *J. rigidus* community. The stands nearer to the lake, being lower in level, are subject to occasional washing and their salt content is relatively low. The plant cover is very high (90–100%) and the associates include *Inula crithmoides, Phragmites australis* and *Tamarix nilotica* and the semi-wild *Phoenix dactylifera*. Local patches of *Inula crithmoides* are recorded in the more sandy areas at the foot of the dunes. As the level of the saline flat rises, salinity, as a result of extensive evaporation and poor washing, increases. This is accompanied by a decrease in vigour and cover of *J. rigidus*. Patches of *Arthrocnemum macrostachyum* (= A *glaucum*) are here scattered among the *Juncus* tussocks. Other associates are *Cressa cretica* and *Nitraria retusa*. Barren patches are common in certain parts, attributable to high salinity associated with soil dryness. "The halophytes are generally found in areas where the salt soils are mostly wet" (Walter, 1961).

Sand Formations

The ecological conditions affecting the vegetation pattern in the sand formations are soil salinity, depth of water-table, thickness of sand sheets and volume of water resources (rainfall, run-off, underground). Three main communities dominated by *Zygophyllum album, Nitraria retusa* and *Tamarix nilotica* may be recognized.

1. *Zygophyllum album* community. *Z. album* is present where soil salinity is relatively low and the surface deposits are friable mixed sand with a low content of

organic matter. It forms an outermost zone that gradually joins the desert plain communities surrounding the Moghra Oasis.

Z. album, in the Moghra Oasis, forms a community with a sparse vegetation (cover is about 5%). In the outermost (higher) fringes, where conditions are less favourable, the plants are small and build no mounds. In the lower (innermost) parts where moisture conditions are high, *Z. album* builds low mounds. Associates include *Alhagi maurorum, Nitraria retusa* and *Tamarix nilotica.* Rare individuals of *Artemisia monosperma* may be present in places adjoining the desert plain.

2. *Nitraria retusa* community. *N. retusa* grows in extensive patches in the sand dunes bordering Moghra Lake where it builds hummocks. It plays an important role in the fixation of sand dunes in the Moghra Oasis.

   The ecological conditions found in the *N. retusa* zone indicate a wide range of tolerance of the dominant (Girgis et al., 1971). Salinity, depth of water-table, volume of water resources etc. vary in the whole stretch of the *Nitraria* scrub. In the outer fringe where salinity is relatively high (3.4–15.2%) and the underground water is deep, associates include *Tamarix nilotica* and *Zygophyllum album.* Towards the inner fringes of the zone where soil salinity is relatively low (0.6%) and the soil is moist below the surface, *Alhagi maurorum* forms the main part of the undergrowth together with *Juncus rigidus, Phragmites australis, Sporobolus spicatus* and *Z. album.*
3. *Tamarix nilotica* community. *T. nilotica* scrub in the Moghra Oasis is restricted to the outer fringes of the dune zone skirting the saline flats. The flora *of Tamarix* scrub includes a few species: *A. maurorum, Cressa cretica, N. retusa* and *Z. album.*

Apart from the above mentioned species recorded by Girgis et al. (1971) the flora of the differeent communities of Maghra Oasis compreses 19 more species namely (Salem and Waseem, 2003.): *Acacia ehrenbergiana, Calligonum comosums, Centropdoia forsskaolii, Cistanche phelypaea, Convolvulus lanatus, Cressa cretica, Cynodon dactylon, Cyperus* sp., *Halocnemum strobilaceam, Hyoscyamus muticus, Imprata cylendrica. Launaea mudicaulis, Minuartia geniculata* (= *Rhodalsine geniculata*), *Moltkiopsis ciliata, Panicum turgidum, Pynocycla tomentosa, Salsola* sp., *Stipagrostis obtusa* and *S. scoparia.*

Economically, most of the flora of Maghara Oasis have multipurpose use, mainly as grazing plants (91%), medicinal plants (54.4%), fuel and wood plants (36.4%) in addition to the other miscellaneous uses e.g., human food, leather tanning, mats etc. Unfortunately, these plants are highly threatened due to their over use (Salem and Wassem, 2006).

(iv) Siwa Oasis

Geomorphology, Climate and History

Siwa Oasis occupies a depression in the northern part of the Western Desert. It trends in an east-west direction with a length of about 80 km and maximum breadth of 26 km. It lies about 300 km south of Mersa Matruh on the Mediterranean coast (Fig. 2.1).

Siwa Oasis (10–17 m below sea level) is bounded on the north by an escarpment that rises to about 100 m above the depression floor and on the south by a low scarp of about 20–25 m covered with belts of NW-SE trending sand dunes. To the east and west the depression floor rises gradually, merging with the general desert level. The deeper portion is occupied by salt marshes (sabkhas). The deep deposits belong to Recent and Subrecent periods. The exposed rocks are Middle Miocene in age (El-Askary, 1968).

The supply of water in Siwa is artesian, flowing from about 150 springs. Shata et al. (1962a) indicate that this water is derived from Miocene aquifers and seems to be about 5,000 years older than Kharga water; the age of both Kharga and Siwa water is recent, dated 30,000–50,000 years (late Pleistocene).

The water of the Siwa springs is warm, with temperatures varying between 26.5 and 30 °C. The total soluble salts in the water range between 1900 and 8200 ppm with conductivity of 3000–5000 mS/cm. Water of springs used for domestic purposes contains lower amounts of soluble salts and lower conductivity (2300–2800 mS/cm, Saleh, 1970).

Data of the *Climatic Normals of Egypt* (Anonymous, 1960) show that in the Siwa Oasis monthly absolute maximum temperature ranges from 19.7 °C in January to 37.9 °C in July. The monthly absolute minimum ranges from 4.1 °C in January to 20.7 °C in July and the average annual temperature is 21.4 °C. Rainfall is negligible: 9.6 mm/year in the period 1931–1960. Heavy cloudbursts are exceptionally rare (28 mm in one day was recorded on 24 December 1936). Frost may occur during December–February. Values of relative humidity recorded in November, December and January (51%, 58% and 53% respectively) are higher than those in May (30%) and June (31%). Evaporation ranges between 17.0 mm/day in June and 3.0 mm/day in December. The prevailing wind blows from the north and northeast during winter and spring with an average velocity of 40 km/h. During summer and autumn, only one or two days of strong winds from a northerly direction may be expected. The values of Emberger's (1951) Pluviothermic Quotient of the Siwa climate is 1.43, indicating severe aridity.

Siwa Oasis enjoyed the fame of its oracle for over three centuries of the classical period, during which envoys from different parts of the world were sent to consult the oracle. The temple of Jupiter Ammon was located in the Siwa Oasis (Belgrave, 1923). The Athenians kept a special galley in which they conveyed questions across the sea to the oasis. Mersa Matruh, sometimes called Ammonia, was the port for Siwa and it was here that the ambassadors and visitors disembarked and started on their desert journey to the Siwa Oasis. It is easy to imagine the riches and gifts that brought the prosperity which Siwa Oasis enjoyed at that time.

Towards the end of the third century BC, the fame of the oracle declined (Belgrave, 1923), although the answers of Ammon were still being given to solve difficult problems until long after the cessation of the oracle at Delphi. During the second century BC the oracle was hardly visited and Siwa Oasis become less popular (Mitwally, 1953).

Vegetation Types

The plant cover of the Siwa Oasis varies according to the topography and other factors. Four vegetation types have been recognized (Zahran, 1972); reed swamp, salt marsh, sand formations and gravel desert.

Reed Swamp Vegetation

The Siwa Oasis embraces 18 lakes which vary in area and depth. The largest (8.9×4.8 km) and deepest (17 m below sea level) is the Siwa Lake, to the eastern side of the oasis. The reed swamp vegetation is well represented in the shallow water of these lakes by a dense growth of *Phragmites australis* and *Typha domingensis*. In some localities *P. australis* grows in the terrestrial borders of the lakes where there is silting of the fringes originally occupied by the reed swamps.

Salt Marsh Vegetation

The salt-affected lands of the Siwa Oasis may be divided into areas adjacent to the lakes where water comes from the lateral seepage of lake water and the underground water, and areas around the springs where the water-table is very shallow (or exposed). Under the prevailing aridity, there is high evaporation of soil water and accumulation of salts in the surface layers of soil.

Five communities have been recognized in the salt marshes, characterized by *Arthrocnemum macrostachyum, Juncus rigidus, Alhagi maurorum, Cladium mariscus* and *Cressa cretica* (Zahran, 1972).

1. *Arthrocnemum macrostachyum* community. This community occupies the zone close to the reed swamp habitat on the north and northeastern sides of the lakes. In Lake El-Maasir and Lake El-Zaitun the sand formations are ill defined and are replaced by salt marsh dominated by A. *macrostachyum*. The plant cover of this community ranges between 70 and 90%, contributed mainly by the dominant species which forms pure stands. In a few stands *Phragmites australis, Juncus rigidus* and *Tamarix nilotica* are present.
2. *Juncus rigidus* community. Dense growth of J. *rigidus* covers more than 50% of the salt-affected lands of the Siwa Oasis. This community is present in the second zone of the marshes adjoining the lakes and in the salt marshes around the springs and wells. Associated species vary according to the habitats. *Phragmites australis* and *Cynanchum acutum* occur in stands with very shallow or exposed water-table, under conditions of low soil salinity, whereas *Arthrocnemum macrostachyum* and *Inula crithmoides* grow in stands with surface crusts (high salinity). Landward stands are covered with a sheet of saline sand. *Alhagi maurorum, Cressa cretica, Nitraria retusa* and *Tamarix nilotica* are present as well as *Phoenix dactylifera,*[2] which is semi-wild here. *J. rigidus* is used by the bedouins for making best-quality mats. Fairly recently it has been found that its green leafy shoots (culms) can be used as material in paper making (Zahran et al., 1972; Zahran and Abdel Wahid, 1982). A project of establishing a unit to produce paper pulp in the Siwa Oasis, using the vegetative parts of *J. rigidus* growing naturally, may play an important role in the environmental development of the oasis.
3. *Alhagi maurorum* community. This community occurs in three types of habitat that vary ecologically. One habitat is the third zone of the salt marshes where

[2]According to Saleh (1970), Siwa Oasis depends economically on the production of dates and olives.

A. maurorum forms pure stands except in a few parts where it is associated with rare specimens of J. *rigidus* and/or *C. cretica*. Plant cover here is usually thin (<5%). A second habitat dominated by *A. maurorum* is the open areas between the rocky hills near the lakes where the soil surface is covered with 10–20 cm of soft, loose and saline sand. Here, *A. maurorum* builds sand mounds of moderate size and the plant cover is 5–10%. *T. nilotica, N. retusa* and *C. cretica* are common associates; *J. rigidus* is rare. A third habitat is the slopes of the rocky hillocks not far from the lakes; the soil here is mainly of coarse material with the lowest amount of soluble salts. In this habitat, *A. maurorum* sends its roots through the crevices of the slopes. It forms pure stands with larger bushes but does not build mounds. The cover ranges between 5 and 10%.

A. maurorum is a widely distributed species in Egypt that seems to have a wide ecological amplitude. It grows in habitats of different salinities (see e.g. Kassas, 1952c; Zahran, 1967). Kassas and Zahran (1967) considered that A. *maurorum* is a desert plant alien to the salt marsh habitat. It has a root system that may extend several metres reaching soil layers that are less saline and permanently wet. But, as it is an abundant species dominating a characteristic salt marsh community, both in the littoral zone (Kassas and Zahran, 1962) and inland salt marshes (Siwa Oasis for example), it is considered to be an apparently cumulative halophyte (Zahran, 1972), shedding excess salts accumulated in the shoot system from the saline soil by the loss of leaves.

4. *Cladium mariscus* community. *C. mariscus* is very rare in Egypt (Täckholm, 1974). Its dominance is recorded only in the Siwa Oasis where it flourishes (plant cover 100%) in the marshy lands about the inner springs (e.g. El-Zein spring) of the oasis where the water level is very shallow or exposed. Moisture-loving species, e.g. *Cyperus laevigatus* and *Phragmites australis*, are very common associates. Less common plants include *A. maurorum* and *J. rigidus*.
5. *Cressa cretica* community. This community is restricted to the dry saline areas of the Siwa Oasis at the far outer edge of the salt marsh ecosystem where the depth of the surface salt crust is 25–40 cm. The underground part of *C. cretica* reaches the water-table at 50 cm depth. No associate species have been recorded in this community and the plant cover is very thin (<5%).

Sand Formation Vegetation

Sand formations are dominant features of the landscape of the Siwa Oasis. They are formed of aeolian materials which settle on wide areas, forming sheets of sand or building sand bars and/or sand dunes.

The sand flats occupy open ground extending from the foot of the highlands to the salt marsh system. The soil is saline as affected by the lateral seepage of the lake water as well as underground water. In these habitats *Alhagi maurorum* dominates with *Nitraria retusa* and *Tamarix nilotica* as common associates. *Juncus rigidus* is rare. The sand flats are also well represented at the downstream parts of wadis draining to the lakes (e.g. Ain Timiera area) where *N. retusa* and *T. nilotica* form open scrubland and the undergrowth is dominated by *A. maurorum*. Both *Nitraria* and *Tamarix* build huge hillocks.

Sandy bars (50–120 cm high) extend along the edges of the salt marsh ecosystem separating it from the non-saline areas. These sandy bars are formed of fine materials (70% including fine sand and silt). *Imperata cylindrica* is present in such habitats with plant cover up to 50%. Associate species include *A. maurorum, C. cretica, J. rigidus* and *T. nilotica*.

On the south and southeastern sides of the lakes (e.g. Siwa Lake, El-Maragi Lake, Timiera Lake) are extensive areas of sand dunes that vary in height (2–3 m), size and shape. The lows among these dunes are characterized by water runnels draining into the lakes. These runnels are lined with homogeneous sand. The vegetation of each runnel affects its features. The plants preserve the shape and form of the runnels to a great extent; they check the movement of sand and trap aeolian sand around their shoots, building hummocks. Naturalized trees *(Populus euphratica*)* are common on the slopes of these dunes. At the foot of the sand dunes, saline sandflats prevail and *T. nilotica* dominates, associated with *A. maurorum* and *C. cretica*. It is noteworthy that trees of P. *euphratica* were introduced to the Siwa Oasis during Roman times (331 BC, Belgrave, 1923). These trees were used as wind-breaks and also to fix sand dunes to protect the Oasis against moving sand. A few trees of P. *euphratica* still grow on these dunes. Mousa et al. (2006) proved that *P. euphratica* could morphologically and physiologically be adapted to adverse environmental conditions which explains that this plant could be successfully exploit for combating sand encroachment towards the agricultural areas.

The runnels between the sand dunes are characterized by two communities. In the downstream parts of these runnels where soil salinity is relatively high *Zygophyllum album* dominates. The cover ranges between 5 and 10%. *Stipagrostis scoparia* is the only associate of this community. The gradual decrease in salinity of soil upstream in these runnels is associated with the disappearance of *Z. album* and the occurrence of *Cornulaca monacantha* which dominates the upstream parts. Associates are S. *scoparia* and *Z. album*. These three plants are sand binders, building sand hummocks.

Gravel Desert Vegetation

The level of the Siwa Oasis rises towards its periphery. Moisture and salt contents of the soil decrease. The salt-tolerant and moisture-loving species gradually thin and the landscape is changed to barren desert except for growth of a few drought-resistant and drought-tolerant plants in two geomorphological units:

1. The depressed areas scattered in the gravelly desert plains; here sand accumulates and maintains an abundant growth of *Zygophyllum coccineum* associated with *Atriplex halimus, Fagonia arabica, Salsola tetrandra* and *Zygophyllum simplex*.
2. The water runnels dissecting the gravelly desert. In the narrow upstream parts of these runnels, where the ground is formed mainly of coarse sediments (>80%), *Fagonia arabica* forms a pure stand with thin plant cover (<5%). In the middle

**P. euphratica (P. illicitana)* is a 10-15 m high tree with spreading branches and glabrous blue-green, broadly rhoraboidal leaves. The name relates to the Euphrates River of Iraq.

(wider) parts of the runnels (40% coarse material and 60% fine material) *Z. coccineum* also forms a pure stand with thin cover (<5%). Shrubs and trees of *Acacia raddiana* grow in the widest downstream parts of these runnels where the soil is deeper and compact and formed mainly of soft sand (>60%) with less than 40% coarse sediments. Trees of *A. raddiana* have a deep root system and so "they require little or in some cases no irrigation" (Migahid et al., 1960). Rare individuals of *Pergularia tomentosa* grow among the trees of *Acacia*.

Vegetation Types of Wadi Timiera

Wadi Timiera is one the wadis of the Siwa Oasis being on its western edge. It cuts across Gebel Timiera and drains to the Timiera Lake. The upstream part of this wadi is narrow and its substratum is of rocks and coarse material (>60%). Its course widens gradually downstream and forms a sort of delta where the soil is mainly of soft material (>80%).

Fagonia arabica dominates most of the upstream part of the wadi, forming green cushions of variable size. The salt content of the soil is low (0.55–0.63%). As the edaphic conditions change downstream in the wadi (soluble salts 0.73–0.80%, associated with reduction in coarse material), *Z. coccineum* occurs as an almost pure stand. The sand formation is well represented in the midstream part of the wadi where an open scrub of *Nitraria retusa* and *Tamarix nilotica* occurs. Both shrubs build huge sand hillocks. *Alhagi maurorum* dominates the undergrowth between these hillocks. In the downstream part of Wadi Timiera where the soil salinity is up to 4.8% in the subsurface layer and 48% in the surface salt crusts, there is salt marsh vegetation. Here, *Juncus rigidus* is the most abundant halophyte. It forms a dense growth over more than 90% of the area. At the fringes of the Timiera Lake the moisture content of the soil is higher and *Phragmites australis* dominates.

It is worth to state that, El-Kholy (2001) i.e. after 30 years of Zahran's Survey (Zahran, 1972), did not find critical changes in the vegetation types of Siwa Oasis. Through using TWINSPA classification and complementary CCA analysis, El-Kholy (2001) recognized 7 vegetation groups dominated by 9 indicator species, namely: *Salsola tetrandra, Atriplex leuc oclada, Alhagi graecorum (= Alhagi maurorum), Nitratia retusa* and *Zygophyllum album* (halophytes), and *Cornulaca monacantha, Fagonia Arabica, Achillea fragrantissima* and *Zygophyllum coccineum* (xerophytes). El-Kholy (2001) did not refer to the presence of *Cladium mariscus* which may be considered extinct. Also he considered *S. tetrandra* and *Atriplex leucoclada* as dominant halophytes, both were recorded by Zahran (1972) as associated species.

The delta of Wadi Timiera is fringed by low hills on its east and west. On the slopes of these hills towards the lake A. *maurorum* dominates.

(v) Wadi El-Rayan Depression

Wadi El-Rayan is a small enclosed and curiously shaped (clover-leaf like) uninhabited depression 25 km southwest of El-Fayium Province (Figs. 2.1. and 3.4). "The

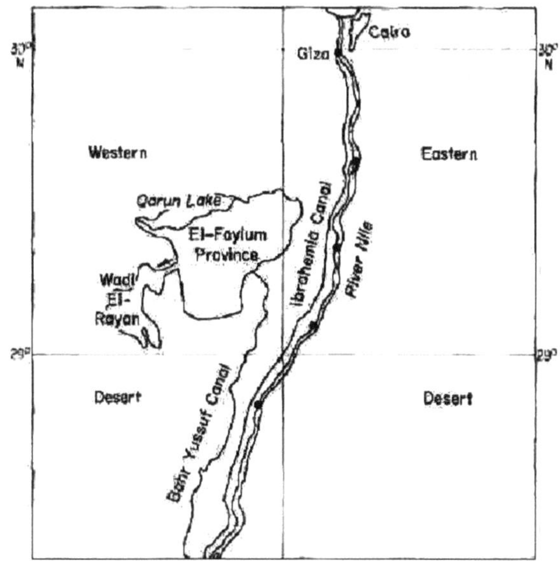

Fig. 3.4 Wadi El-Rayan Depression, Western Desert (*arrow* indicates direction of flow)

depression, discovered by Linant de Bellefonds (1873), is cut out of white limestone of Eocene age, rich in nummulites. The lowest point of the floor of the depression is at 60 m below sea level. The area at the –60 m contour is 22 km^2, at the sea level contour is 301 km^2, and at the 130 m contour about 703 km. Its maximum breadth is 25 km" (Zahran, 1970–1971).

The origin of the word "Rayan" is discussed by Fakhry (1947) – Rayan is Arabic for the "watered one" or the "luxuriant", a suitable name for this wadi which is covered with vegetation at many places and whose subsoil has water at less than 2 m. A bedouin legend gives another explanation. The ruins of ancient buildings are the ruins of the houses of a powerful king called "El-Rayan" and his soldiers who lived here.

Coptic literature gives yet a further interpretation of the name of the wadi. It is stated in the biography of Anba Samuel of Kalamous that he used to go from time to time to worship alone in this wadi and found the word "El-Rayan" in the Arabic text on Abu Salih, the American Worship. The name "Rabana" is a possible one; this is affirmed by its mention in the *Horris Papyrus* in connection with the Libyan war of Rameses (Zahran, 1970–1971).

Ground water is the main source of supply for the Wadi El-Rayan Depression in the south portion of which, and according to Ball (1927), there are three springs deriving their water from the continuous sheet of subterranean water under the Western Desert. Fox (1951) believes that these springs are in the fissured Nubian Sandstone about 600 m beneath the depression. Ball (1927) indicates that "the water of these springs is derived from remote collecting areas and is therefore warm". According to Fakhry (1947), there is evidence that all these springs have long been in use, as their water is drinkable. In the first and second centuries AD the depression was inhabited and a part of it was cultivated.

The vegetation in the Wadi El-Rayan Depression is confined to areas around springs. Besides the trees of *Phoenix dactylifera* and *Acacia raddiana* there are bushes undershrubs and grasses, e.g. *Tamarix* spp., *Nitraria retusa, Zygophyillum album, Desmostachya bipinnata, Alhagi maurorum* and *Fagonia arabica*. Common xerophytes can also be seen in the desert surrounding the depression, e.g. *Calligonum comosum, Cornulaca monacantha, Farsetia aegyptia, Heliotropium luteum, Panicum turgidum* and *Pituranthos tortuosus*.

Saleh et al. (1984) stated that the vegetation of Wadi El-Ryan Depression is confined to three habitats: 1. interdunes, 2. basis of large sand dunes and 3. around the springs. Thirteen plant species have been recognized, four of these (*Cyperus laevigatus, Phragmites australis, Typha domingensis* and *Juncus* sp. are growing only in the wet habitat around the springs. Two species, namely: *Zygophyllum album* and *Calligonum comosum* are inhabiting the interdunes and bases of large dunes. *Nitraria retusa, Alhagi graecorum* and *Desmostachya bipinnata* are restricted to the interdune habitat. Four species, namely: *Tamarix nilotica, Phoenix dactylifera, Sporobolus spicatus* and *Imperata cylindrical* are growing in both interdunes and around springs habitats.

During the last two decades, Wadi El-Rayan Depression has been utilized as a natural reservoir for the drainage of a part of the agricultural land of El-Fayoum Province. Two large lakes have been created. The creation of such large bodies of water in the extreme arid Western Desert have tremendous ecological effects on all living organisms of the depression itself and near areas.

Amin (1998) identified three major ecosystems in Wadi El-Rayan area, namely: desert ecosystem, lake ecosystem and spring ecosystem. Each of these systems has its characteristic environmental features which affect vegetation therein. The desert ecosystem is represented by two distinct localities. The first one is located in the eastern part of Wadi El-Rayan between the lakes shore and the limestone ridge where vegetation is restricted to scattered patches dominated by *Zygophyllum coccineum* with low plant cover (< 5%), Two associate species were recorded: *Kochia indica* and *Fagonia arabica*. The second desert locality is located in the SW part of the depression between the southern shore of the lake in the east and the spring region in the west. The vegetation cover in this locality is relatively high (20–30%) and is distinguished into five communities dominated by: *Alhagi graecorum, Calligonum comosum, Desmostachya bipinnata, Nitraria retusa* and *Zygophyllum album*.

The lake's ecosystem of Wadi El-Rayan Depression comprises the water bodies of the two lakes and the shoreline habitats. Zonation of vegetation in this ecosystem is obvious caused by the local topographic changes, depth of water and soil salinity. Three vegetation types are recognized: aquatic, swampy and terrestrial. The aquatic vegetation is dominated by two hydrophytes: *Myriophyllum spicatum*, (cover= 20–30%) and *Potamogeton crispus* (cover=50–70%). The accociate species are also hydrophytes, namely: *Najas marina, Potamogeton pectinatus* and *Zannichellia pallustris*. The swampy habitat comprises two ecosytems dominated by *Phragmites australis* and *Typha domingensis* (20–30% cover). The associate species are many and include: *Cyperus laevigatus, Tamarix nilotica, Alhagi graecorum, Calligonum comosum, Desmostachya bipinnata, Zygophyllum album, Z. coccineum, Pluchea dioscordis, Salicornia fruticosa, Cynanchum acutum, Phoenix dactylifera, Imperata cylindrical* and *Eichhornia crassipes*.

The spring ecosystem of Wadi El-Rayan Depression is of ecological interest. There is a wide variation between the three springs as regard their water supply and rate of flow. The first (northern spring), has a limited water flow (1.6 L/min) with TSS 3,000–4,000 ppm. and moisten small area (50–150 m). Two plant communities are recognized dominated by *Alhagi graecorum* and *Cressa cretica*. The associate species are: *Phragmites australis, Nitraria retusa, Juncus acutus, Sporobolus spicatus, Tamarix nilotica, Zygophyllum album* and *Phoenix dactyliferea*. The middle spring has a water flow of 4.8 L/min. with 3,500 ppm soluble salts. Four communities dominated by *Tamarix nilotica, Alhagi graecorum, Phragmites australis* and *Nitraria retusa* inhabit the area of this spring. The associate species are: *Calligonum comosum, Desmostachya bipinnata, Juncus acutus, Tamarix nilotica* and *Typha domingensis*.

The southern spring is the largest in Wadi El-Rayan area with an average water flow of 14.4 L/min. and mean salinity of 2.9 g/L. Three communities dominated by *Alhagi graecorum, Phragmites australis* and *Tamarix nilotica* are recognized. *Calligonum comosum* is the only associate species with *A. maurorum* and *T. nilotica* community whereas *P. australis* is associated with *Juncus acutus* and *Phoenix dactylifera*.

El-Henawy (2008) stated that Wadi El-Rayan Depression has been declared by prime ministerial decree No. 943, 1989 as a natural protectorate for ecosystem protection and integrity, environmental education and ecotourism. Also, it is a protected area managed mainly for the sustainable use of natural ecosystem.

(vi) Bahariya Oasis

General Remarks

Bahariya Oasis lies between Lat. 27°28' and 28°30' N and Long. 28°35' and 29°10' E (Fig. 2.1). It differs from the other oases in being entirely surrounded by escarpments and in having many hills (mostly black, a few are reddish, others are white) giving the oasis an entirely different appearance from that of other Egyptian oases. It is an oval depression with its axis trending from NE to SW and an area of about 1,800 km^2. The average depth from the general desert plateau level to the floor of the excavation is less than 100 m.

The Bahariya Oasis is not invaded much by mobile sand dunes as are the other oases to the south, nor is there much sand on its surface. There are, however, a few small dunes around the agricultural land of the villages. Some of these dunes are still advancing at a rate of 15 m/year causing disastrous effects. Water can be obtained by digging shallow wells; the deepest is not more than 7 m (Abu Al-Izz, 1971).

The climate of the Bahariya Oasis is extremely arid. The annual rainfall is 4.3 mm. However, "accidental" cloudbursts may happen at intervals of about five years: 14 mm on 26 November 1936 and 29 February 1940, 13 mm on 19 December 1934 and 16 mm on 18 April 1948. The hottest months are June, July and August, with absolute maximum and mimimum temperatures of 39.8 °C and 19.2 °C respectively. The coldest month is January, with absolute maximum and minimum temperatures of 19.9 °C and 4.7 °C respectively. The relative humidity ranges between 43% in December and 26% in June (Anonymous, 1960).

Main Habitats and Vegetation Types

The main habitats in the Bahariya Oasis favourable for plant growth are the swamps, salt marshes including sand flats and sand formations (W.A. Girgis, 1977, unpublished).

The Swamps

In Bahariya Oasis, the swamps occupy the shallow lands having continuous feed of water either from springs and/or from the irrigated lands. During winter, the water-table of these areas is usually exposed. These swamps are dominated mainly by *Typha domingensis* and rarely by *Phragmites australis* associated with water-loving species, e.g. *Cyperus difformis, C. laevigatus* and *Samolus valerandi;* submerged algae, e.g. *Chara* sp. and free-floating hydrophytes, e.g. *Lemna gibba, L. minor, L. perpusilla, Marsilea minuta* and *Utricularia gibba* ssp. *exoleta. Nymphaea caerulea* v. *aschersoniana*, the blue sacred flower of the ancient Egyptians, a very rare floating hydrophyte in Egypt, is commonly recorded in Bahariya Oasis (Täckholm, 1974).

The swamps may be silted with aeolian sand that becomes dark in colour, enriched with organic matter from decaying rhizomes and fallen leaves. *Typha* withstands this silting process until the soil becomes more saline (>5%) and the water-table becomes deeper. In this habitat *Phragmites australis* dominates. At an intermediate stage *Typha* and *Phragmites* are co-dominant with *Juncus rigidus* as an associate. With further increase in salinity *Typha* is excluded and a *Phragmites* stage results. At this stage salt marsh species invade, e.g. *Aeluropus lagopoides, Cyperus laevigatus, Imperata cylindrica, Spergularia marina, Sporobolus spicatus* and *Tamarix nilotica.*

The Salt Marshes

The salt marshes cover the greater part of the oasis floor. They may be classified into two categories according to the level, nature of surface deposits and depth of ground water: wet and dry salt marshes.

Wet Salt Marshes

These are areas with shallow ground water (<1 m). They are always associated with swamps and springs or irrigated lands, and are concentrated in the northern section of the oasis. These wet saline lands are almost flat; their level rises gradually to the surrounding dry salt marshes. The surface deposits are dark in colour, rich in organic matter and include a high proportion of fine sand and silt. The vegetation of the wet salt marshes may be organized in zones dominated by *Cyperus laevigatus, Juncus rigidus* and *Salicornia fruticosa* in a sequence of increased ground level.

1. *Cyperus laevigatus* community. The *C. laevigatus* community is not common in Bahariya Oasis. Limited lawns are recorded in certain areas. These lawns are inundated during the winter and are sufficiently wet during the summer. This zone follows *Typha* swamps or *Typha* and/or *Phragmites* silted swamps. The soil is muddy with salinity in the range 0.2–0.7%; salinity decreases down the profile. The plant cover of the *C. laevigatus* community is high (up to 100%) and mostly contributed by the dominant. *Cyperus* is extensively grazed by domestic animals so that it forms a carpet-like growth not more than 10 cm high. Associate species are: *Imperata cylindrica, Juncus acutus, J. rigidus, Phragmites australis* and *Typha domingensis*.
2. *Juncus rigidus* community. This community is common in the northern part of the oasis. It follows zones of *C. laevigatus* and *P. australis* upslope. The water-table is shallow. During winter, water creeks traverse the *Juncus* salt marsh, whereas during summer the salt marsh is almost dry. These fluctuations in the level of ground water may be related to agricultural operations. The soil is dark, rich in organic matter and with much fine sand and silt. Soil salinity ranges between 3.6 and 5.3%. *Juncus rigidus* forms dense thickets (cover >70%). Associate species include *Aeluropus lagopoides, Alhagi maurorum, Frankenia revoluta, Imperata cylindrica, Juncus acutus, Salicornia fruticosa, Sporobolus spicatus* and *Tamarix nilotica*.
3. *Salicornia fruticosa* community. This community follows the *J. rigidus* zone upslope where the water-table is deeper. The surface soil layer is sandy, overlying darker or finer layers. In moist areas S. *fruticosa* forms an almost pure population with well-grown plants and high cover (up to 70%). In drier, more saline areas the cover is low (10–15%); plants of *Salicornia* are smaller and the associates include *Aeluropus lagopoides, Desmostachya bipinnata, Imperata cylindrica, Juncus acutus, J. rigidus* and *Tamarix nilotica*.

Dry Salt Marshes

Dry salt marshes represent a process of sand deposition over the original salt marshes, so their level is higher than that of the wet salt marshes. The soil surface is always undulating and yellowish or grey. This type of salt-affected land covers the majority of the oasis floor. The organic carbon content is usually low.

The vegetation of the dry salt marshes may be distinguished into four communities dominated by *Sporobolus spicatus, Alhagi maurorum, Desmostachya bipinnata* and *Tamarix nilotica*. The last three communities represent more than 80% of the natural vegetation of the oasis. No zonation has been recognized in the dry salt marshes of the Bahariya Oasis, but the mosaic pattern is related to the thickness of the sand deposits, depth of ground water and soil salinity.

1. *Sporobolus spicatus* community. The habitat of this community is sand sheets, generally flat or slightly undulating. The surface of the sand is loose, not cemented with salt crusts. The water-table is deeper than 120 cm. Soil salinity is relatively low. S. *spicatus* builds tussocks of considerable size. The cover of this community ranges between 20 and 70%, mostly contributed by the dominant. In areas with dense cover *Sporobolus* forms an almost pure stand, its vigorous tussocks leaving

no place for other plants. *Alhagi maurorum* is the most common associate contributing to the cover. *Stipagrostis scoparia* may be locally co-dominant in parts with deeper sand sheets. There are eight less common associates, namely: *Aeluropus lagopoides, Cressa cretica, Frankenia revoluta, Hyoscyamus muticus, Juncus acutus, Schanginia aegyptiaca, Spergularia marina* and *Tamarix nilotica. Stipagrostis scoparia* and *Hyoscyamus muticus* that grow better in non-saline habitats are restricted in Bahariya Oasis to the *S. spicatus* community. Salinity may not be a limiting factor for non-halophytes in some stands of this community. Salinity may also be non-limiting in the salt marshes of the Red Sea coast (Kassas and Zahran, 1967) and in Wadi El-Natrun (Zahran and Girgis, 1970).

2. *Alhagi maurorum* community. *A. maurorum* is a very common plant in the Bahariya Oasis. Its community is frequent in the areas among P. *nilotica* hillocks forming the undergrowth of this scrubland and ilso among the tussocks of the halfa grassland dominated by *Desmostachya bipinnata.*
   The surface of the ground supporting the growth of the A. *maurorum* ommunity is usually undulating. The sand deposits are loose at the urface but compact below. Salinity is relatively low and the water-able is deeper than two metres.
   A. maurorum usually forms pure stands of dense cover (30–80%). *T. nilotica* and *D. bipinnata* are common associates. *Calotropis procera* is recorded in one of the stands; this species is absent from the oases northwards. The other associates include *Aeluropus lagopoides, Cressa cretica, Frankenia revoluta, Juncus rigidus, Schanginia aegyptiaca, Sporobolus spicatus* and *Zygophyllum album.*
   A. maurorum shows morphological and physiological plasticity according to water stress (Kassas, 1952c). This has been observed in A. *maurorum* growing in the Bahariya Oasis. During winter the plants were straw yellow; in summer they were dark green, with the new growth topping the dry brownish old parts. A few flowering specimens were observed in summer, particularly in moist places.
3. *Desmostachya bipinnata* community. The halfa grasslands dominated by *D. bipinnata* are a feature of the vegetation of the Bahariya Oasis that are usually seen on sand accumulations and low mounds overlying the salt marsh ground. The sand and ground water are deeper than those of the A. *maurorum* community. The soil includes appreciable amounts of fine ingredients; its salinity is more than 2.5%.
   D. bipinnata is a rigid grass that may reach 150 cm, but as it is extensively grazed and cut its height rarely exceeds 30 cm. Its growth varies in relation to water supply and soil salinity. In depressed areas bounded by hills where a hard pan is formed at 20 cm depth, plant cover is low (<20%). At a higher level in the same area, the cover, though distinctly greater (70%), is mostly of dead plants and appears cushion-like. This may be due to increased salinity. On the dunes where salinity is low, the growth of *Desmostachya* is luxuriant. A. *maurorum* and *T. nilotica* occasionally grow in some of these stands.
4. *Tamarix nilotica* community. *T. nilotica* scrub is very common in the Bahariya Oasis. It occurs on sand dunes and parts of the salt marshes with the deepest sand deposits. The shrubs of *T. nilotica* are more vigorous on the sand dunes than in the salt marshes.

T. nilotica scrub represents the main climax of the salt marsh habitat of the oasis. The plant cover is in three distinct layers: a frutescent layer including the dominant, a suffrutescent layer that contributes little to the cover and includes *D. bipinnata* and *Z. album*, and a ground layer formed mainly of A. *maurorum* that is usually rich.

Naturalized trees of *Phoenix dactylifera* are widely distributed in the Bahariya Oasis where date production is the backbone of its economy.

(b) The Southern Oases

These are the oases south of Lat. 26 °N in the Western Desert (Fig. 2.1) including: Kharga Oasis, Dakhla Oasis, Dungul Oasis, Kurkur Oasis in addition to other small Oases eg. Nakheila, Bir Tarfawy, Bir Murr, Bir Kurayim and Bir Kisebi (Bornkamm, 1986). An ecological account of four of these oases – Kharga, Dakhla, Kurkur and Dungul – is given.

(i–ii) Kharga and Dakhla Oases

Landform

Kharga and Dakhla Oases occupy part of a great natural excavation in the southern section of the Western Desert of Egypt. This excavation includes also the slightly elevated plain (140 m above sea level) between them. The depression is open towards the south and southeast. Altitude rises gradually to the southwest, reaching 400 m in the direction of Gebel Uweinat.

The long axis of Kharga Oasis (about 200 km west of the Nile, Fig. 2.1) is in a north-south direction. It is bounded on the north and east by a steep lofty escarpment. To the south and southwest there is no definite boundary, so it is difficult to estimate the area of the depression precisely. It is long and narrow, 185 km from north to south and between 15 and 30 km from east to west (Abu Al-Izz, 1971). In the northwest, its width reaches 80 km. The depression floor is between 300 and 400 m below the surrounding plateau. The lowest point of the oasis floor is almost at sea level whereas the highest is at 115 m above sea level.

The western edge of the Kharga Depression consists of level land with mobile scattered sand dunes; in the northwestern part there is a high scarp. To the south, the depression is open and has no distinctive features. The northwestern wall between Ayun* Um Dabdab and Ayun Amur is strongly dissected by deep wadis flowing from the plateau into the depression. At the mouths of the wadis lines of sand dunes extend for several kilometres. These mobile dunes move across the floor of the oasis with a rate ranging from 20 to 30 m/year. Their movement overwhelms cultivated lands, wells, roads, buildings etc. Some villages are seriously affected, being enclosed from the north, east and west, by bodies of dunes (Beadnell, 1909).

---

* Ayun is Arabic for springs

The northern wall of the Kharga Oasis consists of a steep limestone scarp reaching a height of 371 m in the west. In the east the wall is 355 m. The eastern border of the depression is clearly the highest (400 m).

Dakhla Oasis is about 120 km west of Kharga Oasis. Its long axis is in a W-NW-E-SE direction. Its length is about 55 km and its width varies between 10 and 20 km. Altitudes are in the range 100–400 m above sea level. It is bounded in the north by a precipitous escarpment running more or less irregularly for at least 250 km eastward to Kharga Oasis. To the south of this cliff is a low-lying expanse of sandstone country forming a gentle undulating plain, the general surface of which rises imperceptibly to the south. In the east, there is lowland, covered with sand dunes, extending to the Kharga Oasis. In the west there are also some dunes which make it difficult to trace the western boundary of the oasis. The lowest point of Dakhla Oasis land is 100 ni above sea level and its surface rises gradually towards the rim.

Climate and Water Resources

The Kharga and Dakhla Oases are located in a dry rainless part of the Great Sahara. Rainfall is almost nil whereas the mean annual relative humidity is lower in summer (26–32%) than in winter (53–60%). The mean annual evaporation is 18.4 mm Piche/day and ranges from 25.1 to 9.5 mm Piche/day in summer and winter respectively. Temperature is moderate in winter: absolute minimum 6.0–4.8 °C and maximum 22.1–21.5 °C, but rises very high in summer: absolute minimum 23.4–23.1 °C and maximum 39.2–39.5 °C. There are recorded extreme maxima of about 50 °C in both oases. Mean wind velocity ranges from 6.8 to 11.6 km/h in winter to 11.1–19.8 km/h in summer (Migahid et al., 1960). The Pluviothermic Quotient (Emberger, 1951, 1955) is nearly zero indicating extreme aridity.

The water resource in both oases is, thus, underground. In Kharga Oasis, there are two distinct strata separated by a 75 m band of impermeable grey shale. The upper bed is exposed at the surface and forms the true artesian water sandstone from which the flowing wells of the Kharga Oasis derive their supply. In the Dakhla Oasis, the water supply comes from a bed of white sandstone which corresponds to the surface water sandstone of Kharga.

Plant Cover

The plant life of the Kharga and Dakhla Oases has been studied by Oliver (1930–1931), Hassib (1951), Tackholm & Drar (1954), Migahid et al. (1960), El-Hadidi and Kosinova (1971), Shalaby et al. (1975), Abu Ziada (1980), Girgis et al. (1981) and others.

Seven vegetation types have been recognized in Dakhla and Kharga Oases, namely: hydrophytic vegetation, reed swamp vegetation, halophytic vegetation, psammophytic vegetation, xerophytic vegetation, vegetation of cultivated lands and vegetation of waste lands. The main communities of each of these vegetation types are described.

Hydrophytic Vegetation

The aquatic vegetation is richly developed in the pools, ditches, wells and irrigation canals of Kharga and Dakhla Oases which have permanent fresh or brackish waters. This vegetation is represented in these two oases by five communities.

1. *Utricularia gibba* ssp. *exoleta* is dominant in the freshwater wells in sheltered places. It is associated with *Potamogeton pectinatus.*
2. *Ruppia maritima-Zannichellia palustris* are co-dominant in the brackish shallow pools where the water from irrigation collects. Plants of these two species are submerged and fixed at the bottom of the pools. *Potamogeton pectinatus* is common within this community.
3. *Najas graminea* is dominant in the shallow irrigation channels and is associated with *N. minor.*
4. *Lemna gibba* is a free-floating aquatic plant present in most water bodies of the two oases. It is associated with *Lemna perpusilla.*
5. *Nitella* spp. associated with *Chara* spp. These submerged algae are attached by rhizoids to the mud, forming thick masses at the bottom of the water body. This community occurs in stagnant water, mainly in drains.

Reed Swamp Vegetation

The helophytic vegetation is richly developed in the Dakhla and Kharga Oases. It occurs in marshy areas where the water is at the surface and the soil is waterlogged throughout the greater part of the year. This vegetation, thus, occurs around ditches, swampy ground in rice fields, around wells and in pools and in drains.

Two reed swamp communities dominated by *Typha domingensis* and *Phragmites australis* are recognized.

1. *Typha domingensis* community. *T. domingensis* is widespread in these two oases, particularly in the Dakhla Oasis. It forms extensive patches of dense growth in brackish shallow swamps, shallow ditches and silt swamps. The soils of these habitats are muddy and rich in organic matter with moderate salinity. The water-table is shallow or at the surface. This may be silted with aeolian sand that becomes dark, being enriched with organic matter from decayed rhizomes and leaves. The cover ranges between 60 and 100%. The most common associate species is *Phragmites australis* which may be co-dominant with *T. domingensis* in some swamps. *Tamarix nilotica, Imperata cylindrica* and *Juncus rigidus* are also common on the wet fringes of the swamps. Several water-loving species, e.g. *Cyperus laevigatus, C. longus, C. mundtii* and *C. rotundus*, are present. Other associates that usually occur on the fringes of this community include *Alhagi maurorum, Cynodon dactylon, Desmostachya bipinnata, Panicum repens* and *Phoenix dactylifera* (perennials) and *Conyza linifolia, Euphorbia peplus* and *Sonchus oleraceus* (annuals).
2. *Phragmites australis* community. *P. australis* is a very common strongly rhizomatous reed in Dakhla and Kharga Oases. It is usually dominant in the shallow swamps that result from the flow of water from springs or drainage systems. It thus occupies the

shallower or more saline parts of the swamps than those dominated by *T. domingensis*. *Typha* forms dense thickets in the deeper water with *Phragmites* towards the periphery of the swamps.

These conditions are seen in pools, ditches, wells, irrigation canals and rice fields. This reed community occurs also on silted swamps resulting from deposition of aeolian sand and water-borne fine sediments. The soil moisture content ranges between 25 and 35%.

The plant cover of this community is usually high (50–100%), the main bulk being provided by the dominant reed which forms extensive thickets. The surface runners (legehalme) of this reed may extend for as much as 20 m along the surface of the soil. Also, in some instances the plant produces these aerial creeping culms resembling stolons floating just above the water surface. *T. domingensis* is the most common associate species in the swampy areas of this community whereas *I. cylindrica, J. rigidus* and *T. nilotica* are frequent in the less wet habitat. In the swampy areas C. *laevigatas, C. rotundus* and *Ruppia maritima* are also found. The flora of the fringes of the swamps dominated by *Phragmites* comprise all species recorded on the fringes of *Typha* swamps in addition to *Aeluropus lagopoides, Lagonychium farctum*, and *Zygophyllum coccineum* (perennials) and *Anagallis arvensis, Ambrosia maritima, Kochia indica* and *Rumex dentatus* (annuals).

Halophytic Vegetation

The vegetation of the Dakhla and Kharga Oases is essentially halophytic. Extensive tracts are converted into salinas owing to the flow of springs, irrigation water and ill-drainage. These salt marshes may be also differentiated into:

A. Wet salt marshes where the underground water is shallow, the surface deposits include appreciable amounts of fine sand and silt and the ground is almost flat; and B. Dry salt marshes where the underground water is deeper and the surface deposits are sandy and undulating.

A. Wet Salt Marshes

Four communities may be recognized in this group dominated by *Cyperus laevigatus, Juncus acutus, Schanginia aegyptiaca* and *Suaeda monoica*.

1. *Cyperus laevigatus* community. *C. laevigatus* is widespread in the Dakhla Oasis and of limited distribution in the Kharga Oasis. Its usual habitat is the wet saline flats associated with a shallow water-table. The plant cover of the stands dominated by *C. laevigatus* is high and ranges between 80 and 100%, the dominant providing the main bulk of the cover. Its rhizomes and roots are interwoven in a network covering the water-table (Abu Ziada, 1980). *C. laevigatus* is severely grazed so that it forms carpet-like growth not more than 10 cm high. *Cynodon dactylon, Imperata cylindrica, Juncus rigidus* and *Tamarix nilotica* are the most common associates. Other less common members of this community are mainly water-loving halophytes, namely *Cyperus mundtii, C. rotundus, Diplachne fusca* and *Panicum repens. Typha domingensis* and *Phragmites australis* as well as *Potamogeton pectinatus* have also been found in this community where the

water-table is at the surface. In the less wet areas, *Aeluropus lagopoides, Alhagi maurorum* and *Cressa cretica* have been recorded. Associates include *Launaea capitata* (biennial), *Ammi majus, Anagallis aruensis, Centaurium spicatum, Chenopodium murale, Conyza Unifolia, Cyperus difformis, Echinochloa colona, Melilotus indica, Polypogon monspeliensis* and *Sonchus oleraceus* (annuals). The ground layer of this community is obviously the most conspicuous as it includes the dominant and most of the associate species. *T. nilotica* is the only representative of the frutescent layer.

2. *Juncus rigidus* community. *J. rigidus* is very common rush in the Dakhla and Kharga Oases. It dominates in the saline flats and in the saline-neglected lands formerly under cultivation. Owing to the successive rise in soil salinity, up to 5.5% in certain areas (Girgis et al., 1981), this land became unsuitable for the cultivation of conventional crops. Such habitats are very suitable for the invasion of J. *rigidus* which can grow in the wet, dark-coloured substratum having a thick crust of salts on the surface. The plant cover of the *J. rigidus* community is dense (40–100%) contributed chiefly by the dominant rush. *T. nilotica* is the most common associate, being present in most of the stands of this community. Common associates include *Alhagi maurorum, Cynodon dactylon, Cyperus laevigatus* and *Phragmites australis*. Rarely present associates include *Acacia nilotica, Cressa cretica, Cyperus rotundus, Lagonychium farctum, Phoenix dactylifera, Polygonum equisetiforme, Spergularia media* and *Zygophyllum coccineum. Centaurium spicatum* is the only annual so far recorded.

   The suffrutescent layer of the *J. rigidus* community is dense as it includes the dominant rush and most of the associates. The frutescent layer is restricted and includes *L. farctum, P. dactylifera* and *T. nilotica*. In the ground layer are short perennials, e.g. *Cressa cretica, Cyperus* spp. and *Cynodon dactylon* as well as the ephemerals.

3. *Schanginia aegyptiaca* community. S. *aegyptiaca* is a procumbent annual herb with extremely sappy, linear leaves. The community dominated by this succulent halophyte is widespread in the Kharga and Dakhla Oases on wet salt marshes and lands formerly cultivated.

   The flora of this community includes *Aeluropus lagopoides, Alhagi maurorum* and *Tamarix nilotica* as the most common associate species. Less common ones include *Atriplex leucoclada, Chrozophora obliqua, Cressa cretica, Cynodon dactylon, Hyoscyamus muticus, Juncus rigidus, Phragmites australis, Polygonum equisetiforme, Salsola baryosma* and *Sporobolus spicatus*. Among the therophytes are *Ammi majus, Asphodelus tenuifolius, Centaurium spicatum* and *Kochia indica*. The suffrutescent layer is obviously the most conspicuous as it includes the dominant and most of the associates.

4. *Suaeda monoica* community. S. *monoica* is a halophytic succulent shrub or small tree, 2–4 m high. It flowers all the year round and becomes black when dry. It is extensively grazed and is specially preferred by camels (Long, 1955; Zahran, 1982b). The community dominated by S. *monoica* is not common in the Kharga and Dakhla Oases, being confined to the wet salt flats that are usually dark in colour and have a thin surface crust of salts. The cover of this community

ranges from 5 to 10%. *Tamarix nilotica* is the most common associate. *Aeluropus lagopoides, Alhagi maurorum, Imperata cylindrica, Phragmites australis* and *Zygophyllum coccineum* are frequent associates. Rarely recorded are *Cressa cretica. Polygonum equisetiforme, Salicornia fruticosa* and *Salsola baryosma*.

The S. *monoica* community has a well-developed frutescent layer, including the dominant and the most common associate *(T. nilotica)*. The suffrutescent layer, with most of the other associates, also contributes to the cover. The ground layer is very thin and is represented by *C. cretica* and low-growing individuals of *Alhagi maurorum*.

B. Dry Salt Marshes

Six communities have been recognized in this saline system dominated by *Cressa cretica, Aeluropus lagopoides, Sporobolus spicatus, Alhagi maurorum, Imperata cylindrica* and *Tamarix nilotica*.

1. *Cressa cretica* community. *C. cretica* is a mat-like hairy plant of grey appearance. Its community is uncommon in the Kharga and Dakhla Oases, occupying the dry salt marshes and saline fallow lands with occasional deposits of sand sheets. The cover ranges between 10 and 80% with the main bulk provided by the dominant. *Alhagi maurorum* and *Tamarix nilotica* are common associates. Other species present are *Aeluropus lagopoides, Calotropis procera, Cynodon dactylon, Hyoscyamus muticus, Imperata cylindrica, Lagonychium farctum. Phoenix dactylifera, Phragmites australis, Sporobolus spicatus* and *Zygophyllum coccineum* (perennials), and *Ammi majus, Asphodelus tenuifolius, Cichorium pumilum, Conyza linifolia* and *Kochia indica* (annuals). The ground layer of the *C. cretica* community thus constitutes the prominent part of its phytocoenosis. The suffrutescent and frutescent layers are thin.
2. *Aeluropus lagopoides* community. The *A. lagopoides* community is often associated with dry salt marshes that are usually covered by a thin salt crust. It forms a common halophytic grassland in the Kharga and Dakhla Oases with 20–50% plant cover.

   Tamarix nilotica is the most common associate; less common are *Alhagi maurorum, Cressa cretica, Hyoscyamus muticus, Phragmites australis, Salsola baryosma* and *Sporobolus spicatus*. Other associates include most of the species recorded in the *C. cretica* community in addition to *Atriplex leucoclada, Capparis aegyptia, Chrozophora obliqua, Conyza dioscoridis, Juncus rigidus, Phoenix dactylifera, Scirpus tuberosus, Suaeda monoica* (perennials) and *Centaurium spicatum, Corchorus olitorius, Polypogon monspeliensis* and *Sonchus oleraceus* (annuals). Such flora shows that the suffrutescent layer is relatively dense, including the dominant and most of the perennial associates. The frutescent and ground layers are thin.
3. *Sporobolus spicatus* community. *S. spicatus* is a stiff perennial grass, of pale colour, with creeping stolons which are often several metres long with tufts of leaves and short culms at the rooting nodes (Täckholm, 1974). The community of this grass covers vast areas of the Kharga Oasis. Its cover ranges between 10 and 80%, provided mostly by the dominant that builds sand mounds of moderate size. It is heavily grazed up to a certain height where its tussocks become rigid and spinescent.

Tamarix nilotica is the abundant associate. Common species present include *Alhagi maurorum, Hyphaene thebaica, Imperata cylindrica* and *Phoenix dactylifera*. Other associates of this community include most of the species recorded in the A *lagopoides* community in addition to *Cyperus laevigatus, Suaeda monoica* and *Typha domingensis* (perennials) and *Solanum nigrum* (annual).

In *A. lagopoides* and *S. spicatus* grasslands, the suffrutescent layer is well developed whereas the frutescent and ground layers are thin.

4. *Alhagi maurorum* community. *A. maurorum* is a perennial leguminous plant with spreading spines. It is a valuable fodder for camels and other domestic animals and is known to yield a medically valuable manna. It is also a useful soil binder and as it grows gregariously it protects the soil against erosion. According to Kassas (1952c), A. *maurorum* is dependent for its moisture supply on underground water that may be deeply seated (in desert) or shallow in coastal areas. However, in saline soil, the zone of absorption of the root of *A. maurorum* is found to be located deeper, below the layers of high salinity.

   The A. *maurorum* community is a feature of the vegetation of Kharga and Dakhla Oases as well as other oases and depressions of the Western Desert. It occurs everywhere in saline flats overlying salt marsh beds, dry salt marshes, slopes, hills, plains etc. *Tamarix nilotica* is the most important species of this community. *Aeluropus lagopoides* and *Sporobolus spicatus* are also commonly present. Less common associates are *Acacia nilotica, Capparis aegyptia, Cressa cretica, Francoeuria crispa, Hyoscyamus muticus, Lagonychium farctum. Phoenix dactylifera, Salsola baryosma* and *Zygophyllum coccineum, Phragmites australis* and *Juncus rigidus* are confined to the wet stands. The therophytic associates of this community comprise *Bassia muricata, Chenopodium murale, Kickxia elatine, Melilotus indica, Solanum nigrum* and *Sonchus oleraceus*. The cover of the *A. maurorum* community (40–70%) may be differentiated into four layers, there being a few trees. The frutescent layer is thin and includes the shrubs. The suffrutescent layer is the most conspicuous as it contains the dominant and most of the associates. The ground layer contributes little to the plant cover.

5. *Imperata cylindrica* community. The *I. cylindrica* community is widespread in the Kharga and Dakhla Oases. It is associated with high levels of dry salt marshes, depressions surrounded by dunes and sand accumulation overlying dry salt marshes and canal banks. *Imperata* is a sand-binding grass useful for the fixation of dunes. The cover of this community is high (60–90%), contributed mainly by the dominant grass which is usually fired by the inhabitants to yield a new dense growth for grazing animals.

   The most frequent associates are *Alhagi maurorum, Juncus rigidus, Phragmites australis* and *Tamarix nilotica. Acacia raddiana. Phoenix dactylifera* and *Sporobolus spicatus* are commonly present. Less common associate species are *Calotropis procera. Cassia italica, Chrozophora obiiqua, Francoeuria crispa, Inula crithmoides, Lagonychium farctum, Lotus corniculatus, Maerua crassifolia, Pluchea dioscoridis (Conyza dioscoridis), Scirpus tuberosus* and *Typha domingensis* (perennials) and *Ammi majus, Centaurium spicatum, Kochia indica, Melilotus indica, Phalaris paradoxa, Silene nocturna* and *Sonchus oleraceus* (annuals). The suffrutescent layer is the most conspicuous one; both the frutescent and ground layers are thin.

6. *Tamarix nilotica* community. *T. nilotica* is a desert shrub with a deep root system that reaches the water-table. Chapman (1960) reported that *Tamarix* communities are restricted to well-drained soils.

Tamarix spp. are salt excretives or crinohalophytes (see e.g. Waisel, 1961). The leaves are often covered with a glistening bloom of hygroscopic salt crystals.

T. nilotica is one of the widely present species in the Kharga and Dakhia Oases. It is a very common associate within the communities of the reed swamp and salt marsh vegetation. *T. nilotica* is also a common associate in the communities of xerophytic vegetation of these oases; it is a species of wide ecological amplitude. The community dominated by *T. nilotica* in the Kharga and Dakhia Oases has been observed where sand accumulates and on sand dunes having an underground water-table and also in the dry salt marshes with deep sand deposits that usually have thin crusts of salts on the surface. *Tamarix* scrub, which represents the climax of the salt marsh vegetation, is subject to destructive cutting for fuel and other household purposes. The plant cover ranges between 20 and 50%, but in favourable localities is 70–80%, i.e. almost closed scrub. *Aeluropus lagopoides, Alhagi maurorum, Imperata cylindrica, Juncus rigidus* and *Phragmites australis* are the common associates. Less common perennials of this vegetation include *Calotropis procera, Conyza dioscoridis, Cressa cretica, Cyperus laevigatus, Francoeuria crispa, Hyoscyamus muticus, Polygonum equisetiforme, Salsola baryosma, Sporobolus spicatus, Suaeda monoica* and *Typha domingensis*. The therophytes include *Centaurium spicatum, Frankenia revoluta, Kochia indica* and *Schanginia aegyptiaca*.

The frutescent layer in the *T. nilotica* community is well developed as it includes the dominant shrub. The suffrutescent layer comprises most of the perennial associates and makes a considerable contribution to the cover. The ground layer is, however, thin comprising the matlike perennials (*Cressa* and *Cyperus*); in wet stands ephemerals enrich this layer.

Psammophytic Vegetation

Sand Plains

Sand plains are flat expanses of wind-drifted siliceous sand. In the Kharga Oasis these plains are scattered. In Dakhia Oasis there is a broad sand plain in Wadi El-Akola (Akole is the vernacular name of *Alhagi*). Sand plains also surround most of the villages of the Dakhia Oasis.

Wandering within the sand plains are mobile sand dunes 20–30 m high. Wind action is very severe on these dunes as well as on the sand plains. The surface sand of these plains is derived mainly from the dunes. The presence of sand blowing and sand deposition usually go hand in hand. Any place subject to wind erosion at one time may-receive deposited sand at another. Such continual wind action causes instability of the soil and variable thickness, which in turn affect plant cover. Thus, sand plains are almost bare of vegetation in exposed parts, but may support a sparse cover where sheltered, e.g. in the protection of dunes.

To the north of the sand plains is a series of bare mobile dunes which intersect another series on the eastern side of the plains. The direction of the prevailing wind is northwest-southeast. The eastern dunes are old and stabilized by vegetation. The sand in the low ground at the meeting point of the two dune series is more sheltered and stable than that in the open plain some distance away. Consequently, in the sheltered area a community of *Cressa cretica* has developed in the forward concavity of one of these dunes and a community dominated by *Alhagi maurorum* associated with *Stipagrostis scoparia* is present in other parts of the protected area. The *C. cretica* community forms open and pure growth with thin cover (<5%), the aboveground part of the plant being 10–30 cm high, i.e. there is a ground layer only. The *Alhagi-Stipagrostis* community has a cover of 5–10%.

No living plants exist in the open parts of the plains, only dried stumps of doum trees *(Hyphaene the baica), Tamarix* sp. and others, representing the remnants of a previous plant cover. This vegetation evidently developed on the clay substratum before it was covered by the wind-blown sand dunes. After the sand of the dunes had blown away from the area the dried stumps became exposed again.

In the drier, elevated, sand plains of the Kharga Oasis are such xerophytes as *Alhagi maurorum* (dominant), *Calotropis procera* and *Hyoscyamus muticus*. In the Dakhla Oasis other species may be found, e.g. *Aerva javanica, Stachys aegyptiaca, Suaeda fruticosa* and *Tamarix nilotica*. Individual shrubs of *T. nilotica* reach considerable size and usually build hillocks. In some places, *Calotropis*, a sand non-binding shrub, was found buried under the drifted sand.

Sand Dunes

Sand dunes at different stages of development are numerous in the oases of the Western Desert. The younger dunes are mobile and lack vegetation. They are of the "Barchan" crescentic type with a high, steep forward margin on the leeward side. The medium-sized barchan has a height of about 15 m and a length of about 200 m. Behind the steep leeward part the barchan extends into a tapering tail. These bare dunes migrate slowly in the direction of the wind. Oliver (1930–1931) states that "the projecting slopes travel faster than the main body and move in advance, and these cusps are kept fed with sand which slides along the perimeter on either flank of the dunes".

The travelling barchans (10–20 m/year, Beadnell, 1909) lack vegetation since there is no stable soil on which the plants can settle. On the older stabilized dunes, *Tamarix nilotica* and sometimes *Alhagi maurorum* grow abundantly and cover the summits as well as the slopes. Vegetated dunes usually have their higher, steep margin facing windward, and to leeward the tapering tail of more recently deposited sand is quite barren. "The growing plants act as a stabilizing agent which breaks the velocity of the sand-bearing wind, thus leading to deposition of the burden of sand on their shoots" (Migahid et al., 1960).

Moving sand is a nuisance in the oases not only because it overwhelms dwellings, cultivations and wells but also because it lodges wherever the conditions are stabilized, e.g. by vegetation, thus raising the level. It is usual to see dunes encroaching trees of *Balanites aegyptiaca* and *Hyphaene thebaica* as in Baris Village of Kharga Oasis. The farms of this oasis are protected by successive belts of *Saccharum officinarum, Eucalyptus, Lagonychium* and *Acacia*.

Xerophytic Vegetation

The xerophytic vegetation of the desert ecosystem of the Kharga and Dakhla Oases may be considered under eleven communities dominated by *Citrullus colocynthis, Zygophyllum coccineum, Salsola baryosma, Chrozophora obliqua, Hyoscyamus muticus, Stipagrostis scoparia, Calotropis procera, Lagonychium farctum, Tamarix nilotica, Acacia nilotica* and *Balanites aegyptiaca*.

1. *Citrullus colocynthis* community. In the Kharga and Dakhla Oases this community is not common, but it is present on sand plains covering a substratum of fine sediments. The cover is often dense (average 40%), contributed mainly by the dominant which forms large patches of prostrate growth and carries a wealth of small melon-like fruits with profuse seeds during late summer. The fruits are bitter, highly purgative and used for curing rheumatic pains. Extracts of fruits have antitumour activity against sarcoma 37; the activity is due to the cucurbitacins (Ayensu, 1979).

   The most common associate species of the *C. colocynthis* community include *Alhagi maurorum, Calotropis procera, Chrozophora obliqua, Hyoscyamus muticus* and *Tamarix nilotica*. The other associates include five common, and five rare perennials (Table 3.2) and a few annuals, e.g. *Launaea capitata* and *Tribulus longipetalus*.

   The ground layer is the most conspicuous, including the dominant species. The suffrutescent layer comprises most of the associates whereas in the frutescent layer are a few species of shrubs and trees.
2. *Zygophyllum coccineum* community. Within the desert ecosystem of the Kharga and Dakhla Oases, the *Z. coccineum* community is not common. It is present in sand plains and poorly saline peripheries of the salt marsh ecosystem. The plant cover ranges between 10 and 80%, contributed mainly by the dominant. *T. nilotica* is the most closely associated species with another ten common perennial associates (Table 3.2) and a few annuals e.g. *Asphodelus tenuifolius* and *Schanginia aegyptiaca*. The suffrutescent layer is thus the well-developed one, both frutescent and ground layers being ill-defined.
3. *Salsola baryosma* community. This community occurs in a variety of habitats, ranging from wet compact fine sandy soil of fallow land to moderately saline dry salt marshes covered by sheets of loose sand. It is a common community. The dominant is a desert succulent chenopod that forms low sand mounds 20–50 cm high. *Tamarix nilotica* is the most common associate. Other associates include 22 frequently present perennials (Table 3.2) and the following annuals: *Anagallis arvensis, Asphodelus tenuifolius, Chenopodium murale, Conyza linifolia, Kochia indica, Malva parviflora, Melilotus indica, Polypogon monspeliensis* and *Schanginia aegyptiaca*.
4. *Chrozophora obliqua* community. *C. obliqua* is a desert undershrub that flowers in summer, dries up in winter and its dry stalks are used as mulch and manure. In the Kharga and Dakhla Oases, the *C. obliqua* community is present in the fallow lands of loose sand overlying beds of fine deposits. It is common in the Dakhla Oasis and rare in the Kharga Oasis. Its cover varies between 10 and 80%. The most common associates include *Cynodon dactylon, Lagonychium farctum* and *Tamarix*

Table 3.2 Floristic list of the dominants and perennial associate species of the eleven communities of the xerophytic vegetation of the Kharga and Dakhla Oases, Western Desert

Species	*Communities*										
	1	*2*	*3*	*4*	*5*	*6*	*7*	*8*	*9*	*10*	*11*
The dominants											
1 *Citrullus colocynthis*	D	–	–	C	–	–	C	C	C	C	–
2 *Zygophyllum coccineum*	R	D	C	–	MC	O	–	C	–	–	–
3 *Salsola baryosma*	R	C	D	–	MC	O	–	–	C	–	–
4 *Chrozophora obliqua*	MC	–	–	D	O	–	C	C	C	–	–
5 *Hyoscyamus muticus*	MC	C	C	C	D	MC	C	–	C	C	–
6 *Stipagrostis scoparia*	–	–	C	–	O	D	–	–	C	–	–
7 *Calotropis procera*	MC	C	C	C	MC	C	D	C	C	MC	C
8 *Lagonychium farctum*	R	C	C	MC	–	–	MC	D	C	MC	C
9 *Tamarix nilotica*	MC	Ab	MC	MC	MC	C	MC	MC	D	MC	–
10 *Acacia nilotica*	C	C	–	–	O	–	O	C	C	D	C
11 *Balanites aegyptiaca*	–	–	C	–	–	–	O	–	C	–	D
Associate perennials											
Abutilon pannosum	–	–	–	–	–	–	–	–	C	–	–
Aeluropus lagopoides	C	C	C	–	O	–	–	–	–	–	–
Aerva javanica	–	–	–	–	–	C	–	–	–	–	–
Alhagi maurorum	MC	C	C	C	MC	O	C	C	C	C	C
Astragalus trigonus	–	–	–	C	O	–	–	–	–	–	–
Atriplex leucoclada	–	–	C	–	O	–	–	C	–	–	–
Capparis aegyptiaca	–	–	C	–	–	–	–	–	O	–	–
C. decidua	–	–	–	–	–	–	C	C	–	–	–
Cassia italica	–	–	–	C	–	–	C	C	–	–	–
Convolvulus pilosellifolius	–	–	–	C	O	–	–	–	–	–	–
Cressa cretica	–	C	C	–	O	–	–	C	–	C	–
Cynodon dactylon	C	C	C	MC	O	–	C	C	C	C	–
Cyperus laevigatus	–	–	–	C	–	–	–	–	–	–	–
Fagonia arabica	–	–	C	–	O	–	–	–	–	–	C

Table 3.2 (Continued)

Species	*Communities*										
	1	*2*	*3*	*4*	*5*	*6*	*7*	*8*	*9*	*10*	*11*
F. indica	–	–	–	–	–	–	–	–	C	–	–
Francoeuria crispa	–	–	C	–	–	–	C	C	C	–	–
Haplophyllum tuberculatum	–	–	C	–	–	–	–	–	–	–	–
Heliotropium bacciferum	–	–	–	C	–	–	–	C	C	–	–
Hyphaene thebaica	–	–	–	–	–	–	–	C	C	–	C
Imperata cylindrica	R	–	C	–	–	O	O	C	C	C	–
Juncus rigidus	–	–	C	–	–	–	–	C	–	C	–
Lotus corniculatus	C	–	–	–	–	–	–	–	–	–	–
Maerua crassifolia	–	–	–	–	–	–	C	O	–	–	C
Phoenix dactylifera	C	C	C	C	O	–	O	C	–	MC	MC
Polygonum equisetiforme	–	–	C	MC	O	–	–	–	C	–	–
Saccharum spontaneum	–	–	–	–	–	–	–	–	C	–	–
Sida alba	–	–	–	–	–	–	–	–	C	–	–
Sporobolus spicatus	R	–	C	–	O	O	O	C	C	–	–
Suaeda monoica	–	–	C	–	O	–	–	–	–	–	–
Tamarix aphylla	–	–	–	–	–	–	–	C	C	–	C
Tephrosia apollinea	–	–	–	–	–	–	C	–	–	–	–
Trichodesma africanum	–	–	C	C	O	O	O	C	C	–	–
Ziziphus spina-christi	–	–	–	–	–	–	–	C	O	–	–

Abbreviations: D. dominant; Ab, abundant; MC, most common; C, common; O, occasional; R, rare.

nilotica. Twelve common perennial associates are also present (Table 3.2). The number of associate therophytes is the highest of those of communities of these two oases and include *Amaranthus graecizans, Ambrosia maritima, Ammi majus, Brassica tournefortii, Centaurium spicatum, Conyza linifolia, Dactyloctenium aegyptium, Digitaria sanguinalis, Eragrostis aegytiaca, Kickxia elatine, Kochia indica, Launaea capitata. Lolium rigidum, Plantago lagopus, Portulaca oleracea, Senecio desfontainei, Sorghum virgatum, Tribulus longipetalus* and *Verbesina encelioides*. Stratification of growth is obvious in this community where three layers (frutescent, suffrutescent and ground) are well represented.

5. *Hyoscyamus muticus* community. *H. muticus* is a stout, fleshy, puberulent plant, richly branched from the neck. It is recorded as an associate species in most of the halophytic and xerophytic vegetation types of Kharga and Dakhla Oases. The community dominated by *H. muticus* is rare in these two oases. It usually occurs on the silty deposits overlain by thin sheets of sand. The cover ranges between 20 and 70%, contributed mainly by the dominant. The most common associates are *Alhagi maurorum, Calotropis procera, Salsola baryosma, Tamarix nilotica* and *Zygophyllum coccineum*. Other associates (Table 3.2) include 15 occasionally present perennials and the following annuals: *Anagallis aruensis, Asphodelus fistulosus* v. *tenuifolius, Corchorus olitorius, Kochia indica* and *Sonchus oleraceus*.
6. *Stipagrostis scoparia* community. S. *scoparia* is not a common grass in the Kharga and Dakhla Oases, being confined to sand accumulations and slopes of dunes in a few localities of the Kharga Oasis. It is a sand binder and forms mounds up to 50 cm high. The vegetation is very sparse, covering less than 5% of the ground, with *Hyoscyamus muticus* as the most common associates. Other associates include nine common or occasional perennials (Table 3.2) and only one annual (*Kochia indica*).
7. *Calotropis procera* community. *C. procera* is one of the desert woody shrubs that is often cut for fuel. The fruit is smooth, swollen and spongy-apple like and the seeds bear long hairs used by the local inhabitants for making cushions. The plant is rich in latex that is highly poisonous and causes inflammation of the eyes. Ayensu (1979) states "C. *procera* contains calotacin, calotropin, uscharidin, calctinic acid etc. Its latex has insecticidal activity and abortifacient effects in rats. It is used as purgative, emetic, diaphoretic, expectorant, and can be used also to treat dysentery, colds and elephantiasis".

   The scrubland vegetation dominated by *C. procera* is common in the Kharga Oasis but rare in the Dakhla Oasis. It is stratified into four layers. The upper stratum is very thin and includes the trees *Acacia nilotica, Balanites aegyptiaca* and *Phoenix dactylifera*. The shrub layer is the most conspicuous, containing the dominant, *Lagonychium farctum* and *Tamarix nilotica*. The suffrutescent layer includes most of the other perennial associates (Table 3.2). In the ground layer are *Citrullus colocynthis* and *Cynodon dactylon* (perennials) and *Tribulus longipetalus* and *Verbesina encelioides* (annuals).
8. *Lagonychium farctum* community. *L. farctum* is a small prickly desert shrub that varies in growth form in relation to habitat conditions. It forms either straggling bushes or grows prostrate on the ground. In localities with limited

water resources the plant is dwarf, not exceeding 30 cm. In favourable localities it forms robust growth of considerable size and height. It is a rare desert shrub recorded in the Nile Delta, eastern Mediterranean coast, Oases and the southern region of the Eastern Desert (Täckholm, 1974). Dominance of *L. farctum* has been recorded in Kharga and Dakhla Oases (Shalaby et al., 1975; Abu Ziada, 1980), where it forms scrubland vegetation confined to the silty deposits that may be overlain with sheets of loose sand. The plant cover may be sparse or dense thickets depending on the environmental conditions. In this scrubland the frutescent layer is the most conspicuous, giving the community its character. This layer includes the dominant shrub, *Calotropis procera, Capparis decidua, Maerua crassifolia* and *Tamarix nilotica*. The tree layer is represented by a few specimens of *Acacia nilotica* and *Phoenix dactylifera*. The suffrutescent layer includes most of the other perennial associates (Table 3.2) and a few annuals, e.g. *Ambrosia maritima* and *Schanginia aegyptiaca*. In the ground layer are *Citrullus colocynthis, Cressa cretica* and *Cynodon dactylon* (perennials) and *Anagallis arvensis* and *Launaea capitata* (annuals).

L. farctum in the Kharga and Dakhla Oases is subject to destruction: the pods and foliage are used as fodder and branches as fuel. Local inhabitants burn the shrub, being a carrier for some parasites that infect pomegranate.

9. *Tamarix nilotica* community. In the desert ecosystem of the Kharga and Dakhla Oases, scrub dominated by *T. nilotica* is common on the sand accumulations and low sand mounds that border the cultivated lands and villages of these two oases. Where a water supply is available, *T. nilotica* is a good sand binder and can be used to fix sand dunes. The windward slopes of dunes stabilized by *T. nilotica* are steep while towards the leeward side a tapering tail of more recently deposited sand develops which is quite barren. *T. nilotica* acts as a barrier which breaks the velocity of sand-bearing winds, thus leading to the deposition of their burden of sand on the shoots. *T. nilotica* has the ability of rapid growth and emergence through the deposited sand. By its upward growth the level of the dunes is gradually increased. The underground parts are profusely branched within the mass of accumulated sand and in this way help to fix and stabilize the dunes. Kassas (1952a) states that "the growth of *T. nilotica* scrubland would indicate the presence of underground water reserve".

In the desert ecosystem of Kharga and Dakhla Oases, the *T. nilotica* community is made up of four layers. The tree layer is thin and includes *Acacia nilotica, Balanites aegyptiaca, Hyphaene thebaica* and *Ziziphus spina-christi*. The shrub (frutescent) layer includes the dominant, which forms large patches of thicket, and also a number of associate shrubs such as *Abutilon pannosum, Calotropis procera, Capparis aegyptia* and *Tamarix aphylla*. The suffrutescent layer, however, contains the greater number of species (Table 3.2) but its contribution to the plant cover is limited. The ground layer is very thin and includes *Citrullus colocynthis, Cynodon dactylon, Fagonia indica, Sida alba* (perennials) and *Bassia muricata* (annual). The flora of the *T. nilotica* community of the salt marsh and desert ecosystems is clearly a mixture of halophytes and xerophytes.

10. *Acacia nilotica* community. *A. nilotica* is a common desert tree in Kharga and Dakhla Oases. Gum is secreted on the surface of the trunk and collected by the inhabitants. The crown of this tree provides excellent shade and together with the fallen leaves and other plant remains makes microhabitat conditions favourable for insects that infect the ripe fruits. The trees are subject to destructive cutting for charcoal making, feeding animals etc. In some localities of these oases the undergrowth of A. *nilotica* is cleared for cultivation.

    A. nilotica scrubland represents one of the advanced desert types. It is present in the sand plains overlying fine silty deposits of the Kharga and Dakhla Oases. The average cover of this scrubland ranges between 30 and 40% with *Tamarix nilotica, Phoenix dactylifera, Calotropis procera* and *Lagonychium farctum* as the most common associates. The undergrowth is differentiated into suffrutescent and ground layers, the species of which are shown in Table 3.2.

11. *Balanites aegyptiaca* community. *B. aegyptiaca* is a common tree in the Kharga and Dakhla Oases. Its scrub is confined to areas of deep silty compact deposits, the depth of which may reach 10 m.

    The *B. aegyptiaca* scrub in these two oases, like *Acacia nilotica* scrub, represents an advanced type of the xerophytic vegetation. It is the climax stage of the xerosere of the Egyptian desert. The tree layer is the most conspicuous as it includes the dominant as well as *Hyphaene thebaica, Phoenix dactylifera* and *Tamarix aphylla*. The lower frutescent layer is less important and includes the shrub species *Calotropis procera, Lagonychium farctum*, and *Maerua crassifolia*. The suffrutescent and ground layers are negligible and are represented by *Alhagi maurorum* and *Fagonia arabica* respectively.

Vegetation of Cultivated Lands

Cultivation in the Kharga and Dakhla Oases depends entirely on water flow from deep artesian wells, of which there are several hundred in the area of these oases. Some of the wells date from Roman times while others belong to the Pharaonic period. The water comes up warm and most of the wells are over-flowing. A means for the economic use of the artesian water must, however, be developed since water is most valuable in the desert. The continual flow of water leads to the formation of salt marshes. Areas subject to repeated flooding and drying gradually become suitable for cultivation. The high aridity of the climate enhances evaporation and the deposition of a crust of salt at the surface.

The date palm *(Phoenix dactylifera)* is the main crop of the Kharga and Dakhla Oases. It is grown not only for its fruits but also for fibres in the dried leaves which are used in the manufacture of baskets and other articles. In addition to the date palms, the cultivable land in these two oases is cropped in hay (mainly alfalfa), cereals (mainly sorghum, wheat, rice and barley) and horticulture (mainly olives, grapes, citrus, pomegranate and apricot). Doum trees *(Hyphaene thebaica)* are also common. Their nuts as well as the fruits of handal *(Citrullus colocynthis)* are exported with *Acacia* fruits which are used for tanning. *Acacia* also yields a valuable wood durable for coating the wells.

The most common weeds of winter cultivation in Kharga and Dakhla Oases include *Cynodon dactylon, Melilotus indica* and *Sonchus oleraceus*. Less common winter weeds include *Chenopodium murale, Conyza linifolia, Eruca sativa. Polygonum equisetiforme* and *Sorghum virgatum. Cynodon dactylon* is the most frequent weed of summer crops. Common summer weeds include *Chenopodium murale, Convolvulus arvensis, Echinochloa colona* and *Sorghum virgatum*. El-Hadidi and Kosinova (1971) mentioned several other less common winter and summer weeds of the two oases.

Frequently unfavourable conditions are related to rising of the water-table; salinization and waterlogging develop. Under these conditions, in the Kharga and Dakhla Oases, species that endure saline soil and saturated substrata may invade. These include *Aeluropus lagopoides, Cressa cretica, Cyperus rotundus, Frankenia pulverulenta, Scirpus tuberosus* and *Sphenopus divaricatus*.

Vegetation of Waste Lands

Vast areas of neglected lands are found outside the cultivated areas of Kharga and Dakhla Oases. These are no longer irrigated although in the past they probably were under irrigation. They have been neglected because of a decrease in the supply of irrigation water through wells being overwhelmed by drifted sand, or through the rise of the well-head by constant accretion to the extent that artesian pressure could no longer raise the water to the surface, or else the land may deteriorate through salinity because of continual irrigation and flooding with the weakly saline artesian water. Such flooding, in the absence of drainage and under the prevailing conditions of intense evaporation, causes progressive increase in soil salinity. The waste lands are often intermediate in level and salinity between the salt marshes and the sand plains or desert plateaux. They support a vegetation of weed species, which is sometimes fairly dense with cover up to 60%.

The vegetation of the waste lands of Kharga and Dakhla Oases comprises four communities (Migahid et al., 1960).

1. *Zygophyllum coccineum* community. *Z. coccineum* dominates in areas where the ground level is fairly high but the water-table is so close to the surface that available soil water is within the reach of roots of the dominant species and its xerophytic associates. The plant cover is about 50%. The associates include *Astragalus trigonus* v. *leucacanthus* as the most common and *Asphodelus tenuifolius, Fagonia arabica, Hyoscyamus muticus, Moltkiopsis ciliata* and *Phlomis floccosa* are also common.
2. *Tamarix nilotica-Alhagi maurorum* community. This community occurs in the relatively dry sandy soil. It is associated with *Sporobolus spicatus* only. All these species accumulate wind-drifted sand and so build up sand hillocks and hummocks. Cover is about 50%.
3. *Alhagi maurorum* community. Plant cover is about 60%; the soil is sandy. All plants accumulate sand around their shoots to build up small mounds. *Tamarix nilotica* is the most abundant associate species; others include *Cyperus laevigatus, Sporobolus spicatus* and *Zygophyllum simplex*.

4. *Tamarix nilotica-Zygophyllum coccineum* community. This occurs in areas with dry sandy substratum; plant cover is less than 40%. Associates include *Alhagi maurorum, Aeluropus lagopoides, Cyperus laevigatus, Haplophyllum tuberculatum, Hyoscyamus muticus* and *Polypogon monspeliensis*.

(iii + iv) Kurkur and Dungul Oases

The Nubian Oases

In the Nubian Desert west of the Nile (south-eastern part of the Western Desert) are five small uninhabited Oases, namely: Kurkur, Dungul, Nakheila, Kurayim and Bir Murr. An ecological account on the first two is given here.

Kurkur Oasis is a part of Wadi Kurkur and Dungul Oasis is a part of Wadi Dungul (Fig. 2.1). According to Kassas and Imam (1954), the term wadi means a dried riverbed in a desert area. A wadi may be transformed into a temporary water-course after heavy rain. Each wadi has a main channel and branched affluents. Wadis in Egypt are much more abundant in the Eastern Desert than in the Western Desert. In a typical wadi, the vegetation is subject to seasonal changes resulting from differences in growth form of component species and seasonality of climate. Neither Wadi Kurkur nor Wadi Dungul are typical wadis; they are typical oases since their vegetation is supported by ground water not by rain. However, the upstream parts of these two wadis, where underground water is very deep, represent a typical wadi, vegetation being more dependent on rainfall (Boulos, 1966a; Zahran, 1966).

Wadi Kurkur lies about 62 km southwest of Aswan and about 52 km west of the Nile. It occupies what seems to be the confluence of three wadis joined in the form of a letter Y. These are the upstream part of Wadi Kurkur which extends until it meets the River Nile at Dabud (Boulos, 1966a). Wadi Dungul lies on the top of the Sin-el-Kidab scarp (262 m, Hume, 1908), south of Aswan by about 160 km (Fig. 2.1).

The area of Wadi Kurkur and Wadi Dungul is almost rainless; the mean annual rainfall at Aswan (the nearest meteorological station) is 3 mm. Rainfall is not an annually recurring phenomenon but an "accident" that may happen once every decade. The main source of water for these two oases is seepage along a line which can be indicated, to a large extent, by the plant growth. In Kurkur Oasis there are three wells reported by Butzer (1964) and Reed (1964) and in Dungul Oasis there is the Ain El-Gaw spring, the fossil water of which is fresh and cool (Zahran, 1966). The vegetation associated with these wells is remarkably dense.

Kurkur and Dungul Oases are of ecological and historical interest. They are located in the area of the Nubian Desert (Lat, 23°54' and 23°26'N and Long. 32°19'E and 31°37'E, respectively) where the huge man-made lake of the Aswan High Dam has been constructed. Accordingly, these two oases, and other Nubian Oases, have attracted visits by ecologists, archaeologists, both Egyptian and other. The Peabody Museum of Natural History, Yale University, USA arranged an expedition to study the Oases of this part of the Nubian Desert west of the Nile. This expedition was not only historically oriented, excavating remains of former civilizations, but also concentrated on events of the last half-million years that preceded the rise of the first ancient civilization of Africa. "For a study of the purely local climate, one needs an oasis, where the environment was and is unaffected by the complexities of the Nile's flow. A small oasis (and uninhabited e.g. Kurkur and Dungul) would be sensitive to local changes" (Reed, 1964).

Ecological Characteristics

Wadi Kurkur

Ecologically Wadi Kurkur consists of four units: the oasis proper, northwest wadi, north wadi and south wadi (Boulos, 1966a).

The Oasis Proper

Kurkur Oasis is the part of Wadi Kurkur best covered by dense growth of plants with groves of doum and date palms and patches of reeds around the wells which are mostly silted. Two of these wells are permanently open, each actually being a small shallow pool surrounded by reeds. Within the area of Kurkur Oasis the water-table is high.

The vegetation around the wells is in the form of a ring of reed dominated by *Typha domingensis* and/or *Phragmites australis* followed by a ring of rush *(Juncus rigidus)*, a further and more extensive zone of halfa grass *(Desmostachya bipinnata)* and an outer zone of a mosaic of *D. bipinnata* and *Alhagi maurorum*. In the vicinity of one of the wells is a patch of *Imperata cylindrica*, a halfa grass similar in appearance to *D. bipinnata*, which forms a carpet in the area ranging from complete ground cover in the centre to sparse cover on the periphery. This green carpet is studded by *Hyphaene thebaica* and *Phoenix dactylifera* palms; the former are more numerous. In the central part of the oasis a single bush of *Tamarix nilotica* was present in 1965 near one of the wells.

Northwest Wadi

The northwest wadi may ecologically and geomorphologically be divided into two sections: a downstream area deeply cut across the limestone plateau with a clearly defined channel bounded by cliff sides, and an upstream section with an ill-defined shallow course on the surface of the plateau. In the latter section the course of the wadi is often lost amidst extensive areas of vegetation. The whole runs along what seems to be growth of plants that obviously would not otherwise be found in this rainless area.

The downstream area (near the confluence of the oasis proper) is characterized by a carpet *of Alhagi maurorum*, patches of *Zygophyllum coccineum* and an open scrub of trees of *Acacia raddiana* and shrubs of A. *ehrenbergiana*. This rich growth of *Alhagi* indicates a copious supply of subsurface water. Further up the wadi, the vegetation is an open *Acacia* scrub with undergrowth dominated by *Z. coccineum* associated with

Fagonia arabica. This vegetation is typical of desert wadis which receive occasional rains or run-off from higher ground; the situation seems to indicate a local blockage of the seepage. A distant part of the wadi supports a few palms, showing a surface supply of water deeper than the central part, near the wells, and is also characterized by a number of hillocks formed of sand mixed with dead remains of *Tamarix amplexicaulis*. These are evidently relicts of hillocks built around the growth of *Tamarix*. The presence of *T. amplexicaulis* hillocks in Kurkur Oasis (as well as in Dungul Oasis) may be attributed to former more humid climatic conditions which no longer exist. Similar hillocks of *T. nilotica* and *T. aphylla* are recorded in the deltaic part of Wadi Qena (Girgis, 1965) and other wadis of the Eastern Desert. In the vicinity of *T. amplexicaulis* hillocks is an extensive patch of *Cressa cretica* indicating saline soil.

The upstream section has an eastern part which may be described as halfa grass *(Desmostachya)* country and a western part as camel-thorn *(Alhagi)* country with Acacia scrub. The presence of *Alhagi* and Acacia indicates a subsurface supply of water in contrast to desert plants which might survive on water from sporadic rains (Boulos, 1966a).

North Wadi

The mouth of the wadi is about 600 m to the north of the northern well of the oasis. The vegetation of the central part of the oasis expands, though in a thinner form, into the mouth. In this part there is a rather dense scrub of A. *raddiana* and *A. ehrenbergiana* with occasional doum palms and a rich undergrowth of *Zygophyllum coccineum*. Throughout the major part of this wadi the vegetation is essentially an open scrub of A. *ehrenbergiana*. In the upstream part the vegetation is a thin cover of *Z. coccineum*.

Apart from the vegetation within its mouth, the rest of the north wadi is vegetated typically like the other desert wadis, which receive some water occasionally. This wadi is clearly less favoured by seepage than the northwest wadi.

South Wadi

South wadi is the downstream continuation of the oasis, and is a part of the principal channel of Wadi Kurkur of which the other wadis are affluents. The channel follows a southern direction for a short distance from the wells of the oasis, then turns and cuts its way across the scarp of the Sin-el-Kidab down to the plain across which it proceeds east until it meets the Nile.

The vegetation of this part of the wadi extending southward from the wells is a continuation of that of the oasis: doum and date palms, A. *raddiana* and A. *ehrenbergiana*, with undergrowth *of Desmostachya* and *Alhagi*. On the peripheral parts, *Z. coccineum* is abundant.

At the eastward bend of the course of the wadi bed, the vegetation abruptly changes into a dense growth of *Phragmites australis* and *Juncus rigidus*, with some bushes of *Tamarix amplexicaulis*. This complex is very similar to the vegetation around the wells of the oasis (*Typha* is absent). Though this locality has no apparent well, the water-table is obviously high.

Further eastward is a part of the wadi where the bed is covered By large rounded boulders mixed with other deposits. An area here is covered by dense thickets of T. *amplexicaulis*. Associate species include *J. rigidus* and *D. bipinnata*. The peripheral ridges have lines of *A. maurorum* which follow fissures in the rock. *Z. coccineum* is also present.

The area of *T. amplexicaulis* represents the eastern limit of the influence of ground water. The remainder of the wadi channel eastward is a typical desert habitat with a sparse growth of *Z. coccineum* and a few widely spaced bushes of *Tamarix* and stumps of dead date palms. In this part there are individuals of dry *Schouwia thebaica*, a desert annual crucifer. These are obviously remains of *Schouwia* growth following a previous rain, a rare incident in this nearly rainless desert.

As the wadi crosses the scarp of the Sin-el-Kidab it follows a shallow and ill-defined course across the plain for about 50 km throughout which the vegetation is a sparse growth of *Z. coccineum* and *Fagonia parviflora*.

The vegetation in a small affluent wadi joining the south wadi at its eastward bend is a rich cover of *Alhagi* in the downstream part and an open growth of *Z. coccineum* and *Fagonia thebaica* v. *violacea* in the upstream part. The downstream part receives some of the subsurface supply of water whereas the upstream part is typical desert habitat.

Vegetation of the Kurkur-Dungul Road

The notable feature of the vegetation of the road between Kurkur and Dungul Oasis is the profuse dry remains of a number of species: *Bassia muricata, Fagonia thebaica* v. *violacea, Farsetia ovalis, F. ramosissima, Monsonia nivea, Reseda pruinosa, Schouwia thebaica, Stipagrostis plumosa* and *Tribulus mollis*. These evidently represent a rich ephemeral vegetation which appeared in a rainy year. Living individuals of some desert perennials are present: *Acacia ehren-bergiana, Cornulaca monacantha, Fagonia parviflora* and *Salsola wryosma*. The growth of these plants may depend on the presence of wine underground water.

Wadi Dungul

Ecologically Wadi Dungul consists of two main units:

1. main channel and 2. Dungul and Dineigil Oases (Zahran, 1966).

The main channel of Wadi Dungul runs in an east-west direction, but southwest towards its extremity. In the downstream part there is pure relict community dominated by *Salsola baryosma*. The indvidual bushes are dry because of the extreme drought over a very long riod. The soil is compact and formed mainly of silt mixed with clay covered with scattered pieces of rock detritus. The wadi is barren about 13 km upwards, then there is a community dominated by *Tamarix amplexicaulis*, the bushes of which build huge sandy hillocks and small sand mounds. There are many fossil hillocks with dead plant remains of T. *amplexicaulis*. These hillocks were formed during the Pleistocene period (Said and Issawy, 1964).

The country of *T. amplexicaulis* continues for about 7 km up the wadi as a pure community. Further up is the country of *T. aphylla* which builds and stabilizes huge hillocks of sand. *Stipagrostis vulnerans* is the only associate species. *T. aphylla* scrub thins gradually up the wadi to be replaced by a grassland community dominated by S. *vulnerans*, a sand-binding grass rarely occurring in Egypt (Täckholm, 1974). In Wadi Dungul this grass builds hummocks of moderate size and its community is associated with *T. aphylla* and *Aristida mutabilis* v. *aequilonga* (Zahran, 1966).

The number of plants decreases up the wadi until it becomes barren. The channel of the wadi in this barren part is blocked with very high sand dunes not suitable for plant growth. In the most upstream part are a few depauperate individuals of *Acacia ehrenbergiana* and *Ziziphus spina-christi*.

Dungul and Dineigil Oases have been formed by the blockage of the drainage lines of Wadi Dungul and the formation of local hollows which were later opened by the breaking of these obstructions (Said and Issawy, 1964).

The vegetation of these two oases is generally richer than that of the main channel of Wadi Dungul. This may be due to the relatively high underground water supply of the oases.

Dungul Oasis is located at about 21 km from the mouth of Wadi Dungul. This oasis has ecological and floristic interest in the presence of the ancient palm tree *Medemia argun* (fan palm with unbranched stem up to 10 m high). In Egypt *Medemia* is recorded only in the Dungul Oasis, being absent from the other oases and from elsewher in Egypt (Täckholm, 1974). Täckholm (1956) states "*M. argun* may be present in Nakheila Oasis near Dungul Oasis". The late Professor Vivi Täckholm confirmed the occurence of this palm in the Dungul area in a field trip in November 1963. Outside Egypt, *M. argun* occurs in Senna Wadi Doum and Darfour of Sudan (Täckholm and Drar, 1950). In Dungul Oasis there is only one tree of *M. argun*. Six juvenile palms have been found growing around the mother palm (Zahran, 1996). Other palms recorded are *Hyphaene thebaica* and *Phoenix dactylifera*.

The undergrowth in Dungul Oasis is formed of dense salt-tolerant halfa grass, *Imperata cylindrica*, with 50–70% cover.

Dineigil Oasis is located south of Dungul Oasis and "it is relatively higher in level than Dungul Oasis" (Said and Issawy, 1964). The vegetation of Dineigil Oasis is much richer and thicker. It comprises three communities dominated by *Alhagi maurorum, Juncus rigidus* and *Imperata cylindrica*.

1. *Alhagi maurorum* community. A. *maurorum* is a widespread species in Deneigil Oasis. "It is a valuable indicator of the underground water" (Kassas, 1952c). Its community occurs in three habitats of the oasis: slopes covered with thin sheets of sand, crevices of the bare slopes and water channels at the foot of the slopes. On the slopes and in the crevices, *Alhagi* forms pure stands whereas in the third habitat it is associated with *Juncus rigidus* and *Imperata cylindrica*. Groves of *Hyphaene thebaica* and *Phoenix dactylifera* are also present.
2. *Juncus rigidus* community. In distribution, abundance and cover of *J. rigidus* its community comes next in importance to that of *A. maurorum*. Dominance of J. *rigidus* has been observed in the water channels of the Dineigil Oasis and it is also co-dominant with *I. cylindrica* in the depressed areas. *A. maurorum* and *Hyphaene thebaica* are the associate species.

3. *Imperata cylindrica* community. In Dineigil Oasis, the *I. cylindrica* community is not as well developed as in the Dungul Oasis. Its stands are usually pure, but may be associated with *A. maurorum, J. rigidus* and *H. thebaica.* A shrub *of Acacia tortilis* and few *Phoenix* palms have been recorded in one of the stands.

Vegetation Around Ain El-Gaw Spring

This is a fresh-water spring in Dineigil Oasis at the base of a grove of *H. thebaica* and *P. dactylifera* palms. Its water has a continual underground source. The area around the spring is dominated by *I. cylindrica* and *Sporobolus spicatus.*

Bornkamm et al. (2000) stated that the vegetation of Dungul oasis has not been changed very much since the investigation of Zahran (1966).

3.4 Gebel Uweinat

3.4.1 General Characteristics

In the extreme southwestern part of the Western Desert where boundaries of Egypt, Sudan and Libya meet, lies Gebel Uweinat (the mountain of little springs) with a circumference of 160 km and an area of more than 1500 km^2. Its average elevation is 1907 m above sea level (Lat. 21°54'N, Long. 24°58'E). This mountain was long unknown, until it was explored by Ahmed Hassanein Bey (1925) during his long sojourn in the Western Desert.

The eastern portion of Gebel Uweinat which includes the summit (1934 m, Osborn and Krombein, 1969) is Nubian sandstone. Underlying the sandstone is igneous rock which forms the western part of the mountain. The high southwestern sandstone plateaux have their bases buried in a talus forming the pediment of a broad, sloping alluvial plain. In contrast, on the western extremity of Uweinat, granite cliffs have weathered into piles of exfoliated boulders rising steeply from a level plain (Shaw and Hutchinson, 1934; Said, 1962).

The springs of Uweinat are actually rock basins (guelta) beneath the boulders in the west and a series of springs and small pools in a narrow, winding gorge (Karkur) in the southwest. The springs occur where impervious porphyries hold up water that has percolated through sandstone (Ball, 1928). Rainfall replenishes these sources of surface water every seven or ten years (Peel, 1939). “Rain was recorded at Uweinat in the autumn of 1921, the spring of 1934 and in June 1960” (Williams and Hall, 1965). “According to soldiers at Ain Doua, rain had fallen more recently, probably in 1962, in the SE section of the mountain” (Osborn and Krombein, 1969). Rains in the Gebel Uweinat area are of short duration and violent. “Trees may be uprooted and carried onto the plains”. The basins south and west of Uweinat are known to become filled with water to a depth of 2 m (Kamal El-Din, 1928). The flora of Gebel Uweinat is Indo-Saharan, not Mediterranean nor montane (Shaw and Hutchinson, 1931) and includes genera and species common in the wadis of the granitic Red Sea mountains of Egypt.

The earliest records of the flora of the Uweinat area are those of Hassanein Bey (1924a, b), Kamal El-Din (1928) and Newbold (1928). Shaw and Hutchinson (1931, 1934) produced the first detailed account on the flora of the Gebel Uweinat area, including some ecological observations. They listed 22 species in 12 families of angiosperms. Osborn and Krombein (1969) recorded 55 plant species belonging to 22 families of angiosperms. Of these, 33 species and four varieties were new to the Uweinat area. These families are: Aizoaceae, Amaranthaceae, Asclepiadaceae, Boraginaceae, Capparaceae, Compositae, Convolvulaceae, Cruciferae, Cucurbitaceae, Euphorbiaceae, Gramineae, Juncaceae, Labiatae, Leguminosae, Nyctaginaceae, Orobanchaceae, Palmae, Primulaceae, Tiliaceae, Typhaceae, Urticaceae and Zygo-phyllaceae. Boulos (1982) recorded 73 species in 32 families from the Uweinat area. These families include in addition to the above mentioned: Cistaceae, Cleomaceae, Geraniaceae, Malvaceae, Moraceae, Portulacaceae, Resedaceae, Salvadoraceae, Solanaceae and Tamaricaceae.

3.4.2 The Vegetation

Four vegetation types have been recognized in the Uweinat area, namely: ephemeral vegetation, ephemeral and annual vegetation, perennial vegetation near the wells and perennial vegetation in gorges (Boulos, 1980, 1982).

(a) Ephemeral Vegetation

Ephemeral or annual plants appear after occasional rains, continue to grow and complete their life cycle in a period ranging from few weeks to one year. The length of their life-span usually depends on the amount of rain available to the plants. These constitute the ephemeral vegetation. On the other hand, dry remains of some plants suggest that certain perennials may behave as ephemerals, or potential annuals. Producing their seeds in shorter time (Haines, 1951). "The potential annuals in the Gebel Uweinat area include *Citrullus colocynthis, Trichodesma africanum* v. *abyssinicum* and *Zilla spinosa*", Boulos (1982) states.

According to a NOAA satellite image, it was cloudy on 16–17 December 1977 over an area 5 km southeast of Peter and Paul mountains (some 50 km NW of Gebel Uweinat). As our expedition on 20th October 1978 passed by a small wadi with sandy soil and trachyte boulders, some plants were still growing. The only species which was still rather green, bearing flowers and fruits, though showing some signs of senility, was *Fagonia arabica,* while *Stipagrostis plumosa* was almost dry, *Farsetia ramosissima* and *Trichodesma africanum* v. *abyssinicum* were perfectly dry. Considering that these plants have germinated after the rain of December 1977, ten months before we had seen them, it may be concluded that the perennials *F. arabica* and *T. africanum* v. *abyssinicum* change their growth from perennials into potential annuals to meet these severe and unfavourable conditions of the environment. *S. plumosa* and *F. ramosissima*, which were known to behave as annuals or perennials (Täckholm, 1974), successfully acquired the annual habit producing their seeds which are now kept in soil for the next unpredictable shower.

(b) Ephemeral and Perennial Vegetation

Ephemeral and perennial plants occur mixed together in wadis and water catchments, a community entirely dependent on rainfall, where no perennial groundwater or any other permanent water supply is available. The difference between this vegetation and the previous category, where there were only ephemerals, is the availability of additional moisture in the form of run-off, producing a layer of temporary underground moisture available to the root systems of perennial plants during their life time. This moisture lasts for 3–4 years and unless replenished by further rainfall is depleted by evaporation. Again, the difference between this and the next category, in which perennial vegetation occurs near wells, is the continuous supply of water from wells which permits long-living trees to exist as well as smaller plants, irrespective of rain.

A mixture of ephemeral and perennial plants is exemplified in some wadis of Gilf Kebir, e.g. Wadi Bakht and Wadi Ard El-Akhdar, and includes *Panicum turgidum* and *Zilla spinosa* (perennials) and *Stipagrostis plumosa* (perennial or annual), *Trichodesma africanum* v. *abyssinicum* (potential annual) and *Anastatica hierochuntica* (annual) (Boulos, 1982).

(c) Perennial Vegetation Near Wadis

The conspicuous perennial vegetation in the vicinity of wells comprises trees, shrubs and perennial herbs around five wells (Tarfawi, Tarfawi west, Kiseiba, El-Shab and Kurayim); several species cover small to large areas depending on the abundance of water and its availability to the plants. The most luxuriant growth is certainly that at Tarfawi well where *Phoenix dactylifera* and *Tamarix nilotica* grow in dense groves. In a protected shaded area, under a date palm grove, *Juncus rigidus* occurs in thick tufts up to 1.5 m high. Some of the date palm trees are still bearing rather good quality dates. Huge trunks of *T. nilotica* lie on the ground mixed with the debris of its leaves and branches, suggesting that *Tamarix* nourished in the not too distant past. This may also suggest that the ground water has become depleted and may no longer support large trees as it used to do a hundred years ago (Boulos, 1982). The present living *Tamarix* is in the form of shrubs rather than trees. In the vicinity of other wells, *T. nilotica* grows as a shrub forming sandy hummocks that may coalesce and form enormous crescent-shaped hills of pure stands.

A group of small shrubs *of Acacia ehrenbergiana* exists 25 km northwest of Kiseiba well, on a hummocky sand dune.

The herbaceous perennials are mainly grasses – *Sporobolus spicatus* grows around the wells, very close to the water or where the water level approaches the surface; the soil is usually saline. *Imperata cylindrica* and *Phragmites australis* grow near and around Kiseiba and Kurayim wells. *Desmostachya bipinnata* was recorded by Shaw and Hutchinson (1934), but this is probably a mis-identification based on sterile specimens, and the grass concerned may be *Imperata cylindrica*. Another perennial grass, *Stipagrostis vulnerans*, with stiff sharp spiny leaves, forms a pure stand extending over one km on the sand dunes near El-Shab well. *Alhagi graecorum (A. maurorum*, A. *mannifera*, Täckholm, 1974; Daoud, 1985) is recorded from the area of Tarfawi and Kurayim wells, covering extensive parts of the drier or seemingly dry areas near the wells. It provides excellent fodder, especially for camels – hence its name camel-thorn.

At Kiseiba well, the palm groves are striking and three different species are present: date palm *(Phoenix dactylifera)*, doum *(Hyphaene thebaica)* and another species not previously known in Egypt. The last, called by bedouins delib, may be a species of *Borassus* (El-Hadidi, 1981).

(d) Perennial Vegetation in Gorges

The perennial vegetation in the winding gorges of Gebel Uweinat depends on ground water from seepage. Occasionally rains may feed the ground water and ephemeral vegetation appears shortly afterj showers (Boulos, 1982). The following vegetation types may be recognized in the gorges:

(i) Vegetation near springs
This vegetation may be compared with the perennial vegetation near wells previously described. Ain* Brins at Gorge Karkur provides an example of where water-loving perennials grow, namely *Imperata cylindrica, Juncus rigidus, Phragmites australis* and *Typha domingensis*. *Phoenix dactylifera* is also recorded; however, only small scattered trees occur. Some annuals grow on the muddy borders of the springs – *Eragrostis aegyptiaca, Polypogon monspeliensis* and *Portulaca oleracea*.
(ii) Vegetation dominated by herbs and small shrubs
This vegetation type is commonly present in the gorges of Gebel Uweinat. It comprises perennial herbaceous and shrubby species and very few trees. The following are the most characteristic species: *Aerva javanica* v. *bovei, Cassia italica, Citrullus colocynthis, Cleome chrysantha, C. droserifolia, Crotalaria thebaica, Fagonia thebaica, Francoeuria crispa* and *Pergularia tomentosa*. One or few species may stretch over the greater part of a gorge. Toward the mouth of gorges, where they join the open desert, *Fagonia indica* and *Stipagrostis plumosa* are more frequent.
(iii) Vegetation dominated by shrubs and trees
Trees and shrubs which characterize this vegetation are restricted to four species: *Acacia ehrenbergiana, A. raddiana, Ficus salicifolia* and *Maerua crassifolia*. Usually the two *Acacia* species grow together and are associated with *Panicum turgidum*. In Gorge Talh the conspicuous element of vegetation is the luxuriant growth

---

*Ain is the Arabic name for spring.

of A. *raddiana* (Talh is the vernacular name of A. *raddiana*) which forms, together with the less common *A. ehrenbergiana*, an open thorny forest, with dense tufts of P. *turgidum* covering much of the gorge bed. In Gorge Abdel Malek, however, trees *of Acacia* are very rare, whereas shrubs *of Maerua* are abundant and trees of *F. salicifolia* start to appear at about 850 m (Boulos, 1982).

(iv) Vegetation of higher altitudes

According to Leonard (1969), *Ochradenus baccatus* occurs between 900 and 1400 m, while *Heliotropium bacciferum, Lavandula stricta, Monsonia nivea* and *Salvia lanigera* grow between 1250 and 1850 m.

Apart from the above-mentioned species, Boulos (1982) recorded the following in the Uweinat area: *Argyrolobium saharae, Aristida funiculata, A. mutabilis, Astragalus vogelii, Boerhavia diandra, B. diffusa. Cassia italica, Cistanche phelypaea, Citrullus lanatus, Convolvulus austro-aegyptiacus, C. cancerianus, C. prostratus, Corchorus depressus, C. olitorius, Crotalaria thebaica, Cynodon dactylon, Diceratella sahariana, Eragrostis aegyptiaca, Eremopogon foveolatus, Fagonia bruguien, Helianthemum lippii. Hibiscus esculentus* (cultivated), *Hyoscyamus boveanus, Indigofera arenaria, I. sessiliflora, Juncus subulatus, Lotononis platycarpos, Lycopersicum escuientum* (cultivated), *Morettia philaena, Nicotiana rustica* (cultivated), *Ocimum basilicum* (cultivated), *Parkinsonia aculeata, Pulicaria undulata, Salvia lanigera, Samolus valerandi, Stipagrostis zittelii* and *Tribulus terrestris*. Täckholm (1974) recorded also, e.g. *Acacia arabica* v. *adansoniana, Amaranthus albus, Forsskaolea tenacissima, Indigofera lotononoides* and *Reaumuria vermiculata.*

According to Leonard (2001), the vegetation of Gebel Uweinat area may be classified under 3 main types: hygrophytic vegetation, sandstone vegetation and granitic and geneiss vegetation. Each of these vegetation types consists of plant communities with their characteristic and associated species. The total floristic elements are 89 species: 87 flowering plants and 2 bryophyta (one liverwort and one moss). The hygrophytic vegetation inhabits the slightly humid habitat. Its floristic composition is 34 species: 6 characteristic species (*Crypsis vaginifolia, Eragrostis aegyptiaca subsp. humifusa, Juncus rigidus, Phragmites asutralis, Polypogon monspelnensis* and *Typha domingensis*. 2 bryophyta (*Funaria convexa* and *Riccia cavernosa*) and 26 flowering plants e.g. *Acacia tortilis*, subsp. *tortilis, Aristida adscensionis, Amaranthus gracaecizans, Astragalus vogelii* subsp. *vogelii, Eneopogon desvauxii, Euphorbia granulata, Imperata cylindrica, Lotononis platycarpa, Mollugo cerviana var. cerviana, Polycarpaea robbairea, Sporobolus spicatus* etc.. are the associated species.

The sandstone vegetation is an open forest formed of 56 species as follows: 10 characteristic species (*A. tortilis subsp. tortilis, Aerva javanica var. bovei, Atractylis aristata, Citrullus colocynthis, Crotalaria thebaica, Fagonia indica var. indica, Lavandula antinea, Ochardenus baccatus, Panicum turgidum* and *Stipagrostis acutifolia subsp. acutifolia*, and 44 flowering associates e.g. *Acacia ehrenbergiana, Centropodia forsskaolii, Cleome ambylocarpa, Cocculus pendulus, Convolvulus prostratus, Desmostachya bipinnata, Farsetia stylosa, Forskaolea tenacissima, Francoeuria undulata, (= F. crispa), Hyoscyamus muticus, Lotononis platycarpa,*

Monsonia nivea, Pergularia tomentosa, Senna italica subsp. italica, Tamarix arborea, Tribulus bimucronatus etc.

The granite and gneiss vegetation type of Gebel Uweinat area is formed of 3 communities, two are dominated by *Cleome chrysantha, Tribulus pentandrus var. pentandrus* and the third is co-dominated by *Cleome droserifolia* and *Fagonia thebaica*. The associated species (27 species) include: *Acacia ehrenbergiana, A. tortilis subsp. tortils, Arisida funiculata, A. mutabilis, Boerhavia repens subsp. diandra, B. repens subsp. viscosa, Cistanche phelypaea, Citrullus colocynthis, Corchorus depressus, Cullen plicata, Euphorbia granulata, Fagonia indica var. indica, Indigofera disjuncta var disjuncta, Lotononis platycarpa, Panicum turgidum, Pergularia tomentosa, Salvadora persica, Senna italica subsp. italica, Stipagrostis rigidifolia, Tribulus macropterus var. macropterus, Trichodesma africanum* etc.

3.5 The Gilf Kebir

3.5.1 General Remarks

The Gilf Kebir (= great plateau, great barrier) is a huge residual sandstone plateau located at about 150 km north of Gebel Uweinat on the SW most part of Egypt. Both, Gebel Uweinat and Gilf Kebir, are considered the most prominent landscape of Egypt's Western Desert and are, thus, declaired together by EEAA (Egyptian Environmental Affairs Agency) as a Natural Protectorate (Prime Ministrial Decree No. 10, 2007) with an area of about 47,940 km^2 (Anonymous, 2007). It is thus one of the world's largest conservation area.

The Gilf Kebir Plateau has an area of about 7,700 km^2 and rises about 300 m above the desert floor (1075 m.a.s.l.). Its surface features bear remarkable similarity to features revealed to Mars (El Baz, 1980). The heavely eroded sites of Gilf Kebir* plateau are deeply dissected by wadis that have been penetrated by incredible system.

Alaily et al. (1987) investigated the soil types in Gilf Kebir area and revealed that lithosols, orthic solonchaks and haplic yermosols, mainly from sandstones, buildup the soil on the plateau. Cambic arenosols, eutric regosols (or eutric fluvisols) from alluvial sediments and haplic yermosols from debris as well as takyric yermosols and solonchaks from playa sediment occur in the wadis. All soils are rich in plant nutrients except nitrogen which occurs in high concentration only in the orthic solonchaks.

* It has been so named by the Egyptian explorer, Prince Kamal El- Din on January 1926

3.5.2 Plant Life

The plant diversity of Gilf Kebir area includes 16 species distributed in the different habitats (Kehl and Bornkamm, 1993). On the plateau itself (average 900 m a.s.l.), three vegetation units are recognized. The largest unit occurs on the stony plateau shows the lowest species diversity. Three species grow episodically, namely: *Salsola baryosma,* (high abundance), *Stipagrostis acutifolia* and *Zilla spinosa* (rare). On the central and northern parts of the plateau, with shallow sandy runnels, *Stipagrostis acutifolia* predominates associated with *Fagonia arabica, F. indica* and *Salsola baryosma*. In sandy-filled notches and crevices, plant diversity is relatively high. In addition to the above mentioned species, the flora comprises: *Astragalus vogelii, Farsetia ramosissima, Monsonia nivea, Onobrychis crista-galli, Rumex vesicarius, Schouwia thebaica* and *Trichodesma africanum*.

The vegetation in the wadis and their lower parts are often characterized by relatively high abundance and cover but rare presence of phanerophytes. Patches of green and flowering plants e.g. *Fagonia arabica, Stipagrostits acutifolia* and *Acacia raddiana* indicating a favourable water supply. A large trunk of *Tamarix* sp. was also recorded (Alaily et al. 1987). Wadis with large catchment areas are dominated by *Zilla spinosa* and *Citrullus colocynthis* with rare associates of *Stipagrostis acutifolia* and *Fagonia arabica*. In the pediment area, east and south of the Gilf Kebir (600 a.s.l.), vegetation fades out, only scattered individuals of the grass *Stipagrostis acutifolia* grow. On the other hand, the western part of the Gilf Kebir is characterized by rich growth of *Acacia raddiana* and *Maerua crassifolia*.

Chapter 4
The Eastern Desert

4.1 Geology and Geomorphology

The Eastern Desert of Egypt occupies the area extending from the Nile Valley eastward to the Gulf of Suez and the Red Sea which is about 223,000 km^2, i.e. 21% of the total area of Egypt (Fig. 2.1). It is higher than the Western Desert as it consists essentially of a backbone of high, rugged mountains running parallel to and at a relatively short distance from the coast. The peaks of many of these mountains are more than 1500 m above sea level. These mountains are flanked to the north and west by an intensively dissected sedimentary plateau. The folding and faulting that have occurred during geological history have caused these mountains to be dissected into several blocks of a series of mountain groups (Abu Al-Izz, 1971).

The mountains of the Eastern Desert are of two types: igneous and limestone. The igneous mountains extend southward from about Lat. 28°N to beyond the Sudano-Egyptian border (Lat. 22°N). The highest peak, Gebel Shayeb El-Banat (El-Shayeb, near Lat. 27°N), reaches 2184 m above sea level. To the north of the igneous mountains are the extensive and lofty limestone mountains of South Galala (1464 m), North Galala (1274 m) and Gebel Ataqa (871 m) separated by broad valleys (wadis).

To the west of the Red Sea mountains lie two broad plateaux, parted by the road of Qift to Qusseir (Lat. 26°N). This is a mountainous area which has the badland features typical of arid and semi-arid regions. The northern plateau is of Eocene limestone, the other is of Nubian sandstone and covers a broad area, approximately one-quarter of the total of the Eastern Desert. These plateaux differ from each other not only lithologically, but also geomorphologically. The sporadic rainfall of the Red Sea mountains has a differential influence depending on the nature of the formations on which it falls. On the Eocene limestone narrow wadis are formed similar to canyons (Hume, 1925); in the Nubian sandstone area, running water produces broad wadis. This means that the Eastern Desert is greatly dissected by valleys and ravines and that all its drainage is external. Eastward drainage to the Red Sea is by numerous independent wadis; the westward drainage to the Nile Valley, however, mostly coalesces into a relatively small number of great trunk channels.

The dissection of the Eastern Desert by dense networks of wadis indicates that although the present time is a dry period, this could not have always been the case. Egypt must have witnessed some periods of pluviation.

The range of the Red Sea coastal mountains, thus, divides the Eastern Desert into two main ecological units: the Red Sea coastal land and the inland desert.

4.2 Ecological Characteristics

This section gives an account of the environmental characteristics and vegetation types of the Red Sea coastal land and the inland desert as being the main ecological units of the Eastern Desert of Egypt.

4.2.1 The Red Sea Coastal Land

(a) General Features

The Red Sea is an elongate trough extending NW-SE from the Sinai Peninsula (Lat. 29°50N) to the Bab-El-Mandab Strait (Lat. 12°35N, Long. 43°3E) and separates the Arabian Peninsula from the African continent. Throughout most of its length (2000km), its opposing shorelines are remarkably parallel (Fig. 1.1). In the south the width of the sea is only 175km in the area between Jizan in Saudi Arabia (eastern coast) and Massawa in Ethiopia (western coast). It decreases to a minimum of 30–40km at Bab-El-Mandab Strait. The central region of the Red Sea has a depth of more than 2000km (Zahran, 1977).

Inland of the Red Sea shore-lines, there is a narrow coastal plain which is backed by a prominent escarpment (1500–3000m high). The escarpment marks the uplift of the margins of the Arabian Shields and is the structural edge of the Red Sea Rift area (Chapman, 1978).

The Asian (eastern) coast of the Red Sea extends from Aqaba southwards to the Bab-El-Mandab Strait for about 2600km from Jordan (at the north of the Gulf of Aqaba) to Saudi Arabia (2140km) and northern Yemen (460km). The African Red Sea coast (western) extends from Suez southwards to the Bab-El-Mandab Strait for about 2860km in Egypt (1100km), Sudan (740km), Ethiopia (1020km) and Djibouti (Fig. 1.1).

The Red Sea coastal land of Egypt extends from Suez (Lat. 30°N) to Mersa Halaib (Lat. 22°N) at the Sudano-Egyptian border (Fig. 2.1). The land adjacent to the Red Sea in Egypt is generally mountainous, flanked on the western side by the range of coastal mountains. In the deep trough between the shore-line and the highlands extends a gently sloping plain which varies in width from 8 to 35km. In certain parts of the west coast of the Gulf of Suez (e.g. Khashm El-Galala, 60km south of Suez), there is scarcely any plain, the mountains rising almost directly from the gulf. The coastal plain is covered with sand over which the drainage systems of wadis meander with their shallow courses.

Along the Red Sea coast of Egypt, there are parallel lines of coral reefs between 50 and 100 m wide. They increase in density and width southwards and reach 250 m wide to the south of Mersa Alam (675 km south of Suez). The temperature of the Red Sea water is 21–22°C which is suitable for the formation of reefs and is further enhanced by such factors as the shallowness of the sea (in the coastal area its depth does not exceed 40 m), the high salinity (4%) and the purity of water near the coast.

The climate of the Red Sea coastal land of Egypt is arid. The mean annual rainfall ranges from 25 mm in Suez, 4 mm in Hurghada to 3.4 mm in Qusseir (Fig. 2.1). The main bulk of rain occurs in winter, i.e. Mediterranean affinity, and summer is, in general, rainless. Variability of annual rainfall is not unusual. In Suez, for example, the years 1949 and 1958 were very dry: total annual rainfall was 2 mm and 3.1 mm respectively; whereas in the years 1952 and 1956, which were relatively wet, rainfall was 56 mm and 55 mm respectively (Kassas and Zahran, 1962, 1965). Temperature is high and ranges between 14 and 21.7 °C in winter and 23.1–46.1°C in summer. Relative humidity ranges from 43% in summer to 65% in winter. The Piche-evaporation is higher in summer (13.7–21.5 mm/day) than in winter (5.2–10.4 mm/day).

The effect of topography on precipitation is universal but more pronounced in coastal regions. Within the arid and semi-arid countries, coastal mountains may cause ample orographic rain which is referred 'as "occult precipitation",' "horizontal precipitation" or "fog precipitation" (Moreau, 1938). This orographic rain may produce rich vegetation on slopes of high mountains – "Nebelwald", "Nebeloasen" (Troll, 1935); "mist oases" (Kassas, 1956).

When the whole range of the Red Sea coastal mountains of Egypt considered, the southern blocks (represented by the Elba group) may be seen to receive greater amounts of water from orographic rain in the northern blocks.

(b) Vegetation Types

Apart from the valuable floristic studies by Ruprecht (1849), Schweinfurth (1865a,b, 1896–1899), Ascherson and Schweinfurth (1889a,b) and Drar (1936), the plant ecology of the Red Sea basin has been the subject of several investigations. Schweinfurth (1865a,b, 1896–1899) accumulated valuable ecological observations on the Red Sea coastal land. Ferrar (1914) describes some of the mangrove swamps of the northern Red Sea coast. Troll (1935) gives a more detailed account of the southern part of the Red Sea coastal land. Vesey-FitzGerald (1955, 1957), Batanouny and Baeshin (1982), Younes et al. (1983), Zahran (1982b) and Zahran et al. (1983, 1985b) describe the vegetation types of the Saudi Arabian Red Sea coast. Hemming (1961) gives accounts of the plant cover of the coastal area of northern Eritrea. Kassas (1956, 1957, 1960) presents ecological information on the Red Sea coastal land of Sudan. Montasir (1938) provides an ecological description of the salt marsh vegetation, the mangrove vegetation and the desert vegetation of the Egyptian Red Sea coast. Hassib (1951) gives a life-form spectrum of the flora of the Red Sea region of Egypt, and detailed accounts of the vegetation and flora of the Red Sea coastal land of Egypt have been given by Kassas and Zahran (1962, 1965, 1967, 1971), and Zahran (1962, 1964, 1965, 1977) and Zahran and Mashaly (1991). These studies show that ecologically this coastal area may be considered under three principal ecosystems; (1) coastal salt marsh, (2) coastal desert and (3) coastal mountains. The environmental factors, notably tidal movement, sea water spray, sea water seepage and waves, characters of the substratum, land relief, and local and microclimates, are the main factors that limit the type and extent of the plant cover in each ecosystem. These systems are described below.

(i) The Littoral Salt Marsh

The coastal (littoral) salt marshes comprise areas of land bordering the sea, more of less covered with vegetation and subject to periodic inundation by tides. They have certain features related to the proximity of the sea that distinguish them from inland salt marshes (Chapman, 1974).

The problem of delineating the landward limit of the littoral salt marshes is of ecological importance. These salt marshes are essentially fringes of inland desert, their landward boundary being defined by the desertic qualities. Ecological factors, such as type of terrain and climate, can be used to delimit the littoral marshes. When there is a narrow belt of lowland along the coast that is cut off from the water by a steep barrier of mountains, e.g. along the Red Sea coast of Egypt, the limits are clear, but with a broad plain that stretches inland from the coast there may be no distinct physiographic limit to the littoral formation, and other habitat features including vegetation must be used.

Vegetation characteristics related to physiographic attributes, reflecting both climatic and edaphic factors, provide the best single basis for delimiting littoral salt marshes. These salt marshes may be only a narrow strip within the reach of salt spray, a few hundred metres wide or they may extend inland for many kilometres (e.g. 100–150 km on the Somali Red Sea coast, Meigs, 1966).

In this section the ecological relationships of the halophytic vegetation of the littoral salt marshes of the Red Sea coastal land of Egypt that extends for about 1100 km from Suez southwards to Mersa Halaib are described and evaluated. The plant cover of representative sample areas is indicated and the community types of this ecosystem presented. A short note on the invasive species and the vegetation of one of the off-shore islands are also described.

Sample Areas

Seven sample areas have been selected to represent the patterns of the halophytic vegetation of the Red Sea coastal land of Egypt which includes the eastern coasts of both the Gulf of Suez (400 km) from Suez to Hurghada and the Red Sea proper of Egypt (700) km from Hurghada to Mersa Halaib (Fig. 2.1).

Sample Area 1 (Ain Sokhna Area)

Ain Sokhna is a warm brackish-water spring about 50 km south of Suez on the western coast of the Gulf of Suez. The water of the spring originates at the north-eastern foot of the Galala El-Bahariya mountain and drains toward the shore-line where it forms a pool of warm vater which gives the area its name and which attracts many visitors, making it a popular seaside resort. Figure 4.1 is a sketch map of the vegetation of the Ain Sokhna area. The central region is covered by the green growth of *Juncus rigidus* which forms a pure mat around the source of the spring, along the path of the drainage towards the sea and around its spreading area near the sea-shore. To the south of the spring is a series of more or less circular patches covered by *Juncus* which may indicate subsidiary springs. *Tamarix nilotica* partly surrounds the *Juncus* area and invades its fringes. Its growth varies from isolated bushes to dense tickets. The expanse to the south of the *Juncus* area is saline ground with bushes of abundant *Nitraria retusa* and scarce *T. nilotica*, with undergrowth of *Juncus*. To the north are also scattered bushes of *N. retusa*, with abundant *Limonium pruinosum. Arthrocnemum glaucum* dominates locally.

At the shore-line is a sand bar vegetated with *J. rigidus* and *T. nilotica* and a few clumps of *Phoenix dactylifera*.

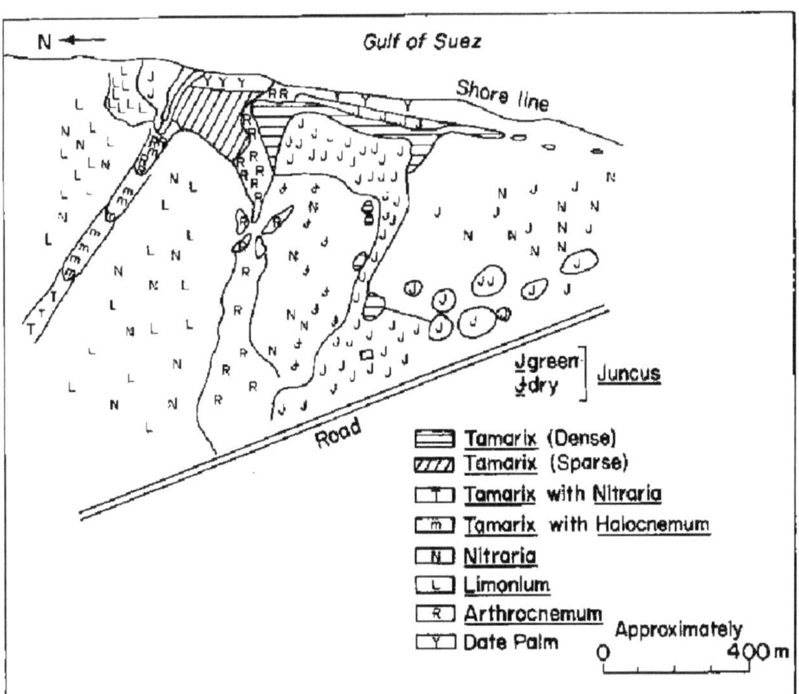

Fig. 4.1 Sketch plan of the vegetation of the Ain Sokhna area, Gulf of Suez (sample area 1)

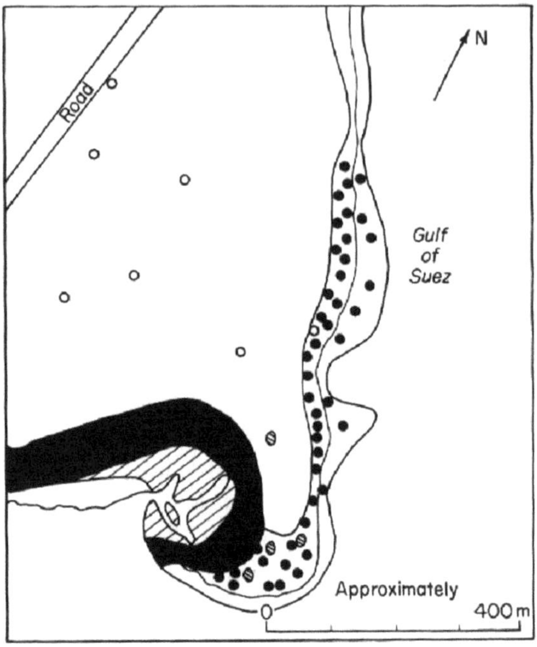

Fig. 4.2 Sketch plan of the vegetation of the area near Hurghada, Gulf of Suez (sample area 2). Black, *Halocnemum strobilaceum*; shaded, *Arthrocnemum glaucum*; O, *Nitraria retusa*; •, *Zygophyllum album*

Sample Area 2

Figure 4.2 shows a sketch map of the vegetation of an area that extends between 12 and 13 km north of Hurghada. It is a part of a littoral salt marsh. Four plant communities are recognized. One is dominated by *Zygophyllum album* which builds low sand mounds parallel to the northwest shore-line. Occasional individuals of *Nitraria retusa* are present. A second community, dominated by *Arthrocnemum glaucum* (= A. *macrostachyum,* Meikle, 1985), covers the tidal mud flat ground that fringes the southern shore-line. A third type is dominated by *Halocnemum strobilaceum.* It covers an inland zone of sand bars which fringes the zone of A. *glaucum.* The inland expanse of dry saline ground is covered by widely spaced mounds of *Nitraria.*

The vegetation in this part is apparently destroyed and many of the mounds are barren, with only dead remnants *of Nitraria.*

Sample Area 3

This area stretches between the mouth of Wadi Ireir and the delta of Wadi Gimal. It covers salt marshes extending from 48 to 54 km south of Mersa Alam.[1] The

[1] Mersa Alam is about 675 km south of Suez (275 km south of Hurghada), at Lat. 25° N.

shore-line is characterized by a group of fresh-water springs that may be recognized at low tide (Beadnell, 1924). A fringing zone of *Avicennia marina* mangrove is separated from the shore-line by a narrow belt of sand bars and mounds covered by *Zygophyllum album*. Small trees of *Avicennia* are present in this part on low sand mounds. The occurrence of *Avicennia* on the inland side of the shore-line may seem strange but similar patches are also recorded in a few localities of the Red Sea coast, e.g. in Mersa Sherm, 31 km south of Mersa Alam, there is a patch of the mangrove extending for about 400 m landwards. The littoral zone also includes *Juncus rigidus* and a discontinuous belt of *Aeluropus* spp.[2] On the inland side of the salt marsh is a zone of *Tamarix nilotica* which extends into the mouth parts of the two wadis. In the mouth of Wadi Ireir *Tamarix nilotica* is associated with *Acacia tortilis*.

The vegetation of the delta of Wadi Gimal shows several interesting features. Within the downstream part is a grove of *Phoenix dactylifera* and *Hyphaene thebaica* palms, looking like a neglected oasis. In the mouth part of the wadi channel is a zone of *T. nilotica* associated with *Zygophyllum coccineum* followed landward by a belt of *T. nilotica* associated with *Panicum turgidum* and *Acacia tortilis*. This is followed by a transitional zone where *T. nilotica* is mixed with *P. turgidum, A. tortilis, Leptadenia pyrotechnica* and *Balanites aegyptiaca*, a vegetation type which extends for several kilometres upstream Wadi Gimal.

Sample Area 4

The vegetation is described from a 13 km stretch of coastal salt marsh extending from 94 to 107 km south of Mersa Alam. The shore-line is fringed by a continuous belt of mangrove vegetation dominated by *Avicennia marina*. The shore-line zone of the salt marsh supports *Arthrocnemum glaucum* which varies from scattered individuals (cover 10%) to dense growth (cover 90%). This is followed by a narrow zone of *Zygophyllum album* represented by discontinuous strips. The inland zone also supports *Limonium axillare*. The salt marsh is bounded inland by high ground of the desert plain which is dissected by a number of small wadis dominated by *Zygophyllum coccineum*.

Sample Area 5

The salt marsh vegetation within the area which extends from 120 to 123 km south of Mersa Alam is described. The shore-line is fringed to seaward by *Avicennia marina* mangrove. On the inland side four distinct zones of vegetation are found. The most seaward is dominated by *Arthrocnemum glaucum* (cover 50–60%), the second by *Zygophyllum album* (cover 15–20%), the third by *Limonium axillare* (cover 20–25%) and the fourth by *Aeluropus* spp. (cover 50–60%).

---

[2] *Aeluropus brevifolius* and *A. lagopoides* often grow together in the Red Sea salt marshes and are referred to here, and subsequently, as *Aeluropus* spp. However, these grasses have been classified together (Cope and Hosni, 1991) under the variable species *A. lagopoides*.

Sample Area 6

The area surrounding a closed lagoon (Walalbab Lagoon), about 28 km north of Mersa Halaib on the Sudano-Egyptian border (Fig. 2.1), is characterized by the encroachment of the sand dunes on the littoral salt marsh. On the inland side is a zone overlying the higher ground of the coastal plain and stretching at its foot. This is covered by a community dominated by *Sphaerocoma hookeri*, followed by a zone of *Sporobolus spicatus* grassland developed on a sandy cover at a lower level. On the northern side the S. *spicatus* grassland extends seaward to the narrow littoral zone covered by *Zygophyllum album*. On the southern side several zones of vegetation may be recognized. The inland high ground characterized by the dominance of S. *hookeri* is followed by a belt of low sand mounds covered by *Z. album*- The midground is occupied by a zone of *Aeluropus* spp. grassland and a zone of *Limonium axillare*. The shore-line is fringed by sand mounds covered by *Z. album*.

Sample Area 7

This area includes 47 km of the Red Sea coast between Mersa Naam and Mersa Abu Fissi North, 48 km and 95 km north of Mersa Halaib respectively. In the salt marsh vegetation here a number of zones may be distinguished:

1. Mangrove zone represented by *Avicennia marina*. This zone ranges from 10 to 100 m wide and its cover varies from 20 to 80%,
2. Shore-line zone of *Arthrocnemum glaucum*. This zone ranges from 10 to 20 m wide and its cover from 20 to 30%.
3. Zone of Halopeplis perfoliata which occupies low ground between the shore-line of tidal mud or sand (with *A. glaucum*) and an inland zone of high saline ground.
4. This zone is covered by the growth of salt marsh grassland dominated by *Aeluropus spp.*
5. A zone of Suaeda monoica occupies the ground between the salt marsh and the desert coastal plain dominated by a *Salsola baryosma* community.

The Community Types

Within the Red Sea coast of Egypt, two halophytic types of vegetation have been recorded: mangrove and salt marsh. The mangrove vegetation is represented by a single community dominated by *Avicennia marina*. However, in the salt marsh vegetation twelve communities dominated by: *Halocnemum strobilaceum, Arthrocnemum glaucum, Halopeplis perfoliata, Limonium pruinosum, L. axillare, Aeluropus* spp., *Sporobolus spicatus, Halopyrum mucronatum, Zygophyllum album, Nitraria retusa, Suaeda monoica* and *Tamarix nilotica* are distinguished. These communities occur in zones following the shore-line where there is mangrove vegetation, each zone being occupied by one of these communities. Within any locality, however, only a few of these communities are represented, and a zone may include a mosaic of more than one community depending on local topography or soil conditions (Kassas and Zahran, 1967).

Distinct communities occupy the salt ground and dunes overlying it. Coastal salt marsh is formed when land rises in relation to the sea; mud accumulates progressively on tidal flats and colonization by plants starts on the exposed mud, increasing its stability. Eight communities belong to this group. The first zone, inland of the mangrove, is occupied by *Halocnemum strobilaceum* and *Arthrocnemum glaucum* communities while a *Tamarix nilotica* community occupies the most landward zone and marks the boundary between the salt marsh and desert plain. The sand mounds are usually covered by *Zygophyllum album* and the sandy hillocks by *Suaeda monoica* or *Nitraria retusa*. These sand formations may be formed within any of the salt marsh zones and the communities dominated by these three species may occur in any part of the salt marsh. The rare grassland community dominated by *Halopyrum mucronatum* is also present.

Mangrove Vegetation

1. *Avicennia marina* community. Mangrove vegetation dominated by A *marina* fringes the shore-line of the Egyptian Red Sea coast from Hurghada southward. It is a notable and common feature of the littoral landscape. However, it does not extend northward to the coast of the Gulf of Suez; the northernmost locality is the bay of Myos Hormos (c.22 km north of Hurghada) (Zahran, 1977).

A *marina* usually grows in pure stands. Within a stretch of about 40 km (Lat. 23°N to 22°40 N) *Rhizophora mucronata* may be mixed with *A. marina* as a co-dominant or as an abundant associate, or it may form pure stands. Where both species grow together *Rhizophora* forms an open layer higher than the thick and almost continuous bushy canopy of *Avicennia*.

The usual habitat of the mangrove of the Egyptian Red Sea coast is the shallow water along the shore, especially in protected areas: lagoons, bays, coral or sand bars parallel to the shore. In a few localities, *Avicennia* grows on the terrestrial side of the shore-line, and in one locality (delta of Wadi Gimal) the bushes are partly covered by sand hillocks. This situation is apparently due to the silting of the shore-line zone originally occupied by the mangrove.

The structure of the mangrove vegetation on the Red Sea coast of Egypt is simple – usually a single layer of A *marina*. In localities where *R. mucronata* is included, this forms a stratum towering over that of *Avicennia*. The ground layer of this community is formed of associate marine phanerogams, e.g. *Cymodocea ciliata, C. rotundata, C. serrulata, Halophila ovalis, H. slipulacea* and *Halodule uninervis*.

The tidal mud of the mangrove vegetation of the Red Sea coast is usually grey or black, and often foul-smelling. The total water-soluble salt content ranges from 1,2 to 4.3%, the organic carbon content ranges from 0.3% to 2.2% and the pH from 8.5 to 9.0. A notable difference between the tidal mud colonized by *Avicennia* and that by *Rhizophora* is the low content of calcium carbonate in the former (4.5–19.5%) as compared with the calcareous mud (80%) in the latter.

Salt Marsh Vegetation

1. *Halocnemum strobilaceum* community. This community is common within the littoral salt marsh of the Gulf of Suez but not in the region further south. The cover is often a pure stand of the dominant. However, nine other species are recorded. *Arthrocnemum glaucum* and *Zygophyllum album* are the most common associates; other less common species include five which dominate other communities, namely: *Alhagi maurorum, Cressa cretica, Limonium pruinosum, Nitraria retusa* and *Tamarix nilotica* as well as *Salsola villosa* and *Zygophyllum simplex*.

   The growth of *H. strobilaceum* occurs in two forms: circular patches on flat tidal mud and sheets of irregular-shaped patches on shore-line bars. This community occupies the inland side of the shore-line bar of sand on rock detritus heaped up by wave and tidal action.

   H. strobilaceum is rarely found south of Hurghada. It is not recorded from the Sudan Red Sea coast (Andrews, 1950–1956; Kassas, 1957) nor in Eritrea (Hemming, 1961). *H. strobilaceum* is also abundant in the northern section of the eastern (Saudi) Red Sea coast and of the Arabian Gulf coast (Halwagy, 1973; Zahran, 1982b).

2. *Arthrocnemum glaucum* (= *A. macrostachyum*) community. This community occurs throughout the area sampled though it is less common in the northern part (coast of the Gulf of Suez). The *A. glaucum* community occupies the same shore-line zone as the *H. strobilaceum* community and shows similar growth habit. Within this community, eleven associate species have been recorded. The common ones are *Halopeplis perfoliata, Limonium axillare* and *Zygophyllum album. H. strobilaceum* is recorded only as a rare associate in the northern 400 km and *Atriplex farinosa* is confined to the shore-line area. Other associates include *Aeluropus* spp., *Avicennia marina, Juncus rigidus, Tamarix nilotica, Sevada schimperi* and *Suaeda volkensii.*

   The communities dominated by *Halocnemum* and *Arthrocnemum* are physiognomically and structurally similar. The dominant species form carpets of a single layer 30–50 cm high. The upper (shrub) layer is usually absent or negligible. Wind- or water-borne sediments are often deposited in the form of mounds on which the cover by the dominant species is frequently patchy, but on the shore-line sand bars *Halocnemum* and *Arthrocnemum* may form almost continuous mantles.

3. *Halopeplis perfoliata* community. Within the 950 km stretch from Suez to Mersa Kilies, *H. perfoliata* is recorded in one locality (55 km north of Ras Gharib*) but is otherwise very rare or absent. Southward of Mersa Kilies *Halopeplis* and its community are common features of the littoral vegetation. *Halopeplis* is also very common further south in the Sudan and Eritrea (Kassas, 1957; Hemming, 1961). Almost the same geographical distribution of *H. perfoliata* has been recorded within the eastern (Asiatic) Red Sea coast (Zahran, 1982b). The cover ranges

---

*Ras Gharib is located at km 240 south of Suez.

from 5 to 40% contributed mainly by the dominant (*Halopeplis*). *Arthrocnemum glaucum* and *Zygophyllum album* are the most common associates. Less common species include eight perennial halophytes {*Aeluropus* spp., *Atriplex farinosa, Cyperus conglomeratus, Halopyrum mucronatum, Nitraria retusa, Sevada schimperi, Sporobolus spicatus* and *Suaeda monoica*) and two annual species alien to this habitat: *Launaea cassiniana* and *Zygophylluni simplex*.

The *H. perfoliata* community has the general features of the salt marsh vegetation: simplicity of structure, limited number of species and notable differences in cover due to minor changes in ground level. It occupies the third zone from the shore-line which is usually lower in level than the second where wave-heaped detritus or wind-deposited sand may form slightly elevated bars. This zone is also lower than the ground further inland, a situation that impedes free drainage. In a few localities sand mounds are formed around *Halopeplis*. The vegetation is usually single layered.

4. *Limonium pruinosum* community. *L. pruinosum* is a non-succulent (salt excretive) semi-shrub which occupies two distinctly different habitats: littoral salt marsh (Kassas and Zahran, 1967) and desert limestone cliffs (Kassas and Girgis, 1964). There are probably two ecotypes.

   In the littoral salt marsh of the Red Sea coast of Egypt, *L. pruinosum* dominates a community common in the region of the Gulf of Suez but not further south. It is not recorded in the Sudan flora (Andrews, 1950–1956).

   The plant cover of the *L. pruinosum* community ranges from 10 to 20%. Associate species include *Cressa cretica, Halocnemum strobilaceum, Nitraria retusa, Suaeda volkensii, Tamarix nilotica* and *Zygophyllum album*. The *L. pruinosum* community occupies a zone on the inland border of the shore-line belt of *Halocnemum strobilaceum* and it rarely has more than one layer.

5. *Limonium axillare* community. *L. axillare* is a non-succulent excretive semi-shrub halophyte rarely recorded in the northern 500 km stretch of the Red Sea coast, but in the southern stretch it is common. The cover of the community dominated by this sand-binder varies between 5 and 50%, the dominant contributing most of the vegetation.

   Aeluropus spp. and *Zygophyllum album* are the most common associates. Common species present are *Halopeplis perfoliata, Salsola baryosma, S. vermiculata, Sevada schimperi* and *Suaeda volkensii*. Less common associates include *Arthrocnemum glaucum, Nitraria retusa* (halophytes), *Anabasis setifera, Cornulaca monacantha, Heliotropium pterocarpum, Panicum turgidum, Polycarpaea repens, Salsola villosa, Sphaerocoma hookeri* v. *intermedia, Taverniera aegyptiaca* and *Zygophyllum coccineum* (xerophytes) and two annuals (*Lotononis platycarpos* and *Zygophyllum simplex*). The mixed population indicates that this community is extending to inland fringes of the marsh where both salt-tolerant and salt non-tolerant species may grow, so containing a richer flora than that of the other strictly halophytic communities. The vegetation of this community has only one layer.

6. *Aeluropus* spp. community. The two morphologically and ecologically similar *Aeluropus brevifolius* and A. *lagopoides* are here combined (as previously

noted). The growth form of the dominant is usually that of a creeping grass, but in one locality (Mersa Alam) it forms peculiar cone-like masses of interwoven roots, rhizomes and sand.

The grassland community dominated by *Aeluropus* spp. occupies the fifth landward zone of the salt marsh. The dominant grass forms patches or mats of dense growth. These are sometimes covered by flakes of salt which denote that the plants may at some time be temporarily wetted by saline water. The cover of this community varies greatly in different stands and ranges from 10 to 80% contributed by the dominant. The most common associates are *Cyperus conglomeratus, Sporobolus spicatus* and *Zygophyllum album* while common ones are *Arthrocnemum glaucum, Halopeplis perfoliata, Limonium axillare, Salsola baryosma, Sevada schimperi* and *Tamarix nilotica*. Other less common species include *Halopyrum mucronatum, Suaeda monoica, S. volkensii* (halophytes), *Sphaerocoma aucheri* (xerophyte) and seven annuals: *Aristida funiculata, A. meccana, Astragalus eremophilus, Crotalaria microphylla, Lotononis platycarpa, Polycarpaea repens* and *Zygophyllum simplex*.

7. *Sporobolus spicatus* community. *S. spicatus* is a halophytic grass common in the southern 100km of the Egyptian Red Sea coast and further south in the Sudan (Kassas, 1957). The grass cover ranges from 20 to 60%, contributed mainly by the dominant which forms an upper layer. The lower layer includes such associates as *Aeluropus* spp. According to Kassas and Zahran (1967), the associate species of this community constitute 42 species: 27 perennials and 15 annuals. This is a large number by salt marsh standards. The fifteen annuals are alien to the salt marsh habitat and are often much more prevalent in the desert wadis and coastal sand sheets and dunes. The perennial associates include halophytes, e.g. *Cyperus conglomeratus, Halopyrum mucronatum, Limonium axillare, Sevada schimperi* and *Zygophyllum album* and xerophytes, e.g. *Panicum turgidum* and *Sphaerocoma hookeri* v. *intermedia. Crotalaria microphylla, Launaea cassiniana* and *Zygophyllum simplex* are among the therophytic associates.

   The S. *spicatus* grassland community occupies a zone inland to that of the *Aeluropus* grassland. In the *Sporobolus* zone the sand deposits are deeper and the soil salinity is much less than in the *Aeluropus* zone. In a few places the two types of grasslands exist in the same zone in a mosaic pattern: the *Sporobolus* in the higher parts, forming island-like patches among an expanse of lower, saline ground covered by *Aeluropus*. Within the stands *of Aeluropus* the water-table is usually shallow (40–100cm), but in the *Sporobolus* stands it is normally deeper than 150cm. The surface deposits are apparently wind-borne sand and are usually loose, especially in areas between the *Sporobolus*-covered patches. This is a notable difference between the soil of this community and that *of Aeluropus* where the soil is frequently cemented by surface accumulation of salts.

8. *Halopyrum mucronatum* community. This grassland community is recorded in a limited area, about 2km long, extending to the south of Mersa Abu Ramad (33km north of Mersa Halaib). It is also recorded by Hemming (1961) in the northern Eritrean Red Sea coast in an area of about 250–450m2 and in the Jizan

coast of the Red Sea in Saudi Arabia in a habitat which appears similar to those of Egypt and Eritrea (Zahran, 1982b). It is a rare grass in both coasts of the Red Sea. *H. mucronatum* grows on much higher sand bars and hillocks than those bearing *Sporobolus* and *Aeluropus*. The sand deposits are usually continuous and the salt marsh ground on which these deposits are formed is usually completely covered by the cover of this community (usually one layer) ranges from 20 to 80%, mostly contributed by the dominant grass. In a few stands *Zygophyllum album* has an appreciable cover (30%) but other associate species contribute little to this. The flora of this grassland includes few halophytes {*Aeluropus* spp., *Limonium axillare, Sevada schimperi* and *Zygophyllum album*) and several xerophytes e.g. *Aerva javanica, Aristida adscensionis, Asphodelus tenuifolius, Heliotropium pterocarpum, Indigofera argentea, Neurada procumbens* and *Panicum turgidum.*

9. *Zygophyllum album* community. *Z. album* is ubiquitous in the sampled coastal area, and the community type dominated by it shows a wide range of ecological conditions. This is indicated by the numerous associates (44 species including six annuals) with differing ecological requirements and by its geographical range which includes, among other regions, the whole stretch of the Egyptian Red Sea coast. The associates include halophytes, e.g. *Aeluropus* spp., *Arthrocnemum glaucum,Atriplex farinosa. Cressa cretica, Halocnemum strobilaceum, Halopeplis perfoliata, Halopyrum mucronatum, Limonium axillare, L. pruinosum, Nitraria retusa, Sevada schimperi, Sporobolus spicatus, Suaeda monoica, S. volkensii*, and xerophytes e.g. *Calotropis procera, Ephedra alata, Hammada elegans, Launaea spinosa, Lygos raetam, Panicum turgidum, Polycarpaea repens, Salsola baryosma* and *Sphaerocoma hookeri*. The dominant contributes mainly to the cover (5–50%).

The phytocoenosis of this community is of several layers. The ground layer includes annuals, e.g. *Aristida meccana, Launaea capitata, Lotus schimperi, Monsonia nivea* and *Zygophyllum simplex*. The suffrutescent layer includes the dominant and several perennial associates e.g. *Hammada elegans* and *Panicum turgidum*. The distantly open frutescent layer includes *Lygos raetam, Nitraria retusa* and *Tamarix nilotica*.

The *Z. album* community may also be classified into two subtypes on the basis of the geographical area of associated halophytic species. In the north, the most common associates are *Limonium pruinosum* and *Nitraria retusa* which are absent in the south whereas *L. axillare* is lacking from the sands in the north and is a common associate in the south.

In the Red Sea coast of the Sudan, *Z. album* is listed among the common associated species of the salt marsh communities (Kassas, 1957). In his second survey of the Red Sea coast of the Sudan M. Kassas (personal communication, 1966) found that *Z. album* is widespread along the whole stretch southward to the Sudano-Ethiopian border. In Tokar (about 160 km south of Suakin and about 80 km north of the Ethiopian border), for example, the *Z. album* community covers vast areas of the salt marshes. In the Asian (Saudi) Red Sea coast, *Z. album* is also very abundant (Zahran, 1982b).

Z. album, a succulent halophyte, tolerates a wide range of soil conditions. Distantly spaced individuals may occur on the ground of dried salt marsh. It may form sand mounds or the mounds may be so crowded that they form continuous sheets of sand.

10. *Nitraria retusa* community. *N retusa* is a halophytic succulent shrub that may form and protect hillocks of sand usually much larger than the mounds of *Zygophyllum album*. Both species are common on the inland wadis of the Eastern Desert (Kassas and Girgis, 1965). The two species are ecologically related. Geographically *Nitraria* is confined to the northern 700 km stretch of the Red Sea coast of Egypt, i.e. from Suez to Mersa Alam. Southward on the Egyptian, Sudanese and Eritrean Red Sea coast *N. retusa* is absent (Kassas and Zahran, 1967; Kassas, 1957; Hemming, 1961). *Z. album* is widespread along the whole African Red Sea coast. The same distribution occurs along the Asian Red sea coast (Zahran, 1982b).

    The *N. retusa* community is present in two habitat types. In one, *Nitraria* forms saline mounds or hillocks that stud the flat ground of be salt marsh. Commonly *Nitraria* covers the north-facing part of the hillocks; the rest is barren. The second habitat comprises sandy bars, actually chains of sandy hillocks, fringing the shore-line. Associate species share the spaces between these sand formations. In this community, *Nitraria* contributes the major cover (5–30%). *Limonium pruinosum* and *Zygophyllum album* are the most common associates. Less common *are Aeluropus* spp., *Halocnemum strobilaceum* and *Tamarix nilotica* (halophytes) and *Hammada elegans, Launaea spinosa, Lygos raetam, Ochradenus baccatus, Tamarix aphylla* and *Zygophyllum coccineum* (xerophytes).

11. *Suaeda monoica* community. S. *monoica* is a succulent halophyte comparable in habit and habitat to *Nitraria retusa*. The two species have an ecological range that extends beyond the limits of the salt marsh to the fringes of the coastal desert plain. Both may form and protect mounds and hillocks of sand, though the S. *monoica* hills may be larger. However, the species seem to have different geographical areas: *N. retusa* occurs in the northern 700 km stretch; S. *monoica*, gradually replaces *N. retusa* within the 300–700 km stretch south of Suez, whereas in the south *N. retusa* is absent and S. *monoica* is a salient feature of the Sudanese and Eritrean Rea Sea coasts (Kassas 1957; Hemming, 1961). The same geographical distribution of *S. monoica* has also been noticed on the Saudi Red Sea coast (Zahran, 1982b).

    The flora of the S. *monoica* community comprises 51 species 9 halophytes, 17 xerophytes and 25 therophytes. The phytocoenosis shows definite layering. *S. monoica* contributes the main part of the frutescent layer. Within the intermediate stretch of the coastal land (300–700 km south of Suez), *Nitraria retusa* may also contribute to this layer with *Tamarix nilotica*. The suffrutescent layer includes *Aeluropus* spp., *Arthrocnemum glaucum, Halocnemum strobilaceum, Halopeplis perfoliata, Hammada elegans, Heliotropium undulatum, Panicum turgidum, Salsola baryosma, Sevada schimperi* and *Zygophyllum coccineum*. Prostrate perennials and annuals, *e.g. Aizoon canariense, Amaranthus graecizans,*

Arnebia hispidissima, Caylusea hexagyna, Launaea capitata, L. cassiniana, Neurada procumbens and *Zygophyllum simplex* form the ground layer. The growth of *Suaeda monoica* within Wadi Di-ib, about 60 km north of Mersa Halaib, shows its wide ecological range. The main channel of this wadi collects surface drainage from the coastal montane country. It crosses the coastal plain with a gradually widening channel forming a deltaic fan within the littoral belt. On the inland side, the delta is choked with sand dunes covered and stabilized by *S. monoica*. In this part the plant forms what seems to be a forest of green dunes with a cover reaching 70–90%. In the spring the spaces between these dunes may be clothed with therophytes. In the middle part of the delta *S. monoica* thins out (cover 20–30%), the hillocks are much smaller and lower than the dunes of the inland part, and the floor is covered by layers of cracked silt which mark the downstream extremities of the torrential floods. In the downstream part of the delta S. *monoica* occupies a salt marsh habitat extending to the shoreline. Associate species of the upstream part are 17 perennials and 18 therophytes, all belonging to the non-saline desert habitat In the middle part, the number of associates is limited (four perennials and five ephemerals). The xerophytic associates of the upstream area are absent from the downstream part and are replaced by a few halophytes.

12. *Tamarix nilotica* community. *T. nilotica* is one of the commons excretive halophytic bushes in the whole African and Asian Red Sea coasts (Andrews, 1950–1956; Cufondontis, 1961–1966; Kassas, 1957; Verdcourt, 1968; Zahran, 1977, 1982b). It grows in a variety of habitats and in various forms. In many parts of the dried salt marsh it forms thickets and gives rise to sand hillocks in the sand-choked deltaic parts of wadis that drain inland country and flow only into the shore-line. Its presence is characteristic of desert wadis with salt- or brackish-water springs that may form sluggish streams of saline water, e.g. Wadi Ambagi 140 south of Hurghada (near Qusseir).

Like other woody plants in arid lands, *T. nilotica* is cut for fuel and other household purposes. Its sparseness in the southern areas of the Egyptian Red Sea coast is, thus, due to its exploitation rather than ecological factors.

In this phytocoenosis. *T. nilotica* contributes the main part of the cover (10–70%). Associates, mostly perennials, include eight halophytes and 17 xerophytes. The most common halophytes are *Juncus rigidus* and *Zygophyllum album*; others are *Aeluropus* spp., *Arthrocnemum glaucum, Halopeplis perfoliata, Limonium axillare, L. pruinosum* and *Nitraria retusa*. The xerophytes include *Acacia raddiana, A. tortilis, Calligonum comosum, Calotropis procera, Convolvulus lanatus, Cornulaca monacantha, Heliotropium arbainense, Hyoscyamus muticus, Leptadenia pyrotechnica, Lygos raetam, Panicum turgidum, Phoenix dactylifera, Tamarix aphylla* and *Zygophyllum decumbens*. Annuals include *Astragalus vogelii*.

The cover of the *T. nilotica* community is in three layers. The frutescent layer includes the dominant and a number of associated trees and shrubs: *Acacia* spp., *Phoenix, Leptadenia, Lygos, Tamarix* spp., etc. The suffrutescent and the ground layers contribute only little to the cover except in the inland wadis with salt-water streams where *Juncus rigidus* forms extensive mats.

Other Communities

Apart from the communities described above, the littoral salt marshes of the Egyptian Red Sea coastal land are characterized by certain communities that are of limited distribution.

1. *Tamarix passerinoides* community. *T. passerinoides* is morphologically and ecologically comparable to *T. nilotica* but the former is rare along the Egyptian Red Sea coast. Its presence is limited to a narrow stretch: El-Mallaha 20–40km south of Ras Gharib (Zahran, 1962), which is an inland depression separated from the shore-line by an elevated raised beach and is fed with seawater through underground seepage. It is 20km long and 5km wide. The bottom of the depression includes several small salt-water lagoons, fringed by extensive saline ground covered by surface crusts. This is, in turn, fringed by salt marsh vegetation.

   T. passerinoides dominates this saline habitat. Individual bushes on sand dunes grow to a considerable size. Associate species are *Arthrocnemum glaucum, Nitraria retusa, Suaeda monoica* and *Zygophyllum album.*
2. *Juncus rigidus* community. *J. rigidus* is a cumulative halophyte which is very tolerant to increased soil water stress and climatic aridity (Zahran, 1982b). Its dominance in the Red Sea coastal marshes is restricted to the Ain Sokhna area (c.50km south of Suez) where its cover is up to 90–100% (Kassas and Zahran, 1962). The associates include: *Cressa cretica, Halocnemum strobilaceum* and *Tamarix nilotica. J. rigidus* is also locally dominant in the coastal plain of Wadi Araba.

   The *J. rigidus* community may also be present in the inland salinas where it is often associated with *Tamarix passerinoides. J. rigidus* chokes the channels of the creeks whereas *T. passerinoides* fringes their banks.
3. *Salicornia fruticosa* community. *S. fruticosa* is morphologically and ecologically comparable to *Arthrocnemum glaucum* and it is very difficult to distinguish between these two species unless they are in flower or fruit (Täckholm, 1974). In the study coastal area, *S. fruticosa* is recorded from one part of El-Mallaha where it forms pure patches with 50–100% cover. The ground is covered with a thick salt crust having 80% soluble salts (Zahran, 1962).
4. *Cressa cretica* community. *C. cretica* is a cushion-chamaephyte excretive halophyte. Its dominance on the littoral Red Sea marshes of Egypt is in the delta of Wadi Hommath of the Gulf of Suez where the surface soil of the saline sand flats has total soluble salts of 60% (Kassas and Zahran, 1962). Associate species include *Alhagi maurorum, Arthrocnemum glaucum, Halocnemum strobilaceum, Imperata cylindrica, Limonium pruinosum, Nitraria retusa* and *Zygophyllum album. Cressa cretica* is also abundant in the delta of Wadi Araba saline habitat (120km south of Suez).
5. *Imperata cylindrica* community

   I. cylindrica is an aggressive rhizomatous grass community known as cogon grass (halfa grass in Egypt). It is rarely recorded in the Egyptian Red Sea coast, however, in the delta of Wadi Hommath there are few pure stands demoinaed by this grass. Eisa (2007) stated that *I. cylindrica* has an allelopathic effect as it produces phenolic compounds. Together with competition, this grass may inhibit the growth and survival of other associated species. This may explain the pure stands in the less saline area of Wadi Hommath.

Invasive Weeds

The Red Sea coast, and since more than 20 years, is being under development programmes for tourist attraction. Establishment of coastal resorts, villages, hotels and new cities are rapidly going on. The soils of these establishments are salt affected and not suitable for the growth of garden plants, sensitive to soil salinity. Thus, soils collected from the agricultural lands, of the Nile Region are carried to the sites of the new establishment of the Red Sea coast to raise gardens. Such soils act as a seed bank for the weeds naturally growing in the field crops and canal banks of the Nile Valley and Nile Delta. The seeds associated with these soils are spread in the gardens of the new establishments soon germinate and produce weed flora new to the Red Sea and are thus considered invasive species. Sheded and Shaltout (1998) investigated the flora of the gardens of eleven coastal resorts distributed along the Red Sea in Suez, Hurghada and Qusseir. A total of 51 species belonging to 42 genera and 18 familes were recorded. Thirty one species of these have been recorded for the first time along the Red Sea coast, e.g. *Avena fatua, Amaranthus hybridus, Chenopodium album, Convolvulus arvensis, Corchorus endivia, Cyprus alopecuroides, Conyza bonariensis, Euphorbia hirta, Echinochloa crus-galli, Gynandropsis gynandra, Hibiscus trionum, Emex spinosa, Polygonum equisetiforme, Panicum repens, Sesbania sesban, Sorghum halepense, Lolium multiflorum, Ricinus communis* etc (Zahran 1962, 1964; Kassas and Zahran 1962, 1965, 1967; Mashaly, 1996).

Off-Shore Islands

The off-shore islands of the Red Sea are close to the shore-line and are in three rows. Each island has an igneous core with a coral reef formed at a time when the core was covered with sea water. Coral development continued until the core appeared above sea level (Abu Al-Izz, 1971).

Abu Minqar Island is 3 km southeast of Hurghada. It is divided by shallow creeks and characterized by three vegetation types: mangrove salt marsh and high ground. The mangrove vegetation is represent by pure thickets of *Avicennia marina* which is well developed within the creeks (cover 80–90%). The mangrove area is fringed inland by salt marsh vegetation dominated by *Arthrocnemum glaucum*. The width of this zone varies in obvious relation to the ground level. Where there is a gradual landward rise in level the zone is wide, whereas where the mangrove habitat is bounded by low cliffs it may be absent or very narrow. In this zone *A. glaucum* often forms pure stands with 60–80% cover. Rare individuals of *Zygophyllum album* may be present. In the high ground, vegetation is mostly a sparse growth of *Z. album* with occasional presence of *Nitraria retusa* and *Suaeda monoica*. The vegetation is confined to isolated locations where some soft deposits cover the underlying rocks. The barren rock surface is usually sterile.

(ii) The Coastal Desert

The vegetation within the salt marsh ecosystem shows zonation which makes the pattern fairly easy to interpret. That is not the situation in the vegetation of the desert ecosystem, which presents a complicated pattern owing to different conditions of topography, characters of the surface deposits and the relationships with the mountain groups.

The desert ecosystem of the Egyptian Red Sea coastal land extends between the littoral salt marsh belt and the coastal range of hills and mountains on the inland side. By reason of this intermediate position, its ecological conditions show, especially on its fringes, transitional characters. Nevertheless, the inland boundaries of the coastal desert are as clear as its seaward ones.

Much of the Red Sea coastal plain is far from the reach of tidal water. The habitat is usually non-saline but climate and soil aridity are the main environmental features. It is essentially a gravel-covered plain traversed by the downstream extremities of the main wadis and is dissected by smaller drainage runnels that may extend from the foot-hills of the coastal front or may not reach the coast. The downstream extremities of the main wadis may form deltaic basins. Superimposed on this pattern, aeolian deposits may form sheets, mounds or hills of various heights and extents. This complex situation produces a similar complex of habitat conditions.

Within the desert plain ecosystem the soil transporting agencies (wind and water) are actively operating. The alluvial deposits range from fine silt to coarse gravel and boulders, and often build terraces on the side of the water course. The building and destruction of these terraces are mainly physical processes that are independent of the vegetation. The aeolian deposits are sandy and are bodies ranging from small mounds to hills. These are, as a whole, built around the plant growth (phytogenic) and their maximum size seems to depend on the species.

The vegetation of the coastal desert is confined to the drainage system (run-off desert). It shows a mosaic pattern and distinct seasonal; aspects mainly due to the preponderant growth of therophytes during the late winter and early spring. This aspect of seasonal phenology is not seen in the salt marsh ecosystem.

The desert ecosystem supports a greater number of species and the floristic composition of the communities is usually much more elaborate than the simple composition of the salt marsh communities. In certain parts, the coastal desert embraces inland saline habitats represented in this study by, for example, the El-Mallaha depression south of Ras Gharib. The vegetation of these saline habitats belongs ecologically and floristically to the salt marsh type. Similar vegetation may occur in certain parts within the main wadis where brackish water springs form local saline habitats.

The vegetational pattern associated with these habitat conditions further complicated by differences in the intensity of human interference (cutting) and wildlife grazing and also by differences in the geographical ranges of the plants.

The vegetation of the Red Sea coastal desert is described under the following titles:

1. The coastal desert wadis; and
2. The communities.

The Coastal Desert Wadis

Kassas and Imam (1954) state "The term wadi designates a dried river bed in a desert which may be transformed into a temporary water course after heavy rain. Each wadi has a main channel and branched affluents or tributaries".

The wadi habitat has distinctive features including a characteristic plant cover. It has the great merit of being a drainage system collecting water from an extensive catchment area. The water supply of a wadi seems to be much greater than the recorded rainfall. Hassib (1951) states: "The whole stream is about 300 m wide and 1–2.5 m deep, rushing continuously for 2–3 days. It sweeps away the vegetation and sometimes men, cattle and roads. A certain amount of water percolates into the soil, forming ground water which is utilized by numerous wells sunk in their course".

This quotation shows that the water supplies of wadis are immense and hence explains the actual, or potential, richness of wadi vegetation. This advantage is counter balanced by two destructive agencies: torrent and grazing. The central part of a wadi bed, which is the waterway, is usually devoid of plants, vegetation being mostly restricted to the sides. In any bend of a wadi meander, the plant cover is very scarce on the outer curve, where the torrent effect is greatest, and well developed on the inner curve. The influence of torrents is partly mechanical, destroying and uprooting the plants, and partly erosional, removing the soil.

Wadis are subject to serious grazing. The most common species are the least grazed. Many palatable species acquire a cushion-shaped or grazed-trimmed growth-form which they do not exhibit if protected. Cutting and lumbering are especially directed towards plants that are valuable for fuel. These destructive processes deprive the soil of its plant cover, render it susceptible to torrential erosion and to deflation, and hinder the natural development of the habitat. The vegetation scarcely attains maturity and is usually kept in a juvenile or deflected stage of development.

The soil of the wadi is usually of rock detritus, ranging in texture from fine silt to gravels and boulders. The wadi bed is often covered with layers of fine material alternating with coarse gravel. As the texture of these sediments is indicative of the transporting capacity of the water bodies contained in the wadi, the alternating layers reflect episodic variation in the water resources of the wadi. The alternation of layers of different texture has a substantial influence on the water available to plants. A gravel bed at the ground surface will be subject to desiccation and will afford few possibilities for seed germination. It may, however, safeguard against run-off as it allows the underground layer of gravel to store greater amounts of free water in its spaces. However, the gravel bed has the least water-retaining capacity.

The soil depth is by far the most important feature. A thin soil is moistened during the rainy season but dried with the approach of the dry one. A deep soil allows for the storage of some water in the subsoil which will provide a continuous supply of moisture for the deeply seated roots of perennials.

The Red Sea coastal land of Egypt is characterized by a numbe of wadis that run eastward to flow into the Gulf of Suez and Red Sea. These include (listed from north to south) Wadi Hommath, Wadi Hagul, Wadi El-Bada, Wadi El-Ghweibba and Wadi Araba (Gulf of Suez) and Wadi Bali, Wadi Ghadir, Wadi Gimal, Wadi Aideib, Wadi Di-ib, Wadi Serimtai, Wadi Shillal, Wadi Laseitit, Wadi Hawaday and Wadi Naam (Red Sea).

The vegetation of three wadis which flow into the Gulf of Suez and Red Sea, Wadi Hagul, Wadi Araba and Wadi Serimtai, is described.

Wadi Hagul Drainage System

This is an extensive wadi occupying the valley depression between Gebel Ataqa to the north and the Kahaliya ridge to the south. Its main channel extends for about 35 km and collects drainage water on both sides. The upstream part cuts its shallow channel into ochreous-coloured marls and grits and carolin-beds of the Upper Eocene. The main channel proceeds in a southeast direction, traversing limestone beds of the Miocene. Further downstream, the wadi widens and cuts its way across recent alluvial gravels before it finally traverses the coastal plain towards the Gulf of Suez (Sadek, 1926).

With reference to the vegetation and geological features of Wadi Hagul, three main sectors may be distinguished: upstream, middle and downstream.

In the upstream sector the plant cover varies in apparent relation to the area drained. The finer runnels here are short, their cover is sparse and their beds are strewn with coarse boulders. The vegetation is usually a community of *Iphiona mucronata*. In the finer runnels the dominant species is *Fagonia mollis*. Other species include *Centaurea aegyptiaca, Gymnocarpos decander, Helianthemum lippii, Heliotropium arbainense, Linaria aegyptiaca, Scrophularia deserti* and *Zygophyllum decumbens*. Runnels with greater drainage are usually characterized by a community dominated by *Zygophyllum coccineum*. The bed is covered with a rock detritus including some soft materials. Associate species include *Artemisia judaica, Cleome droserifolia, Crotalaria aegyptiaca, Fagonia mollis, Heliotropium arbainense, Iphiona mucronata, Launaea spinosa, Lavandula striata, Lygos raetam, Pennisetum dichotomum, Pituranthos tortuosus, Reaumuria hirtella, Scrophularia deserti* and *Trichodesma africanum*. Of particular interest here is *Cleome droserifolia*, a xerophyte common eastward but very rare westward in the Eastern Desert.

Within the upstream part of the main channel of Wadi Hagul, two communities may be recognized. One is dominated by *Zilla spinosa* and is well developed on elevated terraces of mixed deposits. The other is dominated by *Launaea spinosa* and represents a further stage in the building up of the wadi bed: the floor deposits are deeper and include a greater proportion of salt deposits admixed with coarse rock detritus. Associate species of the *Z. spinosa* community include *Artemisia judaica, Asteriscus graveolens, Crotalaria aegyptiaca, Fagonia mollis, Launaea spinosa, Lycium arabicum, Pennisetum dichotomum, Pituranthos tortuosus, Pulicaria undulata, Scrophularia deserti, Trichodesma africanum, Zygophyllum coccineum* and *Z. decumbens*. The same associates have been recorded in the *L. spinosa* community with the addition of *Cleome droserifolia, Lavandula stricta* and *Reaumuria hirtella*. Further downstream this community includes *Acacia raddiana, Lygos raetam* and *Tamarix nilotica*.

In the middle section of Wadi Hagul, *Leptadenia pyrotechnica* occurs; this is common further eastward but rarely found westward. Here the vegetation of the main channel and its tributaries consists of three communities. One is dominated by *Hammada elegans* and occupies the gravel beds that form the raised terraces. The second is co-dominated by *Launaea spinosa* and *Leptadenia pyrotechnica* and occupies the waterways, that is, the areas that are flooded by the occasional torrents. The third is represented by scattered patches of *Tamarix aphylla* occupying relicts of terraces of soft deposits. There are few trees of good size, the vegetation being mostly patchy bushy growth. Associate species include *Artemisia judaica, Cleome droserifolia, Crotalaria aegyptiaca, Fagonia bruguieri, F. mollis, Gymnocarpos decander, Iphiona mucronata, Launaea spinosa, Lavandula stricta, Lycium arabicum, Lygos raetam, Pituranthos tortuosus, Zilla spinosa* and *Zygophyllum coccineum*.

In the downstream part of Wadi Hagul which cuts across the coastal gravel beds, the channel is wide, but the course of the ephemeral streams is ill-defined. The wadi obviously changes its course on different occasions. The result is a reticulum of thin, branching and coalescing water courses. The meshes of this reticulum are patches of raised gravel isles. All are included on both sides by the gravel bed of the coastal plain. The vegetation of the main wadi is dominated by *Hammada elegans* with individuals of *Launaea spinosa* and *Lygos raetam*. The vegetation of the affluent runnels is mainly a grassland with *Panicum turgidum* and/or *Pennisetum dichotomum*. Further downstream, the wadi meets the littoral salt marsh vegetation.

Mashaly (1996) stated that after 34 years of Kassas and Zahran (1962), there are no fundamental changes neither in the physical environment nor in the vegetation types of Wadi Hagul. Actually, no man interference has been observed in this wadi.

Wadi Araba Drainage System

The limestone block of Gebel El-Galala El-Bahariya (north Galala) separated from the comparable block of the Galala El-Qibliya (south Galala) by a valley about 30 km wide (N-S). This valley, mostly Lower Cretaceous sandstone is occupied by the Wadi Araba system which drains the southern scarps of north Galala mountain and northern scarps of south Galala mountain. The main channel of Wadi Araba terminates near Zaafarana lighthouse (120 km south of Suez) and extends westwards to the central limestone country of the Eastern Desert.

The coastal front of the wide valley is dissected by the main channel of Wadi Araba (about 70 km) and a number of smaller wadis which drain the southeast of the north Galala mountain and the northwest of the south Galala mountain. On the northern and southern side are everal independent wadis.

Wadi Araba consists of a main channel and innumerable tributaries (some several kilometres long) draining the two Galala blocks on the northern and southern sides and the fringes of the limestone plateau in the west. The main channel has a well-defined trunk cut across the sandstone beds.

Ecologically, a distinction may be made between (a) the littoral and coastal saline habitats and (b) the desert vegetation of the channels of the drainage system.

Within the former group are the littoral zone of raised beach, the littoral zone of sand hillocks and the inland zone of dry saline plain, i.e. whole seaward belt of the coastal plain is saline and its soil and vegetation are distinctly different from those of the inland desert. The ecology of the drainage system depends on the extent of the catchment areas. The flora, however, shows certain differences between the wadis (or parts of wadis) cutting across the limestone masses of the ridges and those cutting across the sandstone.

The littoral and coastal plains are characterized by simple vegetation types and a limited number of species. The difference between the various communities often reflects differences in relative abundance of species rather than of floristic composition.

The littoral beach of Wadi Araba is represented by Mersa Thelemit, a small bay 7–8 km south of Zaafarana lighthouse. A flat zone fringing the shore-line is bounded inland by a second raised beach which forms the fringe of the inland, narrow coastal plain. The littoral beach is less than one metre above sea level whereas the inland raised beach is 1–4 m above it.

The Mersa Thelemit has a plant cover consisting of *Nitraria retusa* and *Zygophyllum album*. The latter forms pure stands on the flat littoral ground; *Nitraria* is much more abundant on the littoral side. The littoral sand hillocks extend to the north of Zaafarana lighthouse. The surface of these dunes is covered by a saline crust and a rich growth of *Halocnemum strobilaceum* associated with a few plants of *Zygophyllum album* and *Nitraria retusa*.

The coastal plain is transitional between the littoral salt marsh and the inland desert plain. In this *Nitraria retusa* is the most abundant species. In some patches *Cressa cretica* or *Juncus rigidus* is locally dominant.

In the main channel, the wadi is characterized by extensive stands of *Tamarix aphylla*. This species may develop into trees, but in Wadi Araba it forms sand hillocks (1–4 m high and 10–100 m^2 area) covered completely by the growth of *Tamarix*. The *Tamarix* hillocks are present all along the main channel of the wadi but their density varies. Human destruction is the apparent cause of the thinning of these hillocks. In certain localities the hillocks are so crowded that they look like thickets.

Associated with *T. aphylla* and forming hillocks of smaller sizes are *Calligonum comosum* (locally dominant), *Ephedra alata* and *Leptadenia pyrotechnica*. Near the mouth of Wadi Araba *Tamarix amplexicaulis* is a common associate. The spaces between these hillocks and mounds are the habitat of a number of species: *Aerva javanica, Artemisia judaica, Centaurea aegyptiaca, Convolvulus hystrix, Farsetia aegyptia, Francoeuria crispa, Hammada elegans, Heliotropium luteum, H. pterocarpum, Hyoscyamus muticus* and *Taverniera aegyptiaca*. Near the mouth of the wadi *Nitraria retusa* and *Zygophyllum album* are also present.

Distantly dispersed within the main channel of the wadi and its main tributaries is *Acacia raddiana*. These few trees and shrubs represent relicts of a dense population, destroyed by lumbering.

The community dominated by *Hammada elegans* is obviously the most common. It seems to replace *T. aphylla* scrubland, wherever the latter is destroyed. The ground cover is often coarse sand and gravel. Growth of *H. elegans*, which may build small sand mounds, is also abundant within the tributaries and runnels of the drainage system. The associates of this community include *Artemisia judaica, Calligonum comosum, Ephedra alata, Francoeuria crispa, Pergularia tomentosa* and *Zilla spinosa* (Sharaf El-Din and Shaltout, 1985).

In the upstream extremities of the main channel is an area where *Anabasis articulata* is locally dominant. The vegetation here is of a few species: *Ephedra alata, Hammada elegans* and *Lygos raetam.*

Within the western affluents of Wadi Araba which traverse the limestone plateau of the Eastern Desert a community dominated by *Launaea spinosa* may be recognized. Associate species include *Acacia raddiana, Achillea fragrantissima, Artemisia judaica, Asteriscus graveolens, Atractylis flava, Echinops galalensis, Launaea nudicaulis, Lygos raetam, Matthiola livida, Ochradenus baccatus, Pituranthos tortuosus, Pulicaria undulata, Trichodesma africanum* and *Zilla spinosa.* Grassland vegetation dominated by *Lasiurus hirsutus* is also recognized within the upstream affluents. Associates include *Acacia raddiana, Asteriscus graveolens, Calligonum comosum, Centaurea aegyptiaca, Cleome droserifolia, Echinops spinosissimus, Fagonia mollis, Heliotropium arbainense, Iphiona mucronata, Launaea nudicaulis, L. spinosa, Linaria aegyptiaca, Lygos raetam, Panicum turgidum, Pergularia tomentosa, Plantago ciliata* and *Pulicaria undulata.*

Within the affluents of the middle part of Wadi Araba, apart from the common *Hammada elegans* community, two other communities may be recognized. One is dominated by *Artemisia judaica*, the other is grassland dominated by *Panicum turgidum* which is severely affected by grazing. Associates of the *A. judaica* community are *Calligonum comosum, Centaurea aegyptiaca, Cleome droserifolia, Farsetia aegyptia, Hammada elegans. Pergularia tomentosa, Taverniera aegyptiaca* and *Zilla spinosa.* In the *P. turgidum* community the associates are *Artemisia judaica, Calligonum comosum, Ephedra alata, Farsetia aegyptia, Hammada elegans and Zilla spinosa.*

Within the upstream parts of the main tributaries that cut across the limestone blocks of the two Galala mountains the plant cover is characterized by the preponderance of *Zygophyllum coccineum* which is present within the main channel of Wadi Araba. Common associates include *Asteriscus graveolens, Erodium glaucophyllum, Fagonia mallis, Iphiona mucronata* and *Zilla spinosa.*

The main channel contains a number of brackish-water springs, e.g. Bir Zaafarana, and Bir Buerat. Around these springs are neglected groves of *Phoenix dactylifera* and patches of halophytes – *Juncus rigidus, Nitraria retusa* and *Zygophyllum album.*

Wadi Serimtai Drainage System

This is one of the biggest drainage systems in the southern section of the Egyptian Red Sea coast. It is about 50 km north of Mersa Halaib. The plant cover of this wadi and of the other wadis of the southern section does not vary greatly.

The littoral downstream belt of the delta of Wadi Serimtai supports halophytic vegetation comprising two communities; one is dominated *Aeluropus brevifolius* and the other by *Zygophyllum album.* The *Aeluropus* community occurs in the low ground of wet brown sand covered by a silty saline layer. The plant cover is 70%, contributed by the dominant grass. The associates, mainly halophytes, include *Arthrocnemum glaucum, Halopeplis perfoliata, Limonium axillare, Suaeda pruinosa* and *Zygophyllum album*. The *Z. album* community is present on flat soft sand sheets on which the dominant builds small unlocks of sand. The plant cover ranges from 30 to 40%. Being nearer to the desert habitat of the wadi, associates of this community are a mixture of xerophytes and halophytes. The xerophytes (perennials and annuals) include *Asphodelus tenuifolius, Euphorbia scordifolia, Launaea cassiniana, Monsonia nivea, Panicum turgidum Polycarpaea repens, Stipagrostis hirtigluma* and *Zygophyllum simplex*. The halophytes are *Limonium pruinosum, Salsola vermiculata Sporobolus spicatus* and *Suaeda monoica*.

The westward (upstream) change in the soil characteristics and land relief is associated with a change in the vegetation of the wadi. West of the area dominated by *Zygophyllum album* is a narrow zone (200 m) dominated by *Salsola vermiculata* (29–30% cover) that builds hummocks of 2 × 1 m area and 60 cm high. Associate species are mainly salt non-tolerant desert perennials and annuals, e.g. *Acacia tortilis, Asthenatherum forsskaolii, Calotropis procera, Cyperus conglomeratus, Eragrostis ciliaris, Euphorbia scordifolia, Glossonema boveanum, Monsonia nivea, Panicum turgidum, Stipagrostis hirtigluma* and *Zygophyllum simplex*.

In the midstream part of Wadi Serimtai the main channel is dominated by *Panicum turgidum*. The substratum is of coarse sand mixed with scattered rock detritus. Plant cover is 10–15% and increases to 60–65% during the winter season with dense growth of therophytes. *Acacia raddiana* is the abundant associate perennial. *Aristida adscensionis* and *Stipagrostis hirtigluma* are the abundant ephemerals. Other associates include *Abutilon pannosum, Aerva javanica, Arnebia hispidissima, Calotropis procera, Caylusea hexagyna, Cucumis prophetarum, Stipagrostis ciliaris, Leptadenia pyrotechnica, Monsonia nivea, Neurada procumbens, Polycarpaea repens* and *Rumex simpliciflorus*.

The dominance of *P. turgidum* continues for a few kilometres in the main trunk of the wadi, but then its abundance decreases gradually until it is replaced by a community dominated by *Acacia tortilis*. The plant cover of this community is about 40–50%: 10–20% perennials and 30–40% annuals. The most common associates are *Aristida adscensionis, Stipagrostis hirtigluma* and *Panicum turgidum*. Other associates include *Abutilon pannosum, Aerva persica, Aizoon canariense, Asphodelus tenuifolius, Caylusea hexagyna, Cenchrus pennisetiformis, Cleome brachycarpa, Cucumis prophetarum, Dipterygium glaucum, Farsetia longisiliqua, Heliotropium pterocarpum, H. strigosum, Ifloga spicata, Launaea massauensis, Neurada procumbens, Panicum turgidum, Polycarpaea repens, Salsola vermiculata Tephrosia purpurea, Tragus berteronianus* and *Zygophyllum simplex*

The vegetation of the upstream part of the wadi is an open scrubland dominated by *Acacia raddiana*. The ground is of coarse sand mixed with gravel and rock detritus. Between the two *Acacia* communities (A. *tortilis* in the east and A *raddiana* in the west) is a transitional zone co-dominated by both *Acacia* spp. Apart from the above-mentioned associates, other species recorded in the upstream part of the wadi include *Amaranthus graecizans. Antirrhinum orontium, Aristida meccana,*

Astragalus eremophilus, Chenopodium murale, Euphorbia granulata, Ochradenus baccatus, Robbairea delileana, Sisymbrium erysimoides, Spergula fallax, Trianthema salsoloides, Tribulus pentandrus and *Trichodesma ehrenbergii.*

The Communities

The vegetation of the Egyptian Red Sea coastal desert may be classified into two main types: ephemeral and perennial. These two types of vegetation and their communities are described below.

Ephemeral Vegetation

Ephemeral vegetation indicates soil conditions that allow for no everyear storage moisture: soil wetness is maintained during only a part of the year. This may be due either to scantiness of rainfall or also to surface deposits being too shallow. The ephemeral nature is the distinctive feature of this vegetation and not the habit of the species since many perennials may acquire ephemeral growth form under these conditions.

During the rainy season, the desert plain may show green patches of ephemerals. These areas may be independent of the drainage system. A rainy season is not an annual recurring phenomenon within a particular area. The rainfall is mostly inconsistent in time and space; cloudbursts that bring the desert rain are often of limited extent.

Ephemeral vegetation usually forms a mosaic of patches each of which may be dominated by one or several species. One type of ephemeral growth may recur in the same patch for several years. This may be due to the availability of seeds where parent plants have existed. Depending on the growth form of the dominant (or most abundant) species, three types of ephemeral vegetation are recognized in the Red Sea coastal desert: one dominated by succulent plants, a second by grasses and a third by herbaceous species (Zahran, 1964).

The succulent type of ephemeral vegetation may be dominated by *Zygophyllum simplex, Trianthema crystallina, Tribulus pentandrus* or *Aizoon canariense*. The tissue of these succulents can store water that may be used later in the growing season. They are characterized by shallow roots and survive throughout a season longer than other ephemerals. In exceptionally wet years or highly favoured localities *Zygophyllum simplex* may extend its life-span for a whole year or more. Also included in this group may be the ephemeral growth of *Spergula fallax* and *Spergularia marina* which appear in exceptionally wet seasons or in areas fed by springs.

Ephemeral grasses include several species of *Aristida, Bromus, Cenchrus, Eragrostis, Schismus, Stipagrostis*, etc. This vegetation is of special importance for the nomadic herdsmen for whom it is valuable pasture.

The herbaceous ephemeral type of vegetation may be dominated by one of a great variety of species or may be mixed growth with no obvious dominant. Within the Egyptian Red Sea coastal desert, patches of ephemerals dominated by one of the following species are recorded: *Arnebia hispidissima, Asphodelus tenuifolius, Astragalus eremophilus, A. vogelii, Filago spathulata, Ifloga spicata, Malva parviflora, Neurada procumbens, Plantago ciliata, Schouwia thebaica, Senecio desfontainei, S. flavus, Tribulus longipetalus* and *T. orientalis*. The growth of these plants may also provide valuable grazing.

The patches of ephemerals are associated with soft deposits, usually shallow sheets of sand, or especially favoured localities. This type of habitat provides a briefly sustained water supply during the rainy season; the shallow surface deposits eventually dry almost completely. The vegetation may be quite dense and the number of individual plants several hundred/km^2. Plant cover varies from 5 to 50% and the patches of ephemerals look like micro-oases amidst the dry desert plain.

Perennial Vegetation

The perennial xerophytic vegetation of the Red Sea coastal desert may be classified into two main types – suffrutescent and frutescent. The vegetation of the suffrutescent type consists of an upper layer (30–120 cm) which includes the dominant species and a ground layer (<30 cm) including the associate annuals and cushion-forming perennials, e.g. *Cleome droserifolia* and *Fagonia mollis*. Woody plants of 120–150 cm are sparse in the frutescent type. Here there are three main layers – the two of the suffrutescent type and a higher one including the dominant. This vegetation includes some trees of more than 5 m, e.g. *Acacia raddiana* and *Balanites aegyptiaca*.

The Suffrutescent Perennial Vegetation

The suffrutescent perennial vegetation is widespread in the Egyptian Red Sea coastal desert. It is distinguished by a permanent framework of perennial xerophytes. This vegetation is of two layers: a suffrutescent and a ground layer. In the majority of communities the suffrutescent layer characterizes the vegetation; the ground layer is of dwarf or trailing perennials, enriched by the growth of ephemerals during the rainy season.

The suffrutescent perennial vegetation includes the following units which may be recognized:

(a) Succulent half-shrub forms
   1. *Zygophyllum coccineum* community
   2. *Salsola baryosma* community
   3. *Hammada elegans* community

(b) Grassland forms
   4. *Panicum turgidum community*
   5. *Other communities*

(c) Woody forms
   6. *Zilla spinosa community*
   7. *Launaea spinosa community*
   8. *Cleome droserifolia community*
   9. *Sphaerocoma hookeri v. intermedia community*
   10. *Other communities*

Succulent Half-Shrub Forms

1. *Zygophyllum coccineum* community. *Z. coccineum* is a leaf and stem succulent that remains green all through the year. Yet this xerophyte seems to have an age limit of several years and in this differs from *Z. album* which does not appear to have a limited life. The regeneration of *Z. coccineum* is confined to rainy years when numerous crowded seedlings may appear but are eventually thinned to a limited number (1–2/m^2). In exceptional localities where the Z. *coccineum* population may have a higher density, the individuals are usually small, being far below the normal size of about 0.1 m^2. The most abundant species is *Zilla spinosa. Cleome droserifolia* and *Indigofera spinosa* are commonly present.

   Within the Z. *coccineum* community, the frutescent layer is very open and may be of one or more of the following: *Acacia raddiana, A. tortilis, Leptadenia pyrotechnica, Lycium arabicum, Lygos raetam*, and *Tamarix nilotica*. The suffrutescent layer contains the main bulk of the perennials including the dominant and its most common associates. The ground layer may include such dwarf or prostrate species as *Fagonia mollis, Heliotropium arbainense, Robbairea delileana* and *Cyperus conglomeratus*. The species population may be enriched during the rainy season by the profuse growth of therophytes.

   The presence of Z. *coccineum* in the Egyptian Red Sea coastal desert is not very widespread, being confined to the limestone country. It occurs from the seaward fringes of the coastal desert plain to the inland mountain country. This wide range of ecological conditions is reflected by the presence of associates, including halophytes, e.g. *Limonium axillare* and *Nitraria retusa*, and plants of the mountain habitat e.g. *Moringa peregrina.*

   In the Red Sea coastal lands of Saudi Arabia (Asian coast) and the southern section of the African coast (Eritrean coast), Z. *coccineum* is one of the halophytic plants that forms a widespread community (Hemming, 1961; Younes et al., 1983). There might be two Z. *coccineum* ecotypes, one intolerant of soil salinity and the other a salt-tolerant form.

   The Z. *coccineum* community of the Red Sea coast is present within the drainage systems. Within the main channels of the larger wadis its growth is usually confined to the parts flushed by torrents and community dominated by Z. *coccineum* is rarely found on the terraces higher than the water course.

2. *Salsola baryosma* community. Within the Egyptian Red Sea coastal desert the community dominated by S. *baryosma* is common within the southern section of the coastal plain extending south of Mersa Alam. Within the perennial framework of this community, S. *baryosma* contributes the main cover (10–50%). *Panicum turgidum* is abundant associate and *Salsola vermiculata* and *Acacia tortilis* commonly present. Less common associates include *Sevada shemperi* and *Sporobolus spicatus*. The growth of therophytes is notably. *Aristida adscensionis, A. meccana, Astragalus eremophilus, Eragrostis ciliaris, Stipagrostis hirtigluma* and *Zygophyllum simplex* cover about 50% of the spaces between the perennials. The parasite *Cistanche tinctoria* is usually present on the bushes of *S. baryosma*. The frutescent layer of this community is made up of *Acacia tortilis* and *Lycium arabicum* and a few individuals of

Calotropis procera. The suffrutescent layer includes the prostrate perennials and is enriched during the rainy season by the ephemerals.

Unlike the *Zygophyllum coccineum* community which is usually confined to the water courses flushed by torrents, the S. *baryosma* community spreads within the whole of the wadi channel and extends over the sides of the wadi delta. It is particularly common in the area between Lat. 23°10′N and 22°20′N. In this part the coastal plain is wide (15–25 km). The downstream extremities of the wadis flowing into the plain may have ill-defined courses that are often lost among sheets of sand. The S. *baryosma* community grows on these sheets of sand. However, like *Z. coccineum, S. baryosma* does not build mounds and hillocks.

Within the Saudi Red Sea coastal land, the community dominated by *S. baryosma* occupies the transitional zone between the littoral salt marsh ecosystem and the coastal desert (Zahran et al., 1985b).

3. *Hammada elegans* community. Within the Red Sea coastal plain the community dominated by *H. elegans* is confined to the coastal plain of the Gulf of Suez, i.e. the 400 km stretch from Suez to Hurghada. From Hurghada southwards to the Sudano-Egyptian border, *H. elegans* is absent. A comparable community dominated by *H. elegans* has also been recorded within the Saudi Red Sea coast (Zahran, 1982b). In this community, *H. elegans* is consistently the most abundant species though its cover may not exceed 20% and may be much lower (<5%). The most common associate is *Launaea spinosa. Artemisia judaica, Francoeuria crispa, Panicum turgidum* and *Zilla spinosa* are common associates. *Lygos raetam, Pennisetum dichotomum* and *Zygophyllum coccineum* are less common. Other perennial xerophytes include *Anabasis articulata* and *Cleome droserifolia. A. articulata* is species similar in habit to *H. elegans*; both are desert succulent xerophytes of the family Chenopodiaceae. *Anabasis* dominates a particular community with close physiognomic similarities to the *Hammada* community. It is abundant in several parts of the Egyptian deserts: Cairo-Suez desert (El-Abyad, 1962), Mediterranean coastal desert (Tadros and Atta, 1958b). Within the Red Sea coastal land, *A. articulata* is restricted to a few localities, e.g. delta of Wadi Hommath (33 km south of Suez) and delta of Wadi Araba (Zahran, 1962). *A. articulata* is not recorded in the Saudi Red Sea coastal land (Migahid, 1978).

The vegetation of the *H. elegans* community is represented by three layers. The frutescent layer is of little significance and includes the shrubs: *Acacia raddiana, Calotropis procera, Tamarix aphylla* and *T. mannifera*. The suffrutescent layer contains the dominant and numerous associated perennials. The ground layer includes dwarf and prostrate perennials, e.g. *Centaurea aegyptiaca, Erodium glaucophyllum, Fagonia mollis. Paronychia desertorum* and *Polycarpaea repens*. During the rainy season this layer is enriched with the growth of therophytes.

Within the coastal land of the Gulf of Suez, the *H. elegans* comunity is abundant on the gravel terraces of the wadis. In a few localities, e.g. Wadi

Hommath, *Hammada* builds hillocks that may reach considerable size, but the formation of such sand hillocks by *Hammada* is exceptional. Usually, the growth of *H. elegans* in the Egyptian desert is taken to indicate conditions where the softer deposits are gradually removed leaving the coarse lag materials (El-Abyad, 1962).

Perennial Grassland Forms

4. *Panicum turgidum* community. The *P. turgidum* community is one of the most common within the desert of Egypt, especially on sandy formations. It is also one of the most extensively grazed. *P. turgidum* is a good fodder plant. There is also evidence that it is one of the wild grain-plants which may be used as a supplementary food resource for desert inhabitants. Zohary (1962) states "Durkop (1903) reports that the grain of *P. turgidum* is collected in the Sahara by the natives of southern Algeria, ground and baked into bread."

   P. turgidum is a tussock-forming grass that may acquire an evergreen habit, especially in favorable environments. Under less favorable conditions it may have a strictly deciduous growth form and remain dry and look dead for a prolonged period (a few years) but regain its green habit after the rain.

   The vegetation, though distinctly a grassland type, shows the usual three layers. The frutescent layer includes bushes (*Acacia tortilis, Leptadenia pyrotechnica, Lycium arabicum* and *Maerua crassifolia*) and trees e.g. *Acacia raddiana, Balanites aegyptiaca* and *Calotropis procera*. The suffrutescent layer is made up of the dominant grass and a range perennial associates, e.g. *Hammada elegans, Launaea spinosa, Pituranthos tortuosus, Salsola baryosma, Sporobolus spicatus* and *Zilla spinaosa*. The ground layer includes several dwarf or prostrate perennials e.g. *Citrullus colocynthis, Fagonia mollis* and *Polycarpaea repens*. Therophytes, e.g. *Aristida adscensionis, Brachiaria leersioides, Caylusea hexagyna, Eragrostis ciliaris, Euphorbia scordifolia, Stipagrostis hirtigluma* and *Zygophyllum simplex* form dense populations in especially favoured localities or in years with high rainfall, with cover, between the perennials, that may reach 40–60%.

   Widespread in the Red Sea coastal desert, the *P. turgidum* community is made up of a rather large number of associate species having different geographical ranges within this coastal desert. For instance, *Acacia tortilis, Maerua crassifolia* and *Salsola baryosma* are present in the southern section of the coastal plain only whereas *Hammada elegans* is confined to the northern section.

   P. turgidum is mostly a sand-dwelling grass. Its growth, if not excessively grazed, may build up sand mounds and hillocks. It is an effective sand-binding xerophyte which also grows on mixed sand and gravel deposits but is uncommon on limestone detritus.

   The *P. turgidum* grassland is widespread throughout the African Sahara (Zohary, 1962). In the arid desert of the Arabian Peninsula P. *turgidum* is one of the most common grassland types (Zahran, 1982b).

5. Other grassland communities. Reference has been made to two grassland types: the *Pennisetum dichotomum* community and the *Lasiurus hirsutus* community.

The former is abundant in the wadis of the limestone country and is associated with salt terraces. A detailed description of this community is given in the next section of the inland part of the Eastern Desert. Within the Red Sea coastal desert, a few stands of this grassland type are recorded in the wadis dissecting limestone formations. *Pennisetum dichotomum* is a tussock-forming grass similar in habit to *Panicum turgidum*. Without the inflorescence it is often difficult to distinguish between them. Both are grazed though *Pennisetum* seems to be less palatable.

Lasiurus hirsutus is apparently more drought-tolerant than both *P. turgidum* and *P. dichotomum*. It grows in smaller runnels with smaller catchment areas and hence less water resources. Within the Red Sea coastal desert, there are a few stands dominated by *L. hirsutus* associated with small runnels dissecting the sand gravel formations of the coastal plain.

There is a fourth grassland type dominated by *Hyparrhenia hirta* which is very rare in the Red Sea coastal desert and in the inland desert as well. The dominance of *H. hirta* is confined to some of the fine upstream runnels dissecting the limestone formations. Associated species are chasmophytic and lithophytic species such as *Fagonia mollis* and *Helianthemum kahiricum*. Rattray (1960) in his survey of the African grassland, described *H. hirta* as one of the Cape species which is also found along the coast of North Africa. *H. hirta* is of considerable importance in the southwestern Red Sea coastal mountains of Saudi Arabia. Its dense growth and dominance have been recorded in nost of the protectorates of these mountains (Zahran and Younes, 1990). It is also recorded in other parts of the Middle East (Zohary, 1962).

Woody Perennial Forms

6. *Zilla spinosa* community. *Z. spinosa* is a spinescent woody undersrrub. It is normally a perennial xerophyte which under favourable conditions grows as an evergreen that flowers throughout most of the year. Under less favourable conditions it acquires a deciduous growth form. It may behave as an annual under extreme conditions. Individuals of *Z. spinosa* may reach considerable size with an area of 1–1.5 m^2 and height up to 80–170 cm. It may also be a small woody plant not exceeding 20–30 cm. A species with a wide range of habit and size is a valuable indicator of habitat conditions.

   Within the Red Sea coastal desert, the plant cover of the *Z. spinosa* community is usually low (5–20%) and its individuals are normally small. Within the southern section, *Z. spinosa* very often acquires a deciduous or an annual growth form. *Artemisia judaica* and *Cleome droserifolia* are the most common associates. The abundance of *C. droserifolia* is a character that distinguishes the *Z. spinosa* community of the Red Sea coastal land from the comparable community of the inland Eastern Desert where *C. droserifolia* is a rarity. *Aerva javanica, Francoeuria crispa, Leptadenia pyrotechnica, Pulicaria undulata* and *Zygophyllum coccineum* are common associates. Less common species include *Acacia flava, A. raddiana, Hammada elegans, Iphiona mucronata, Lasiurus hirsutus, Launaea spinosa, Lycium arabicum, Lygos raetam, Panicum turgidum, Pennisetum dichotomum* and *Zygophyllum decumbens*.

The vegetation shows the usual three layers. The dominant species and a great variety of associated perennials form the frutescent layer which contributes the main bulk of the plant cover. The frutescent layer is often very thin and may include one or more of the following bushes: *Acacia flava, A. tortilis, Leptadenia pyrotechnica, Lycium arabicum, Lygos raetam, Maerua crassifolia* and *Moringa peregrina* & trees, e.g. *Acacia raddiana* and *Balanites aegyptiaca*. The ground layer includes prostrate perennials such as *Citrullus colocynthis, Cucumis prophetarum, Fagonia parviflora, Monsonia nivea* and *Robbairea delileana*. The ephemerals enrich this layer during the rainy season. The community dominated by *Z. spinosa* is confined to the channels of the wadis and is absent from other smaller runnels. It usually occurs in parts of the wadi bed that are covered with alluvial deposits. The size and growth form of the dominant vary in obvious relationships with texture and depth of these surface deposits. In contrast to some other communities described, this community is made up of xerophytes only and none of the halophytes.

7. *Launaea spinosa* community. *L. spinosa* (Compositae) is a milky sap xerophyte having a growth form comparable to that of *Zilla spinosa* (Cruciferae); both are spinescent with individuals ranging from small woody plants to bushy undershrubs of considerable size. Also, both may develop an evergreen growth form under favourable conditions and a summer deciduous growth form under less favourable conditions. The two communities dominated by *Z. spinosa* and *L. spinosa* are confined to the channels of the wadis and their main tributaries. However, the community dominated by *L. spinosa* is restricted to the northern part of the Red Sea coastal desert, i.e. the coastal desert of the Gulf of Suez. It does not extend westward into the Eastern Desert.

   The plant cover of the *L. spinosa* community ranges between 10 and 25% and is contributed mainly by the dominant and the abundant associate, *Zilla spinosa*, which is recorded in all stands (Zahran, 1962). Other most common associates are *Artemisia judaica, Cleome droserifolia, Crotalaria aegyptiaca, Echinops galalensis, Iphiona mucronata, Lasiurus hirsutus, Lavandula stricta, Lygos raetam, Matthiola livida, Pennisetum dichotomum, Pituranthos tortuosus, Tamarix nilotica* and *Zygophyllum coccineum*.

   The three layers typical of the desert ecosystem communities are represented here. The most important layer is the suffrutescent one which contains the dominant species and the most common associates, The ground layer includes a number of perennials, e.g. *Asteriscus graveolens, Citrullus colocynthis* and *Helianthemum lippii* and the therophytes. The frutescent layer is usually thin and includes one or more of the following trees and shrubs: *Acacia raddiana, Lycium arabicum, Lygos raetam* and *Tamarix nilotica*.

8. *Cleome droserifolia* community. *C. droserifolia* is a low, much branched, densely glandular-hispid undershrub with small orbicular three-nerved leaves. It forms low cushions of spreading growth that may not exceed 30 cm in height; it occurs in the ground layer.

In Egypt, *C. droserifolia* is present in all regions except the Mediterranean coastal desert (Täckholm, 1974). It is recorded in the southern section of the Saudi Red Sea coastal land (Zahran, 1982b).

The dominance of *C. droserifolia* in the Egyptian Red Sea coastal desert is confined to the larger runnels; it does not occur in the channel of the main wadis covered with very coarse detritus including large boulders. The dominant may accumulate pads of soft material and it grows in isolated patches confined to particular positions among the coarse boulders.

Ecologically, the community dominated by *C. droserifolia* is associated with the limestone country of the Red Sea coastal desert (Zahran, 1964). *Zilla spinosa* and *Zygophyllum coccineum* are the most common associates. *Fagonia mollis* and *Francoeuria crispa* are frequently present whereas less common associates include *Acacia flava, A. raddiana, A. tortilis, Artemisia judaica, Hammada elegans, Iphiona mucronata, Launaea spinosa, Leptadenia pyrotechnica, Lycium arabicum, Lygos raetam, Panicum turgidum, Pituranthos tortuosus* and *Zygophyllum decumbens*.

Stratification of vegetation is clear in this community. The frutescent layer is extremely thin. The suffrutescent layer is made up of a variety of species including the most common associates mentioned above. The dominant is in the ground layer which includes also *Citrullus colocynthis, Corchorus depressus, Cucumis prophetarum* and *Fagonia mollis*. This layer is enriched by therophytes during the rainy season.

9. *Sphaerocoma hookeri* community. S. *hookeri* is a glabrous, blue-green, densely branched shrub (60–70 cm high), with knotty branches, and fleshy, terete, opposite or whorled leaves. Its presence in Egypt is confined to the Red Sea coastal desert: S. *hookeri* ssp. *intermedia* is rarely recorded in the southern region of the inland part of the Eastern Desert (Täckholm, 1974).

   The community type dominated by *S. hookeri* is present in the southern part of the Egyptian Red Sea coastal desert (south of the Tropic of Cancer). It grows in a special type of habitat – sand dunes formed at the boundaries between the littoral salt marshes and the coastal desert plain. These dunes may also form embankments covering the edges of the desert plain in places where a low plateau rises abruptly at the border of the low ground of the littoral belt. Thus, the associate species of this community include halophytes, e.g. *Cyperus conglomerate, Limonium axillare, Salsola vermiculata, Sevada schimperi* and *Sporobolus spicatus* and xerophytes, e.g. *Acacia tortilis, Asthenatherum forsskaolii, Heliotropium pterocarpum, Lycium arabicum, Monsonia nivea, Panicum turgidum, Polycarpaea repens* and *Salsola baryosma* and also ephemerals, e.g. *Aristida* spp.

   The three common layers of vegetation can easily be recognized in this community. The suffrutescent layer includes the dominant and most of the associates, but the frutescent layer is thin, including the shrubs e.g. *Acacia tortilis*. The ground layer comprises *Aristida, Cyperus, Eragrostis, Monsonia, Sporobolus* etc.

10. Other communities. There are a number of communities dominated by woody perennials that are less widespread in the Red Sea coastal desert. These include

communities dominated by each of: *Iphiona mucronata, Artemisia judaica, Pituranthos toriuosus* and *Calligonum comosum.*

The community dominated by *Iphiona mucronata* is confined to smaller affluent wadis or the upstream extremities of greater tributaries traversing the limestone country. The most abundant associates include: *Gymnocarpos decander, Launaea spinosa, Zygophyllum coccineum* and *Z. decumbens.*

The *Artemisia judaica* community is present in a few localities within the wadis. It is apparently confined to conditions where the surface deposits are a mixture of alluvial limestone detritus and aeolian sand. Such mixtures are particularly those which traverse and drain the limestone and sandstone formations. The most common associates of this community include *Calligonum comosum, Ephedra alata, Farsetia aegyptia, Hammada elegans* and *Zilla spinosa.* This community is widespread in the mountainous country of southern Sinai (Migahid et al., 1959).

The community dominated by *Pituranthos tortuosus* occurs in a few of the runnels dissecting the gravel plain. Associate species include *Artemisia judaica, Fagonia mollis, Iphiona mucronata, Lavandula stricta* and *Lindenbergia sinaica.*

The *Calligonum comosum* community is associated with sand aeolian deposits. The dominant forms sandy mounds and hillocks. Associate species include: *Artemisia judaica, Francoeuria crispa, Hammada elegans, Tamarix aphylla* and *Zilla spinosa.*

Frutescent Perennial Forms

The frutescent perennial vegetation includes the scrubland types of the desert vegetation. The plant cover is in three layers: a frutescent (120–500 cm), suffrutescent (30–120 cm) and a ground layer (<30 cm). The frutescent layer is here dense enough to give the vegetation characters that distinguish it from the previous categories. Two main forms may be recognized in this type: the succulent shrub form and the scrubland form. The former is not represented in the Red Sea coastal desert but a type dominated by a succulent xerophytic bush, *Euphorbia candelabra*, is recorded from one part of the Red Sea coastal desert of the Sudan some 50–70 km south of the Egyptian coast (Zahran, 1964). The scrubland form is represented by the communities dominated by *Acacia raddiana, A. tortilis, Lycium arabicum* and *Tamarix aphylla* and several less common types. In these communities the vegetation is well developed as is evident from the long list of associate species. All these communities are present in the channels of the main wadis and their larger tributaries. The distribution of the *Acacia* scrubland types extends over wadis traversing the coastal plain and their upstream parts across the mountain area. The description of these communities, given below, depends on stands representing their growth within the two ecosystems. The vegetation of these communities, especially that dominated by *A. raddiana*, are (and for a long history have been) subject to extensive lumbering for fuel and charcoal manufacture. Camel breeding and charcoal manufacture are the main industries of the local inhabitants, especially in the Gebel Elba district (Zahran, 1964).

11. *Lycium arabicum* community. *L. arabicum* (= *L. shawii* var. *shawii*) is a spinescent shrub which may develop an evergreen growth form under favourable conditions of water resource. Growing in less favourable habitats it is deciduous, shedding its leaves early in the season and the main part of its shoot may also dry up. Within the thick growth of Gebel Elba scrubland, *L. arabicum* may show a climbing habit.

    Within the Egyptian Red Sea coastal desert the *L. arabicum* community is particularly common in the southern part. In the coastal plain extending from mountain ranges to the littoral salt marsh belt this community is one of the most common of the wadis and the deltas of the main channels. It may also dominate the channels of the large tributaries. The plant cover is in the ranges of 10–50%. The dense cover is due to the rich growth of the therophytes which almost completely cover the gaps between the perennial vegetation. It is estimated that the cover in some stands is 70–90% contributed by the ephemerals and 20% by the perennials. In other stands the cover of the ephemerals is slightly higher (40–50%) than that of the perennials (30–40%). In some other stands the cover is mainly perennial and the ephemerals contribute very little to this.

    In the *L. arabicum* community the abundant associates are *Acacia tortilis, Panicum turgidum* and *Salsola vermiculata* (perennials) and *Euphorbia scordifolia* and *Zygophyllum simplex* (annuals). Less common species include *Acacia raddiana, Aristida adscensionis, A. funiculata, A. meccana, Asphodelus tenuifolius, Balanites aegyptiaca, Calotropis procera, Launaea cassiniana, Leptadenia pyrotechnica, Lotononis platycarpa, Polycarpaea repens, Salsola baryosma* and *Stipagrostis hirtigluma.*

    Stratification of vegetation is clear in this community. The frutescent layer includes the dominant and several shrubs and trees, e.g. *Acacia raddiana, A. tortilis. Balanites aegyptiaca* and *Lygos raetam.* This layer contributes the bulk of the perennial cover and forms the main part of the permanent framework which gives the character to the vegetation. The suffrutescent layer, though made up by the majority of the perennials, including some of the most common associates, contributes little to the cover. In the ground layer there are several perennials and almost all the ephemerais. In years of good rainfall ephemerais form patches of dense growth between the components of the other layers.

12. *Acacia tortilis* community. *A. tortilis* is a flat-topped or umbrella-shaped spinescent shrub. The size varies from dwarf shrubs to much larger ones. The *A. tortilis* community is a most common scrubland type within the desert area extending to the south of Mersa Alam but it is absent from the stretch from Mersa Alam northward to Suez (c.700 km). There is, however, a limited locality along the Suez-Ismailia desert road (16–17 km north of Suez) on the Suez canal west bank where there are a few patches of *A. tortilis* scrubland.

    The *A. tortilis* community occurs in a variety of habitats. It may be seen on the slopes of the low hills at the northern and eastern foot of the mountains, e.g. Gebel Elba, and the similar coastal mountains. The most common habitat is the channels of the main wadis of the larger tributaries. Unlike the *A. raddiana* community which occurs on the softer deposits of the wadi channels, the A. *tortilis* community is present on the coarse deposits.

The vegetation cover of the A. *tortilis* community is primarily formed by the canopy of the dominant. In a few instances, the most common perennial associates, e.g. *Lycium arabicum* and *Panicum turgidum*, and therophytes, e.g. *Aizoon canariense, Aristida adscensionis* and *Zygophyllum simplex*, contribute substantially to the cover of the permanent framework. Other associates include: *Acacia raddiana, Aerua javanica* (*A. persica*, Täckholm, 1974), *Balanites aegyptiaca, Indigofera spinosa, Maerua crassifolia, Moringa peregrina, Salsola vermiculata* and *Zilla spinosa*.

By virtue of the growth form of the dominant species, the frutescent layer is the most notable part of the vegetation. This layer includes trees, e.g. *Acacia raddiana* and *Balanites aegyptiaca*, and shrubs, e.g. *Leptadenia pyrotechnica, Lycium arabicum* and *Maerua crassifolia*. Most of the associates, e.g. *Aerva javanica, Panicum turgidum, Salsola vermiculata* and *Zilla spinosa*, are in the suffrutescent layer. The main bulk of the ephemerals are in the ground layer.

13. *Acacia raddiana* community. A. *raddiana* is one of the most common and widespread plants of the Eastern Desert. It is recorded in almost all the main wadis of this desert, including the Red Sea coastal desert. Comparison of the distribution of the two *Acacia* species, *A. raddiana* and *A. tortilis* shows that the former is much more widespread all over the Eastern Desert whereas the latter is mostly confined to the southern part. But, where the two species grow together, there is evidence that A. *tortilis* is the more drought tolerant. A study of the wadis of the Abu Ghusson district (Lat. 24°20′N, Long. 35°10′E) showed that most of the very abundant A. *raddiana* shrubs were dry, dead or almost dead whereas shrubs of A. *tortilis* were thriving almost normally. This was repeatedly observed in a number of wadis where the habitat was subject to a spell of rainless years (Zahran, 1964).

    In the main wadis draining the Gebel Elba area, the two *Acacia* scrubland types may be present. A *tortilis* shrubland covers the gravel terraces and the A. *raddiana* open forest is in the channels which contain ephemeral streams and where the deposits are much softer. In these localities A. *raddiana* shows some features of forest growth such as the presence of lianas (*Cocculus pendulus*, and *Ochradenus baccatus*) and parasites, e.g. *Loranthus acaciae* and *L. curviflorus*.

    It may be added that A. *raddiana* scrub and open forest is confined to the channels of the main wadis and is not present on the hills and mountain slopes where other *Acacia* spp. (A. *etbaica, A. laeta, A. mellifera* and A. *tortilis*) may grow. This has been also observed in the Saudi Red Sea coastal desert (Zahran, 1993a).
14. *Tamarix aphylla* community. *T. aphylla* is a species that normally grows as a tree that may reach considerable size. It may form a dense open forest that represents one of the main climax communities of the desert wadis. Under the influence of cutting, grazing and other destructive agencies it acquires a bushy growth form that covers the ground in patches. In localities subject to the accumulation of wind-borne sand it may form dunes which it covers.

In the *T. aphylla* community of the Egyptian Red Sea coastal desert the cover ranges between 5 and 50%. It is mainly contributed by the dominant and the ephemeral vegetation during the rainy years. The most common associates include *Acacia raddiana, Balanites aegyptiaca, Calligonum comosum, Farsetia aegyptia, Francoeuria crispa, Hammada elegans, Iphiona mucronata, Lycium arabicum, Ochradenus baccatus, Salvadora persica, Zilla spinosa* and *Zygophyllum coccineum.*

The flora of this community is of a variety of species with different; ecological amplitude. For instance, *Acacia raddiana, Balanites aegyptiaca and Tamarix aphylla are* trees that require ample water resources and deep valley-fill deposits and that represent the climax stage in the wadi-bed development. *Hammada elegans* and *Zygophyllum coccineum* are succulents that withstand different conditions. *Gymnocarpos decander* and *Iphiona mucronata* represent pioneer stages in the wadi-bed development and occur in localities where the surface deposits covering the bed rocks are very thin.

T. aphylla is present in the whole stretch of the Egyptian Red Sea coastal desert. Because its wood is softer than that of the other trees (*Acacia* and *Balanites*) of the area, its scrublands are apparently the first vegetation to be destroyed by cutting. This may be one of the main causes for the limited dominance of *T. aphylla* in the coastal desert. Relict patches of *T. aphylla* forest are usually seen in the main wadis. Climatic changes affect trees of *T. aphylla* more adversely than other trees.

15. Other scrubland communities. Reference may be made to four scrubland types that are occasionally found within the Red Sea coastal desert. One is dominated by *Acacia ehrenbergiana.* It is usually associated with valley-fill deposits that include considerable proportions of soft material that make the deposits compact and difficult to excavate.

A second community is dominated by *Capparis decidua.* It is recorded in the wadis of the southern region where the dominant may form large patches of bushy growth that may build sand hummocks. *C. decidua* may also have a tree growth-form. This community is usually associated with soft deposits that may be alluvial or aeolian.

The *Leptadenia pyrotechnica*, community is more widespread in the Egyptian Red Sea coastal desert than the two previously mentioned communities. Like all the other scrubland types, this type is confined to the channels of the main wadis. It is usually associated with wadi terraces of mixed deposits.

The *Salvadora persica* community is recorded in a few of the main wadis, e.g. Wadi Bali. The dominant forms patches of growth that show the influence of repeated cutting. S. *persica* is the tooth-brush tree; if saved from cutting, it forms trees with upright trunks. In the mid-part of Wadi Bali, *S. persica* is associated with *Artemisia judaica, Cleome droserifolia, Pulicaria undulata, Zygophyllum coccineum* etc. (Zahran, 1964). In the Saudi Arabian Red Sea coastal desert, *S. persicA* (Arak in Arabic) grows densely in Wadi Arak (Zahran, 1982b). The dense growth of S. *persica* in the Sudanese Red Sea coastal desert has also been mentioned by Kassas (1957).

A fifth community is an open forest dominated by *Balanites aegyptiaca*. Patches of this are present in a few of the main wadis especially within the mountain ranges. These patches are clearly relicts of a much more widespread growth. Trees of *B. aegyptiaca* are recorded in almost all the wadis of the southern section. The fleshy fruits are collected and eaten by the bedouins.

(iii) The Coastal Mountains

General Features

The Red Sea coastal land of Egypt is bounded inland by an almost continuous range of mountains and hills. This range, as already mentioned, has a number of high peaks. It forms the natural divide between the eastward drainage (to the Red Sea) and the westward drainage (across the Eastern Desert to the Nile). The presence of this coastal range has apparently influenced the climate and the water resources of the Eastern Desert. In the words of Murray (1951):

Regular rainfall ceased over Egypt below the 500-m contour sometime about the close of the Plio-Pleistocene period, three-quarters of a million years ago and, though torrents from the Red Sea hills have been able to maintain their courses to the Nile through the foothills of the Eastern Desert, the Libyan (Western) Desert has ever since been exposed to erosion by the wind alone.

The influence of orographic precipitation is hardly noticeable within the coastal mountains of the Gulf of Suez. The expanse of water of the Gulf supplies the dry winds crossing from Sinai with appreciable amounts of moisture. By contrast, the coastal mountains of the Red Sea proper show the influence of orographic rain. The amount of this rain, judging by the type of vegetation, varies according to altitude and distance from the sea. "So abundant the vegetation in all the wadis draining from Elba, that it is impossible to approach the mountains very close with loaded camels, owing to the closeness of the trees" (Ball, 1912). He adds "of the ten days I remained on the summit in April and May 1908 only three days were clear".

The Red Sea coastal mountains of Egypt may be categorized into:

1. Mountains facing the Gulf of Suez
2. Mountains facing the Red Sea proper.

The northern mountain is Gebel Ataqa and the southernmost is Gebel Asotriba. The highest is Gebel Shayeb El-Banat (2187 m) some 40–50 km west of Hurghada (Lat. 27°14′N) at the mouth of the Gulf of Suez.

Mountains Facing the Gulf of Suez

Geology and Geomorphology

The coastal chain of mountains included within the west coast of the Gulf of Suez, between Suez and Hurghada (400 km), is made up of the predominantly limestone plateau blocks of Ataqa, Kahaliya, Akheider, the Galala El-Bahariya, the Galala El-Qibliya and the chain of basement complex bounded on the north by Gebel Gharib and on the southern border by the mass of Gebel Abu Dukhan, Gebel Qattar and Gebel Shayeb El-Banat.

The limestone block is a table-like plateau dissected by a number of the drainage systems. The vegetation is clearly associated with the type of landform but is, except for lichens, confined to the drainage runnels.

The basement complex mountains are jagged masses of igneous rocks with peaks rising to 1700 m or more. The landform-types are different and the slopes are dissected by shallow runnels with precipitously sloping channels. The beds of these channels are as a rule covered with massive blocks and boulders. These runnels flow into the drainage lines at the foot of the mountains (Sadek, 1926, 1959).

Gebel Ataqa (c.817 m) is a mountain block covering an area about 300 km^2. It is bounded on its north and east margins by precipitous cliffs dropping abruptly for 300–400 m. To the west and south it slopes gradually. This rocky massif is dissected by water runnels of drainage systems each with a main channel receiving water from branching affluents. The drainage runnels of the northern side flow into Wadi Bahara which runs parallel to the Cairo-Suez desert road. The runnels of the south-western and western slopes drain into Wadi Hagul. In the southwest there are a few wadis representing a gradual increase in the catchment area, e.g. Wadi Hommath.

Gebel Kahaliya Ridge is of limestone made up of Gebel Kahaliya (660 m) and Gebel Umm Zeita (250 m). The drainage runnels of the southwest slopes of the former feed Wadi El-Bada whereas those of the second go to Wadi Hagul.

Gebel Akheider Ridge forms an eastward extension of the Eocene plateau of the Eastern Desert. It comprises Gebel Noqra (436 m), Gebel Akheider (367 m) and Gebel Ramiya (300 m). The topographic pattern causes the main bulk of the drainage of this ridge to feed the affluents of Wadi El-Ghweibba.

Gebel El-Galala El-Bahariya (Galala-north) is the greatest massifi block which forms one of the most notable topographical features on the west side of the Gulf of Suez. The north edge is bounded by steep cliffs extending for about 60 km from east to west and rising to 977 m near its eastern (seaward) end and about 700 m near its western extremity. Most of this mountain is of Middle Eocene limestone. On the east the Galala cliffs face the Gulf of Suez. The plain separating them is very narrow. The runnels of the southern side of this mountain drain into Wadi Araba.

At the foot of Khashm El-Galala issues a spring of slightly warm sulphuretted water of very brackish nature – “Ain Sokhna”. Most probably it owes its origin to the fault at the foot of the mountain block.

Gebel El-Galala El-Qibliya is a mass of Eocene limestone which rises to over 1200 m in the east and slopes gradually to merge into the limestone plateau of the Eastern Desert. The runnels of the northern side of this mountain drain into Wadi Araba.

The Gebel Shayeb El-Banat group constitutes four mountains, namely, from north (Lat. 27°20′N) to south (Lat. 26°5′N): Gebel Abu Dukhan (1705 m), Gebel Qattar (1963 m), Gebel Shayeb El-Banat (2187 m) and Gebel Umm Anab (1782 m). Gebel Shayeb El-Banat is the highest peak within the Red Sea coastal mountains. These are the igneous blocks that form a range extending some 40–50 km to the west of Hurghada. The Gebel Shayeb group thus form hills facing the southern part of the Gulf of Suez and the northeast part of the Red Sea.

Plant Life

The notable difference between the limestone plateau and the basement complex mountains is that the water resources available for plants in the former is mainly the run-off water of the convectional rainfall, whereas that in the latter includes also orographic condensation of cloud moisture. The difference causes the pattern of the vegetation within the limestone plateau country to be such that the lower the altitude of the habitat the less arid it is, as it receives a greater proportion of drainage. This is not necessarily the case within the jagged mountains of the basement complex: high up the mountains, the vegetation may indicate habitat conditions less arid than those lower down the slope. Reference may be made in this respect to Troll (1935) and Kassas (1956, 1960).

Within the coastal hills of the Gulf of Suez, the vegetation is confined to the upstream part of the drainage system and to the slopes of these hills. The water courses are usually well defined in the hill country. Across the coastal plain, on the other hand, the courses are usually ill-defined runnels within the much wider courses of the wadi.

Plant cover varies in relation to the extent of the area and the texture of the bed cover. Several plant communities may be recognized; some are common in both coastal ecosystems (desert and mountains) and some are confined to the hill ecosystem. *Acacia raddiana. Anabasis articulata, Artemisia judaica, Hammada elegans, Launaea spinosa, Leptadenia pyrotechnica, Lygos raetam* and *Panicum turgidum* belong to the first category. The *Zilla spinosa* and *Zygophyllum coccineum* communities arc common in the wadis of the limestone hills.

There are a few plants that characterize the cliffs and dry waterfalls that intercept the courses of the wadis traversing the hills. These are *Capparis cartilaginea, C. spinosa, Cocculus pendulus* and *Ficus pseudosycomorus*. Reference may also be made to the slopes of the Cretaceous limestone hills which include *Anabasis articulata, Halogeton alopecuroides, Heliotropium pterocarpum* and *Salsola tetrandra* together with such common species as *Hammada elegans, Ochradenus baccatus* and *Pergularia tomentosa*.

The Communities

Within the drainage system of the mountains facing the Gulf of Suez, two main communities may be recognized. One is dominated by *Zilla spinosa* and is widespread within the channels of the wadis and the second is characterized by the preponderance of *Moringa peregrina* and is confined to the upstream parts of the wadis draining the slopes of the higher mountains.

1. *Zilla spinosa* community. *Z. spinosa*, the most abundant plant in the majority of the wadis, acquires, in this district, a distinctly deciduous growth form. The shoot is dry and plants often appear to be dead. In rainy years, which are not of regular occurrence, plants are profusely regenerated. It is suspected that

Z. spinosa is here a particular variety (*Z. spinosa* v. *microcarpa*; Täckholm, 1974) or an ecotype (potential annual, Zahran, 1964). This requires further ecological and taxonomic studies.

Common perennial associates are *Aerva javanica, Artemisia judaica, Calligonum comosum, Cleome droserifolia, Fagonia mollis, Leptadenia pyrotechnica, Solenostemma arghel* and *Zygophyllum coccineum*. Abundant and common ephemerals are *Aizoon canariense, Arnebia hispidissima, Asphodelus tenuifolius, Ifloga spicata, Lotus arabicus, Reichardia orientalis, Robbairea delileana* and *Senecio flavus*.

In certain parts of these wadis *Acacia raddiana* is locally dominant: there are patches of A. *raddiana* scrubland which are apparently relicts of better growth that has been destroyed. *Acacia* scrubland is presumed to represent the natural climax vegetation of these wadis.

2. *Moringa peregrina* community. *M. peregrina* is one of the most interesting plants in the mountain ranges of the Red Sea coastal land. It is a 10–15 m high tree, [with white bark] usually destitute of leaves. These, when present, consisting of 3 pairs of long, slender, junciform pinnae, looking like opposite virgate branchlets. The pendulous pods ripen in October, the angled nutlike, white seeds [behen-nuts] are of a bitter-sweet nauseous taste and rich in oil (ben-oil) (Täckholm, 1974). The behen-nuts are collected by the local natives and sold at a good price. The ben-oil of these seeds is used for special lubrication purposes. This particular attribute has saved this plant which is too valuable to be cut for fuel.

   The *Moringa* scrub is represented by patches that cover limited areas of the upstream runnels of the drainage systems. These are runnels collecting water at the foot of the higher mountains. A survey of *Moringa* within the Red Sea mountains extending from Lat. 27°20′N to 22 °N (Table 4.1) shows that this species is confined to the foot of the mountains that are higher than 1300 m. Lower mountains and hills have almost no *Moringa* at their foot. The ground where *Moringa* grows is usually covered with coarse rock detritus, a character typical of the upstream runnels at the foot of the mountains.

   The usual association of *Moringa* with coastal mountains more than 1300 m high is not a sharp limit, since mountains nearer to the coast are better favoured than those further from the coast. Within the mountain range of Hurghada, the *Moringa* community contains the following xerophytic associates: *Acacia raddiana, Aerva javanica. Artemisia judaica, Capparis cartilaginea, C. decidua, Chrozophora plicata, Cleome droserifolia, Fagonia mollis, Francoeuria crispa, Hyoscyamus muticus, Launaea spinosa, Lavandula stricta, Leptadenia pyrotechnica, Lindenbergia sinaica, Lycium arabicum, Ochradenus baccatus, Periploca aphylla, Zilla spinosa* and *Zygophyllum coccineum*.

The Nakkat Habitat

The convectional rainfall of the Gulf of Suez coastal area, according to the average rainfall of Hurghada, is 3 mm a year (Anonymous, 1960). But the vegetation and

Table 4.1 Mountains within the Red Sea coastal land of Egypt showing altitude and the presence (+) or absence (–) of *Moringa peregrine*

Gebel	Altitude (m) and occurence of *Moringa*	Gebel	Altitude (m) and occurence of *Moringa*
Abu Harba	1705 (+)	Umm Laseifa	1210 (–)
Abu Dukhan	1661 (+)	Nugrus	1504 (+)
Abu Guruf	1099 (–)	Hafafit	857 (–)
Qattar	1963 (+)	Zabara	1360 (+)
Shayeb El-Banat	2187 (+)	Miqif	1198 (–)
Umm Anab	1782 (+)	Hashānib	1133 (–)
Abu Fura	1032 (–)	Abu Hamamid	1745 (+)
Weira	1035 (–)	Samiuki	1486 (+)
Mitiq	1112 (–)	Hamata	1977 (+)
El-Sibai	1484 (+)	Elba	1428 (+)
Abu Tiyur	1099 (–)	Shindodai	1426 (+)

the water resources in the mountain area indicate greater precipitation. In the wadis running at the foot of the mountains there are several shallow wells of fresh water. On the slopes or cliffs of the mountains there are cracks from which a continuous trickle of water oozes ("nakkat" is Arabic for "dropper") and runs down the slopes, water collecting in a pot-hole at the foot of the slope forming a "bir" (well) or in some parts forming hollows (gelts) along the slope. The source of the nakkat is often a fissure in the solid basement complex rocks of the mountains, usually situated near the top. The courses of the runnels dissecting the slopes of the mountains may contain pot-holes that are periodically filled with water. These pot-holes are usually lined with calcareous skin material dissolved in the water that collects in the pot-hole. In this peculiar habitat of these trickles, ferns e.g. *Adiantum capillus-veneris*, mosses and algae, alien strangers of the desert environment, grow. Associates are such water-loving plants as *Imperata cylindrica, Phragmites australis, Solanum nigrum* and *Veronica beccabunga*.

The nakkat habitat is also typical for *Ficus pseudosycomorus*. Stunted individuals of *Phoenix dactylifera* occur in many nakkats, hanging from the top or near the source. The wet areas that fringe the birs and gelts are often covered with a rich growth of *Cynodon dactylon, Cyperus laevigatus, Imperata cylindrica* and *Juncus rigidus*.

The restriction of *Moringa* to the foot of the higher mountains and the presence of nakkats as features peculiar to such mountains indicate that the high altitude leads to greater water resources.

In one of the runnels across the eastern slope (facing the sea) of Gebel Shayeb El-Banat, the following plants have been recorded: *Acacia raddiana, Aerva javanica, Artemisia inculta, A. judaica, Capparis cartilaginea, Chrozophora oblongifolia, Citrullus colocynthis, Cleome droserifolia, Fagonia mollis, Francoeuria crispa, Hyoscyamus muticus, Lavandula stricta, Lindenbergia abyssinica, L. sinaica, Moringa peregrina, Periploca aphylla, Pulicaria undulata, Solenostemma arghel, Teucrium leucocladum, Zilla spinosa* and *Zygophyllum coccineum*.

Mountains Facing the Red Sea Proper

These include the groups of mountains that extend between Lat. 24°50′N and 22°N on the Sudano-Egyptian border and comprise: Gebwl Nugrus group, Gebel Samiuki group and Gebel Elba group (Kassas and Zahran, 1971).

Gebel Nugrus Group

This group, which extends between Lat. 24°40′N and 24°50′N, faces the full stretch of the Red Sea. The highest of this group is Gebel Nugrus "... a great boss of red granite rising to a height of 1505 m among schist and gneisses" (Ball, 1912). This group includes Gebel Hafafit (857 m), Gebel Migif (1198 m), Gebel Zabara (1360 m) and Gebel Mudargag (1086 m).

In the district of Gebel Nugrus the growth of ephemerals in the wadis and on the rills across the mountain slopes is usually rich during the wet years.

The growth of *Moringa* characterizes the foot of the high mountains (above 1300 m) but not the lower ones. The main wadis draining this group are the habitat of various types of open scrub dominated by one of the following: *Acacia raddiana, A. ehrenbergiana, Balanites aegyptiaca, Leptadenia pyrotechnica* and *Salvadora persica*. The *A. raddiana* scrub is confined to wadis draining westward. The *B. aegyptiaca, L. pyrotechnica* and S. *persica* scrublands are less common and are mostly confined to the channels of the main wadis.

Gebel Samiuki Group

This group comprises Gebel Abu Hamamid (1745 m), Gebel Samiuki (1283–1486 m) and Gebel Hamata (1977 m). The last is the nearest to the sea (40 km) whereas Gebel Abu Hamamid is furthest away (65 km).

The vegetation in the Gebel Hamata area is much richer in species and in plant cover than that in the Gebel Samiuki area. However, the flora of this group as a whole is richer than that of the Gebel Nugrus group to its north which is again richer than that of the Gebel Shayeb group still further northward.

Gebel Elba Group

This is an extensive group of granite mountains situated on the Sudano-Egyptian border (Lat. 22°N) and includes Gebel Elba (1428 m), Gebel Shindeib (1911 m), Gebel Shindodai (1426 m), Gebel Shillai (1409 m), Gebel Makim (1871 m) and Gebel Asotriba (2117 m). This group, especially Gebel Elba, is particularly favoured by its position near the sea. The richness of the vegetation of the Gebel Elba area is so notable, compared to the rest of the Egypt, that this is considered as one of the main phytogeographical regions of the country (Drar, 1936; Hassib, 1951). The flora of the Gebel Elba group is much richer than that of the other coastal mountain groups. The number of species collected (Zahran, 1962, 1964) within the areas of the four mountains are: 33 species in the area of Gebel Shayeb group, 92 species in the area of Gebel Nugrus group, 125 in the area of Gebel Samiuki group and 458 species in the area of Gebel Elba group.

Though not the highest of its group, Gebel Elba is nearest to the sea (20–25 km). The whole group faces a northeast bend of the shore such that Gebel Elba faces northward to an almost endless stretch of water. Gebel Elba is the most northerly of its group, the rest being in its shadow from the north winds.

Within the block of Gebel Elba, the vegetation on the north and northeast flanks is much richer than that on the south and southwest. The difference is equal to that between rich scrubland or open parkland on one side and desert vegetation on the other.

The northern and northwest slopes of Gebel Elba are drained by Wadi Yahameib and Wadi Aideib. These wadis are densely covered with *Acacia* thickets the only place in the Eastern Desert where the vegetation looks like a forest. The north and northeast slopes of Gebel Elba are richly vegetated. Three latitudinal zones of vegetation may be recognized: a lower zone of *Euphorbia cuneata*, a middle zone of *E. nubica* and a higher zone of moist habitat vegetation. In this higher zone are stands *of Acacia etbaica, Dodonaea viscosa, Ficus salicifolia, Pistacia khinjuk* and *Rhus abyssinica*. Within these higher zones ferns, mosses and liverworts are present.

Gebel Karm Elba is one of the main foot-hills of Gebel Elba which lies on its east. The north and northeast slopes of this block are characterized by the abundance of *Delonix elata.*

The southern slopes of Gebel Elba drain into Wadi Serimtai, one of the most extensive drainage systems within the whole district. The *Acacia* scrub of this wadi is much more open than that of Wadi Aideib. The southern slopes are notably drier: the vegetation is mostly confined to the rills and runnels of the drainage system. The most common type of vegetation within these runnels is a community dominated by *Commiphora opobalsamum*. On the higher altitudes, some shrubs of *Acacia etbaica* and *Moringa peregrina* may be found.

The western slopes of Gebel Elba are even drier than southern ones. The vegetation is of therophytes (mostly *Zygophyllum simplex*) which appear in rainy years.

The differences in vegetation on different slopes are also notable in the eastern foot-hills of the Gebel Elba group. The higher hills are 175 m rises in the coastal plain to the east of Gebel Elba. On the north and northeast slopes the vegetation is characterized by the preponderance of *Euphorbia cuneata*. On the southern slopes *Aerva javanica* is dominant with only rare plants of *E. cuneata*. On one of the foot-hills (nearer to the shore-line) the northern and eastern slopes are covered by a rich growth of *Acacia nubica* whereas on the southern and western slopes there is an open growth of *Aerva javanica.*

The vegetation of Gebel Shindodai and Gebel Shillal is also richer than that of Gebel Shindeib. The three mountains lie along the same latitude but Gebel Shindodai and Gebel Shillal are on the east, nearer to the sea than Gebel Shindeib. It will similarly be noted that the eastern side of Gebel Asotriba (Soturba) & (Schweinfurth, 1865a) has also a richer flora than the inland slopes.

The eastern flanks of Gebel Shindodai and northern slopes of Gebel Shillal feed the upstream tributaries of Wadi Shillal which is covered by dense *Acacia* scrub. The vegetation of the northeast slopes of Gebel Shindodai show four main zones from base to top: (1) a zone characterized by the abundance of *Caralluma retrospiciens*, (2) a zone characterized by the abundance of *Delonix elata*, (3) a zone of *Moringa peregrina* and (4) a zone with bushes of *Dodonaea viscosa, Pistacia khinjuk* v. *glaberrima* and *Euclea schimperi* and with numerous bryophytes and ferns including *Ophioglossum polyphyllum* and other moisture-loving species such as *Umbilicus botryoides*. The northeast slopes of Gebel Shillal are richly vegetated with a great variety of species. A number of zones may be recognized from base to top: (1) a zone of *Acacia tortilis* and *Commiphora opobalsamum*, (2) a zone of *Acacia etbaica* and *A. mellifera*, (3) a zone with patches of *Cordia gharaf, Dodonaea viscosa, Maytenus senegalensis, Rhus oxyacantha*, and a number of bryophytes and ferns.

Flora of the Red Sea Coastal Mountains

The vegetation on the slopes of the mountains, especially Gebel Elba, is delimited into altitudinal zones, the lower of which shows recognizable characters of community structure. The vegetation of the upper zone is obviously influenced by minor differences of habitat. The individual plants are crowded in patches, forming a mosaic that makes the recognition of clearly defined communities difficult. With such ill-defined pattern, the relationship between the habitat condition and vegetation may be interpreted on the basis of moisture requirements of species. This interpretation is supported by studies carried out on the mountain groups further south in the Sudan (Troll, 1935; Kassas, 1956, 1957, 1960) and in the East African territories (Gillilan 1952; Keay, 1959). The flora of the Egyptian Red Sea coastal mountai includes over 400 species (Zahran and Mashaly, 1991). Table 4.2 gives information on the distribution and range of abundance of a number of selected species of four mountain groups, namely: Shayeb group, Nugrus group, Samiuki group and Elba group (Kassas and Zahran, 1971). The flora included in the table is classified into three main growth form categories: (a) trees, shrubs and undershrubs, (b) persistent herbs and herbaceous herbs and (c) ferns and bryophytes. The species of categories (a) and (b) are subdivided into two classes according to their drought tolerance and moisture requirements. This character is deduced from field observations on distribution and growth within the surveyed area. Within the Elba group distinction is made between the coastal hills (CH), hills at the foot of mountains (FH) and mountains. The habitat types of the coastal hills are classified into the north- and east-facing slopes (NS) and the south- and west-facing slopes (SS). A similar classification is made in the foot-hills, but a distinction is made here between the slopes (NS and SS) and runnels dissecting the north-facing slopes (NSr) and those dissecting the south-facing slopes (SSr).

The species of *Acacia* may be classified into a group with relatively low water requirements and a group with higher water requirements. The former group includes *Acacia ehrenbergiana, A. nubica, A. raddiana* and A. *tortilis*. These are all species of arid and semi-arid habitats and are not confined to the mountain country. A *nubica* is reported within the southern part of the coastal belt (Kassas and Zahran, 1971). It is a widespread species within the semi-arid parts of the north Sudan (Kassas, 1960).

Tabel 4.2 Abundance estimates of some selected species in the different habitat types within the mountain groups

Selected species	Elba group												Sami-uki group	Nug-rus group	Sha-yeb group
	CH		FH				Mountains								
	NS	SS	NSr	NS	SSr	SS	NS_1	NS_2	NS_3	NS_4	SSr	SS			
A. *Trees,shrubs and undershrubs*															
1. Lower water requirements															
Acacia tortilis	4–5	+–1	1–3	2–4	2–4	+	3–5	1–3	+–2	–	2–4	–	4–5	2–3	–
A. raddiana	–	–	–	–	–	–	–	+	+	+	+	–	+	+	+
A. nubica	+–6	–	+–2	–	–	–	+–2	+–2	–	–	–	–	–	–	–
A. ehrenbergiana	–	–	–	–	–	–	–	–	–	–	–	–	–	+	–
Leptadenia pyrotechnica	–	–	–	–	–	–	+	–	–	–	–	–	–	+–1	+–1
Ochradenus baccatus	+–1	–	+–2	–	+–1	–	–	+–2	+–1	+	+	–	+	+	–
Ziziphus spina-christi	–	–	–	–	–	–	–	–	+	+	–	–	–	–	–
Commiphora opobalsamum	+–3	–	+–2	+	+–3	–	+–3	+	+	–	3–5	+	–	–	–
Salvadora persica	–	–	–	–	–	–	–	–	+	–	–	–	–	–	+
Lycium arabicum	+–2	+–1	+–3	+–2	+	–	–	–	–	–	+–2	–	+	+	+
Ephedra alata	+–1	–	–	–	–	–	–	–	–	–	–	–	–	–	–
Grewia tenax	+–2	–	+–3	+	–	–	–	+–2	+–3	+–2	–	–	–	–	–
Indigofera oblongifolia	–	–	–	+–2	+–1	–	+	+	+	+	–	–	–	–	–
Balanites aegyptiaca	–	–	–	–	–	–	–	–	+	+	–	–	+	+	–
Maerua crassifolia	–	–	+–1	+–1	–	–	–	–	–	–	–	–	–	+	–
Cadaba farinosa	–	–	–	+	–	–	–	–	+–2	+–2	–	–	–	–	–
C. rotundifolia	–	–	+	–	–	–	–	–	+	+	+–1	–	–	–	–
Capparis decidua	+	–	–	–	–	–	–	+	–	–	–	–	+	+	–

Tabel 4.2 (Continued)

Selected species	Elba group												Sami-uki group	Nug-rus group	Sha-yeb group
	CH		FH				Mountains								
	NS	SS	NSr	NS	SSr	SS	NS_1	NS_2	NS_3	NS_4	SSr	SS			
2. Higher water requirements															
Moringa peregrina	–	–	–	–	–	–	–	–	+–5	+	–	–	+–5	+–5	+–5
Ficus pseudosycomorus	–	–	–	–	–	–	–	–	–	+	–	–	+–4	+–4	+–5
Dracaena ombet	–	–	–	–	–	–	–	–	–	+–5	–	–	–	–	–
Euphorbia cuneata	5	+–1	5–6	5	2–4	–	5–6	2–4	+–3	–	+–1	–	–	–	–
E. nubica	–	–	+	–	–	–	+–2	+–6	+–2	+	–	–	–	–	–
Acacia etbaica	–	–	–	–	–	–	–	–	5	2–4	–	–	–	–	–
A. mellifera	–	–	–	–	–	–	+–4	+–3	+–3	+–3	–	–	+	–	–
A. laeta	–	–	+	–	–	–	–	–	–	2–4	–	–	–	–	–
Delonix elata	–	–	+–5	–	–	–	–	+–5	+	–	–	–	–	–	–
Euclea schimperi	–	–	–	–	–	–	–	–	–	3–5	–	–	–	–	–
Dodonaea viscosa	–	–	–	–	+	–	–	–	1–4	3–5	–	–	–	–	–
Jasminum floribundum	–	–	–	–	–	–	–	–	–	3–5	–	–	–	–	–
J. fluminense v. *blandum*	–	–	–	–	–	–	–	–	–	3–5	–	–	–	–	–
Olea chrysophylla	–	–	–	–	–	–	–	–	–	+–3	–	–	–	–	–
Ephedra foliata	+	–	–	–	–	–	–	–	+–2	+–3	–	–	+	–	–
Rhus abyssinica	–	–	–	–	–	–	–	–	–	3–5	–	–	–	–	–
R. abyssinica v. *etbaica*	–	–	–	–	–	–	–	–	–	3–5	–	–	–	–	–
R. oxyacantha	–	–	–	–	+	–	–	–	–	3–5	–	–	+–3	–	–
Ficus salicifolia	–	–	–	–	–	–	–	–	–	+–2	–	–	+	–	–
Pistacia khinjuk v. *glaberrima*	–	–	–	–	–	–	–	–	–	+	–	–	–	–	–
Withania obtusifolia	–	–	–	–	–	–	–	–	–	+–2	–	–	–	–	–
Maytenus senegalensis	–	–	–	–	–	–	–	–	–	+–2	–	–	–	–	–
Lantana viburnoides	–	–	–	–	–	–	–	–	–	+–3	–	–	–	–	–

Tabel 4.2 (Continued)

Selected species	Elba group												Sami-uki group	Nug-rus group	Sha-yeb group
	CH		FH				Mountains								
	NS	SS	NSr	NS	SSr	SS	NS_1	NS_2	NS_3	NS_4	SSr	SS			
B. *Persistent herbs and herbaceous herbs*															
1. Lower water requirements															
Aerva persica	+–3	5	2–3	+–3	2–3	5	+–4	+–2	+	–	2–4	+	2–3	2–3	2–3
Launaea spinosa	–	–	–	–	–	–	–	–	–	–	–	–	–	–	2–3
Cleome droserifolia	–	–	–	–	–	–	–	–	–	–	–	–	2–3	3–4	+–2
Fagonia bruguieri	–	–	–	–	–	–	–	–	–	–	–	–	–	+	+
F. tristis v. *boveana*	–	–	+	–	–	–	–	–	–	–	–	–	+	+	+
Pulicaria crispa	–	–	–	–	–	–	+	–	–	–	+	–	+	+	+
Zilla spinosa	–	+	–	–	–	–	–	–	–	–	–	–	2–5	2–5	2–5
Echinops galalensis	–	–	–	–	–	–	–	–	–	–	–	–	+–2	+–2	+–2
Solenostemma argel	–	–	–	–	–	–	–	–	–	–	–	–	+–2	+	+–2
Salsola vermiculata	+–2	+–2	+	+	+	–	2–3	2–3	+	–	+	–	–	–	–
Solanum dubium	1–3	–	3–2	+–2	–	–	+	+	–	–	–	–	–	–	–
Seddera latifolia	1–2	+	1–2	–	–	–	+–3	+	–	–	–	–	–	–	–
Convolvulus hystrix	1–2	1–2	1–2	+	1–2	+	–	+	–	–	–	–	–	–	–
Farsetia longisiliqua	–	–	–	+–1	–	–	–	–	+	+	+	–	+	+	–
2. Higher water requirements															
Lindenbergia abyssinica	–	–	–	–	–	–	–	–	–	+–2	–	–	–	–	+
L. sinaica	–	–	–	–	–	–	–	–	–	+–2	–	–	+–1	+–1	+–1
Parietaria alsinifolia	–	–	–	–	–	–	–	–	–	+–1	–	–	+–1	+–1	–
Solanum nigrum	–	–	–	–	–	–	–	–	–	+–1	–	–	–	–	+
Leucas neufliseana	–	–	–	+–1	–	–	–	+	+	+–2	–	–	–	–	–
Veronica beccabunga	–	–	–	+	–	–	–	–	–	+–1	–	–	–	–	+
Ruellia patula	–	–	–	+–1	+	–	+	+	+	+	–	–	–	–	–

Tabel 4.2 (Continued)

Selected species	Elba group												Sami-uki group	Nug-rus group	Sha-yeb group
	CH		FH				Mountains								
	NS	SS	NSr	NS	SSr	SS	NS_1	NS_2	NS_3	NS_4	SSr	SS			
Micromeria biflora	–	–	+	+–1	–	–	+	+	+	+–2	–	–	–	–	–
Galium setaceum	–	–	–	+	–	–	–	–	–	+–1	–	–	–	–	–
G. tricorne	–	–	–	–	–	–	–	–	–	+–1	+	–	–	–	–
G. spurium v. *tenerum*	–	–	–	–	–	–	–	–	–	+–1	–	–	–	–	–
Scrophularia arguta	–	–	–	+–1	–	–	–	–	–	+–2	–	–	–	–	–
Pancratium tortuosum	–	–	–	+–1	–	–	–	–	–	+–1	–	–	–	–	–
Echinops hussoni	–	–	–	–	–	–	–	–	–	+–2	–	–	–	–	–
Ocimum menthaefolium	–	–	–	–	–	–	+	+	+	1–2	–	–	–	–	–
Commelina benghalensis	–	–	–	+–1	–	–	–	–	–	1–3	–	–	–	–	–
C. forsskalei	–	–	–	–	–	–	–	–	–	1–3	–	–	–	–	–
C. latifolia	–	–	–	–	–	–	–	–	–	1–3	–	–	–	–	–
Bidens bipinnata	–	–	–	–	–	–	–	–	–	1–3	–	–	–	–	–
B. schimperi	–	–	–	–	–	–	–	–	–	1–3	–	–	–	–	–
Oxalis anthelmintica	–	–	–	–	–	–	–	–	–	1–3	–	–	–	–	–
Priva cordifolia v. abyssinica	–	–	–	–	–	–	–	–	–	1–3	–	–	–	–	–
Osteospermum vaillantii	–	–	–	–	–	–	–	–	–	1–3	–	–	–	–	–
Melhania denhamii	–	–	+	+	+	–	–	–	–	1–3	–	–	–	–	–
Umbilicus botryoides	–	–	–	–	–	–	–	–	–	+	–	–	–	–	–
C. *Ferns and bryophytes*															
Adiantum capillus-veneris	–	–	–	–	–	–	–	+	1–3	+–1	–	–	+	+	+
Actiniopteris australis	–	–	+	–	–	–	–	–	1–3	+–2	–	–	–	–	–
Cheilanthes coriacea	–	–	–	–	–	–	–	–	+	+–2	–	–	–	–	–
Notholaena vellea	–	–	–	–	–	–	–	–	–	+–3	–	–	–	–	–
Onychium melanolepis	–	–	–	–	–	–	–	–	–	1–3	–	–	–	–	–

Tabel 4.2 (Continued)

Selected species	Elba group												Sami-uki group	Nug-rus group	Sha-yeb group
	CH		FH				Mountains								
	NS	SS	NSr	NS	SSr	SS	NS_1	NS_2	NS_3	NS_4	SSr	SS			
Ophioglossum polyphyllum	–	–	–	–	–	–	–	–	–	+	–	–	–	–	–
Funaria mediterranea	–	–	–	–	–	–	–	–	–	+–2	–	–	–	–	–
F. pallescens	–	–	–	–	–	–	–	+	+	1–2	–	–	–	–	–
F. pulchella	–	–	–	–	–	–	–	+	+	1–2	–	–	–	–	–
Gymnostomum calcareum	–	–	–	–	–	–	–	+	+	1–2	–	–	–	–	–
Hyophila laxitexta	–	–	–	–	–	–	–	–	+	1–2	–	–	–	–	–
Mannia androgyne	–	–	–	–	–	–	–	–	+	1–2	–	–	–	–	–
Plagiochasma rupestre	–	–	–	–	–	–	–	–	+	1–2	–	–	–	–	–
Riccia aegyptiaca	–	–	–	–	–	–	–	–	+	1–2	–	–	–	–	–
R. atromarginata	–	–	–	–	–	–	–	–	+	1–2	–	–	–	–	–
R. crystallina	–	–	–	–	–	–	–	–	+	1–2	–	–	–	–	–
Timmiella barbula	–	–	–	–	–	–	–	–	+	1–2	–	–	–	–	–

CH, hills near the coast; FH, foot-hills of the mountains; NS, north-facing slopes; NS_1 lowest zone of the north-facing slopes; NS_2, middle zone of the north-facing slopes; NS_3 and NS_4, upper zones of the north-facing slopes; NSr, runnel/north-facing slopes; SS, south-facing slopes; SSr, runnel/south-facing slopes; +–10, abundance estimate according to a modified Domin scale; – = not recorded.

After Kassas and Zahran (1971).

A *tortilis* forms open thickets on the north slopes of the coastal hills. It is also very abundant on all of the slopes of the foot-hills of Gebel Elba and on the lower parts of the north slopes of the same mountain. In these habitats it is associated with *Euphorbia cuneata* which is often dominant. This is very similar to the growth of A *tortilis* on the foot-hills of the Red Sea coastal mountains of the Sudan. A *raddiana* is almost absent from the coastal hills and foot-hills. It is only occasionally found on the north slopes of Gebel Elba within the runnels dissecting the south slopes. A *raddiana* is less drought tolerant than A. *tortilis*.

A. *nubica* has a limited distribution within the Red Sea coastal mountains. This drought-deciduous species occurs in a few types of habitat within the Elba district. It dominates the north slopes of a few of the coastal hills that are covered by surface sheets of sand but is absent from the north slopes of the small hills. It is occasionally found within the runnels of the north slopes of the foot-hills and within the lower levels (NS! and NS_2) of the slopes of the Elba mountains.

The other group of *Acacia* species include A. *etbaica, A. laeta* and *A. mellifera* (Table 4.2). A. *etbaica* is a mountain species that forms rich growth within the higher zone of the mountain slopes of Elba (NS_4). It is also present on the north slopes of the coastal mountains (Shindodai, Shillal and Asotriba) but not on the slopes of the inland ones. A. *mellifera* has a wider range within the northern slopes of Gebel Elba (from base to top). A. *laeta* is less common but does well on the highest zones of the north slopes of Gebel Elba (NS_4). It is recorded from one locality within the foot-hills type (NSr). It is also present in the Gebel Hamata foot-hills (Samiuki group).

Other species within the group trees, shrubs and undershrubs with low water requirements include several which are widespread within the desert habitats of the Egyptian Red Sea coast. Most of these species are either absent or very rare on the southern slopes of the coastal hills and the foot-hills. *Leptadenia pyrotechnica* is much more abundant within the wadis draining the hills and mountains than on the slopes. *Ochradenus baccatus*, which is also common in the wadis, does better on the slopes. It is the most abundant bush within the runnels of the slopes of Gebel Elba (SSr) where it may form groups of thickets. *Salvadora persica*, which forms patches of rich growth in parts of the main wadis, is rare on the mountain slopes. *Lycium arabicum*, a species that dominates a community on the coastal desert plain, is common on the coastal hills and the foot-hills. It is extremely rare on the north slopes of the Elba mountains but locally abundant within some of the runnels of the south slopes of the other mountain groups. *Ephedra alata* is present on the north slopes of the coastal hills but not in the other habitats of the mountains. *Grewia tenax* is common on the north slopes of the coastal hills, foot-hills and the mountains but absent from the south slopes. *Indigofera oblongifolia* is common in several localities of the north slopes of the foot-hills and the runnels of the south slopes of these hills. It is also present on the north slopes of the Elba mountains. *Balanites aegyptiaca* occurs on the higher zones of the north slopes of the mountains and the runnels of the south slopes of the foot-hills. It is also present on the Gebels of the Samiuki and Nugrus groups. *Maerua crassifolia, Cadaba farinosa* and *C. rotundifolia* are recorded within the foot-hills and the north slopes of the Elba mountains.

Trees and shrubs that are less drought tolerant include, apart from the *Acacia* spp. mentioned above, several species which are dominant within other communities. *Moringa peregrina* is present on the higher zones of the north-facing slopes of the mountains, especially Gebel Shindodai. It is also present within the mountains of the Samiuki, Nugrus and Shayeb groups. *Ficus pseudosycomorus* is associated with dry water-fall habitat types and is especially common in the northern mountain groups: Samiuki-Shayeb.

Dracaena ombet is recorded in the highest zones of the north and east slopes of Gebel Elba In several localities there are limited groves of this tree; otherwise there are isolated individuals. Reference may be made to the studies on the growth of *D. ombet* within the Sudanese coastal mountains including the mist oasis of Erkwit (Kassas, 1956, 1960). The occurrence of *Dracaena* in Gebel Elba is its most northern limit within the Red Sea coastal mountains (Kassas and Zahran, 1971).

Euphorbia cuneata is one of the most abundant species within the coastal hills, the foot-hills and the base zone of the Elba mountain (Table 4.2). It dominates a special community on the north slopes but is only occasional on the south slopes of the coastal hills. On the south slopes of the foot-hills it may be common on the runnels but very rare on the slopes outside the runnels. It forms a rich growth that characterizes the lower zone (NS) of the north slopes of the Elba mountain, but its growth is gradually reduced up these slopes. *E. cuneata* is not recorded from the northern mountain groups: Samiuki-Shayeb. *E. nubica*, apparently a species with higher water requirements, is only occasional on the north slopes of the coastal hills and the foot-hills, absent from their south slopes and occasional within the lower zone of the north slopes of Gebel Elba (NS) where *E. cuneata* dominates. *E. nubica* dominates a middle zone (NS_2) of these slopes, thins up the slopes and is absent from the south slopes. The distribution of the two communities dominated by these two *Euphorbia* species is comparable to their distribution within the coastal hills and mountains of the Sudan (Kassas, 1960).

Delonix elata is abundant within certain localities – the Karm Elba group of the foot-hills and the zone NS_2 of the Gebel Shindodai of the Elba group.

There is a group of species that are the least resistant to drought and are confined to the highest zones of the north slopes of the Elba group (NS_4). These include *Dodonaea viscosa, Ephedra foliata, Euclea schimperi, Jasminum fluminense, J. floribundum, Lantana viburnoides, Maytenus senegalensis, Olea chrysophylla, Pistacia khinjuk, Rhus abyssinica, R. abyssinica* v. *etbaica* and *Withania obtusifolia*. Most of these are species that are dominant or very abundant within the wettest zone of the mist oasis of Erkwit, Sudan (Kassas, 1956).

The persistent herbs and clearly herbaceous herbs may also be classified into two groups, one having lower water requirements than the other. The first group includes *Aerva javanica* which dominates the vegetation on the southern slopes of the coastal hills and the foothills and which is common in the highest (wettest) zone of the north slopes of the Elba mountains (NS_4).

Cleome droserifolia, Fagonia bruguieri, F. tristis v. *boveana* and *Launaea spinosa* seem to be confined to the northern mountain groups (Shayeb-Nugrus-Samiuki) and are not recorded in the Elba group. Except for *Cleome droserifolia*, these species are not recorded in the Sudanese flora (Andrews, 1950–1956) and appear to be geographically confined to the northern parts of the Red Sea coast.

Francoeuria (Pulicaria) crispa occurs on the coastal hills, the lower zone of the south slopes of the mountains and also in the runnels of their south slopes. *Zilla spinosa* is rare within the Elba group but is one of the most abundant plants within the northern mountain groups. *Echinops galalensis* and *Solenostemma arghel* are similarly rare within the Elba group and are common within the northern mountain groups. *Convolvulus hystrix, Salsola vermiculata, Seddera latifolia* and *Solanum dubium* are commonly found within the habitats of the coastal hills, foot-hills and the lower zones of the north slopes of the Elba mountains but are rare in the northern mountain groups. *Farsetia longisiliqua* is occasional on the north slopes of the foot-hills and Elba group. It is also present in the Samiuki and Nugrus groups.

The second group of herbs is much more restricted in distribution as they are confined to the less arid localities. The 25 species listed in this group (Table 4.2) are all recorded in the higher zones of the north slopes of the Elba group. In this habitat they grow better than in any of the other habitats.

The ferns and hydrophytes are mostly confined to less arid habitats and all grow best on the upper zones of the north slopes of the mountains. *Adiantum capillus-veneris*, the most widespread fern in the Egyptian desert, is present on the north slopes of the Elba mountains and their foot-hills. It is also common within the pot-hole and nakkat habitats of the northern groups. Other ferns and bryophytes are confined to the Elba group (Table 4.2).

Selected Communities

On the slopes of the mountains the vegetation reflects different habitat conditions on the different slopes or in zones of the same slope. The moisture regime and air temperature are the foremost ecological factors. A brief description of two communities of these habitats is given below (Zahran, 1964).

1. *Aerva javanica* community. *A. javanica* (*A. persica*, Täckholm, 1974) is a white woolly undershrub with alternate entire leaves and with flowers in dense woolly spikes. It is a xerophyte common in the upstream parts of the wadis of the Red Sea coastal desert. Its community occurs in the montane country. *A. javanica* is also common in the Red Sea coastal desert of Saudi Arabia and abundant in the southwest (Hejaz) mountains where the bedouins use its woolly branches instead of cotton for making pillows (Zahran, 1982b).

   The *Aerva* community is usually found on the south slopes of both the coastal hills and foot-hills. Its plant cover is usually thin, not exceeding 10%, forming an open vegetation. This community has all the features of desert (xerophytic) vegetation, e.g. thin cover and notable seasonal variations owing to the growth of ephemerals during the rainy season. *Acacia tortilis* and *Blepharis edulis* are the abundant associates. Other common perennials include *Abutilon fruticosum. Convolvulus hystrix, Cucumis prophetarum, Euphorbia cuneata, Heliotropium strigosum, Indigofera spinosa, Pupalia lappacea* and *Tephrosia purpurea.*

One of the notable features of this community is the abundance of therophytes, the cover of which is several times that of the perennials. *Aristida adscensionis* and *Stipagrostis hirtigluma* are abundant annuals. *Asphodelus tenuifolius, Euphorbia granulata, Kohautia caespitosa* and *Melanocenchris abyssinica* are also common.

2. *Euphorbia cuneata* community. *E. cuneata* is a small non-succulent tree up to 3 m. Old branches are covered with bark and younger ones are often spiny-tipped. The *E. cuneata* community is the most common vegetation on the north slopes of the foot-hills and the lower zones of the north slopes of the Elba group. It may also be present, though in a depauperate form, in some of the runnels dissecting the south slopes of the foot-hills.

   The perennial plant cover of this community ranges from 5 to 50%, the main part of this being *E. cuneata. Abutilon pannosum* and *Indigofera spinosa* are the abundant perennial associates. *Acacia tortilis, Aerva javanica* and *Salvia aegyptiaca* are also common. Less common species include *Blepharis edulis, Euphorbia nubica, Farsetia aegyptia, Forsskaolea tenacissima* and *Panicum turgidum. Actiniopteris australis* is the only fern recorded.

   The ephemeral plant cover may be up to 70%. *Aristida adscensionis* and *Cenchrus pennisetiformis* are the abundant associates. *Pimpinella etbaica* and *Stipagrostis hirtigluma* are common.

4.2.2 The Inland Desert

The inland part of the Eastern Desert lies between the range of the Red Sea coastal mountains in the east and the Nile Valley in the west, an area of about 223,000 km^2 (Hassib, 1951). It is a rocky plateau dissected by a number of wadis. Each wadi has a main channel with numerous tributaries. The whole forms a drainage system collecting the run-off water, i.e. the Eastern Desert is divided piecemeal into catchment areas of these drainage systems. Most of the wadis drain westward into the Nile, only a few extending northeast, terminating near Belbis on the border of the Nile Delta.

The inland part of the Eastern Desert can be divided into four main geomorphological and ecological regions:

a. The Cairo-Suez Desert
b. The Limestone Desert
c. The Sandstone (Idfu-Kom Ombo) Desert
d. The Nubian Desert.

(a) The Cairo-Suez Desert

(i) Geomorphology and Climate

The Cairo-Suez Desert traverses the country between the Mokattam hill (300 m) in the west and the Ataqa mountain (817 m) in the east. It is limited on the south by the fringe of the limestone plateau of the Eastern Desert (Lat. 30°N) and to the north by fringes of the cultivated land of the Ismailia irrigation canal (Lat. 30°30′N). An area of such extent (c.3000 km^2) naturally represents a number of geomorphological, geographical and ecological features. Kassas and Imam (1959) state "the Cairo-Suez Desert is a mosaic of Eocene limestone, Oligocene, Miocene and Pliocene formations". El-Abyad (1962) wrote "In the Cairo-Suez Desert there are two main types of sedimentary formations: limestone and fluviatile-deposits. The limestone rocks form table-like landforms ranging in size from vertical cliffs of the extensive plateaux to small buttes. The fluviatile gravels and sands form series of rolling low and gravel-covered basins."

The southern ridge of the Cairo-Suez Desert is represented by the escarpment of the limestone plateau extending from Gebel Ataqa near Suez to Gebel Turra to the east of Cairo. The second limestone ridge is a discontinuous series of hills including Gebels Mokattam, El-Nassuri and Angabiya and a chain of Eocene limestone low hills extending eastwards towards the foot of Ataqa. The third ridge includes Gebels Eueibed and El-Gafra and extends westwards in the form of a series of low small hills. The fourth ridge includes Gebels Gineifa, Umm Kuteib, Gibra and Umm Raqm. The strip between the first and second ridges is occupied by Oligocene gravel forming the huge gravel hills of Gebels Yahmum, El-Asmar, El-Royesat and El-Khashab (the petrified forest). The strip between the second and the third ridges is a rolling gravel country traversed by the Cairo-Suez desert road. The area between the third and fourth ridges is similarly covered with gravels and is crossed by the Cairo-Suez railway. To the north of the Gebel Gineifa, the Gebel Umm Raqm ridge extends in an almost flat plain of Plio-Pleistocene gravels and includes the silt terraces of Wadi Tumilat in the far north.

The general topography of the whole Cairo-Suez Desert shows a NW slope. Consequently the main drainage wadis extend in a north-west direction terminating near Belbis. The area is traversed by a number of wadis with their upper stream parts cutting across the northern limestone ridge. The northern scarps of Ataqa are drained by runnels feeding Wadi Bahara which extends eastwards and drains into the plain west of Suez.

The drainage lines cutting across the limestone are usually well defined and their deeply cut channels may be a barren rock surface or may have a deep layer of valley-fill including coarse limestone rounded boulders and softer sand and silt. The sides of the floor may be fringed by strips of silt terraces.

The channels cutting across the gravel and sand formations are usually shallow and ill-defined. Their beds may expose limestone surfaces of the underlying formations or may have a deep valley-fill of flint gravels and sand.

In the Cairo-Suez Desert, rainfall is characterized by its scantiness (25–30 mm year) and its great variability from year to year. Except for the winter months, the main part of the year is rainless, i.e. the climate is very arid. Cloudbursts may cause abundant rainfall within a limited area. The climate of this desert is classified as the Saharan-Mediterranean type of Emberger (1951).

(ii) Vegetation Types

The vegetation of the Cairo-Suez Desert may be described under the following titles.

1. Vegetation of Wadi El-Gafra system
2. Vegetation of Ataqa Scarp
3. Vegetation of Gebel Asfar Dunes
4. Vegetation of the Gravel Desert
5. The plant communities.

1. Vegetation of Wadi El-Gafra System

Wadi El-Gafra is one of the main drainage systems of the Cairo-Suefl Desert. Its basin extends for about 70 km from the mountain ranges on the northern edge of the Eocene limestone plateau downwards until it terminates near Belbis on the eastern border of the Nile Delta. This system drains the central part of the Cairo-Suez Desert. It receives tributaries extending southwards across the plateau of the Eastern Desert. These tributaries include: Wadi Umm Seyal which drains the northwest of Ataqa mountain, Wadi Umm Gerfan and Wadi Etheily which drain Gebel Umm Rayahat and Wadi Gendali which receives tributaries draining Gebel El-Katamiya and Gebel Abu Shama.

Wadi Umm Seyal

This is the eastward affluent of the Gafra system. In its upstream area *Zygophyllum coccineum* gains dominance. The middle part is characterized by the *Z. coccineum-Panicum turgidum* community. The downstream part supports *Lygos raetam* scrubland.

Wadi Umm Gerfan

The floor of the upstream part of this wadi, which cuts in the limestone plateau, is barren rock with little or no veneer of rock detritus. This is the habitat of a vegetation dominated by *Stachys aegyptiaca* which is not palatable to grazing animals and is only frequently nibbled by sheep and goats (Kassas and Imam, 1954). The characteristic species of this community include *Asteriscus graveolens, Erodium glaucophyllum, Fagonia kahirina, F. mollis, Gymnocarpos decander, Helianthemum kahiricum, Iphiona mucronata, Limonium pruinosum, Pituranthos tortuosus, Reaumuria hirtella* and *Zygophyllum coccineum.*

Wadi Etheily

The upstream part of this wadi extends across the limestone plateau whereas its middle and downstream parts cross the gravel country. The vegetation of the middle part is of three communities dominated by *Artemisia monosperma, Ephedra alata* and *Panicum turgidum*. In the downstream part *Hammada elegans* dominates.

Wadi Gendali

Within the channel of Wadi Gendali two communities dominated by *Anabasis articulata* and *Nitraria retusa* are found. The former is widespread within the course of the wadi, extending downstream for About 10 km north of its emergence from the edge of the plateau. The *N. retusa* is here restricted to the silt terraces that are well developed on the sides of the wadi. In the Bir (well) Gendali environs, *N. retusa* grows in patches and the associate species occur between them. The abundant species is *Zygophyllum album. Atriplex halimus* is also common. The dominant and these two associates are halophytes. Common xerophytic associates include *Achillea fragrantissima, Artemisia monosperma, Gymnocarpos decander, Launaea capitata* and *L. nudicaulis*. In certain areas of Wadi Gendali small patches of *Tamarix nilotica* occur. These are apparently relicts of much richer thickets which have been destroyed by cutting.

Near its emergence from the plateau, Wadi Gendali receives its tributary Wadi El-Katamiya, which also extends across the limestone plateau. In this wadi are *Achillea fragrantissima, Pennisetum dichotomum* and *Zygophyllum coccineum* which are limestone country species.

The *Anabasis articulata-dominated* part of the wadi is followed by vegetation dominated by *Ephedra alata*. Within this part, the wadi receives many affluents that drain the gravelly country which it traverses. These affluents are mostly lined with sand and are vegetated with *Panicum turgidum* grassland. The finer runnels support *Lasiurus hirsutus* grassland.

A little before the junction between Wadi Gendali and the Cairo-Suez Desert road, the *Ephedra alata* community is replaced by the *Hammada elegans* community in the course of the wadi to its mouth where it joins the eastern confluent of Wadi El-Gafra.

Hassan et al. (2005) reported that *Anabasis articulata* and *Nitraria retusa* are still the main dominants of Wadi Gendaly after more than 40 years of El-Abyad (1962). They recorded 60 species belonging to 23 families having the following life forms: 35% are chamaephytes, 23.3% are hemi-cryptophytes, 11.7% are phanerophytes, 3.3% are geophytes and 1.7% are parasites.

Wadi El-Gafra Confluents

All the previously mentioned wadis drain into the eastern confluent of Wadi El-Gafra, namely Wadi El-Fool which receives a number of affluents that are less extensive than those of the east confluent. The channel of the confluent bears vegetation which contains a central zone (fringing the water channel) and a terrace zone in its peripheries. The central zone is usually covered with soft deposits and supports *Lygos raetam* and *L. raetam—Panicum turgidum* scrubland vegetation. The peripheral terraces are usually gravelly and bear *Hammada elegans* vegetation.

Main Channel of Wadi El-Gafra

The main channel of Wadi El-Gafra is mostly occupied by *Lygos raetam* scrubland. The notable character is the preponderance of *Calotropis procera* and the abundance of *Acacia raddiana*. As it emerges from the Gafra limestone plateau, the wadi crosses the Cairo-Suez railways and extends across the northern gravel country. Its vegetation is here a mixture of *Lygos raetam* and *Hammada elegans*. The two branches of the wadi bear similar vegetation.

2. Vegetation of Ataqa Scarp

The north and northeast scarps of Gebel Ataqa are dissected by drainage lines of different extent. They range from minor runnels to extensive wadis. As they emerge from the hills the wadis and runnels cut across the erosion pavement and the plain which extends into the foot of Gebel Ataqa. The wadis are enriched by the *Launaea spinosa-Zygophyllum coccineum* community. The plant cover of the larger runnels is mostly the *Z. coccineum* community. The minor runnels are characterized by a vegetation in which *Fagonia mollis* is the dominant with thin cover (3.5%). Associate species include *Anastatica hierochuntica, Blepharis edulis, Hyparrhenia hirta* and *Phagnalon barbeyanum*.

The larger wadis cutting across the scarps of Ataqa are often deep ravines with their floors covered with coarse rock detritus. The vegetation comprises a number of communities dominated by *Cleome droserifolia, Launaea spinosa, Zygophyllum coccineum* and *Z. decumbens*.

3. Vegetation of Gebel Asfar Dunes

In the northwest of the Cairo-Suez Desert (southern fringe of the Nile Delta) is a body of sand dunes known as Gebel Asfar district. They cover an area of about 40 km^2 and form a huge crescent with one arm extending in a north-south direction and the other an east-west direction. This formation is of groups of coalescent sand ridges each of which ranges from 80 to 120 m long and 25 to 40 m high. El-Beheiri (1950) showed that the sand is mostly derived from the Nile alluvial deposits and is only partly of continental desert origin. This mass of sand dunes overlies a plain of non-marine Miocene gravel and sand. The district receives drainage from several wadis on the south, north and east sides.

The sand dunes are mostly devoid of plant cover, but some parts are vegetated by an open grassland with *Stipagrostis scoparia* as the dominant species. Characteristic associates include *Asthenatherum forsskaolii, Calligonum comosum, Cornulaca monacantha* and *Moltkiopsis ciliata*; all are sand dwellers. The lows between the sand are often covered by a rich vegetation. The species, in addition to those mentioned above, include *Anabasis articulata, Convolvulus hystrix, Hammada elegans, Lasiurus hirsutus, Pituranthos tortuosus, Polycarpaea repens* and *Stipagrostis plumosa*.

The wadis and other runnels of the drainage systems that skirt the dunes are comparable in their ecological characteristics to the other drainage systems of the

Cairo-Suez Desert. *Anabasis articulata* and *Hammada elegans* are abundant. This is, perhaps, the only part of the Cairo-Suez Desert where *A. articulata* and *H. elegans* are frequently found together.

4. Vegetation of the Gravel Desert

The gravel desert provides several landform types that may be genetically related. In the first place the parent fluviatile deposits are mixtures of particles ranging from gravels to fine silt and clay. Under the transportation agencies of wind and water, the finer material is removed and the coarser material left to accumulate on the surface as lag deposits. This is the process of natural sorting or sieving which is gradual; stages intermediate between the original mixed deposits and the mature stage of gravel "armour" may be recognized. The mature armour, closely packed gravels, once established, protects the underlying deposits against further transportation. The gravel armour provides a surface impenetrable to roots and the undersurface layer of gypsum and salts provides an added resistance to the growth of roots. The result is that the mature gravel surface is usually barren (Kassas and Imam, 1954).

The vegetation of the gravel desert of the Cairo-Suez Desert has been discussed by Kassas and Imam (1959) under three types of habitats: gravel slopes, affluent runnels and main channels.

Gravel Slopes

The plant cover on the gravel slopes varies according to five features:

1. The maturity of the gravel armour. This is indicated by the compactness of the surface gravels which may be so closely packed that they form a mature impenetrable armour or may be loosely strewn and the softer deposits below may still be subject to transportation.
2. The angle of slope. The more gentle the slope the less is the rate of run-off and the better favoured the vegetation.
3. The direction of the exposure (aspect). South-facing slopes are consistently warmer than north-facing ones.
4. Extent and depth of surface soft deposits. The gravel armour may become buried under wind-borne sands. The sand may be patchy or continuous and may be very shallow or up to 20 cm deep. This depends on the exposure in relation to the sand-bearing wind and the local features of the ground.
5. The nature of the deposits below the gravel surface. This concerns the profile pattern of the deposits and their geological nature.

Tal-El-Mokatat (striped hill by water runnels) is a huge body of Oligocene gravel at 35 km on the Cairo-Suez Desert road. Nine quadrats were monitored on the slopes of the hill (Table 4.3). These permanent quadrats were of 100 m^2, a large size being necessary to sample the open vegetation adequately. Study of species:area relationships (Kassas, 1953b) shows that even this size falls considerably short of the "minimal area" for very open desert vegetation (which may be greater than 1500 m^2), especially

Table 4.3 The floristic composition of nine permanent quadrats within the gravel slopes of Tal-El-Mokatat locality, Cairo-Suez Desert (after Kassas and Imam, 1954)

Species	Quadrats								
	North-facing slopes					South-facing slopes			
	1	2	3	4	5	6	7	8	9
A. Perennials									
Allium desertorum	+	+	+	+	+	.	.	+	.
Atractylis flava	+	.	+	+	+	.	+	+	+
Dipcadi erythraeum	.	+	+	+	+	.	+	.	.
Diplotaxis harra	+	.	.	.	+	.	.	.	+
Echinops spinosissimus	.	.	+	.	.	.	.	.	.
Ephedra alata	.	+	+	+	+	.	+	.	.
Euphorbia retusa	.	.	.	+	.	+	.	.	+
Fagonia glutinosa	.	.	.	+	.	.	.	.	.
Farsetia aegyptia	.	+	+	+	+	.	+	.	.
Gagea reticulata	.	+	+	.	+	.	.	.	.
Gymnocarpos decandrum	.	+	+	.	+	.	.	.	.
Halogeton alopecuroides	.	+	+	+	+	.	.	+	.
Hammada elegans	+	+	.	+	+	.	.	.	+
Heliotropium luteum	.	.	+	.	.	.	.	.	.
Lasiurus hirsutus	.	+	+	+	.	.	.	.	.
Launaea nudicaulis	.	.	.	.	+	.	.	.	.
Linaria aegyptiaca	.	.	.	.	+	.	.	.	.
Onobrychis ptolemaica	.	.	+	+	+	.	.	.	.
Pancratium sickenbergeri	.	+	+	+	+	.	.	+	.
Panicum turgidum	.	.	.	+	+	.	.	.	.
Pergularia tomentosa	.	+	.	.	+	.	.	.	.
Pituranthos tortuosus	.	+	+	+	+	.	.	.	.
Reaumuria hirtella	+	.	.	.	+	.	.	.	.
Urginea undulata	.	.	+	.	+	.	.	.	.
Total perennials	5	12	15	14	19	1	4	4	4

Table 4.3 (Continued)

Species	Quadrats								
	North-facing slopes					South-facing slopes			
	1	2	3	4	5	6	7	8	9
B. Annuals									
Anastatica hierochuntica	·	·	·	·	·	+	+	·	+
Calendula micrantha	+	+	+	+	+	·	+	+	+
Centaurea aegyptiaca	+	+	+	+	+	+	+	·	+
Centaurea pallescens	+	+	+	+	+	+	+	+	+
Erodium pulverulentum	·	+	·	+	+	·	·	+	·
Fagonia latifolia	+	·	·	·	+	+	+	+	+
Filago spathulata	·	+	·	·	+	·	·	·	·
Gastrocotyle hispida	·	·	·	·	·	·	·	·	+
Gymnarrhena micrantha	·	·	·	·	·	·	·	+	·
Ifloga spicata	·	+	+	+	+	·	·	·	+
Linaria haelava	·	·	·	·	+	·	·	·	·
Malva parviflora	·	·	·	+	·	·	·	·	·
Matthiola livida	+	·	·	·	+	·	·	·	·
Medicago laciniata	+	·	+	+	+	·	+	·	·
Mesembryanthemum forsskalei	+	·	·	·	·	+	·	·	+
Plantago ovata	+	+	+	+	+	+	+	+	+
Polycarpon succulentum	·	+	+	+	+	·	·	·	·
Pteranthus dichotomus	+	·	·	+	+	+	·	+	+
Reichardia orientalis	+	+	+	+	+	·	+	+	+
Roemeria dodecandra	·	·	·	·	+	·	·	·	·
Rumex vesicarius	+	·	+	+	+	+	+	+	+
Schismus barbatus	+	+	+	+	+	+	+	+	+
Silene villosa	·	+	+	·	·	·	·	·	·
Stipa capensis	+	+	+	+	+	+	+	+	+
Total annuals	13	12	12	14	18	10	11	11	14

+, present; ·, absent.

when seasonal growth of ephemerals is also involved. In the investigations at Tal-El-Mokatat quadrats 1–5 were situated on north-facing slopes and represent conditions of similar exposure and increasing thickness of the sandy cover overlying the gravel surface. Quadrats 6–9 were on south-facing slopes and represent increasing thickness of the sandy surface cover (Kassas and Imam, 1959). The first obvious feature of the records of these quadrats is the preponderance of perennials on the north-facing slopes. Within the quadrats of the same aspect the number of perennials increases as the thickness of the sandy cover increases. The ephemerals do not exhibit such marked differences. A second point is that in the quadrats of the south-facing slopes devoid of any surface cover of sand, the number of ephemeral species in each quadrat is double – or more – the number of perennial species. In quadrat 6 there are ten ephemerals and only one perennial.

The monthly records of periodicity show that ephemerals germinate, grow, flower and dry up on south-facing slopes 3–4 weeks earlier than on north-facing ones. The perennials which occur on both slopes show earlier phenology on the south-facing slopes.

Gravelly Hillocks

The substrata of the slopes of the gravelly hillocks are of non-marine Miocene and Oligocene gravels devoid of any surface mantle of sand (Kassas and Imam, 1959). The abundant species of the slopes (not water runnels) are *Mesembryanthemuni forsskaolii* and *Stipa capensis. Centaurea pallescens, Pteranthus dichotomus* and *Zygophyllum simplex* are somewhat less frequent. *Centaurea aegyptiaca* is the common perennial on both types of gravel. *Fagonia glutinosa* and *Hammada elegans* are more common on the non-marine gravel than on the Oligocene gravel. *Allium desertorum, Gypsophila capillaris* and *Stipagrostis plumosa* are more frequent on the Oligocene gravel.

Affluent Runnels

The slopes of the gravel hills are usually dissected by runnels which contain and guide the run-off water. They are, in fact, the finer tributaries of the surface drainage systems. A clear distinction must be drawn between the vegetation of the runnels transecting the Pliocene gravel on the one hand and the Miocene and Oligocene gravels on the other. The vegetation in the former is essentially ephemeral with *Mesembryanthemum forsskaolii* as the abundant plant, whereas the latter supports a grassland community dominated by *Lasiurus hirsutus*. The distinction is seen where the runnels are lined by soft material. In those runnels devoid of such a sandy lining the vegetation, in both classes of gravel, is a depauperate cover of *Hammada elegans*. Thus, three communities dominated by *M. forsskaolii, L. hirsutus* and *H. elegans* may be recognized in these affluents.

In the *H. elegans* community, the type of habitat is water runnels devoid of sand cover. The plant cover is usually less than 5%. Common perennial associates include *Farsetia aegyptia, Panicum turgidum* and *Stipagrostis plumosa*. Less common perennials are *Atractylis flava, Echiochilon fruticosum, Heliotropium luteum,*

Lasiurus hirsutus, Moltkiopsis ciliata, Tribulus alatus and *Zilla spinosa*. Rare perennials include *Fagonia glutinosa, Gypsophila capillaris, Lygos raetam. Paronychia desertorum* and *Pergularia tomentosa*. Therophytic associates include *Centaurea aegyptiaca, C. pallescens, Ifloga spicata, Neurada procumbens, Plantago ovata, Schismus barbatus, Silene villosa* and *Zygophyllum simplex*.

In the *L. hirsutus* community, the type of habitat is water runnels across Oligocene and non-marine Miocene gravel. It is an open grassland with a relatively large number of associate species. *Atractylis flava, Euphorbia retusa, Fagonia glutinosa, Farsetia aegyptia, Launaea nudicaulis, Panicum turgidum, Pituranthos tortuosus* and *Zilla spinosa* are the abundant perennial associates. Less common perennials include *Artemisia monosperma. Astragalus spinosus, Colocynthis vulgaris, Convolvulus lanatus, Dipcadi erythraeum, Echinops spinosissimus, Ephedra alata, Gymnocarpos decander, Hammada elegans, Pancratium sickenbergeri* and *Salvia aegyptiaca*. The abundant annuals are *Centaurea pallescens, Launaea cassiniana* and *Schismus barbatus*. Less common therophytes include *Gastrocotyle hispida, Linaria haelava, Lotus villosus, Medicago laciniata, Monsonia nivea, Neurada procumbens. Ononis reclinata, Plantago ovata, Polycarpon succulentum. Reseda arabica, Silene villosa* and *Zygophyllum simplex*. The water runnels across Pliocene gravel are characterized by annual plant cover dominated by *Mesembryanthemum forsskaolii*. The abundant perennial associate is *Zilla spinosa* while the common ones include *Fagonia bruguieri, F. glutinosa, Farsetia aegyptia, Panicum turgidum* and *Stipagrostis plumosa*. Less common perennial associates include *Atractylis flava, Atriplex halimus, Cleome arabica, Diplotaxis harra, Echiochilon fruticosum, Erodium glaucophyllum, Fagonia arabica, Hyoscyamus muticus, Heliotropium luteum* and *Linaria aegyptiaca. Centaurea aegyptiaca, C. pallescens, Matthiola livida, Schismus barbatus, Senecio desfontainei* and *Zygophyllum simplex* are the common therophytes. Other annuals include also *Aizoon canariense, Astragalus mareoticus, Erodium laciniatum, Reichardia orientalis* and *Sonchus oleraceus*.

Main Channels

The main channels of the wadis cutting across the gravel desert are characterized by four chief communities dominated by *Hammada elegans, Panicum turgidum, Zilla spinosa* and *Artemisia monosperma*.

1. *Hammada elegans* community. Within the gravel desert system, the *H. elegans* community is one of the most common. It is well developed on the sandy sides of the main drainage channels. The sand is usually built into mounds protected by the bushes *of Hammada*. These mounds are actually relict segments of old terraces of similar alluvial deposits. They differ from the freshly formed mounds, such as those built around the tussocks of *Panicum turgidum*, by being softer in a texture, darker in colour and richer in organic matter content. Between the *Hammada* mounds the surface sand cover has, in the Cairo-Suez desert, a variety of habitats but *Hammada* usually grows best in areas losing their softer material. Here, the species which dominates elsewhere fail to gain ascendancy.

The *H. elegans* community is recognized, in the first instance, by the abundance of this perennial, the growth of which provides the apparent homogeneity of the stand. In this community four layers may be distinguished. The frutescent layer includes *Lygos raetam* is of little significance. The suffrutescent layer is of prime importance in the framework of the community as it includes the dominant and the common associates *Lasiurus hirsutus* and *Panicum turgidum*. The third layer includes *Convolvulus hystrix, Farsetia aegyptia, Gymno-carpos decander, Pituranthos tortuosus* and *Zilla spinosa* and it is richest in number of perennial associates though its total cover is usually less than 5%. The ground layer includes a number of prostrate and dwarf perennials, e.g. *Moltkiopsis ciliata. Paronychia desertorum* and *Polycarpaea repens*. This layer is enriched in the spring by the ephemeral vegetation including *Ifloga spicata, Monsonia nivea, Plantago ovata, Polycarpon succulentum, Schismus barbatus, Silene succulenta* and *Zygophyllum simplex*. In years of good rainfall the cover of this layer may reach 50%.

2. *Panicum turgidum* community. This community is restricted to the drainage lines of the gravel desert. It is well developed in localities receiving sand. Surface accumulation of sand is usually into mounds around the tussocks of *P. turgidum* and *Lasiurus hirsutus*. Between the mounds, the sand cover is a continuous sheet. The texture of the sand varies in alternate layers but it is usually coarse wind-borne material with occasional thin layers of soft water-borne silt.

   Within the permanent framework of the *P. turgidum* community four layers have been recognized. The first includes *Acacia raddiana* and *Lygos raetam*: it is of little importance. The suffrutescent layer includes the dominant, *Hammada elegans* and *Lasiurus hirsutus*; all are sand binders. The third layer includes e.g. *Farsetia aegyptia* and *Zilla spinosa*. The ground layer contains a number of prostrate perennials and all ephemerals mentioned in the *H. elegans* community.

3. *Zilla spinosa* community. The *Z. spinosa* community is not common in the gravel desert ecosystem, being restricted to the drainage lines of Pliocene gravels. The close relationship between *Mesembryanthemum forsskaolii* and *Z. spinosa* is obvious. The former occupies the runnel tributaries of the drainage system, the latter the main channels.

   In this community the frutescent (150–300 cm) and suffrutescent (100–150 cm) layers make up a minor portion of the cover. The third layer (30–100 cm) includes the dominant and many other perennials. It constitutes the main part of the permanent framework of the community. The ground layer includes several dwarf perennial species and is enriched by the therophytes during spring. The floristic composition of the *Z. spinosa* community shows that the common perennial associates include *Convolvulus lanatus, Echinops spinosissimus, Fagonia arabica, F. glutinosa, Farsetia aegyptia, Hammada elegans, Heliotropium luteum, Launaea nudicaulis, Moltkiopsis ciliata, Pergularia tomentosa* and *Pituranthos tortuosus*. It may be noted that *Lasiurus hirsutus* and *Artemisia monosperma*; which are dominant in other communities are scarce or absent here.

The annual associates include those recorded in the other communities of the gravel desert.

4. *Artemisia monosperma* community. This community is confined to one drainage system and its tributaries at 40 km from Cairo along the Cairo-Suez Desert road. This wadi collects the run-off water of an Oligocene area including extensive basalts.

   The flora of the *A. monosperma* community includes *Convolvulus lanatus, Echinops spinosissimus, Hammada elegans, Panicum turgidum* and *Polycarpaea repens* as common perennials. Among less common ones are *Atractylis flava, Cleome arabica, Colocynthis vulgaris, Echiochilon fruticosum, Fagonia bruguieri, F. glutinosa, Farsetia aegyptia, Gymnocarpos decander, Gypsophila capillaris, Lasiurus hirsutus, Pituranthos tortuosus* and *Zilla spinosa*. Annuals include *Mesembryanthemum forsskaolii, Launaea cassiniana, Neurada procumbens, Plantago ovata, Schismus barbatus, Stipa capensis* and *Trigonella stellata*.

5. The Plant Communities

The vegetation of the Cairo-Suez Desert comprises fourteen communities dominated or co-dominated by the following species: *Panicum turgidum, Hammada elegans, H. elegans-P. turgidum, Zygophyllum coccineum, Z. coccineum-P. turgidum, Z. coccineum-Launaea spinosa, Lygos raetam-P. turgidum, L. raetam-Anabasis articulata, Ephedra alata, Artemisia monosperma, Zygophyllum decumbens, Lasiurus hirsutus* and *Acacia tortilis* (El-Abyad, 1962; Kassas and El-Abyad, 1962).

1. *Panicum turgidum* community. The grassland community dominated by *P. turgidum* commonly covers the parts of the main channels of wadis which receive wind-borne sand. It accumulates around the tussocks of the grass, forming isolated mounds that gradually enlarge and eventually coalesce to form sandy patches that cover the originated gravelly or stony bed of the channels. The plant cover of this grassland community varies from 5 to 50%; *P. turgidum* contributes most of the cover. Its growth aspect varies considerably. It has been recorded as flowering during December, July and October, in dry conditions during December, February, March and June, in full foliage during December, January and July and in new sprouting stage during] January and November. The growth aspect of this grass varies in apparent independence of seasonal conditions. Such growth aspects of *P. turgidum* are, in fact, indications of local and temporal habitat conditions. When the conditions are optimal, the plant grows as an ever green grass. Under less favourable conditions it shows a deciduous habit; its aerial parts dry up and the plant looks dead – a mass of dry tussocks and branches. This aspect may be maintained for a number of consecutive years. Whenever the locality receives some water, new sprouts actively resume their green appearance. The receipt of water is of local occurrence because rainfall is usually only local and because the water revenue is often in the form of local torrents or restricted floodsheets. With two different amounts of water supply, the plant cover may exhibit

different aspects. The growth behaviour of the closely associated grass, *Lasiurus hirsutus*, is similar. *Pituranthos tortuosus* and *Lasiurus hirsutus* are the abundant associates. The former maintains its green and flowering aspects all through the year. It has a deeply penetrating root which may facilitate its evergreen character. *Echinops spinosissismus, Farsetia aegyptia, Lygos raetam*, and *Zilla spinosa* are commonly present. *L. raetam* is a bush which forms, together with the less common *Acacia raddiana, Lycium arabicum* and *Tamarix nilotica*, the sparse frutescent layer. Other less common associates are *Artemisia monosperma, Convolvulus lanatus* and *Hammada elegans*. Rarely recorded perennial species include *Anabasis articulata, Asteriscus graveolens, Cleome arabica. Ephedra alata, Pennisetum dichotomum (P. divisum*, Täckholm, 1974) and *Zygophyllum coccineum*. Among therophytes are *Centaurea aegyptiaca* (widespread), *Cotula cinerea, Emex spinosus, Ifloga spicata, Filago spathulata, Lotus arabicus, Plantago ovata, Trigonella stellata* and *Zygophyllum simplex*.

2. *Hammada elegans* community. *H. elegans* is widespread in the Cairo-Suez Desert. It is mostly confined to the water runnels of the gravel formation where the bed is apparently receiving little wind-borne sand or is losing its cover of soft material. This vegetation is also found in wadis where the bed is covered with coarse gravels. In certain wadis *Hammada* covers and evidently protects raised patches of sand and silt mounds with the spaces between veneered with gravel. The exposed sections of the mounds show layering that indicate deposition by water.

   The total cover of the *H. elegans* community in the Cairo-Suez Desert varies from 10 to 70% but is usually 15%. *H. elegans* is the species with greatest density and giving the main part of the cover. Being a stem succulent, *Hammada* maintains its green aspect all through the year. It flowers in the autumn and fruits early in winter. *Panicum turgidum* is the abundant associate. *Farsetia aegyptia, Pituranthos tortuosus* and *Zilla spinosa* are also common. Less common associates include *Artemisia monosperma, Echinops spinosissimus, Ephedra alata*, and *Lasiurus hirsutus. Acacia raddiana* and *Lygos raetam* represent members of the frutescent layer which is widely spaced in this community. The therophytes include *Aizoon canariense, Amberboa lippii, Asphodelus tenuifolius, Astragalus bombycinus, Bassia muricata. Calendula aegyptiaca, Centaurea pallescens, Cleome arabica, Erodium laciniatum, Ifloga spicata, Malva parviflora, Medicago laciniata, Mesembryanthemum forsskaolii, Plantago ovata, Reichardia orientalis* and *Schismus barbatus*.

3. *Hammada elegans-Panicum turgidum* community. This is a transitional community with ecological features of the *H. elegans* and the *P. turgidum* communities. In the floristic composition of this co-domination *H. elegans* and *P. turgidum* are equally abundant (El-Abyad, 1962). *Lasiurus hirsutus, Pituranthos tortuosus* and *Zilla spinosa* are very common perennial associates. The annuals include *Asphodelus tenuifolius, Filago spathulata, Hordeum marinum, Ifloga spicata, Medicago laciniata, Plantago ovata, Stipa capensis* and *Zygophyllum simplex*.

4. *Zygophyllum coccineum* community. The community dominated by *Z. coccineum* is absent throughout the area extending from the eastern border of Cairo at the

beginning of Cairo-Suez Desert road until some 70 km to the east of Cairo, i.e. midway between Cairo and Suez. Z. *coccineum* is present only as rare individuals. Within the eastern part, which includes the drainage systems of Gebel Iweibed and the northern parts of the Ataqa range, this community is, however, widespread. Here, Z. *coccineum* is one of the most common species of the limestone desert, its dominance being associated with the prevalence of the limestone formations. Again within the upstream affluents of the main drainage system which cut across the southern Eocene limestone plateau, Z. *coccineum* is a common plant and in certain localities is dominant. The Z. *coccineum* community does not occur in the gravel desert.

The plant cover of *Z. coccineum* in the Cairo-Suez Desert varies from 5 to 70% in obvious relationship with the water resources and the soil conditions. The most common perennial associate is *Zilla spinosa* which acquires an evergreen habit and may flower all through the year. Common perennial associates are *Fagonia bruguieri. Linaria aegyptiaca* and *Panicum turgidum. Hammada elegans*, one of the most abundant species in the western part of the Cairo-Suez Desert, is here rare. *Artemisia monosperma* also, though dominant elsewhere, is here rare too, as are *Ephedra alata, Fagonia mollis, Launaea spinosa* and *Lygos raetam. Acacia raddiana* is an associate species that deserves special note. It is a relict of a rich growth *of Acacia* scrub which is presumably the natural climax vegetation of the wadis. There are, however, still a few trees of A. *raddiana* in the wadis of the eastern part of the Cairo-Suez Desert. The most common therophytes are *Cleome arabica* and *Zygophyllurn simplex. Mesembryanthemum forsskaolii*, the most common annual in the western part (e.g. in the *Hammada elegans* community), is rare in the *Z. coccineum* community. Other annual associates are *Arnberboa lippii, Anastatica hierochuntica, Astragalus bombycinus, Erodium laciniatum, Ifloga spicata, Reichardia orientalis, Reseda pruinosa* and *Spergularia diandra.*

5. *Zygophyllurn coccineum-Panicum turgidum* community. This community is geographically and ecologically transitional between that dominated by *P. turgidum* and *Z. coccineum.* The former is a widespread community in the western part of the Cairo-Suez Desert whereas the latter is confined to the eastern part. In this co-domination *Z. coccineum* and *P. turgidum* are obviously the two main species. The most common associates are *Farsetia aegyptia, Lasiurus hirsutus, Lygos raetam, Pituranthos tortuosus* and *Zilla spinosa.* The shrub *L. raetam* is a conspicuous feature of this co-domination. *Fagonia bruguieri*, which is a common associate in the *Z. coccineum* community, is here rare. *Acacia raddiana* is an occasional associate. The common annuals include *Erodium laciniatum, Filago spathulata, Ifloga spicata, Malva parviflora, Mesembryanthemum forsskaolii, Rumex vesicarius, Schismus barbatus, Trigonella stellata* and *Zygophyllurn simplex.*
6. *Zygophyllurn coccineum-Launaea spinosa* community. This community is restricted to the extreme eastern part of the Cairo-Suez Desert. Its most evident feature is the abundance of *L. spinosa* which attains a cover equal to that of *Z. coccineum.*

Echinops spinosissimus, Zilla spinosa and *Zygophyllurn decumbens* are the most common associate perennials. Other common perennials are *Achillea fragrantissima, Iphiona mucronata, Panicum turgidum* and *Pituranthos tortuosus*. Less common ones are *Blepharis edulis, Cleome droserifolia* and *Robbairea delileana. Cleome arabica, Ifloga spicata, Lotus halophilus, Malva parviflora* and *Zygophyllurn simplex* are the common annual associates.

7. *Lygos raetam-Panicum turgidum* community. This is an open scrubland type found within the main channels of the larger wadis of Cairo-Suez Desert. It usually occurs in elevated areas in the wadis that may sometimes be flooded. The central part of the channel is normally occupied by this community whereas the bordering sand-and-gravel terraces are covered by a *Hammada elegans* community.

   In the *L. raetam-P. turgidum* co-domination, the cover estimates of both species are equally high. *Zilla spinosa* is the most common associate but its cover is relatively low. *Pituranthos tortuosus* and *Farsetia aegyptia* are common perennial associates: less common ones include *Anabasis articulata, Artemisia monosperma, Convolvulus lanatus, Hammada elegans, Lasiurus hirsutus, Pennisetum dichotomum* and *Zygophyllum coccineum, Fagonia mollis* and *Zygophyllum decumbens*, which are dominant elsewhere, are rarely recorded here. The abundant annuals are *Cleome arabica* and *Plantago ovata*. Other annuals are *Asphodelus tenuifolius, Astragalus bombycinus, Erodium laciniatum, Malva parviflora. Mesem bryanthemum forsskalei* and *Zygophyllum simplex*.

8. *Lygos raetam-Anabasis articulata* community. This is an open scrubland common in the upstream parts of the wadis, including some tributaries dissecting the Oligocene gravels and the Upper Eocen marl. The *Lygos-Panicum* co-domination, on the other hand, occupies the downstream parts of the main wadis. *Panicum* is a favoured fodder plant and provides the main source of animal feed within the Cairo Suez Desert. *Anabasis* and *Lygos* are not usually grazed.

   In the *Lygos-Anabasis* community, the former dominates th frutescent layer and the latter the undergrowth.

   Farsetia aegyptia is the most common associate species but its cove is low. Common perennials here include *Echinops spinosissimus, Gymnocarpos decander, Panicum turgidum, Pituranthos tortuosus* and *Zilla spinosa*. Other perennial associates are *Artemisia monosperma, Astragalus spinosus, Atriplex halimus, Ephedra alata, Hammada elegans, Zygophyllum coccineum* and *Z. decumben* Ephemerals include *Erodium laciniatum, Filago spathulata, Gastrocotyle hispida, Mesembryanthemum forsskaolii, Polycarpaea succulentum* and *Trigonella stellata*.

9. *Anabasis articulata* community. This community is widespread in the southern part of the Cairo-Suez Desert where the fringes of the limestone plateau adjoin the gravel formations of post Eocene age. It is also found, though of limited extent, in the northern extremities of this desert, e.g. Gebel Asfar area. It is mostly confined to wadis of different sizes and is particularly abundant in Wadi Gendali, Wadi Umm Deram and Wadi Abu Derma and their tributaries.

Panicum turgidum is the most frequent associate and *Farsetia aegyptia, Lygos raetam, Pituranthos tortuosus* and *Zilla spinosa* are common. *L. raetam* usually forms an open frutescent layer that includes *Lycium arabicum, Artemisia monosperma* and *Hammada elegans* occasional whereas *Zygophyllum coccineum* is not recorded in the stands of this community. Therophytes include *Anthemis melampodena, Arnebia linearifolia, Bassia muricata, Calendula aegyptiaca, Cleome arabica, Cutandia memphitica, Erodium cicutarium, Gastrocotyle hispida, Matthiola livida, Plantago ovata, Reichardia orientalis, Senecio desfontainei, Trigonella stellata* and *Zygophyllum simplex.*

10. *Ephedra alata* community. This is a well-represented community in Wadi Gendali, Wadi Etheily and limestone parts of other wadis that cut across the middle gravel plain of the Cairo-Suez Desert. In these parts the channels of the wadis are often cut across the whole depth of the gravel beds exposing the underlying limestone or marly limestone at the floor. *Ephedra* builds up sandy mounds that may reach considerable sizes (up to 150cm high) with more or less complete cover. This growth forms a special layer (150–200cm above ground). The spaces between these isle-like, mounds are partly occupied by other species. *Lygos raetam*, when present, is associated with the upper layer. *Panicum turgidum* and *Lasiurus hirsutus* build smaller mounds around those of *Ephedra* forming a second layer (60–120cm). A third layer is formed by the sparse growth of the suffrutescent perennials. The ground layer includes such prostrate and low-growing perennials as *Atractylis flava* and *Fagonia mollis* and it is occasionally rich in ephemerals, e.g. *Asphodelus tenuifolius, Savignya parviflora* and *Trigonella stellata.* Other perennial associates include *Anabasis articulata, Artemisia monosperma, Farsetia aegyptia, Hammada elegans, Pennisetum dichotomum, Pituranthos tortuosus, Zygophyllum coccineum* and *Z. decumbens.*
11. *Artemisia monosperma* community. This community is represented by scattered localities within the Cairo-Suez Desert. It is always within the main channels of wadis. In contrast to many of the other dominants, *A. monosperma* does not seem to build mounds. Though not much grazed, it is subjected to destructive cutting for fuel and other household purposes. The plant shows prolific regeneration by seedlings, a character enjoyed by only a few of the desert perennials.

    Panicum turgidum is the most common associate; it has been recorded in all of the studied stands of the A. *monosperma* community (El-Abyad, 1962). *Lygos raetam, Pituranthos tortuosus* and *Zilla spinosa* are common associate perennials. Species such as *Anabasis articulata, Ephedra alata, Hammada elegans, Zygophyllum coccineum* and *Z. decumbens,* which dominate other communities, are here of minor standing. *Tamarix nilotica* and *Lycium arabicum,* members of the frutescent layer of the desert vegetation, are sparsely or rarely found. Ephemeral associates include *Plantago ovata, Schismus barbatus* and *Zygophyllum simplex.*
12. *Zygophyllum decumbens* community. Within the Cairo-Suez Desert the *Z. decumbens* community is restricted to the northern and eastern foot of the Ataqa range. It usually occupies the narrow and shallow runnels that dissect the low limestone ground extending at the foot of the range.

Z. decumbens often builds small sand mounds that may not exceed 50 cm high, the soil of which is usually mixed sand with some limestone detritus. The cover of this community ranges from 2 to 10%, contributed by Z. *decumbens* which is characterized by succulent leaves that are only scarcely grazed.

Iphiona mucronata is the abundant associate. Other common perennials present include: *Asteriscus graveolens, Gymnocarpos decander, Lasiurus hirsutus, Launaea spinosa, Linaria aegyptiaca* and *Zilla spinosa*. Less common associates include: *Lygos raetam, Panicum turgidum* and *Zygophyllum coccineum. Anabasis articulata* and *Hammada elegans*, dominant elsewhere, are absent here. *Cuscuta pedicellata* is one of the common annuals parasitizing *Iphiona mucronata*. Other annuals include *Asphodelus tenuifolius, Erodium laciniatum, Trigonella stellata* and *Zygophyllum simplex*.

13. *Lasiurus hirsutus* community. This is a desert grassland vegetation present in the minor runnels of the gravel formation whose water revenue is very limited. The stands of this community are usually of limited size: short, narrow and shallow runnels lined with thin or discontinuous sheets of sand (El-Abyad, 1962).

    The most common associate species is *Panicum turgidum* which is closely similar in habitat to *L. hirsutus*. Both are tussock-forming grasses that may develop a deciduous growth form with dry aerial parts. The plants may look dead for several years, but soon after rain they sprout new foliage and regain their green appearance.

    Farsetia aegyptia, Gymnocarpos decander, Hammada elegans and *Pituranthos tortuosus* are common associate perennials. *Artemisia monosperma, Ephedra alata, Lygos raetam, Zygophyllum coccineum* and *Z. decumbens*, which dominate other communities, are here of minor significance. Associate annuals include *Filago spathulata, Ifloga spicata, Medicago hispida, Mesembryanthemum forsskaolii, Plantago ovata, Trigonella stellata* and *Zygophyllum simplex*.

14. *Acacia tortilis* community. Small patches of A. *tortilis* scrubland are found within a limited area 10–17 km north of Suez, some 4–5 km west of the Suez-Ismailia fresh-water canal. Nowhere else in the Cairo-Suez Desert is this type recorded. These patches are apparently relicts of much more widespread vegetation.

    The A. *tortilis* community is confined to runnels traversing sand and gravel country. The survival of these remarkable patches may be due to the seepage of underground water or to the presence of storag ground water.

    In this community, *A. tortilis* forms a definite frutescent layer. The common associates are the low-growing plants, *Convolvulus hystrix, Haplophyllum tuberculatum* and *Heliotropium luteum*, which are the abundant members of the ground layer of this vegetation. *Panicum turgidum, Pituranthos tortuosus, Zilla spinosa* and *Zygophyllum coccineum* are the components of the suffrutescent layer. The annuals which enrich the ground layer include *Atractylis flava, Emex spinosus, Erodium laciniatum, Malva parviflora* and *Schismus barbatus*.

(b) The Limestone Desert

To the west of the Red Sea mountains, north of Lat. 28 °N, extends the limestone Maaza Plateau that makes up most of the inland section of the Eastern Desert

(Abu Al-Izz, 1971). It is an almost uniform flat-topped plateau bordered by huge scarps on all sides and dissected by numerous wadis draining westward to the Nile.

Ecologically, this limestone desert may be divided into two sections: the northern section (Helwan Desert) that includes many wadis, e.g. Bahr-Bila Maya, Digla, Hof, Gibbu, Garawi and Rishrash. The southern section is characterized by the following wadis: Sanur, Tarfa, El-Dir, Tihna, Baathran, El-Mashagig, Hashas, El-Assiuti and Qena (Girgis, 1962; Kassas and Girgis, 1970, 1972; El-Sharkawi and Ramadan, 1983).

(i) The Northern Section (Helwan Desert)

Geomorphology and Climate

Helwan Desert is essentially an expanse of limestone desert fringed on its north and west by sand and gravel formations belonging to the Upper Eocene and Middle Eocene. The Upper Eocene series includes sands, marls, clays and rarely limestone. The Middle Eocene series includes beds for building stone and for the cement industry limestone (Schweinfurth, 1883; Sandford, 1934; Said, 1954; Farag and Ismail, 1956).

The northern front of Helwan Desert is marked by the Bahr-Bila Maya (waterless wadi) which is bounded on the south by the Upper Eocene beds of yellow sandy limestone. On the north are the extensive gravel beds of Gebel El-Khashab (the fossil wood mountain). The mouth of this wadi is bounded on the north by the extremities of Gebel El-Mokattam. To the south, in an almost parallel direction, is Wadi Digla which terminates near the Maadi district; it extends eastward for about 30 km.

Wadi Hof terminates at a point 3 km north of the Helwan district. Its course extends eastward for some 30 km to the foot of Gebel Abu Shama. The main channel of Wadi Hof is a deep ravine cutting across the white and grey limestone of the Middle Eocene.

Wadi Gibbu terminates at a point 2 km south of the western and southern foot of Gebel Abu Shama. The downstream part of Wadi Gibbu is the site of extensive limestone quarrying for the cement fectory of Helwan.

Wadi Garawi discharges at Ettabin (6 km south of Helwan) where an iron and steel factory is built. Its downstream part (6–7 km) is a zone of high fossiliferous Pliocene marly limestone and sandy marl and Plio-Pleistocene terraces of clays and sands.

Schweinfurth (1883) discovered an old dam built across Wadi Garawi. The dam, later described by Murray (1947), was apparently meant to store the torrential waters of the wadi. Its construction is referred to the third or fourth Dynasty.

Further south is Wadi Rishrash, the site of the old garden and an ibex reserve. The downstream part of this wadi cuts across the gravel-and-sand terraces of the Nile Valley, Its eastward extension cuts a deep ravine across the plateau. It receives some of the drainage of the western extremities of the Galala El-Bahariya plateau.

Wadi terraces extend, though discontinuous, on the sides of the floors of the main channels. Terraces are often of alternating bands of alluvium of various textures, which is taken to indicate variations in the volume and velocity of the occasional torrents. Blackenhorn (1921) describes three terraces in the downstream part of Wadi Digla: high (pluvial) terraces (c.7 m thick), low (middle) terraces (c.3 m thick) and wadi flow (lower) terraces.

Helwan Desert is a part of the North African Desert. It lies at the southern boundary of the winter rainfall belt that fringes the area with Mediterranean climate (the Saharan-Mediterranean Climate, Emberger, 1951).

The rainfall is characterized by its scantiness, its seasonality and its inconsistency. Its average annual total is 31 mm, 70% within the December-March period. The June-October period is almost rainless. However, remarkable annual variations in the rainfall have been observed in the years 1900–1961 (Girgis, 1962). Years of the double annual average are recurrent: 1908 (91 mm), 1917 (61 mm) and 1952 (62 mm). Years of less than half the annual average are also recurrent: 1910 (14 mm), 1915 (12 mm), 1922 (13 mm), 1928 (7 mm), 1931 (9 mm), 1932 (6 mm), 1933 (15 mm), 1935 (1 mm) and 1953 (7 mm). The erratic rainfall is local and is caused by cloudbursts giving considerable differences in rainfall from place to place. Sutton (1947) states "Heavy but sporadic storms in desert are usually of the thunderstorm type and apt to cause great floods in otherwise dry wadis".

The Vegetation

The first botanical report on the Helwan Desert is given by Schweinfurth (1901). He lists species "found in the desert east of Helwan within a distance of 10–12 miles from the Nile". He also makes a number of interesting observations on plant life. Stocker (1926–1927) presents an ecological enquiry into the desert of Helwan. Other contributions are by Montasir (1938), Tadros (1949), Kassas and Imam (1954), Girgis (1962), Batanouny (1963) and Kassas and Girgis (1964, 1965).

The vegetation of the Helwan Desert is made up of a permanent framework of perennials and complementary assemblages of therophytes. The structure of the vegetation varies in relation to three main physical features: (a) the extent of the catchment areas; (b) the nature of the surface deposits; and (c) the microclimatic conditions. As the wadis of the Helwan Desert are easily accessible to bedouins and their domestic animals, the vegetation is subject to grazing and cutting. Most of the shrubs are cut for fuel and the few trees of *Acacia raddiana* and *Ziziphus spina-christi* are relicts of previous rich growth. The most common species are, as a rule, the least grazed.

Thirteen communities named after the dominants have been recognized in the Helwan Desert. These are *Stachys aegyptiaca, Zygophyillum decumbens, Z. coccineum, Z. album, Anabasis setifera, A. articulata, Zilla spinosa, Pennisetum divisum, Lycium arabicum, Nitraria retusa, Atriplex halimus, Tamarix nilotica* and *Hammada elegans* (Girgis, 1962).

1. *Stachys aegyptiaca* community. S. *aegyptiaca* is a woolly bush about 1 m tall that usually grows on calcareous rocks (Täckholm, 1974).

   In the Helwan Desert the S. *aegyptiaca* community covers parts of the wadi beds where the rock is exposed and the limestone creviced along joints.

Discontinuous patches of shallow sediments may accumulate in places. This situation occurs in parts of the wadi bed adjacent to, and upstream of, a sudden drop in the level (steps or waterfalls), a situation enabling the ready removal by water action of all or the main part of the surface deposits. The patches of surface sediments are protected by small depressions, boulders etc. As the habitat of the S. *aegyptiaca* phytocoenosis lacks the surface deposits that form the desert soil, this plant assemblage may be a pioneer stage in the development of vegetation in the wadi bed. This habitat allows the growth of rock plants, mostly chasmophytes, rooting in the cracks and deriving their water from the moisture stored in the crevices or in the porous limestone.

S. *aegyptiaca* appears to flower all through the year. The flora of its community comprises *Asteriscus graveolens, Gymnocarpos decander, Iphiona mucronata, Pituranthos tortuosus* and *Zygophyllum coccineum* as common associates. Less common perennials include *Achillea fragrantissima, Alhagi maurorum, Anabasis setifera, Capparis spinosa, Diplotaxis harra, Echinops spinosissimus, Erodium glaucophyllum Fagonia kahirina, F. mollis, Farsetia aegyptia, Hammada elegans, Limonium pruinosum, Pennisetum divisum, Phagnalon barbeyanum* and *Zygophyllum decumbens*. Annual associates include *Anastatica hierochuntica, Diplotaxis acris, Plantago ovata, Trigonella stellata* and *Zygophyllum simplex*.

The species of this community includes a number of chasmophytes growing on the otherwise bare rock surface. Some are strictly chasmophytes with genuine affinity to rocky habitats, e.g. S. *aegyptiaca Gymnocarpos decander, Fagonia mollis* and *Reaumuria hirtella*. Others are facultative chasmophytes, e.g. *Asteriscus graveolens* and *Iphiona mucronata* and there is a third group of non-chasmophytes e.g. *Pennisetum divisum* and *Zilla spinosa*.

2. *Zygophyllum decumbens* community. *Z. decumbens* is a species of peculiar geographical boundaries within the Eastern Desert. It seems confined to a triangle with its apex at a point near El-Saff (c.40 km south of Helwan) in the Nile Valley and its base extending along the coast of the Gulf of Suez from Gebel Ataqa to Gebel El-Galala El-Qibliya. Occasional individuals may be found along the Red Sea coastal desert as far south as Lat. 24 °N. But it is only abundant or dominant within the limits of the above-mentioned triangle (Kassas and Girgis, 1965).

   Within the Helwan Desert, *Z. decumbens* is confined to the southern part. It is absent from Wadis Digla, Hof and Gibbu, poorly represented in Wadi Garawi but well represented in the wadis to the south of Wadi Garawi, e.g. Wadi Rishrash.

   The *Z. decumbens* community is restricted to affluent runnels cutting across the erosion pavement of the lowest level that is the pavement that forms the beds of the wide valley containing the channels of the main wadis. The main wadi cuts its channel across this pavement that extends on its sides at a higher level. The beds of *Z. decumbens* runnels are usually covered with rounded limestone fragments (alluvial) derived from the limestone detritus formed *in situ*, whereas the surface layer includes some transported sandy materials which build around the plants.

In the community dominated by *Z. decumbens* in the Helwan Desert, *Gymnocarpos decander* is the most common associate. *Iphiona mucronata* and *Zilla spinosa* are also common. Other associates are: *Achillea fragrantissima, Anabasis articulata, Atriplex inamoena, Fagonia kahirina, F. mollis, Farsetia aegyptia, Hammada elegam Helianthemum lippii, Heliotropium luteum, Launaea nudicaulis, Limonium pruinosum, Lycium arabicum, Stachys aegyptiaca* and *Zygophyllum album.*

3. *Zygophyllum coccineum* community. Within the part of the Eastern Desert north of Lat. 28′N *Z. coccineum* is confined to the Middle Eocene country and is rarely found within the sand-and-gravel formations of the Oligocene and post-Oligocene.

   In the Helwan Desert, the *Z. coccineum* community is common in the channels of the wadis and their main affluents. The bed is usually covered with a continuous veneer of valley-fill material, mixed with alluvial deposits with particles ranging from fine silt to coarse boulders. The parts of the wadi bed where this community occurs are subject to the scouring effect of torrents. The deposits have a fresh appearance and are light yellow to white. Higher terraces, where the surface deposits are greyish or brownish, do not support this community.

   Reaumuria hirtella and *Zilla spinosa* are the abundant associate perennials of this community. Other perennials include *Achillea fragrantissima, Alhagi maurorum, Anabasis setifera, Atriplex inamoena, Centaurea aegyptiaca, Diplotaxis harra, Erodium glaucophyllum, Fagonia mollis, Farsetia aegyptia, Gymnocarpos decander, Iphiona mucronata, Linaria aegyptiaca, Lycium arabicum, Lygos raetam, Pennisetum divisum, Stachys aegyptiaca* and *Zygophyllum album.* The common annuals include *Ifloga spicata, Plantago ovata, Trigonella stellata* and *Zygophyllum simplex.*

   The cover of the *Z. coccineum* community includes a permanent framework of perennials which is distinguished into a widely open frutescent layer including *Lycium arabicum* and *Lygos raetam*, a suffrutescent layer of the dominant and other bushes and a ground layer including *Atriplex inamoena, Fagonia mollis* (perennials), *Caylusea hexagyna* (biennial) and the ephemerals.

4. *Zygophyllum album* community. *Z. album* is a species of a wide ecological range. It is one of the most abundant of the coastal and inland salt marsh vegetation.

   The *Z. album* community may seem alien to the desert. Within the Helwan Desert, this community occurs in certain wadis, e.g. Wadi El-Warag, and in areas around Helwan Spring where the flowing brackish Water has transformed the area into a saline habitat. This community is also present in parts of the drainage system where alluvial deposits are overlained by aeolian sand, in other words, where the drainage passage is choked by the accumulation of wind-borne sand. The sand is mostly derived from the Nile Valley deposits and is carried eastward by the westerlies. The *Z. album* community may replace that of *Z. coccineum* on the elevated gravel terraces of the main channel of Wadis El-Hay and El-Warag. This wide range of habitat conditions signifies the substantial range of tolerance of *Z. album. Z. coccineum* is the most common associate of the *Z. album* community. Common perennial associates include *Anabasis articulata, Cleome arabica, Fagonia bruguieri, Farsetia aegyptia, Gymnocarpos decander, Hammada elegans, Helianthemum lippii, Iphiona mucronata, Juncus rigidus, Lycium arabicum. Nitraria retusa, Panicum turgidum, Pennisetum divisum, Polycarpaea repens* and *Tamarix nilotica*. Ephemerals include *Bassia muricata, Cotula cincrea, Ifloga spicata Matthiola livida, Plantago ovata, Silene linearis* and *Zygophyllum simplex*. Some of these species deserve special mention. *Nitraria retusa* which is rarely present in this vegetation, dominates a community which represents an advanced stage in the vegetation development of the wadi bed of the limestone desert. *Juncus rigidus* is a salt-tolerant rush. *Pennisetum divisum is* a grassland species which characterizes the silt terraces of the wadis. *Panicum turgidum*, a common grassland species in the gravel-and-sand formation of the gravel desert, is here a rare grass.

5. *Anabasis setifera* community. In several areas of Helwan Desert, notably Wadi Digla, the A. *setifera* community replaces that of *Z. coccineum. A. setifera* grows in the channel of the wadi where the bed is covered with a mantle of coarse rock detritus.

   The structure of this community bears some resemblance to that of *Z. coccineum*. Both of the dominants are succulent perennials but *A. setifera* (Chenopodiaceae) is a winter deciduous plant with the whole of its shoot drying up during the winter. *Z. coccineum* (Zygophyllaceae) is an evergreen. Both species are not grazed when green, but dry A. *setifera* is collected by the bedouins as a palatable dry matter for their livestock.

   The flora of the *A. setifera* community includes *Reaumuria hirtella, Zilla spinosa* and *Zygophyllum coccineum* as the abundant associates. Common associates are *Atriplex halimus, Erodium glaucophyllum, Gymnocarpos decander, Iphiona mucronata* and *Lycium arabicum*. Less common associates are *Artemisia inculta, A. judaica, A. monosperma, Cocculus pendulus, Hammada elegans, Limonium pruinosum, Lygos raetam, Pennisetum divisum* and *Stachys aegyptiaca. Diplotaxis acris, Plantago ovata, Trigonella stellata* and *Zygophyllum simplex* are the common annuals.

   The *A. setifera* community is by no means confined to the channels and tributaries of Wadi Digla. Extensive stands of *A. setifera* are also found in some parts of Wadi Hof and elsewhere. The *Z. coccineum* community appears to occupy the beds of the wadis where Middle Eocene nummulitic hard limestone forming the floor is covered by shallow alluvial detritus. The *A. setifera* community occurs in comparable parts of wadis where marly limestone (mostly Upper Eocene) forming the floor is covered by a veneer of alluvial rock detritus.

6. *Anabasis articulata* community. *A articulate* is one of the desert succulents that is more robust than *A. setifera*. It is capable of building mounds.

   Within the Helwan Desert, the *A. articulata* community is confined to a belt transitional between the Eocene limestone desert to the south and the Oligocene sand-and-gravel desert to the north. This community may be subdivided on ecological and floristic grounds into two subtypes. One subtype occurs in the limestone area and includes several calcicolous species not recorded in the other subtype. The latter is seen on the fringes of the sand-and-gravel desert;

the plants occurring include several arenicolous species not recorded in the other subtype. Both subtypes have enough ecological, floristic and structural affinities to justify being considered subdivisions of one type. Ecologically they both occupy comparable parts of the drainage systems and seem to require a deeper soil (surface deposits) than the previously described communities. The stratification of vegetation is similar in both subtypes: the frutescent layer is well represented by *Lygos raetam* and the suffrutescent one dominated by *A. articulata*.

The flora of the *A. articulata* community in the Helwan Desert has been studied in two types of habitats – within the drainage system of the limestone country, and within the drainage system of the sand-and-gravel formation. In both habitats, *Centaurea aegyptiaca, Farsetia aegyptiaca, Gymnocarpos decander, Lygos raetam, Pituranthos* tortuosus and *Zilla spinosa* are almost equally abundant. The notable difference between the two habitats is the abundance of *Atriplex halimus* on the limestone but its poor representation on the sand-and-gravel habitat. *Reaumuria hirtella* and *Diplotaxis harra* are better favoured on the limestone than on the gravel. *Anabasis setifera* is apparently more selective: on the limestone it is one of the common species whereas it is absent from the gravel desert. *Astragalus spinosus, Pennisetum divisum* and *Urginea undulata* are more favoured in the gravel desert than on the limestone.

7. *Zilla spinosa* community. Within the limestone country of Helwan Desert the *Z. spinosa* community is widespread. It occurs in parts of the wadi bed where the alluvial deposits have a high proportion of soft ingredients. These deposits are not only fine but also deeper than those covering the area occupied by the *Zygophyllum coccineum* community. The depth of these surface deposits is more than 50 cm and alternation of layers of different texture can be seen, a usual character of the wadi-fill material. The *Z. spinosa* community is always confined to the main wadis.

Unlike the *Z. coccineum* community which is restricted to the Eocene limestone country and is virtually absent from the sand-and-gravel desert, the Z. *spinosa* community, though especially common within the limestone area, is also found in parts of the gravel desert. The Z. *spinosa* dominated vegetation seems to be of two types: one in the limestone country and the other in the gravel. The widely spaced species may form other types of communities, elsewhere in the desert.

In the drainage system of the Helwan Desert, *Z. coccineum* is seen to occupy tributary affluents, whereas Z. *spinosa* grows in the main channels receiving the drainage from these affluents. The demarcation between the two types is sometimes very clear. Within the channels of the main wadis *Z. coccineum* occupies the parts of the bed with shallower and coarser deposits whereas *Z. spinosa* is present where the deposits are deeper and softer. The habitat of the *Z. coccineum* community no doubt has a more severe water regime than that of the *Z. spinosa* community.

Z. coccineum is the most closely associated species in the *Z. spinosa* community in the Helwan Desert. Common associates include *Achillea fragrantissima,*

Alhagi maurorum, Farsetia aegyptia, Gymnocarpos decander, Iphiona mucronata, Pennisetum divisum and *Pituranthos tortuosus*; all are species better favoured in the limestone desert except *P. tortuosus* which is very widespread. Other associates include *Anabasis setifera, Asteriscus graveolens, Atriplex halimus, A. leucoclada, Citrullus colocynthis, Echinops spinosissimus, Hammada elegans, Iphiona mucronata, Limonium pruinosum, Lygos raetam, Panicum turgidum, Reaumuria hirtella, Stachys aegyptiaca, Zygophyllum album* and *Z. decumbens*.

8. *Pennisetum divisum* community. The *P. divisum* community is a grassland vegetation confined to the silt terraces that may fringe the channels of the main wadis. *P. divisum (P. dichotomum)* is subject to extensive destruction by grazing and cutting. This reduction of grassland cover is usually followed by the destruction of the silt terraces and hence prevention of its regeneration. The result is that this; grassland, which is perhaps the main grassland of the limestone desert, is represented by limited patches or narrow strips, many of which are obviously relicts of earlier greater patches.

   The habitat of the *P. divisum* community (silt terraces) represents one of the advanced stages of wadi bed development. These terraces; are confined to the main wadis, with extensive catchment areas and considerable water revenue. The silt terraces provide conditions for the storage of moisture in deeply seated layers. As the terraces are protected by the grassland vegetation they are stabilized.

   Zilla spinosa is the abundant associate species occupying the gaps between the tussocks of the dominant. *Launaea nudicaulis* and *Zygophyllum coccineum* are common associates. Less common species include *Achillea fragrantissima, Asteriscus graveolens, Cleome arabica, Cynodon dactylon, Gymnocarpos decander, Hammada elegans, Lavandula stricta, Lycium arabicum, Stachys aegyptiaca* and *Zygophyllum decumbens*.

9. *Lycium arabicum* community. This community represents one of the scrublands that may be present in parts of the main channels of the wadis of the Helwan Desert. It occupies strips of silt terraces fringing the channels of the main wadis. The roots of the dominant extend horizontally for considerable distances. This type of root growth gives effective protection of the underlying silt deposits. The woody stems *of Lycium* are often cut for fuel. The destruction of the scrub entails the destruction of the terraces, damage which is often irreparable.

   The cover of the *L. arabicum* community is in three main layers. The frutescent layer is dominated by *L. arabicum*. Associate shrubs include *Atriplex halimus* and rarely *Lygos raetam*. The suffrutescent layer contains the greater number of perennials including *Achillea fragrantissima, Anabasis setifera, Pennisetum divisum, Zilla spinosa* and *Zygophyllum coccineum*. The ground layer is made up of dwarf and prostrate perennials, e.g. *Cynodon dactylon, Erodium glaucophyllum* and *Fagonia mollis*. During spring this layer is enriched by the growth of the therophytes, e.g. *Erodium laciniatum, Malva parvi-flora, Schismus barbatus, Trigonella stellata* and *Zygophyllum simplex*. Apart from the above-mentioned species, *Artemisia judaica, Farsetia aegyptia, Halogeton alopecuroides* and *Zygophyllum album* are also common.

10. *Nitraria retusa* community. *N. retusa* is a common shrub of the coastal and inland salt marshes of Egypt. It is also common on the silt terraces of the wadis of the limestone desert where it may form scrubland subject to destruction for fuel but not to serious grazing. The bushes *of N. retusa* form patches close to the ground and provide effective protection. This shrub is capable of building hillocks, the woody shoots, when covered with sand, producing adventitious roots and new shoots that help to trap the accumulating sand. The fleshy, sweet and red fruits *of Nitraria* are sought by bedouins and birds.

    In the limestone desert, the cover of this community is often dense (about 60%). *Alhagi maurorum* and *Zygophyllum coccineum* are the most common associates. Less common species include *Limonium pruinosum, Tamarix nilotica, Zilla spinosa* and *Zygophyllum album.* The other perennial associates are *Artemisia judaica, Cynodon dactylon, Diplotaxis harra, Francoeuria crispa, Juncus acutus, Lavandula stricta, Pennisetum divisum, Pituranthos tortuosus* and *Polygonum equisetiforme*. The annuals include *Diplotaxis acris, Schismus barbatus, Trigonella stellata* and *Zygophyllum simplex.*

    The *N. retusa* community includes a number of associate halophytes: *Juncus acutus, Limonium pruinosurn, Tamarix nilotica* and *Zygophyllum album*, indicating that the soil is fairly saline.
11. *Atriplex halimus* community. Within the Helwan Desert *A. halimus* scrub is confined to the drainage system of Wadi Digla and its tributaries. In this respect the dominant is comparable to *Anabasis articulata.* These two species are also associates in the region bordering the Western Mediterranean coastal belt (Tadros, 1953).

    A. *halimus* covers strips of the silt terraces fringing the channels of the main affluents and also forms island-like patches within the channels. In both instances it overlies and protects bodies of alluvial silt. The cover of this community (20–60%) is contributed mainly by A. *halimus. Anabasis setifera* and *Halogeton alopecuroides* are the abundant associates whereas *Farsetia aegyptia, Lycium arabicum, Pituranthos tortuosus* and *Reaumuria hirtella* are common. Less common associates include *Lygos raetam, Pennisetum divisum, Zilla spinosa* and *Zygophyllum coccineum.* Rarely recorded are *Anabasis articulata, Euphorbia kahirensis, Hammada elegans, Haplophyllum tuberculatum* and *Zygophyllum album.* The annuals include *Malva parviflora, Plantago ovata, Reichardia orientalis* and *Trigonella stellata.*
12. *Tamarix nilotica* community. The *T nilotica* scrubland is considered one of the climax types of vegetation in the desert wadis. It has been subject to destruction by cutting for centuries. The result is its almost complete eradication. In the Helwan Desert it is represented by a few relict stands especially in Wadi Rishrash which has, "for several decades, been protected as a private reserve (ex Royal Reserve for the desert Ibex)" (Girgis, 1962).

    The phytocoenosis dominated by *T. nilotica* includes *Alhagi maurorum. Ochradenus baccatus* and *Zygophyllum coccineum* are common associates. Other associates are *Achillea fragrantissima, Artemisia judaica, Atriplex halimus, Francoeuria crispa, Nitraria retusa, Panicum turgidum, Pituranthos tortuosus, P. triradiatus, Zilla spinosa* and *Zygophyllum album.*

13. *Hammada elegans* community. The previously described communities form the plant life of the beds of the limestone desert. The *H. elegans* community occurs in the wadis of the sand-and-gravel desert which separates the limestone plateau from the Nile Valley.

    The flora of the *H. elegans* community includes only four of the dominant species of the twelve previously mentioned communities: *Pennisetum divisum, Zilla spinosa, Zygophyllum album* and *Z. coccineum*. Other associates include *Asthenatherum forsskaolii, Atractylis flava, Cleome arabica, Convolvulus lanatus, Cornulaca monacantha, Diplotaxis acris, Fagonia arabica, Farsetia aegyptia, Heliotropium undulatum, Moltkiopsis ciliata, Panicum turgidum, Polycarpaea repens* and *Stipagrostis plumosa*. This community is also characterized by the abundance of annuals, e.g. *Cotula cinerea, Eremobium aegyptiacum, Ifloga spicata, Matthiola livida* and *Zygophyllum simplex. Cistanche tinctoria* is a common parasite, its host being *H. elegans*.

The Drainage Runnels

These are the minor affluents of the drainage system which receive run-off water from limited catchment areas. Their floors are either devoid of alluvial detritus or covered with coarse boulders and huge blocks (alluvial detritus).

In the Helwan Desert there are two main types of affluent runnels: (a) runnels cutting backwards: (1) rill-line across rocky slopes, (2) precipitous cliffs, (3) stepped cliffs, (4) stepped runnels and (b) runnels dissecting erosion surfaces: (5) short shallow runnels across erosion pavement, (6) long (shallow) runnels across erosion pavement and (7) long (deep) runnels across erosion pavement.

The vegetation of these runnels shows differences of note, but is here much simpler than that of the main wadis already described. Communities are recognized on the basis of the consistent presence of a single species and the character of the seven habitat types.

Rill-Runnels Across Rocky Slopes

A few isolated plants may be present in this habitat. *Fagonia kahirina* is the most common and is often the only one. Rare individuals of *Asteriscus graveolens* and *Reaumuria hirtella* may occur. In years of good rainfall, plants of, for example *Diplotaxis harra, Plantago ovata* and *Schismus barbatus*, may be present. This sparse growth is associated with rill-lines across rocky slopes facing north, northwest and northeast. Rill-lines across south-facing slopes are barren.

Precipitous Cliffs

The most characteristic species of this habitat is *Capparis spinosa*. Associates include *Iphiona mucronata* and *Zygophyllum coccineum. Lycium arabicum* is an "accidental" species not normally in this habitat. Saxicolous associates include *Anabasis setifera, Fagonia mollis, Limonium pruinosum* and *Stachys aegyptiaca*.

Stepped Cliffs

In this habitat, *Limonium pruinosum* is the abundant species and gives the character to the vegetation. *Zygophyllum coccineum* is the most common associate. Common ones include *Fagonia kahirina, F. mollis, Farsetia aegyptia, Gymnocarpos decander* and *Reaumuria hirtella. Capparis spinosa* is here less common. Other associates include *Asteriscus graveolens, Erodium glaucophyllum* and *Helianthemum kahiricum.*

Stepped Runnels

In these runnels, *Erodium glaucophyllum* is the dominant. Common associates are *Asteriscus graveolens, Farsetia aegyptia, Reaumuria hirtella* and *Zygophyllum coccineum.* Less common are *Achillea fragrantissima* and *Lasiurus hirsutus* (non-chasmophytes) which are recorded in this vegetation but not in the previously mentioned type. Other common associates of the stepped cliffs type are here rarely present.

Short Shallow Runnels

These runnels are characterized by the abundance of three species: *Fagonia mollis* dominates the smallest runnels, *Asteriscus graveolens* the medium runnels and *Gymnocarpos decander* is dominant in the larger ones. *Erodium glaucophyllum, Reaumuria hirtella* and *Zygophyllum coccineum* are common associates. Species with low presence values include *Lycium arabicum, Paronychia desertorum* and *Zilla spinosa.*

Long Shallow Runnels

These runnels across hamada (xeric) type of erosion pavement are dominated by *Iphiona mucronata.* Associate perennials are *Asteriscus graveolens, Diplotaxis harra, Erodium glaucophyllum, Fagonia mollis, Farsetia aegyptia, Gymnocarpos decander, Reaumuria hirtella* and *Zygophyllum coccineum.* Annuals, e.g. *Diplotaxis acris* and *Plantago ovata*, occur during rainy years.

Long Deep Runnels

Zygophyllum coccineum is the dominant of these runnels with *Asteriscus graveolens, Erodium glaucophyllum, Farsetia aegyptia, Reaumuria hirtella* and *Zilla spinosa* as common associates. This community is similar to that of the main wadis.

(ii) The Southern Section (Beni Suef-Qena Desert)

These sections of the limestone part of the Eastern Desert extend between Lat. 28°50N and 25°30N, covering the desert areas of five provinces of Upper Egypt, namely: Beni Suef, Minya, Assiut, Sohag and Qena (Fig. 2.1). It is bounded on the north by the Helwan Desert and on the south by the non-calcareous sandstone (Idfu-Kom Ombo) desert.

This section is in the extremely arid part of Egypt which is almost rainless. According to the *Climatic Normals of Egypt* (Anonymous, 1960), the average annual rainfall during the period 1946–1960 was: Beni Suef 8.5 mm, Minya 5.3 mm, Assiut 0.4 mm, Sohag 1.0 mm and Qena 5.3 mm., mainly in winter. These mean values are not due to recurrent annual rainfall but to "accidental" cloudbursts. Sutton (1947) notes that rain may occur once every several years. Dewfall may be a vital source of water for the vegetation, especially the therophytes. The annual mean minimum and annual mean maximum temperatures are: Beni Suef 13.3 °C and 29.8 °C, Minya 13.1 °C and 29.8 °C, Assiut 15.4 °C and 30.4 °C, Sohag 14.5 °C and 31.4 °C and Qena 16.4 °C and 33.5 °C. The lowest absolute minimum and the highest absolute maximum temperatures are: Beni Suef 3.3 °C and 45.7 °C, Minya –0.4 °C and 47.5 °C, Assiut 0.4 °C and 47.7 °C, Sohag 0.0 °C and 46.5 °C and Qena 0.0 °C and 48.2 °C, in winter and summer respectively.

Ecologically, this section of the Eastern Desert may be divided into:

1. Beni Suef-Minya Desert
2. Assiut-Qena Desert

Beni Suef-Minya Desert

Geomorphology

This area is between Lat. 28°50′N and 27°30′N. It is characterized by seven major wadis, namely (from north to south): Sanur, Tarfa, Garf El-Dir, Tihna, Baathran, El-Mashagig and Hashas. The drainage of surface run-off waters received by such wadis leads to the turbidity of the Nile for a considerable distance because of the alluvium which it carries. The turbidity may be expected to have some ecological effects on the vegetation of at least the major drainage trunks of the wadis (El-Sharkawi and Ramadan, 1983).

The area generally slopes towards the NNW. Consequently the lowest points are in the northern part with an average elevation of 50 m above sea level. The wadis of this area are subparallel (except Wadi Baathran) and join the Nile at approximately the same acute angle. The main courses of some of the wadis reach a few kilometres in width, e.g. Wadis Garf El-Dir and Tihna are approximately 3 km wide. The floors of the wadis are covered with alluvium ranging between a few centimetres to several metres thick. These deposits are mainly of nummulitic fragments, quartz, sands and limestone rock fragments.

Vegetation

According to El-Sharkawi and Ramadan (1983), the vegetation of this limestone part of the Eastern Desert is characterized by: 1. Alliance *Zygophyllaeion coccini* 2. Four communities and 3. Companion species.

Alliance Zygophyllaeion Coccini

Three species – *Zygophyllum coccineum, Zilla spinosa* and *Fagonia arabica* – are represented in most of the stands in this area. They are referred to as members of the Alliance *Zygophyllaeion coccini*. The presence values of the three species are: 75%,

70% and 55% respectively. Such relatively high presence values indicate a wide range of tolerance to adverse conditions characterizing such habitats, particularly severe climatic aridity. Phenologically, each of the three species has its own means of tolerance. *Z. coccineum*, being a perennial succulent undershrub, can withstand the scantiness of rainfall, which is not a recurrent event. *Z. spinosa*, a perennial spinescent shrub, can change its mode of growth according to the amount of water available in the soil. It can adopt an annual growth form when soil moisture is available (Kassas and Girgis, 1970). *F. arabica*, also widely distributed, shows considerable tolerance to poor water resources.

The Communities

Four communities are recognized in the wadis of this desert area. Each is co-dominated by two species.

1. *Farsetia aegyptia-Salsola kali* community. This vegetation is characterized by the presence of *Alhagi maurorum, Imperata cylindrica, Launaea cassiniana* and *Trichodesma africanum* as associate species.
2. *Francoeuria crispa-Salsola volkensii* community. In addition to the dominant species, this community is characterized by the following associates: *Artemisia judaica, Astragalus sieberi, A. trigonus, Atriplex inamoena, Ochradenus baccatus, Pituranthos triradiatus, Salsola delileana, Salvia aegyptiaca, Suaeda vera* and *Tamarix amplexicaulis*.
3. *Echinops spinosissimus-Hammada elegans* community. The associate species of this vegetation are *Ochradenus baccatus, Panicum turgidum* and *Pennisetum divisum*.
4. *Echium rauwolfii-Pulicaria undulata* community. The characteristic species of this community include *Capparis aegyptia, Hyoscyamus muticus* and *Pergularia tomentosa*. This vegetation is poorly represented as its members are subject to destructive effects.

The Companion Species

These species, other than the above mentioned, of the wadis are *Anabasis setifera, Aristida adscensionis, Calligonum comosum, Heliotropium digynum, Launaea nudicaulis* and *Monsonia nivea*. The catchment areas play an important role in influencing the amount of plant cover and the phenology of the species present. The density of vegetation varies in the main wadis from that in the small affluents and ravines. In the main wadis, the vegetation is much denser and the vigour of species is greater as observed in Wadis Tihna, Tarfa, El-Mashagig and the western section of Wadi Hashas. In contrast, some affluents of Wadis Baathran, Hashas and Garf El-Dir are characterized by thin plant cover and low abundance value of species.

The distribution of certain species in specific ecologically defined (edaphic) habitats seems to substantiate the view that such species are useful as indicators for their habitat characters even under adverse conditions of high disturbance. Among the species that could be referred to as indicators are *Imperata cylindrica, Launaea cassiniana, Francoeuria crispa* and *Alhagi maurorum*, which are well established in the delta of Wadi El-Mashagig. *Hammada elegans* is widespread in these wadis. *Panicum turgidum* and *Pennisetum divisum* are two grasses of the loose sandy terraces of Wadi Tihna. *Capparis aegyptia* grows in the rocky base of the desert plateau of these wadis.

Assiut-Qena Desert

Geomorphology

This is the part of the inland Eastern Desert between Lat. 27°30′N and 25°30′N. It represents, from a lithological point of view, the transition between the limestone plateau (in the north) and the non-calcareous desert (in the south). It is dissected by several main wadis, e.g. El-Assiuti, Bir El-Ain, Qassab, Qena, Zaidun, El-Matuli and El-Qarn that flow into the Nile.

Wadi El-Assiuti drains into a rectangular plain covered by alluvial gravel. This depression is continuous with the River Nile and its deposits include Pliocene and Post-Pliocene terraces of the Nile Valley. The main channel of Wadi El-Assiuti has its head on a high part of the limestone plateau (above 700 m) which forms the divide between Wadi Qena in the east and wadis of the limestone plateau on the west. The main channel of Wadi El-Assiuti runs east-west and its downstream part traverses the gravel plain before it joins the Nile Valley a few kilometres south of Assiut city. Through its course it receives numerous affluents including Wadis Hubara, Marahel and Habib. Wadi El-Assiuti drains a part of the rainless desert of Egypt and has no connection with the Red Sea mountains which may provide other drainage systems of the Eastern Desert with some water.

Wadi Bir El-Ain is one of the longest wadis crossing the plateau to the east of Sohag (90 km south of Assiut). It extends in a NE-SW direction for approximately 40 km. The wadi walls rise about 130–150 m above its floor. From the mouth of the wadi to Bir El-Ain (spring) (7 km NE), the floor is almost bare and flat, whereas the rest is covered by large boulders and cobbles from the surrounding rocky walls. The width of the wadi varies considerably from 20 to 50 m.

Wadi Qassab is one of the few wadis of the Eastern Desert which extends for about 80 km in a nearly north-south direction. Because of this orientation, it appears to receive less run-off waters compared to those of wadis running east-west. Few ravines, however, drain into the main wadi course. The mouth of the wadi is about 55 km south of Sohag town, east of the Nile Valley in an area which lies between Lat. 26°20′ and 26°45′N and Long. 31°50′ and 32°50′E (El-Sharkawi et al., 1984). Along its course this wadi varies in width from 150 m to about 1 km.

Wadi Qena (Lat. 26°10′N) is the greatest of the Eastern Desert, being about 300 km long. The north-south course of its channel is one of its main characteristics; other principal wadis of the Eastern Desert run mostly E-W or W-E. On the eastern side it receives numerous tributaries collecting the westward drainage of the Red Sea hills. On the western side it receives a number of small tributaries draining the eastern scarps of the limestone plateau which contribute very little to the water resources of the wadi. The downstream part traverses a wide valley which joins a broad gravel-covered deltaic plain bordering the Nile Valley. The alluvial gravels of these downstream parts overlie Pliocene deposits. "Massive hills of these Pliocene deposits appear on the fringe of the mouth of Wadi Qena" (Sandford, 1934).

The bottom of this wadi is filled with debris of different sizes. This means that the amount of running water varies from time to time and the coarse material in the floor of the wadi retards evaporation, but there is no water on the surface since the gravel is unretentive. Near its mouth, Wadi Qena is joined by a major tributary, from the north through the limestone plateau. The central part of Wadi Qena is almost without vegetation because the high speed of the water in that section removes the soil. The only plants in that area are on the sides of the wadi. The influence of grazing has left its mark on the wadi, the areas with the richest vegetation being the least grazed.

Along the slopes of Wadi Qena, gravel deposits form terraces of Pleistocene age. This indicates that the wadi experienced several stages of erosion and deposition related to change in climatic conditions during the Pleistocene.

Further south of Qena is a deltaic plain into which the El-Laqeita drainage system flows. Wadi Zaidun is the main part of this system. The middle part of this wadi traverses sandstone country, whereas the upstream part extends across the basement complex formations of the Red Sea hills.

Wadi El-Matuli is a tributary of Wadi El-Qarn and the latter comprises the deltaic part (El-Sharkawi et al., 1982a). Both wadis run in an area extending about 15 km east of the Nile Valley in the vicinity of Qift (about 30 km south of Qena). The two wadis are rather wide (2 km in some parts), with a flat floor which is mostly exposed to solar radiation at daytime and lacking microhabitat shelters for shade plants, as are present in Wadi Bir El-Ain.

Plant Cover

The plant cover of the southern section of the limestone desert may be considered as follows:

1. Vegetation of the wadis
2. The communities

Vegetation of the Wadis

a. Wadi El-Assiuti. Wadi El-Assiuti may be divided ecologically into three main parts:

1. The downstream deltaic plain, covered by deep alluvium of gravel and sand and dissected by a network of ill-defined water courses;
2. The middle part which has a clearly defined channel, but is often choked with sand embankments bounded by limestone cliffs;
3. The upstream tributaries and runnels dissecting the limestone plateau. Their beds are often covered by coarse deposits of limestone detritus.

Reference may also be made to Wadi Habib, the main tributary flowing into the deltaic plain of Wadi El-Assiuti (Kassas and Girgis, 1972).

The vegetation of the deltaic plain of Wadi El-Assiuti is mostly confined to shallow courses that dissect the sand-and-gravel beds of this plain. The larger courses are the habitat of scrubland vegetation of evergreen growth of *Leptadenia pyrotechnica* associated with rich undergrowth of *Calligonum comosum*. Other associates include *Artemisia judaica, Cornulaca monacantha, Cotula cinerea, Echium rauwolfii, Francoeuria crispa, Phaeopappus scoparius, Poiycarpaea repens, Zilla spinosa* and *Zygophyllum coccineum*. In the smaller courses the vegetation is mostly dominated by *Zilla spinosa* which shows a distinctly deciduous growth form. Associate species include *Calligonum comosum, Cornulaca monacantha, Francoeuria crispa* and *Phaeopappus scoparius*.

The main water course of Wadi El-Assiuti occupies the bed of a well-defined channel which is cut deep through the massive limestone formation. The east-west channel intercepts wind-blown sand which may be deposited as embankments over the northern cliffs bounding the channel, or as small sand dunes or mounds. Some of these dunes are partly stabilized by plants.

There are three main types of vegetation in this part of Wadi El-Assiuti. *Cornulaca monacantha* dominates in areas where limestone detritus of valley-fill is associated with a thin cover of aeolian sand. *Calligonum comosum* forms low sand mounds. *Tamarix aphylla* grows on sand hillocks. In the rainy years ephemeral growth of *Eremobium aegyptiacum* may appear on these sandy deposits.

The affluents and runnels which form parts of the drainage system of Wadi El-Assiuti vary in size as do their catchment areas and water resources. Beds of these affluents are usually covered by coarse alluvium. The vegetation differs in relation to the size of the catchment. In smaller runnels, the plant cover includes decidous growth forms of *Zilla spinosa* and *Artemisia judaica* together with *Anabasis setifera*. In larger affluents the vegetation is of patches *of Calligonum comosum* with *Artemisia judaica, Atriplex leucoclada, Cornulaca monacantha* (occasional bushes) and rare presence of *Acacia raddiana, Eremobium aegyptiacum, Heliotropium ramosissimum* and *Leptadenia pyrotechnica*.

The vegetation of Wadi Habib is of only a few plants of *Leptadenia pyrotechnica* and *Calligonum comosum*. The growth of the former occupies the water course of the wadi whereas *C. comosum* forms sand mounds on the terraces fringing the water course. Apart from these two species there are dead remains of others. Plant populations may be substantially enlarged in the rainy season.

b. Wadi *Bir El-Ain*. According to El-Sharkawi and Fayed (1975), the distribution of plants in Wadi Bir El-Ain seems to be controlled by: first the water requirement of species, second the depth of the water-table and third, the light requirement. Light may be important since the orientation of the wadi changes abruptly, resulting in various degrees of shading. Species with a high chlorophyll content, such as *Capparis spinosa*, are found mainly in areas of lowest light intensity and this species flowers in these areas in particular. On the other hand, species with a low chlorophyll content, e.g. *Zygophyllum coccineum*, are present mainly in exposed areas. However, plant distribution seems to depend primarily on water relations and species may be classified into three types in this respect as follows:

1. Plants of wide distribution, of moderate water requirements. These include *Atriplex dimorphostegia, Capparis spinosa, Fagonia thebaica, Pulicaria undulata* and *Zilla spinosa. C. spinosa* grows on the rocky sidewalls of the wadi;

2. Plants with high water requirements, e.g. *Acacia raddiana, Desmostachya bipinnata, Gnaphalium luteo-album, Juncus rigidus, Ochradenus baccatus, Phoenix dactylifera, Phragmites australis* and *Tamarix aphylla*. These plants are limited to sites of water accumulation or crack seepage;
3. Plants with low water requirements, e.g. *Fagonia bruguieri, Forsskaolea tenacissima, Leptadenia pyrotechnica* and *Moringa peregrina*.

Other species recorded in Wadi Bir El-Ain are *Alhagi maurorum, Artemisia judaica, Cynanchum acutum, Dactylis glomerata, Phaeopappus scoparius, Salsola rigida* (=*S. orientalis*), *Salvia aegyptiaca* and *Ziziphus spina-christi*.

Moringa peregrina is a xerophytic shrub which is plentiful at the foot of the high mountains (>1300 m) of the Red Sea coast of Egypt where relatively high amounts of water accumulate (Kassas and Zahran, 1971). El-Sharkawi and Fayed (1975) consider it a plant with low water requirements but further investigations on its water relations are needed.

c. Wadi Qassab. Of the 48 species recorded in the vegetation of Wadi Qassab 14 are of wide distribution. The rest fall into two groups both of limited distribution (El-Sharkawi et al., 1984). Widely occurring perennials include *Fagonia indica, Forsskaolea tenacissima, Leptadenia pyrotechnica, Morettia philaena, Pulicaria undulata, Trichodesma africanum, Zilla spinosa* and *Zygophyllum coccineum*. The annuals include *Oligomeris linifolia, Schouwia thebaica* and *Trigonella stellata*. Habitats favourable for these species are apparently the relatively dry up- and midstream parts of the wadi. However, in areas of the wadi where there are differences in soil characteristics, these are reflected in changes in vegetation. In areas of alluvial loose non-saline soil of high moisture content a community of annuals co-dominated by *Zygophyllum simplex* and *Frankenia pulverulenta*, associated with *Anastatica hierochuntica, Convolvulus* sp., *Diplotaxis acris* and *Lotononis platycarpa* occurs. Local ecological conditions favourable for such a community are apparently fulfilled at low midstream locations just near the deltaic sedimentary part of the wadi. The appearance of this community in such habitats is temporary and it nourishes only after sufficient surface run-off water has been absorbed. Other plants of this community are *Acacia raddiana, Cleome droserifolia* and *Filago spathulata*.

A *Salsola baryosma-Ochradenus baccatus* community occupies the relatively wet habitats, and its presence indicates slight salinization of at least the surface soil. Common associates include *Artemisia judaica, Francoeuria crispa, Matthiola livida, Plantago ovata* and *Rumex vesicarius*. Three species show affinity to this community, although they grow in dry habitats-*Acacia raddiana, Cleome droserifolia* and *Filago spathulata*. Stands of this community are distributed widely in the wadi course except for the extreme upstream dry part. Apart from the above-mentioned species, the following are also recorded within these stands: *Amberboa lippii, Bassia muricata, Citrullus colocynthis, Fagonia boulosii, F. bruguieri, F. kassasii, Iphiona mucronata, Kickxia aegyptiaca, Launaea capitata, L. mucronata. Paronychia arabica, Phoenix dactylifera, Robbairea delileana, Schismus barbatus, Sonchus oleraceus, Spergularia diandra, Tamarix nilotica* and *Tribulus kaiseri*.

d. Wadi Qena. The drainage system of Wadi Qena may, ecologically, be divided into three areas: the deltaic plain, the principal channel and the affluent tributaries and their runnel feeders.

The deltaic plain that extends for 40 km is of two sections: the downstream one which is a part of the plain that fringes the Nile Valley, and the upstream one which is bounded on the west by Gebel Aras and on the east by Gebel Abu Had and Gebel Qreiya. In the downstream section the channel of Wadi Qena cuts its shallow course across sand-and-gravel beds of the 50-foot terrace of the lower Palaeolithic (Sandford, 1929). The vegetation is a thin cover of *Aerva javanica, Artemisia judaica, Cleome droserifolia, Francoeuria crispa, Hammada elegans, Leptadenia pyrotechnica, Ochradenus baccatus, Zilla spinosa* and *Zygophyllum coccineum*. The upstream section is characterized by clearly defined terraces, the higher ones (Pliocene gravel) and the 100-foot terraces being usually barren. Terraces at a lower level, fringing the 100-foot ones, are characterized by hillocks that are relicts of the growth *of Tamarix aphylla*. In certain parts these fossil *Tamarix* hillocks are so covered and extensive that they seem, to indicate former dense growth of *Tamarix* forest. The lower terraces are covered in parts with an open *Tamarix* scrub of varying density. The beds of the network of the water courses are the habitat of desert vegetation with *Zygophyllum coccineum* and *Hammada elegant* as the abundant plants. Apart from the above-mentioned species, the flora of the deltaic part of Wadi Qena includes also: *Acacia ehren-bergiana, Calligonum comosum, Capparis spinosa, Heliotropium bacciferum, Pergularia tomentosa* and *Tamarix nilotica*.

Within the deltaic plain of Wadi Qena are a few wells in the wadi floor where the water is apparently due to seepage from the Nile through the porous deposits of the deltaic plain.

Reference has been made to the extensive relicts of *Tamarix* growth on the intermediate terraces of Wadi Qena. These are hillocks of sandy material admixed with remains of wood and branches of *Tamarix*. On the lower terraces (3–4 m lower) are patches of rich growth of *T. aphylla* in the part upstream of Bir Aras.

The principal channel of Wadi Qena upstream of the deltaic part is a well-defined course bounded on both sides by a gently sloping plateau. The wadi-fill deposits are obviously deep and mostly compact, apparently due to the incorporation of soft silt. The vegetation is mostly confined to the fringes of the water course and is mainly an *Acacia ehrenbergiana* scrub of various density. Associate species include *Aerva javanica, Artemisia judaica, Francoeuria crispa, Leptadenia pyrotechnica, Salsola baryosma* and *Zilla spinosa*.

The main trunk of Wadi Qena receives, throughout its long course, numerous tributaries on the eastern side but only a few on the west. The eastern tributaries collect the westward drainage of the Red Sea chain of mountains and form the main feeds of Wadi Qena. The western ones drain the rainless limestone plateau.

Wadi Gurdi is the main westward tributary. Apart from its downstream confluence with the principal channel of Wadi Qena, where there is an open *Acacia ehrenbergiana* scrub, the wadi has a very thin suffrutescent growth. In the other parts of its main channel, *Zilla spinosa* is plentiful but bushes of *A. ehrenbergiana* are extremely scarce. In affluent runnels of Wadi Gurdi a thin growth of *Zygophyllum coccineum* is present.

Wadi El-Atrash is one of the main eastern tributaries. It has its head affluents on the slope of Gebel Dokhan and Gebel Attar of the Red Sea mountains. In the main trunk of this wadi are patches of *Leptadenia pyrotechnica* open scrub. The course of some of its affluents may be choked with sheets of sand, which are the habitat of a rich growth of *Calligonum comosum*. The downstream part of Wadi El-Atrash has an ill-defined course traversing a gently sloping erosion pavement covered with basement complex detritus of mosaic appearance and with sand sheets. On these sheets the plants are mainly the annuals *Fagonia mollis* and *Morettia philaena*. Wadi Fatira is a twin of Wadi El-Atrash. Its head parts drain the slopes of Gebel Shayeb El-Banat and Gebel Abu Hamr of the Red Sea coast. In the main channel of this wadi the vegetation is mostly *of Zilla spinosa*. The heads of these wadis are associated with mountain country and their vegetation indicates less arid conditions than do their downstream parts. In these head parts *Acacia raddiana* and *Moringa peregrina* may form local patches of scrubland.

Wadi El-Qreiya joins the deltaic part of Wadi Qena at Bir Aras. It receives two principal tributaries: Wadi El-Markh and Wadi Hamama. These wadis drain some of the Red Sea hills. In the upstream part of Wadi El-Markh are patches of *Leptadenia pyrotechnica* scrub with occasional trees *of Acacia raddiana*. In the middle and downstream parts *Leptadenia* is very scarce and the vegetation is essentially of *Zilla spinosa* associated with *Artemisia judaica* and *Aerva javanica*. The vegetation of the principal course of Wadi Hamama is essentially an open *Acacia ehrenbergiana* scrub, and of the affluents of this wadi mostly of the *Zilla spinosa-Zygophyllum coccineum* type.

e. Wadi Zaidun. This drainage system may be divided into four sections: deltaic plain, main channel, tributary wadis and affluents in the montane country. The deltaic plain is extensive and dissected by a network of shallow courses studded by massive bodies of sand dunes covered by a rich growth *of Tamarix nilotica*. There are only a few of the fossil *Tamarix* dunes of the type that characterizes the downstream part of Wadi Qena. The water course channels are the habitat of vegetation dominated by *Zygophyllum coccineum*. Associates include *Aerva javanica, Crotalaria aegyptiaca* and *Zilla spinosa. Salsola baryosma* is the most abundant species on the flat ground amidst the *Tamarix* dunes and on the island-like patches between the network of water courses. Dead remains of *Schouwia thebaica, Tribulus longipetalus* and *Zygophyllum simplex* are abundant. These remains are indicative of a rich growth of ephemerals that may appear in rainy years.

In the eastern part of El-Laqeita plain, and in the sandstone country further eastward, are a number of clearly defined channels including the main channel of Wadi Zaidun. This part of the drainage system is intermediate between the deltaic plain on the west and the numerous tributary wadis dissecting the basement-complex country on the east. The vegetation in this part is essentially an open scrub of *Leptadenia pyrotechnica* with extensive patches *of Crotalaria aegyptiaca* or *Salsola baryosma*. Associate species include *Aerva javanica* and *Fagonia bruguieri. Zygophyllum coccineum* is either very rare or absent. The rarity of *Z. coccineum* is one of the notable features of the vegetation as it is a most common species in the downstream plain and in the upstream tributaries and affluents of the montane country.

The main channels of the tributaries of Wadi Zaidun drainage system traverse the hills of the basement complex country. The vegetation of these wadis is mainly an *Acacia ehrenbergiana* scrub of various density. Associate shrubs include *A. raddiana* and *Capparis decidua*. The undergrowth is formed by *Aerva javanica, Cassia senna, Citrullus colocynthis, Crotalaria aegyptiaca, Fagonia bruguieri, Francoeuria crispa, Pulicaria undulata, Zilla spinosa* and *Zygophyllum coccineurn*. In certain localities there may be one or both of the two climbers *Ochradenus baccatus* and *Pergularia tomentosa*.

The upstream tributaries of the drainage system collect run-off of the western slopes of the chain of mountains facing the main watershed. The vegetation of these wadis is mainly open forest of *Acacia raddiana* with a decreasing population *of A. ehrenbergiana*; patches of *Salvadora persica* are present in some of these affluents.

f-g. Wadi El-Matuli and Wadi El-Qarn These two wadis, though limited in catchment areas relative to other wadis draining the Eastern Desert into the Nile Valley, are rich in vegetation cover. Twenty-four species have been recorded (El-Sharkawi et al., 1982a) which fall into three major communities. The local geographical distribution of these communities seems to be governed by two major factors:

1. The relative aridity of their habitats; upstream stands catch little water and contain xerophytic species whereas downstream and deltaic habitats shelter more annuals as well as species of relatively high water requirements;
2. The degree of silting or sedimentation. Silting is effective in the deltaic part of the wadis.

One community is characterized by species of wide ecological amplitude, the presence of which ranged between 50 and 100% among stands studied by El-Sharkawi et al. (1982a). These include *Cotula cinerea, Launaea capitata, Pulicaria undulata, Salsola baryosma, Schouwia thebaica, Tribulus pentandrus, Zilla spinosa, Zygophyllum coccineum* and *Z. simplex*.

A second community is limited in distribution and occurs in the upstream parts of the wadis. The species include *Astragalus vogelii, Citrullus colocynthis, Crotalaria aegyptiaca, Fagonia bruguieri, Lotononis platycarpa* and *Morettia philaena*.

A third community is characterized by species downstream and near the deltaic parts of the wadis. Members of this community are generally of higher water requirements and include *Astragalus eremophilus, Malva parviflora, Rumex vesicarius* and *Trichodesma africanum*. Other species showing affinity to this community are *Arnebia hispidissima, Fagonia indica, Forsskaolea tenacissima, Hammada elegans* and *Plantago ciliata*.

The Communities

The principal communities of the Assiut-Qena Desert may be categorized into: (1) suffrutescent woody types, (2) suffrutescent succulent types and (3) scrubland types.

Suffrutescent Woody Types

1. *Zilla spinosa* community. In this part of the inland Eastern Desert the *Z. spinosa* community is one of the most common. Its cover is usually thin (5–20%), contributed mainly by the dominant. The growth aspect of *Z. spinosa* may vary from one locality to another within the same season: in some stands the individuals are green in full flower whereas in others they are dry, probably depending on local differences in rainfall.

   Zygophyllum coccineum is the abundant associate. Other common associates include *Aerva javanica, Artemisia judaica, Citrullus colocynthis, Fagonia bruguieri, Francoeuria crispa, Leptadenia pyrotechnica* and *Pulicaria undulata*. Less common associates include *Acacia ehrenbergiana, A. raddiana, Cassia senna, Chrozophora oblongifolia, Heliotropium bacciferum* and *Polycarpaea repens*. Annual species include *Arnebia hispidissima, Astragalus eremophilus, Euphorbia granulata, Lotononis platycarpa, Senecio flavus* and *Zygophyllum simplex*.

   Three layers may be recognized in these communities: a thin frutescent layer including *Acacia* spp. and *Leptadenia*, a most notable suffrutescent layer as it includes the dominant and the main bulk of associate species, and a ground layer which includes prostrate forms, e.g. *Fagonia* spp. and *Citrullus*, and is enriched by annuals in rainy years.
2. *Aerva javanica* community. This community is present in some of the channels of wadis associated mostly with coarse alluvial deposits. Here, the frutescent layer is very thin and includes *Acacia ehrenbergiana, A. raddiana* and *Leptadenia pyrotechnica*. The suffrutescent layer includes the dominant species and numerous associates, e.g. *Artemisia monosperma, Chrozophora oblongifolia, Cornulaca monacantha, Francoeuria crispa, Ochradenus baccatus, Pergularia tomentosa, Pulicaria undulata* and *Zilla spinosa*. The ground layer includes two of the most common associate species – *Citrullus colocynthis* and *Polycarpaea repens* – and the less common *Fagonia bruguieri*. Ephemerals enrich this layer during rainy years.
3. *Calligonum comosum* community. This community is well represented in the Wadi El-Assiuti system and in some of the eastern tributaries of Wadi Ghuzzi. The dominant species is a much-branched undershrub capable of building sand mounds and small hillocks. It is a winter deciduous plant, remaining in the form of barren shoots throughout winter and early spring and producing flowers and foliage in late spring: *Cornulaca monacantha*, the most common associate, is also capable of building sand mounds. Other common associates include *Acacia raddiana, Anabasis setifera, Artemisia judaica* and *Zilla spinosa*. Less common species are *Atriplex leucoclada, Farsetia aegyptia, Francoeuria crispa, Heliotropium ramosissimum, Leptadenia pyrotechnica, Pergularia tomentosa, Phaeopappus scoparius, Pulicaria undulata* and *Tamarix aphylla*. *Eremobium aegyptiacum* is the common annual. The vegetation of the *C. comosum* community is usually in three layers. There is a thin frutescent layer including *Acacia raddiana, Leptadenia pyrotechnica* and *Tamarix aphylla*. The height of the dominant plant may exceed the 150 cm limit of the suffrutescent layer

but is mostly within this. This layer also includes the most common perennial associates *Artemisia judaica, Cornulaca monacantha* and several other perennials. In the ground layer are *Heliotropium luteum* and all the annuals.

4. *Crotalaria aegyptiaca* community. This community is confined to the non-calcareous country and is absent from the Wadi El-Assiuti system and the wadis of the limestone plateau to the north. *C. aegyptiaca* is one of the xerophytic cryptophytes that shows variation in growth form according to habitat conditions. Here, the frutescent layer is thin and is represented by *Acacia ehrenbergiana, A. raddiana* and *Leptadenia pyrotechnica*. The suffrutescent layer includes the dominant and the bulk of the associates, e.g. *Aerva javanica* (most common), *Artemisia judaica, Cornulaca monacantha, Francoeuria crispa, Pergularia tomentosa, Pulicaria undulata* and *Zygophyllum coccineum*. The ground layer includes *Fagonia bruguieri* and all of the ephemerals in rainy years.

Suffrutescent Succulent Types

This type of vegetation is represented in this part of the inland Eastern Desert by a community dominated by *Zygophyllum coccineum*, one of the widespread communities within the limestone country. It is plentiful in the affluents of the drainage systems and in the parts of the main channels where the deposits are shallow and coarse. It is less common in the basement complex and is absent from the sandstone country. *Calligonum comosum*, which dominates a community associated with wind-borne sand, and *Crotalaria aegyptiaca*, which dominates another community confined to the sandstone country, are rare associates in the limestone area. The most common associate is *Zilla spinosa. Aerva javanica* and *Francoeuria crispa* are common. Less common species include *Acacia ehrenbergiana, A. raddiana* and *Leptadenia pyrotechnica* (frutescent layer), *Artemisia judaica, Cassia senna, Hammada elegans, Pergularia tomentosa, Salsola baryosma* and *Trichodesma africanum* (suffrutescent layer) and *Citrullus colocynthis* and *Fagonia bruguieri* (ground layer).

Scrubland Types

1. *Acacia ehrenbergiana* community. This community is confined to the principal channels of the wadis and is associated with deep valley-fill materials. The soil is compact and many of its layers include some of the soft ingredients of the alluvial deposits that seem to cement the coarse material. It is well represented in Wadi Qena and in the wadis to its south but not in the limestone country of Wadi El-Assiuti and the wadis to its north. The plant cover of this scrub varies from 10 to 20% and is contributed mainly by the dominant. The most common associates are *Francoeuria crispa, Leptadenia pyrotechnica, Salsola baryosma* and *Zilla spinosa*. Other associates include *Acacia raddiana, Aerva javanica, Capparis decidua, Fagonia bruguieri, Ochradenus baccatus, Pergularia tomentosa, Pulicaria undulata* and *Zygophyllum coccineum*. The frutescent layer is well represented as it includes the dominant shrub as well as the most common species (*L. pyrotechnica*). The suffrutescent layer contributes

only little to the vegetation and is usually in widely spaced patches. This layer includes the main bulk of species. The ground layer includes *Fagonia bruguieri* and the ephemerals.

2. *Leptadenia pyrotechnica* community. This community is well represented in this part of the Eastern Desert. In certain localities, e.g. in the deltaic part of Wadi El-Assiuti, the bushes of *L. pyrotechnica* are evergreen. But elsewhere, e.g. Wadi El-Atrash, it may show a deciduous growth-form, with its branches drying up and the whole bush looking dead. The evergreen growth-form may be taken as an indicator of availability of underground water to deeply seated roots as its occurrence is confined to the deltaic part of the wadi where alluvial deposits of sand and gravel allow lateral seepage of the Nile water. In other parts of this system (e.g. Wadi Habib), where the bedrock prevents penetration of the underground water, *Leptadenia* has the deciduous growth-form or may be absent (Batanouny and Abdel Wahab, 1973). The cover of this community varies from 10 to 20%, mostly contributed by *Leptadenia*. The abundant associates are *Acacia raddiana, Artemisia judaica, Francoeuria crispa* and *Zilla spinosa*. Less common are *Acacia ehrenbergiana, Aerva javanica, Atriplex leucoclada, Capparis decidua, Citrullus colocynthis, Cleome droserifolia, Crotalaria aegyptiaca, Fagonia bruguieri, Hammada elegans, Ochradenus baccatus, Pergularia tomentosa, Pulicaria undulata, Salsola baryosma, Trichodesma africanum* and *Zygophyllum coccineum*. Shrub, undershrub and ground layers are all represented.
3. *Tamarix* spp. community. This part of the inland Eastern Desert is characterized by two *Tamarix* scrubs: one dominated by *T. aphylla* present in Wadi El-Assiuti and the other by *T. nilotica* present in Wadi Zaidun. In Wadi Qena both communities occur.

   T. aphylla is one of the desert plants that may acquire a tree habit or its bushy growth may form hillocks of sand. The latter growth-form is abundant in this area. The plant cover of this community ranges between 25 and 40% and common associates are *Anabasis setifera, Calligonum comosum, Cornulaca monacantha, Hammada elegans* and *Zygophyllum coccineum*. Other associates include *Acacia ehrenbergiana, A. raddiana, Artemisia judaica, Capparis decidua, Eremobium aegyptiacum* and *Leptadenia pyrotechnica*.

   The plant cover of the *T. nilotica* community ranges between 30 and 60%, often thinner than that of *T. aphylla*. Common associates are *Aerva javanica, Francoeuria crispa, Hammada elegans* and *Zygophyllum coccineum*. Less common species include *Artemisia judaica, Calligonum comosum, Crotalaria aegyptiaca, Heliotropium bacciferum, Leptadenia pyrotechnica, Ochradenus baccatus, Polycarpaea repens, Salsola baryosma* and *T. aphylla*.

   T. nilotica is usually associated with soils that are more saline than those of *T. aphylla* (Kassas and Zahran, 1967; Kassas and Girgis, 1972).

   The *Tamarix* spp. community shows the usual three layers of the desert vegetation. The shrub layer contributes the main part of the plant cover; it includes the dominant species together with *Leptadenia* and *Ochradenus*. The suffrutescent layer includes the main bulk of the associate perennials and is mostly dominated by *Z. coccineum*. The ground layer includes few perennials, e.g. *Heliotropium bacciferum* and is enriched by the therophytes during the rainy years.

(c) The Sandstone Desert (Idfu-Kom Ombo Desert)

(i) Geomorphology

This region includes the part of the Eastern Desert extending between Lat. 25°30′N and 24 °N. The area is made up of an eastern basement complex section and a western Nubian sandstone section. Within the Nubian sandstone country there are a few patches of Cretaceous limestone. The westward drainage from the Red Sea hills is contained in three principal wadis: Abbad, Shait and El-Kharit (Girgis, 1965).

Wadi Abbad joins the Nile Valley near the village of Ridisiya (opposite Idfu, Fig. 2.1). The drainage system is of a number of wadis, e.g. Wadis El-Shalul, El-Miyah and Barramya. The west part of the Idfu (in the Nile Valley)-Mersa Alam (on the Red Sea coast) desert road passes through the main channel of Wadi Barramya which contains the Barramya gold-mines. These tributary wadis, in their turn, receive drainage from affluent tributaries. The whole system is about 7000 km^2.

Wadi Shait originates on the Red Sea mountains. Its total length is about 200 km and it joins the Nile Valley a little north of Kom Ombo; its total basin is 10,000 km^2. It receives several large tributaries draining the country to its north. These tributaries include Wadis Bizah and Mueilhe. The main trunk of Wadi Shait contains a number of wells e.g. Bir Qubur, Bir Helwat and Bir Salam.

Wadi El-Kharit, described by Ball (1912) as one of the greatest trunk wadis in Egypt, has its principal head at Gebel Ras El-Kharit and debouches on the Kom Ombo plain. It reaches the Nile at the same point as Wadi Shait with a length of about 260 km. It drains an area of more than 23,000 km^2. The principal tributaries of Wadi El-Kharit are Wadis Natash, Antar, Khashab, Abu Hamamid, Garara, Rod El-Kharuf and AH Mikan.

The main parts of Wadi Abbad, Wadi Shait and Wadi El-Kharit traverse the rainless plateau of the Eastern Desert. The water resources are primarily dependent on the chance occurrence of rain on the Red Sea mountains; run-off water will then collect in the upstream tributaries and if the discharge is voluminous it may reach the main trunks of the wadis and may flow downstream, which is a rare occurrence.

(ii) The Vegetation

Erratic rainfall, one of the characters of the desert climate, results in erratic changes in the vegetation in time and space. Differences in seasonal aspects are seen in the growth of ephemerals and the appearance of seedlings of perennials in the wet season which may or may not continue to grow, depending on the volume of water resources. The vegetation of Idfu-Kom Ombo desert is described as follows:

(1) plant cover of the main wadis – Abbad, Shait and El-Kharit; and
(2) the main communities.

Wadi Abbad

Main Trunk

The vegetation of the main trunk of Wadi Abbad was studied by Girgis (1965) in three seasons: February 1961, October 1963 and March 1965.

February 1961

The vegetation of the main trunk of Wadi Abbad in February 1961 was very different from that on other occasions. In that year the downstream part of the wadi was covered by a rich growth of ephemerals. Three areas at different distances east of Idfu City were studied.

Area *1*: *10–20 km east of Idfu.* The plant cover ranged from 30 to 50%. The ephemeral *Schouwia thebaica* contributed a substantial part of this area. *Zygophyllum simplex* formed localized patches of rich growth. This area includes the locality of Bir Abbad where there are a few large trees of *Acacia nilotica* and *Phoenix dactylifera* (both cultivated) and *Balanites aegyptiaca*. Other species recorded in the area include *Acacia ehrenbergiana, A. raddiana, Aerva javanica, Arnebia hispidissima, Astragalus vogelii, Calotropis procera, Citrullus colocynthis, Fagonia bruguieri, Francoeuria crispa.*

Area 2: *50–60 km east of ldfu.* The plant cover in this area was very thin (5–10%), only a few perennials being present, namely *Acacia ehrenbergiana, A. raddiana, Aerva javanica, Cornulaca monacantha, Crotalaria aegyptiaca*, and *Zilla spinosa.* The difference in vegetation between this and that of area 1 shows the striking variation within the same wadi, due to the sporadic and local nature of the rainfall.

Area 3: *100–110 km east of Idfu.* Plant cover ranged between 20 and 50%. In certain localities, *Cassia senna* dominated and in others, e.g. the mouth of Wadi Salatit (104 km east of ldfu), *Cleome chrysantha* was the most abundant plant. Other species in this area have been mentioned in area 1 with the addition of *Acacia tortilis. Astragalus eremophilus, Cleome arabica, Cotula cinerea, Eremobium aegyptiacum, Fagonia parviflora, Farsetia ramosissima, Forsskaolea tenacissima, Morettia philaena, Pergularia tomentosa, Stipagrostis plumosa* and *Trichodesma africanum.* Rare individuals of *Acacia tortilis* were recorded but this species is absent from the western part of Wadi Abbad.

October 1963

The vegetation during this season was very different from that in February 1961. None of the ephemeral growth of *Schouwia thebaica* of the downstream part was present. The vegetation was of the hardy perennials.

Certain species had become prominent: *Aerva javanica, Crotalaria aegyptiaca, Salsola baryosma* and *Zilla spinosa.* These species, together with other perennials, germinate in the rainy season but the early stages of their development are much slower than that of the ephemerals. In consequence, the perennials appear luxuriantly, whereas the ephemerals are still essentially at the seedling stage. Later the ephemerals will have completed their life cycle and eventually dry up while the perennials may exist for only a year, or for several years depending on the volume of moisture stored in the soil. Then they die and the vegetation is of *Acacia* scrub (Girgis, 1965).

In October 1963, the area 10–20 km east of Idfu had a vegetation cover of less than 10% (30–50% in February, 1961). The species recorded were *Acacia ehrenbergiana, A. raddiana, Aerva javanica, Citrullus colocynthis, Francoeuria crispa, Salsola baryosma* and *Zilla spinosa*. The area 100–110 km east of Idfu had an even thinner cover and the species recorded included *A. ehrenbergiana, A. raddiana*, A *tortilis, Aerva javanica, Cassia senna* and *Zilla spinosa. Cleome chrysantha* had almost completely disappeared.

March 1965

The vegetation had changed only very little but for the worse. The *Acacia* scrubs were still there but *Zilla spinosa* was mostly dead and dry. *Aerva javanica, Citrullus colocynthis* and *Francoeuria crispa* were represented by only a few depauperate plants.

Affluent Wadis of Wadi Abbad

Wadi El-Shaghab (March 1965)

The vegetation was an *Acacia ehrenbergiana* open scrub with a plant cover of less than 5%. There were also a few individuals of A. *raddiana*. The undergrowth included a rich growth of dead *Zilla spinosa* and *Salsola baryosma* together with remnants of dry *Schouwia thebaica*.

Wadi El-Muyah

The vegetation of this wadi in February 1965 was similar to that of Wadi El-Shaghab: an open scrub of A *ehrenbergiana* with occasional individuals of *A. raddiana*. The undergrowth included a sparse population of green *Cassia senna, Aerva javanica* and *Citrullus colocynthis* together with dry *Zilla spinosa* (Girgis, 1965).

According to El-Sharkawi et al. (1982b), in Wadi El-Muyah there are three distinct communities, each a subordinate of a larger phytocoenosis dominated by *Zilla spinosa* and *Aerva javanica*. Other species recorded in this wadi are *Acacia raddiana, Artemisia judaica, Capparis aegyptia, C. decidua, Citrullus colocynthis, Fagonia bruguieri, Francoeuria crispa, Gypsophila capillaris, Linaria haelava, Pergularia tomentosa, Pulicaria undulata, Salsola baryosma, Schouwia thebaica* and *Zygophyllum coccineum*. The last named succulent xerophyte is present in great abundance and high vigour.

Wadi Barramya

This wadi is the continuation of the main trunk of Wadi Abbad. It may be divided into two parts – a downstream part traversing the Nubian sandstone country and an upstream one crossing the basement complex country. In both parts the vegetation is essentially an *A. ehrenbergiana* scrub. In the downstream part the undergrowth is of extensive patches of dry *Fagonia bruguieri* and *Zilla spinosa* together with living plants of *Aerva javanica, Cornulaca monacantha* and *Crotalaria aegyptiaca*. In the upstream part a rich growth of *A javanica* and *Cassia senna* forms the undergrowth, with smaller amounts of *Fagonia bruguieri*.

Wadi Shait

The vegetation of this drainage system as a whole is essentially an *A. ehrenbergiana* scrub. The density of the shrub layer varies considerably. In parts where this layer is thin the vegetation may be dominated by one of the several suffrutescent species, some of which behave like ephemerals.

Ecologically, Wadi Shait may be considered in three main sections: downstream, main water channel and upstream.

Downstream Section

In its downstream 30 km part, Wadi Shait occupies a wide course with a number of channels bounded by gravel terraces. The vegetation of this section is a well developed, though open, *A. ehrenbergiana* scrub. *A. raddiana* is occasionally present. The flora here includes *Asphodelus tenuifolius, Astragalus eremophilus, A. vogelii, Cassia italica, Citrullus colocynthis, Fagonia parviflora, Farsetia ramosissima, Lotononis platycarpos, Morettia philaena, Robbairea delileana, Salsola baryosma, Schouwia thebaica, Sonchus oleraceus, Tribulus longipetalus, Trigonella hamosa, Zilla spinosa, and Zygophyllum simplex.*

Main Water Channel

The 30 km part of the course of Wadi Shait that extends between Long. 33′21′E and 33°40′E, cuts across a low sandstone plateau (about 200 m). The scrubland growth is very thin, made up of a few individuals *of Acacia ehrenbergiana* and *A. raddiana*. Near the east extremity of this part there is a single *Balanites aegyptiaca* tree, forming one of the landmarks of the area.

The vegetation is a suffrutescent type dominated mostly by *Zilla spinosa* with very distinct areas which may be considered societies dominated by *Salsola baryosma* and *Francoeuria crispa*. The flora of this part of the wadi also includes *Aerva javanica, Cassia italica, Chrozophora oblongifolia, Crotalaria aegyptiaca, Fagonia arabica, Leptadenia pyrotechnica* and *Morettia philaena*.

Wadi Bizah, one of the extensive tributaries of Wadi Shait, joins in this part. The vegetation of the downstream part of Wadi Bizah is a well developed *A. ehrenbergiana* scrub associated with the same species as in the downstream section of Wadi Shait with addition of *Calligonum comosum, Eremobium aegyptiacum, Forsskaolea tenacissima, Launaea cassiniana, Schismus barbatus* and *Trichodesma africanum*. The upstream extremities of Wadi Bizah are traversed by the Idfu-Mersa Alam desert road. In these areas the growth of *A. ehrenbergiana* is thin. The suffrutescent layer is mostly dominated by *Zilla spinosa*, which usually forms a thin cover (5–10%.) But is enriched by ephemerals during rainy years. Other species recorded include *Acacia raddiana, Aerva javanica, Citrullus colocynthis, Cleome arabica, Cotula cinerea, Fagonia parviflora, Farsetia aegyptia, Tribulus. longipetalus, Stipagrostis plumosa, Trigonella stellata* and *Zygophyllum simplex.*

The main channel of Wadi Shait east of Long. 33°40′E traverses a higher plateau (above 300 m). The channel occupies a narrow, well-defined valley, where *A. ehrenbergiana* is very thin. The narrow terraces fringing the course of the wadi may be covered by the growth of *Leptadenia pyrotechnica.* Bushes of *Calotropis procera* are locally abundant. Otherwise the vegetation is a mosaic of patches dominated by *Citrullus colocynthis, Francoeuria crispa* or *Aerva* javanica. The flora of this part of Wadi Shait includes, in addition to, species elsewhere in this wadi, *Acacia raddiana, Arnebia hispidissima, Hippocrepis constricta, Launaea capitata, Monsonia nivea, Orobanche muteli* and *Pulicaria undulata.*

In this part, Wadi Shait receives its large tributary – Wadi Muweilha – where the scrubland of *Acacia ehrenbergiana, A. raddiana* and *Leptadenia pyrotechnica* is well developed. The undergrowth is much thinner than that of the main wadi and fewer species are present: *Aerva javanica, Citrullus colocynthis, Crotalaria aegyptiaca, Francoeuria crispa* and *Zilla spinosa.* In the upstream extremities of Wadi Muweilha the vegetation is essentially a distantly open scrub of *A. ehrenbergiana* and *A. raddiana.* The undergrowth is dominated by *Crotalaria aegyptiaca* and/or *Zilla spinosa.* Patches of *Panicum turgidum* grassland may be present on the sheets of sand fringing the courses of the affluent wadis.

Upstream Section of Wadi Shait

In the head part of Wadi Shait (Long. 34°30′E) *Acacia ehrenbergiana* scrub is well developed. Associated trees and shrubs include *A. raddiana, A. tortilis, Balanites aegyptiaca* and *Maerua crassifolia.* Other species recorded are *Aristida adscensionis,Arnebia hispidissima, Asphodelus tenuifolius. Cassia italica, Citrullus colocynthis, Echium horridum, Fagonia parviflora, Launaea capitata, Lotononis platycarpos, Panicum turgidum, Pulicaria undulata, Trichodesma ehrenbergii, Zilla spinosa* and *Zygophyllum coccineum.*

Wadi El-Kharit

The downstream part of Wadi El-Kharit was studied in March 1961 and February 1964 (Girgis, 1965). In March 1961 this part was richly covered by green growth, *Schouwia thebaica* being the most abundant ephemeral. Seedlings *of Crotalaria aegyptiaca* were also very abundant, especially in the terraces of the wadi. In February 1964 the green growth was very thin but there were patches of dried *Salsola baryosma* and *Crotalaria aegyptiaca.* The richness of ephemerals in March 1961 was comparable to that noted in the downstream part of Wadi Abbad at the same time. Rainfall in the 1961 season was so ample that the torrents filled the whole course and reached the downstream part of the main wadis. The flora of this part of Wadi El-Kharit includes *Acacia ehrenbergiana, A. raddiana, Aerva javanica, Astragalus eremophilus, Citrullus colocynthis, Cornulaca monacantha, Farsetia ramosissima, Lupinus varius* ssp. *orientalis, Pulicaria undulata, Orobanche muteli, Stipagrostis plumosa, Tamarix* sp., *Trichodesma africanum* and *Zygophyllum simplex.*

The vegetation of the main channel of Wadi El-Kharit is an open scrub of *Acacia ehrenbergiana*. Mounds covered by dead remains of *Salsola baryosma* are obviously relicts of a previous rich growth.

Wadi El-Kharit has three main tributaries: Wadis Natash, Khashab* and Abu Hamamid. Wadi Natash is mostly dominated by *Crotalaria aegyptiaca, Zilla spinosa* or *Citrullus colocynthis*. Other associate species include *Acacia ehrenbergiana, A. raddiana, Aerva javanica Cassia italica, Fagonia bruguieri, Francoeuria crispa* and *Leptadenia pyrotechnica*.

Wadi Khashab derives its name from its relatively well wooded character (Ball, 1912). In its upper parts "it presents the appearance of a broad valley, in which trees are so numerous as to give very pleasing contrast to the dreary wastes on the other side of it". In the main channel of this wadi the vegetation is an open forest of *Acacia raddiana*. The undergrowth is dominated by *Zilla spinosa*. The other species recorded are *Aerva javanica, Caylusea hexagyna, Chrozophora oblongifolia, Citrullus colocynthis, Cleome arabica, Francoeuria crispa, Panicum turgidum* and *Pulicaria undulata*. The vegetation of the affluent wadis differs according to the size of the affluent. In larger affluents it is an open scrub of A. *ehrenbergiana* with rare individuals of A. *raddiana*. This is in obvious contrast with the vegetation of the principal Wadi Khashab where A. *raddiana* is dominant whereas A. *ehrenbergiana* is rare. The vegetation of the smaller runnels is a suffrutescent type dominated by *Zilla spinosa, Crotalaria aegyptiaca* or *Fagonia bruguieri* or grassland dominated by *Panicum turgidum*.

The vegetation of Wadi Abu Hamamid is a well developed *Acacia raddiana* open forest. The trees reach considerable size. *A. ehrenbergiana* is very rare in the principal wadi. The undergrowth of the A. *raddiana* open forest is mostly a rich growth of *Zilla spinosa* associated with *Aerva javanica, Cassia senna, Chrozophora oblongifolia, Fagonia bruguieri, Lotononis platycarpos, Panicum turgidum, Stipagrostis plumosa* and *Zygophyllum simplex*. The vegetation of the affluent wadis includes rare individuals of A. *raddiana* and is dominated by one of A. *ehrenbergiana, Cassia senna, Panicum turgidum* and *Zilla spinosa*.

The Main Communities

The main communities of the Idfu-Kom Ombo desert may be divided into the ephemeral communities and the perennial communities which include suffrutescent woody types, suffrutescent succulent types and scrubland types.

Ephemeral Communities

The ephemeral vegetation in the Idfu-Kom Ombo desert appears after rainy seasons. This vegetation includes ephemeral and also perennial species that behave as ephemerals. This type of vegetation may also include some perennial shrubs and undershrubs but they are often scarce and contribute very little to the cover. The

* Natash is the vemacular name of *Crotalaria*; Khashab is the Arabic word for wood.

ephemeral vegetation of this part of the Eastern Desert is of two main communities: a *Schouwia thebaica* community and a *Fagonia bruguieri* community.

1. *Schouwia thebaica* community. The notable feature exhibited by the stands of this community in spring is the richness of the flora (about 43 species) and the substantial plant cover (20–50%, Girgis, 1962). In summer the cover is usually thin (<5%) and later in the year becomes even sparser due to the dryness and disappearance of ephemerals.

   The abundant associates of this community include one perennial (*Morettia philaena*) and four ephemerals (*Astragalus eremophilus, Cotula cinerea, Tribulus pentandrus* and *Zygophyllum simplex*). Common associates include three ephemerals: *Astragalus vogelii, Lotononis platycarpos* and *Orobanche muteli* and four perennials that seem to behave as ephemerals (*Fagonia parviflora, Farsetia ramosissima, Francoeuria crispa* and *Zilla spinosa*), one perennial (*Citrullus colocynthis*) in which almost only the root remains, and one perennial (*Crotalaria aegyptiaca*) that may survive for more than a year. Other associates include species of various growth forms. Perennials that survive throughout the year include *Acacia ehrenbergiana, A. raddiana, Aerva persica, Calotropis procera, Cornulaca monacantha* and *Salsola baryosma*.

   The ephemeral vegetation shows the usual three layers, but the ground layer is the most notable as it includes the dominant and all of the consistently present associates. The growth *of Schouwia thebaica* may, in certain localities, be tall enough as to be included with the suffrutescent species. In the stands of this community most of the species dominating the perennial communities are either absent (e.g. *Leptadenia pyrotechnica*) or rarely present (e.g. *Acacia* spp., *Cassia senna* and *Salsola baryosma*). Dominants of other communities that are commonly recorded here include *Crotalaria aegyptiaca, Francoeuria crispa* and *Zilla spinosa*.
2. *Fagonia bruguieri* community. *F. bruguieri* is a perennial that behaves as an ephemeral under certain conditions. The floristic composition of this community has been recorded in two different seasons: spring 1961 and autumn 1963 (Girgis, 1965). The perennials *Aristida plumosa, Citrullus colocynthis, Fagonia bruguieri* (the dominant), *Farsetia aegyptia* v. *ovalis, F. ramosissima, Morettia philaena, Pulicaria undulata, Salsola villosa* and *Zilla spinosa* were recorded in full foliage or in full flower and fruit in spring. During autumn, these species were dead and dry or had disappeared. The dry remains of *Fagonia bruguieri* were the most notable feature. Other perennials included the common *Acacia ehrenbergiana, A. raddiana, Aerva javanica* and *Crotalaria aegyptiaca*. Ephemeral associates include *Astragalus vogelii, Cotula cinerea, Eremobium aegyptiacum, Euphorbia granulata, Schismus barbatus, Tribulus longipetalus* and *Zygophyllum simplex. Schuowia thebaica* is rarely recorded in this community.

Perennial Communities

Suffrutescent Woody Types

1. *Zilla spinosa* community. *Z. spinosa* is one of the common species of the Eastern Desert. In years with rainfall it may cover extensive tracts of the wadis,

surviving for a few years before it dries up and remains completely dry until the recurrence of a rainy year.

The plant cover ranges from 5 to 30%, mostly derived from the growth of the dominant. *Acacia ehrenbergiana, A. raddiana* and *Citrullus colocynthis* are the abundant associates. *Aerva javanica, Crotalaria aegyptiaca, Fagonia bruguieri* and *Francoeuria crispa* are common associates. Less common perennials include *Balanites aegyptiaca, Cassia italica, C. senna, Caylusea hexagyna, Fagonia arabica, Farsetia ramosissima, Leptadenia pyrotechnica, Morettia philaena* and *Pulicaria undulata*. The ephemeral associates include rare individuals of *Arnebia hispidissima, Echium rauwolfii, Euphorbia granulata, Lotononis platycarpa, Monsonia nivea* and *Zygophyllum simplex*. Dead remains of *Schouwia thebaica* and *Arnebia hispidissima* are also found in the stands of this community in Wadi Abbad.

The frutescent layer contributes very little to the cover of this community though it includes several of the most common species-*Acacia* spp. together with *Balanites* and *Leptadenia*. The suffrutescent layer is the most notable as it includes the dominant together with such common associates as e.g. *Aerva, Crotalaria* and *Francoeuria*. The ground layer includes *Caylusea, Citrullus* and *Fagonia* spp.

2. *Crotalaria aegyptiaca* community. This is one of the communities that characterizes the Idfu-Kom Ombo desert. Though *C. aegyptiaca* is present all over the Eastern Desert the community which it dominates is widespread in this part only, and not in the Nubian Desert to the south or in the limestone desert to the north. The habitat of this community is usually the sheets of sandy deposits that are mostly aeolian. The cover is thin (5–20%), contributed mainly by *C. aegyptiaca. Acacia raddiana, Citrullus colocynthis* and *Zilla spinosa* are the most common associates. Other perennials include A. *ehrenbergiana, Aerva javanica, Fagonia bruguieri, Francoeuria crispa, Leptadenia pyrotechnica, Stipagrostis plumosa* and *Tamarix* spp. *Leptadenia* and *Tamarix* spp. which dominate scrubland communities are here less frequent. The ephemeral vegetation is represented by *Asphodelus tenuifolius*.

   The usual three layers of desert vegetation are present here. The suffrutescent layer is the most notable as it includes the dominant, and the most common undershrubs together with other perennials.

3. *Cassia senna* community. *C. senna* is a common species in the southern part of the Eastern Desert and is rare in the northern part. It is one of the most extensively collected desert species as it is a famous medicinal plant. Ayensu (1979) reported that *Cassia acutifolia* (*C. senna*) leaves and pods contain glycosidin, sennosides (a mixture of anthraquinone glycosides), rhein and related compounds and can be used as a laxative. Its distribution and density are probably influenced by the intensity of collection by man.

   Dominance of *C. senna* is usually seen in Wadi Abbad and in affluents of the two principal tributaries of Wadi El-Kharit. The plant cover of this community ranges between 5 and 15% contributed mainly by *C. senna. Acacia ehrenbergiana, Aerva javanica* and *Zilla spinosa* are the common associates.

Other associates are *Acacia raddiana, Cassia italica, Chrozophora oblongifolia, Fagonia parviflora, Farsetia ramosissima, Morettia philaena, Panicum turgidum, Polycarpaea repens* and *Stipagrostis plumosa*. Ephemerals include *Arnebia hispidissima, Astragalus eremophilus, A. tribuloides, Euphorbia granulata* and *Zygophyllum simplex*. The ephemerals enrich the ground layer which, like the frutescent layer, is usually thin. The suffrutescent layer is the most notable.

4. *Francoeuria crispa* community. *F. crispa* is a common xerophyte in many of the wadis of the Eastern Desert. Its growth is greater in the rainy season. The plant community dominated by this composite is common in the middle part of the main channels of Wadi Shait but otherwise is infrequent. The cover ranges between 10 and 25%, formed mostly by *F. crispa*. Common associates are *Citrullus colocynthis* (ground layer), *Aerva javanica, Chrozophora oblongifolia, Crotalaria aegyptiaca* and *Zilla spinosa* (suffrutescent layer) and *Acacia ehrenbergiana, A. raddiana, Calotropis procera* and *Leptadenia pyrotechnica* (frutescent layer). In certain localities of this community, *Zilla spinosa* or *Citrullus colocynthis* may co-dominate with *F. crispa*.
5. *Citrullus colocynthis* community. *C. colocynthis* is a common desert plant in Egypt. Its bitter fruits are used in native medicine. Oil extracted from the seeds is used by the bedouins for finishing the tanning of water bags made of goat skin (Girgis, 1965). In easily accessible parts of the Egyptian desert this plant is becoming less common. It has a deeply penetrating (perennial) tap-root which is often thick and which produces an annual crop of trailing stems. Ayensu (1979) reported that the fruit, seed, stem, root and leaf of this plant contain tannin, alkaloids and cucurbitacin. Extracts of the fruits have anti-tumour activity against sarcoma 37, due to cucurbitacins.

   In the Idfu-Kom Ombo desert, the *Citrullus* community is found in the main channels of Wadis Shait and Natash. The abundant associates are *Aerva javanica, Crotalaria aegyptiaca* and *Zilla spinosa* (members of the suffrutescent layer). The shrub layer includes *Acacia ehrenbergiana, A. raddiana, Calotropis procera* and *Leptadenia pyrotechnica*. The ground layer contains the dominant together with *Fagonia bruguieri* and ephemerals during the rainy years.

Suffrutescent Succulent Type

This is represented by one community dominated by *Salsola baryosma*, one of the common species in this part of the Eastern Desert. Wadi El-Kharit is named after its vernacular name. This community is common in Wadis Shait and El-Kharit but not in Wadi Abbad.

The growth of S. *baryosma* follows a cycle which may be outlined as follows. In rainy years the seeds germinate. In the seedling stage (spring) the plant is not a conspicuous member of the cover which may be dominated by the growth of ephemerals. Subsequent growth and development follow during the next summer. The plant reaches its mature size and flowers in the following autumn. At this stage the plant assumes dominance; the ephemerals will have dried earlier in the summer and by the autumn their dry remains may have been blown away. Plants survive for a few years during which sand or silt may be collected around them. Finally the plants dry and dead relicts remain on these mounds. The cycle is resumed whenever a rainy year recurs. In this respect *S. baryosma* follows the cycle of many of the suffrutescent perennial species of this and similar areas, but being a succulent it seems to survive for longer periods than the non-succulent *Zilla spinosa* etc.

Acacia ehrenbergiana and *A. raddiana* are consistently present. Very common associates are *Citrullus* and *Zilla*. Other less common associates include *Aerva, Crotalaria, Francoeuria, Leptadenia* and *Tamarix* spp. The three usual layers of desert vegetation are present in this community.

Scrubland Types

1. *Leptadenia pyrotechnica* community. Limited patches and strips of scrubland dominated by *L. pyrotechnica* are found in the channels of the main wadis and on the sandy terraces that may fringe their courses. The cover of this community ranges from 20 to 30%, contributed mostly by the dominant. In the rare incidence of rich epnemeral growth the cover of the ground layer may become notable. Three layers of vegetation may be recognized in this community. The shrub layer includes the dominant and the common associates *Acacia raddiana* and *Calotropis procera*. The suffrutescent layer includes three of the consistently recorded perennials (*Aerva, Francoeuria* and *Zilla*) together with *Cassia senna, Chrozophora oblongifolia, Crotalaria aegyptiaca, Farsetia ramosissima, Salsola baryosma* and *Trichodesma africanum*. The ground layer includes the trailing *Citrullus* and *Fagonia* spp. and is enriched during rainy years with therophytes, e.g. *Asphodelus tenuifolius, Astragalus eremophilus,A. vogelii, Cotula cinerea, Ifloga spicata, Launaea capitata* and *Zygophyllum simplex*.
2. *Acacia ehrenbergiana* community. The open scrub dominated by *A. ehrenbergiana* is widespread within the wadis of this extremely arid part of the Eastern Desert. The cover ranges between 5 and 25%, mostly of the dominant. *A. raddiana*, though recorded in all of the stands of this community, makes very little contribution to the cover (Girgis, 1965). *Zilla* is one of the most common associates and often forms patches of rich undergrowth. *Citrullus* and *Salsola* are also common. Other associates are as those of the *L. pyrotechnica* community with the addition *of Balanites aegyptiaca, Echium horridum, Lotononis platycarpos, Maerua crassifolia, Panicum turgidum, Schouwia thebaica* and *Stipagrostis hirtigluma*.
3. *Acacia raddiana* community. *A. raddiana* is present in the main wadis of the inland part of the Eastern Desert, but stands dominated by it are not very common. It is the main raw material for charcoal and hence is subject to continued destruction.

   Within the Idfu-Kom Ombo desert, open forest of A. *raddiana* is present in the upstream parts of the main tributaries of Wadis Shait and El-Kharit. On these parts that drain the basement complex hills and mountains, the main channels have deep wadi-fill deposits where some water is stored. A notable feature of these localities is the presence of wells that are holes dug through the alluvium and where the water-table is often reached at depths ranging from 6 to 15 m. Deeply penetrating roots of desert shrubs and trees frequently reach such depths (Girgis, 1965).

The open forest growth dominated by A. *raddiana* is one of the most highly organized types of desert vegetation. Four layers may be recognized. The tree layer includes the dominant and provides the most distinctive character of this type. Trees of *Balanites aegyptiaca* are rarely present. The shrub layer includes *A. ehrenbergiana.* The suffrutescent layer comprises the most common associates: *Aerva jauanica, Salsola baryosma* and *Zilla spinosa* as well as the less common ones, e.g. *Cassia senna, Chrozophora oblongifolia, Farsetia ramosissima, Francoeuria crispa* and *Panicum turgidum.* The ground layer includes *Citndlus colocynthis* and *Cleome arabica* together with the ephemerals, e.g. *Caylusea hexagyna* and *Zygophyllum simplex.*

(d) The Nubian Desert

(i) Geomorphology

The Nubian Desert of Egypt, east of the Nile Valley, comprises the southwest part of the Eastern Desert between Lat. 24°N and 22°N. It is an example of an extremely arid desert. According to the *Climatic Normals of Egypt* (Anonymous, 1960), at Aswan the total annual rainfall is 1.4 mm, the annual mean temperature is 27°C, the absolute maximum ranges between 37°C in December and 50°C in June and the absolute minimum between 1.7°C in January and 22.4°C in September. The mean annual evaporation is 15.4 mm/day (Piche) and the mean annual relative humidity ranges between 20 and 30%.

The Nubian Desert includes a sandstone region fringing the Nile Valley and a basement complex part extending eastward to the Red Sea mountains. The basement complex country is drained by the Wadi Allaqi system; the sandstone country is drained by numerous short wadis. The Nile forms the base line of all these drainage systems (Kassas and Girgis, 1969–1970).

Wadi Allaqi is the most extensive drainage system in the Egyptian desert. Ball (1902) estimates the basin of this wadi as no less than 44,000 km^2. The upstream tributary of Wadi Allaqi drains some of the mountains that form the natural divide between the Eastern Desert and the Red Sea coastal land. These tributaries may receive occasional rainfall and their drainage may accumulate in the main channel of Wadi Allaqi forming "accidental" torrents that are the main source of its water.

The wadis of the sandstone country are two groups of independent systems: a group north of Wadi Allaqi and a second south of it. All are short wadis (20–40 km) that receive "accidental" rainfall, an incident that may recur once every decade. The mouths of these wadis are in -direct contact with the Nile and may be inundated during the flood season. Kassas and Girgis (1969–1970) state "It is expected that the establishment of the High Dam south of Aswan will raise the water level in the river in this area some 60 cm and transform the downstream extremities of these wadis into lagoons of fresh water. The deltaic mouth of Wadi Allaqi will become a fresh-water lake."

These expectations have proved to be well founded. The High Dam (Nasser) Lake inundation has changed the ecology of the downstream part of Wadi Allaqi which has become a habitat for mesophytes (Springuel and Ali, 1990). *Francoeuria crispa* and *Tamarix nilotica* are abundant and have reached a considerable size. All other species recorded (21) in this part of Wadi Allaqi are typical riverain elements.

(ii) Plant Cover

The vegetation of the Nubian Desert may be considered as (1) the plant life of the wadis; and (2) the main communities.

Plant Life of the Wadis

Wadi Allaqi

Wadi Allaqi had a more important role in the history of the country than has been hitherto realized. In almost every tributary there are remains of settlements, including ancient gold mines. One of these mines, the Umm Qureiyat, has a history extending from Ancient Egyptian time until the early decades of this century.

The main channel of Wadi Allaqi has its head in the mountain region of the Red Sea (alt. 1740 m). It follows a westward direction among the rugged country of some mountains (alt. 1148–1230 m). From this area it traverses a plateau country with hills ranging from 300 to 700 m. Then it flows in a northwest direction across country with low hills until it meets its southern affluent (Wadi Gabgaba) and proceeds towards its delta.

The course may, ecologically, be divided into four main sections: a mountainous east section, a middle hilly section, a low plateau section and a deltaic one.

East Section

The east section of the montane region has the richest vegetation. The mountains are within the maritime influence of the Red Sea and have some rain. In certain localities the vegetation may be described as a desert forest, with trees of *Acacia raddiana* and *Balanites aegyptiaca* and lianas such as *Cocculus pendulus* and *Ochradenus baccatus. Salvadora persica* is also common. In this part, Wadi Allaqi receives a few tributaries from the south and numerous tributaries from the north e.g. Wadis Mirikwan and Eqat. Wadi Eqat is taken as a representative of this group. The vegetation of this wadi is of three types of scrubland, characterized by *Acacia raddiana, A. tortilis* and *Balanites aegyptiaca. A. tortilis* scrub occupies affluents of the wadi and gravelly terraces of the main channel. *A. raddiana* scrub occupies the main channel. *B. aegyptiaca* scrub is represented by a few patches in especially favoured parts of the main channel. In this order, these three communities represent increasing requirements of moisture.

Within the whole region of the upstream part of Wadi Allaqi and its tributaries, *Panicum turgidum* grassland is represented by what appear to be relict patches. This grassland is subject to widespread destruction and overgrazing. The dominance of *Acacia tortilis*, the abundance of *Salvadora persica* and the rarity of *Acacia ehrenbergiana* in the main channels of Wadi Allaqi and its tributaries are notable features.

Middle Section

In the middle section, Wadi Allaqi receives numerous large affluents on both the north and south sides. The vegetation of the main channel is essentially a distantly open scrub of A. *ehrenbergiana* with associate trees of A. *raddiana*, but A. *tortilis* is absent. Patches of *Salsola baryosma* are associated with the mouths of the tributary wadis. The appearance of these patches and the extent of their growth largely depend on the amount of rainfall and hence volume of water discharged by these tributaries.

The main channel of Wadi Allaqi is here characterized by several terraces and numerous fossil hillocks with remains of extensive thickets of *Tamarix aphylla* and *Salvadora persica*. Some of the hillocks are of considerable size, up to 12 m high and 500 m^2 area.

The north side affluents of this section of Wadi Allaqi are represented by Wadis Seiga, Murra and Umm Rilan. Wadi Seiga has a main channel about 80 km long and its downstream 40 km traverses an almost flat plain of sand and gravel. Growth of A. *ehrenbergiana* is in discontinuous patches. In certain parts of the wadi, the channel is not clearly defined and the vegetation is mostly herbaceous: *Aerva javanica, Cassia senna, Fagonia indica, Indigofera argentea, Morettia philaena* and *Stipagrostis plumosa*. This vegetation type seems to appear in rainy years. Wadi Murra has a more clearly defined channel bounded by high ground on both sides. It has a brackish well and its channel is a part of the main camel-caravan track traversing this country. Growth of *A. ehrenbergiana* is richer here than in Wadi Seiga. In the downstream part A. *raddiana* is present in the scrub. Wadi Umm Rilan is a smaller one (30 km) with a few short affluent runnels. The vegetation is distantly open A. *ehrenbergiana* scrub. *Aerva javanica* is a very common associate in the wadi and dominates the affluent runnels.

The south side affluent runnels are represented by Wadis Muqsim, Abu Fas, Ungat and Neiqit. The vegetation of Wadi Muqsim is sparse.

Beds of the main channel and its runnel tributaries are mostly covered with coarse sediments. There are a few bushes of A. *ehrenbergiana*. On the sandy patches of the main channel there may be *Aerva javanica, Dipterygium glaucum, Morettia philaena, Panicum turgidum, Solenostemma arghel* and *Stipagrostis plumosa*. In the smaller runnels dissecting the slopes and other higher ground *Fagonia indica* is abundant. In Wadi Abu Fas the vegetation is a mixed scrub of A. *ehrenbergiana* and A. *raddiana*. The latter seems to increase in the parts of the wadi nearer to the mountains. The channel of Wadi Ungat has a group of fresh-water wells which forms one of the main water points along the camel-caravan road. This wadi is characterized by the dominance of A. *raddiana* scrub together with A. *ehrenbergiana* and *Balanites aegyptiaca.* The undergrowth includes a rich growth *of Cassia senna* associated with *Aerva javanica, Astragalus vogelii, Citrullus colocynthis. Euphorbia granulata, Fagonia bruguieri, Indigofera argentea* and *Morettia philaena. Fagonia indica* is common in the runnels dissecting the slopes of the hills. The scrubland growth of Wadi Neiqit is thinner than that of Wadi Ungat and is mostly A *ehrenbergiana. A. raddiana* is rare in the downstream parts but increases nearer to the mountain areas. The undergrowth includes *Aerva javanica, Cassia senna, Fagonia indica, Francoeuria crispa* and *Indigofera argentea.*

Low Plateau Section

The main channel of this section of Wadi Allaqi supports a distantly open scrub of A. *ehrenbergiana* with occasional specimens of A. *raddiana.* Fossil hillocks of *Tamarix aphylla* are occasionally found but not of *Salvadora persica.* The channel is dotted with green patches of vegetation associated with the mouths of some of the affluent wadis. These patches are mostly dominated by *Salsola baryosma* associated with *Cassia senna, Cistanche tinctoria, Citrullus colocynthis, Convolvulus prostratus, Francoeuria crispa, Haplophyllum obovatum, Psoralea plicata* and *Trianthema crystallina.*

On the north side, Wadi Allaqi receives numerous affluents including a number of extensive wadis. The vegetation of these wadis is dominated by *Cassia senna* in the downstream parts and by *Aerva javanica* upstream.

Within Wadi Umm Qureiyat four types of vegetation may be distinguished. In the downstream part where the deposits are soft *Citrullus colocynthis* dominates. In the middle part there are extensive stretches of dense growth of *Cassia senna.* In the upstream part *C. senna* is much thinner. In the affluent runnels of this wadi the vegetation is mostly an ephemeral growth of *Morettia philaena.*

Wadi Haimur is one of the extensive tributaries of the north side of Wadi Allaqi. It contains a group of wells. A *ehrenbergiana* scrub occurs within the whole system of this wadi and the presence of *Cassia senna* is common. *A. raddiana* is rare in the downstream parts but increases nearer to the mountain areas. A part of the main channel of Wadi Haimur is characterized by thickets of *Tamarix nilotica* covering silt terraces. In the part of Wadi Haimur near the wells extensive regeneration of *Cassia senna* was shown in February 1963; this is taken to indicate that some rain had fallen (Kassas and Girgis, 1969–1970). In finer runnels of this wadi, the vegetation is mostly ephemeral comprising *Morettia philaena* and *Fagonia indica.*

Deltaic Section

Wadi Allaqi flows onto a wide deltaic plain covered with a mixture of gravel and sand. The surface deposits of this delta are coarse at the mouth of the channel and soft near the fringes of the Nile. The delta is now essentially part of the reservoir lake of the High Dam.

The vegetation of this deltaic part, during field study in February 1963, was very sparse. There was no *Acacia* scrub that characterizes the defined course of the main channel. Individuals of the following species were recorded: *Aervajavanica, Citrullus colocynthis, Crotalaria aegyptiaca* and *Hyoscyamus muticus.*

Springuel et al. (1997) identified the vegetation of Wadi Allaqi, as a major North African desert system of international importance in the context of conserving desert biodiversity. They investigated the influence of 30 years human influence on the flora produced by periodic flooding of the lower section of the Wadi by Lake Nasser (High Dam lake) and compared the results obtained with those of Kassas and Girgis (1965, 1969–1970, and 1972) and Springual et al. (1991). They recorded *Tamarix nilotica* shrubs in the downstream part of the main channel of Wadi Allaqi. The dominant shrub and the ground layer vegetation co-dominated by *Glinus lotoides* and *Pulicarea crispa* varied in cover and density with the distance from the lake shore to form a zoned distribution up to 2 km in length and 2 km width. In the further downstream section of Wadi Allaqi, the dense stands of *T. nilotica* and its associated species appear to be supported by shallow ground water retained in the wadi fill deposits after the annual Nile flood events which usually raise the level of Lake Nasser by 10–15 cm. Springuel et al. (1997) stated "the *T. nilotica* group identified in this study is a new vegetation for the Allaqi area". The community co-dominated by *T. nilotica* (*T. mannifera*) – *Pulicaria crispa* (= *Francoeuria crispa*) described by Kassas and Girgis (1969–1970) differs from the new vegetation group because the former comprises a relict vegetation community in one part of Wadi Haimur where the ground water level is usually close to the soil surface. Also Springuel et al. (1997) recognized the group of *Cullen plicatum* (*Psoralea plicata*) previously described by Kassas and Girgis (1969–1970) as a part of the community dominated by *Salsola imbricata* (= *S. baryosma*).

The flora of Wadi Allaqi comprises 79 species: 55 perennials, including the dominant species, namely: *Acacia ehrenbergiana, Acacia tortilis* subsp. *tortilis (=A. tortilis), Cullen plicatum* and *Tamarix nilotica* in addition to 24 annuals. The largest number of associated species occur in *A. tortilis* community (57 species) followed by *A. ehrenbergiana* community (40 species), *C. plicatum* community (25 species) and the lowest is that of *T. nilotica* community (11 species). Eight associates, namely: *Panicum turgidum, Citrullus colocynthis, Acacia raddiana, Pulicaria crispa, Fagonia indica, Morettia philaena, Senna alexandrina* and *S. indica* occur in the four communities, whereas 10 species occur in three communities, these are: *Aerva javanica, Crotallaria aegyptiaca, Alhagi graecorum, Salsola imbricata, Dipterygium glaucum, Monsonia nivea* (perennials) and *Glinus lotoides, Astragalus eremophilus, Euphorbia granulata* and *Shouwia thebaia* (annuals). It is worth to mention that three of the four dominant species (*A. ehrenbergiana. A. tortilis* and *T. nilotica*) are not recorded as associate species in any of the other communities but *C. plicatum* is recorded with *A. ehrenbergia* community.

Wadis of the Sandstone Desert

The wadis of the sandstone formation traverse a very dry country. The presence of rich vegetation recorded in certain wadis, e.g. Wadi Kurusku, is apparently exceptional. The downstream parts of these wadis form, at their confluence with the Nile, small khors (channels) that were filled by the Nile water during the flood season before the establishment of the Aswan High Dam. The extent of these khors was dependent on the height of the flood. Nowadays, the khors have been expanded and new habitat conditions created.

The vegetation of these wadis includes a few specimens of *A. ehrenbergiana* and A. *raddiana*. Patches, both of suffrutescent species and of ephemerals, dominated mostly by *Fagonia indica*, were present. Associates are *Aerva javanica. Cassia senna, Citrullus colocynthis, Morettia philaena* and *Solenostemma arghel.*

Wadi Kurusku has a great catchment area. On the occasion of field studies carried out by Kassas and Girgis (February 1963), the bed of this wadi was covered by a rich growth of the ephemerals *Fagonia indica* and *Morettia philaena*. Associate species included *Cassia senna* and *Francoeuria crispa*. In the middle and downstream parts of the main channel of Wadi Kurusku, *Acacia ehrenbergiana* forms an open scrub associated with ephemerals dominated by *Fagonia indica*. The khor formed at the mouth of Wadi Kurusku is fringed by *Acacia nilotica, Alhagi maurorum, Calotropis procera, Glinus lotoides* and *Tamarix nilotica*. The upstream parts of this wadi are often choked with sand embankments which may be the habitat of a thin ephemeral growth *of Fagonia indica, Morettia philaena, Stipagrostis plumosa* and *Tribulus longipetalus*.

Wadi Qar flows into the Nile a few kilometres south of Abu Simbil. In the downstream part of this wadi (now inundated by the water of the reservoir of the High Dam) there are patches of scrub growth including a few specimens of *Acacia raddiana* with the undergrowth *of Astragalus vogelii, Cassia senna, Crotalaria thebaica, Fagonia indica, Morettia philaena, Pulicaria undulata, Stipagrostis plumosa, Tribulus longipetalus* and *T. ochroleucus*.

Main Communities

Though the flora of the Nubian Desert is limited, four groups of vegetation may be recognized: (1) ephemeral types; (2) suffrutescent woody types; (3) suffrutescent succulent types; and (4) scrubland types.

Ephemeral Types

This vegetation is of communities of ephemeral species and perennials behaving as ephemerals.

1. *Morettia philaena* community. The *M. philaena* community characterizes two types of habitat: the smaller runnels of Wadi Allaqi and of some of the wadis of the Nubian sandstone country. The habitat conditions are severe, the plant cover is thin and the species represented are limited. Associates include *Acacia ehrenbergiana, A. raddiana, Aerva javanica, Cassia senna, Fagonia indica, Francoeuria crispa* and *Stipagrostis plumosa*.
2. *Fagonia indica* community. This community is associated with two types of habitat: sheets of sand covering the plains among the hills, and wadis of the sandstone country. *Morettia philaena* is invariably present and *Cassia senna* is a common associate. Other associates are those of the *M. philaena* community.

Suffrutescent Woody Types

1. *Cassia senna* community. *C. senna* is a widespread xerophyte within the Nubian Desert. In Wadi Allaqi the plant cover of this community ranges from 5 to 25%. Of the three layers the shrub layer is thin and includes two common associates, *Acacia ehrenbergiana* and A. *raddiana* together with the less common *A. tortilis* and *Balanites aegyptiaca*. The suffrutescent layer is well developed as it includes the dominant together with the common associate *Aerva persica* and the less common *Crotalaria aegyptiaca, Dipterygium glaucum, Francoeuria crispa, Heliotropium arbainense, Panicum turgidum, Salsola baryosma, Solenostemma arghel* and *Stipagrostis plumosa.*
2. *Aerva javanica* community. *A. javanica* is a xerophyte throughout the whole Eastern Desert. Stands of vegetation dominated by this species in the Nubian Desert are found in a number of the smaller affluents of Wadi Allaqi. The growth of this plant is usually associated with coarse deposits. *Cassia senna* is a consistent associate species. *Acacia ehrenbergiana* and *A. raddiana* are common associates that form the shrub layer. Other associates are *Indigofera argentea, Fagonia indica, Francoeuria crispa* and *Salsola baryosma*. The ground layer is represented by *Euphorbia granulata.*
3. *Indigofera spinosa* community. *I. spinosa* is a silvery leguminous shrub with patent, glossy or yellow, needle-like, sharp spines. In the Nubian Desert it is dominant in a number of the smaller tributaries of the Wadi Allaqi system. The cover is usually low (<5%), no doubt because the dominant is grazed. *Aerva javanica* and *Fagonia indica* are consistently present. *A. ehrenbergiana* (common) and *A. raddiana* (rare) represent the shrub layer. No ground layer has been recognized in this community. Other associates are in the suffrutescent layer and include *Cassia senna, Salsola baryosma* and *Stipagrostis plumosa.*

Suffrutescent Succulent Types

This type is represented by only one community, that dominated by *Salsola baryosma*, a common succulent xerophyte in the Nubian Desert. Its dominance is confined to the main channel of Wadi Allaqi. Dry growth of S. *baryosma* is present all over the channel but patches of its green growth are confined to the confluence areas. In this community, the shrub layer is thin and includes the common *A. ehrenbergiana, A. raddiana, Balanites aegyptiaca, Leptadenia pyrotechnica* and *Salvadora persica*. The suffrutescent layer is the most obvious part of the vegetation as it includes the dominant and a number of common associates: *Aerva javanica, Cassia italica, C. senna, Francoeuria crispa, Panicum turgidum* and *Psoralea plicata*. The ground layer is represented by a thin cover of *Citrullus colocynthis, Convolvulus prostratus* and *Fagonia indica* and may be enriched by the ephemerals during rainy years.

Scrubland Types

These include the communities dominated by large shrubs or trees and so there is at least one vegetation layer higher than 150 cm. These types represent the permanent vegetation of the country, in that these shrubs and trees survive periods of rainless years. Other types of vegetation appear in rainy years and may survive part but not the whole of the subsequent rainless episode.

The distribution of the scrubland types follows a geographical distribution between the eastern part of the Nubian Desert (east of Long. 34°35′E) and a western part. Ecological differences are shown between the requirements of the different dominants.

1. *Acacia ehrenbergiana* community. This is the most widespread scrub in the Nubian Desert. It does not extend into the montane country that forms the eastern border of this desert. The density of A. *ehrenbergiana* and the local pattern of its distribution are variable. In certain wadis there is a thin growth, more or less evenly distributed, over an extensive area; in other wadis it grows strongly in patches. The usual growth form of A. *ehrenbergiana* is that of a much branched bush, but it may be tree-like.

   Vegetation of this community is organized into three layers. The frutescent layer comprises the dominant species and the most common associate *A. raddiana*. The suffrutescent layer contains the consistently present associate *Cassia senna* and the common associates (*Aerva javanica* and *Francoeuria crispa*). The ground layer includes such prostrate and low-growing perennials as *Citrullus colocynthis* and *Fagonia indica* and the ephemerals. Other rarely present species are *Cleome droserifolia, Crotalaria aegyptiaca, Indigofera argentea, Salsola baryosma* and *Stipagrostis plumosa.*

   This scrubland is present in the Wadi Allaqi drainage system west of Long. 34°30′E, but is not recorded further eastwards. It is more frequent in the tributaries than in the main channel. This is probably due to the history of the vegetation (Kassas and Girgis, 1969–1970). The main channel was obviously populated by a rich growth of *Tamarix aphylla* and *Salvadora persica* thickets whereas the *A. ehrenbergiana* scrub was confined to the affluent tributaries with deep wadi-fill deposits that are usually compact.

   A. ehrenbergiana is not subject to widespread lumbering for charcoal manufacture. In this respect it is different from A. *raddiana* which has been the main raw material for charcoal throughout historical times.
2. *Acacia tortilis* community. This type of scrubland is common within the eastern part of the Wadi Allaqi system, east of Long. 34°30′E. West of this line A. *ehrenbergiana* occurs.

   The *A. tortilis* scrub is common in the affluents, and is usually associated with gravel alluvial deposits that are not very deep. The size of the bushes seems to vary according to the habitat conditions; the bushes are as low as 1 m in the smaller runnels and as high as 5 m in the larger affluents. The cover of this community ranges from 15 to 40%. The dominant provides most of the cover. In certain localities, *A. raddiana*, which is present in all of the stands, may also contribute to the cover. The vegetation may be recognized in four layers. The thin tree layer includes A. *raddiana, Balanites aegyptiaca, Calotropis procera* and *Maerua crassifolia*. Many individuals of these species do not reach the height of trees. The shrub layer includes the dominant together with *Leptadenia pyrotechnica* and dwarf individuals of *Calotropis procera, Maerua crassifolia, Salvadora persica* and *Ziziphus spina-christi.* The suffrutescent layer includes most of the other associates, e.g. *Aerva persica, Panicum turgidum* and *Solenostemma arghel.* Rare associates are *Ochradenus baccatus*, which may show a climbing growth form, and *Acacia albida*, which is mostly confined to the banks of the Nile, and may have been cultivated and can thus be taken to indicate previous human occupation. The ground layer is represented by sparse *Cleome droserifolia.*

3. *Acacia raddiana* community. *A. raddiana* is present all over the wadis of the Nubian Desert, but the stands of its open forest are mostly confined to the montane country of the eastern part of the Wadi Allaqi system. In this respect it is different from A. *tortilis* scrub which is frequent in the affluent tributaries and smaller runnels.

   The cover of the A *raddiana* community varies from 10 to 40%. The conspicuous tree layer includes the dominant species, having the greatest cover, together with several common associates such as *Balanites aegyptiaca* and *Ziziphus spina-christi* (which may have tree or shrub form). *Ochradenus baccatus* and *Cocculus pendulus* are two climbers. *Calotropis procera, Capparis decidua, Maerua crassifolia* and *Salvadora persica* may be of tree form but as a result of repeated cutting may form bushy growth. These four species are included in this community in the shrub layer which also contains *Acacia ehrenbergiana, A. tortilis, Leptadenia pyrotechnica* and *Lycium arabicum. Solenostemma arghel* is the most common member of the suffrutescent layer which also includes *Aerva javanica, Cassia senna, Indigofera argentea* and *Salsola baryosma.* The ground layer is represented by rare individuals of *Cleome droserifolia* and *Citrullus colocynthis* and may be enriched by ephemerals during rainy years. This community includes also a few specimens of *Ricinus communis* (semi-wild) which may be planted by the bedouins.

4. *Leptadenia pyrotechnica* community. The vegetation dominated by *L. pyrotechnica* is confined to the montane country of the Wadi Allaqi system east of Long. 34°30′E, where the dominant forms patches which accumulate heaps of soft deposits. Associate species occur between these patches. *Leptadenia*, as do most of the members of the Asclepiadaceae, provides stem fibre that may be used for household purposes. It is consequently subject to partial destruction by cutting.

   Three species which are consistent associates in this community are *Acacia raddiana. A. tortilis* and *Balanites aegyptiaca. Maerua crassifolia, Salvadora persica* and *Solenostemma arghel* are very common associates, whereas *Ochradenus baccatus* and *Stipagrostis plumosa* are common. *Capparis decidua* and the semi-wild *Ricinus communis* are rare.

5. *Tamarix nilotica* community. Reference has been made to the fossil *Tamarix aphylla* hillocks that are common within the main part of the principal channel of Wadi Allaqi. In a part of Wadi Haimur, near the well, is a limited area where the silt terraces are covered by a rich growth of *T. nilotica*. On these terraces *T. nilotica* grows in pure stands with 30–60% cover. As already mentioned, *T. nilotica* forms one of the typical thickets of the salt marshes of the Red Sea

coast. The notable feature of the soil supporting the growth of the *T. nilotica* community in the Wadi Allaqi area is the relatively high salt content: 3.5% in the surface 10 cm layer and 12–30% in the subsurface layer (Kassas and Girgis, 1969–1970). These are the highest records of soil salinity in the Nubian Desert. The area around Haimur well is an inland salt-affected land. No associate xerophytes have been recorded in this community.

6. *Salvadora persica* community. Living *S. persica* is confined to the wadis of the montane country of the eastern Wadi Allaqi system. Fossil remnants of its growth are widespread in the main part of the principal channel of Wadi Allaqi.

   S. *persica* may grow into a tree but in most instances it forms patches covering mounds of soft deposits. This is probably due to the repeated cutting of the young branches which are used by the bedouins as tooth-brushes (Arak). Ayensu (1979) states "A 50% ethanol-water extract of the stem-bark of S. *persica* exhibited *in vitro* antispasmodic activity. It is also used for gonorrhoea, spleen, boils, sores, gum disease, and stomach ache".

   The *S. persica* community is represented in the Nubian Desert by a few stands on the silt terraces of the upstream part of the main channel of Wadi Allaqi. *Acacia raddiana, A. tortilis* and *Leptadenia pyrotechnica* are consistently present. *Calotropis procera* and *Solenostemma arghel* are the common associates. Less common associates are *Balanites aegyptiaca, Capparis decidua, Cocculus pendulus, Lycium arabicum, Ochradenus baccatus* and *Ziziphus spina-christi.*

7. *Balanites aegyptiaca* community. Rare individuals of *B. aegyptiaca* are recorded in a few of the tributaries of Wadi Allaqi west of Long. 34°30′E. Stands of an open forest dominated by *B. aegyptiaca* are reported in a number of the wadis of the eastern Wadi Allaqi montane country.

   The structural organization of this community is very similar to that of the open forest of A. *raddiana*. Both types are also very similar in floristic composition and their ecological relationships. In this community A. *raddiana* and *Leptadenia* are consistently present. In certain localities A. *raddiana* may be co-dominant. The abundant associates are A. *tortilis, Calotropis procera* and *Salvadora persica*. Common ones are *Maerua crassifolia, Ochradenus baccatus* and *Solenostemma arghel*. Other less common associates are *Aerva javanica, Cleome droserifolia, Cocculus pendulus, Salsola baryosma* and *Ziziphus spina-christi* and the semi-wild *Ricinus communis*.

Chapter 5
The Sinai Peninsula

5.1 Geomorphology

The Sinai Peninsula is a triangular plateau in the northeast of Egypt with its apex, in the south, at Ras Muhammed, where the eastern coast of the Gulf of Suez meets the western coast of the Gulf of Aqaba (Lat. 27°45′N). Its base, in the north, is along the Mediterranean Sea (the eastern section of the Egyptian Mediterranean coast that extends for about 240 km between Port Said and Rafah (Lat. 31°12′N)). The area of the Sinai Peninsula (61,000 km^2) is about 6% of that of Egypt. More than half the peninsula is between the Gulfs of Aqaba and Suez (Fig. 5.1).

Geographically, the Sinai Peninsula is separated from the other regions of Egypt by the Gulf of Suez and the Suez Canal. It is continuous with the continent of Asia for over 200 km between Rafah on the Mediterranean and the head of the Gulf of Aqaba which separates it from Saudi Arabia (Fig. 2.1). The core of the peninsula, situated near its southern end, consists of an intricate complex of high and very rugged igneous and metamorphic mountains. The northern two-thirds of the peninsula is occupied by a great northward-draining limestone plateau which rises from the Mediterranean coast, extends southwards and terminates in a high escarpment on the northern flanks of the great igneous core (Said, 1962).

Geomorphologically, the Sinai Peninsula region may be divided into three subregions: southern, central and northern. The southern subregion has an area of about 20,000 km^2 or one-third of the area of Sinai (Shata, 1955, 1956). Its basement complex, which appears on the surface, has the form of a horst, an upstanding block of the earth's crust bordered on all sides by long fault lines. This horst is composed of igneous and metamorphic rocks constituting the foot of old mountains which have been undergoing erosion since their formation.

The southern horst of Sinai was subject to severe crustal disturbances during Tertiary and Quaternary times. This ultimately produced the Gulfs of Suez and Aqaba as well as a number of fault blocks (Beadnell, 1927).

Fig. 5.1 The Sinai Peninsula

The eastern and western edges of the Sinai horst are different from one another. The western coastal plain, known as El-Qaa, is wide. It borders the Gulf of Suez, which has a depth of 100 m and a length (from El-Shatt[1] southwards to Ras Muhammed) of about 400 km and the eastern coastal plain that extends from Aqaba southwards to Ras Muhammed for about 235 km.

The mountains which form the igneous core of the Sinai Peninsula rise to considerably greater heights than any of those of the African part of Egypt. The highest peak, Gebel St Katherine, is 2641 m above sea level. Many other peaks and crests rise above the 2000 m contour, conspicuous among which are Gebel Umm Shomer (2586 m), Gebel Musa (2285 m), Gebel Al-Thabt (2439 m) and Gebel Sebal Pile (2070 m). "The core of the peninsula is highly dissected; its gaunt mountains and deep rocky gorges form one of the most rugged tracts on the earth's surface" (Said, 1962).

Because of its high altitude, the southern section of Sinai receives ample rainfall which has produced wadis. Most of the wadis run in long hollows and appear as hanging valleys. Some wadis flow to the Gulf of Aqaba, e.g. Wadis Ghayib, Nasb and Watir, all of which are steep valleys. Running to the El-Qaa plain, in the west, are for example Wadis Feiran, Sidri, Sudr and Gharandal, all of which are wide and have relatively rich vegetation.

The higher part of the limestone plateau which flanks the igneous core to the north is called El-Tih (Fig. 5.2). At the southern end of Gebel El-Tih is the Ugma Plateau which is 1620 m above sea level. The central portion of the plateau forms fairly open country draining to the Gulfs of Aqaba and Suez.

The western coastal plain of Sinai is relatively broad, extending in western Sinai from the Gulf of Suez to the great western El-Tih escarpment and its continuation southwards in the granitic ridges of southern Sinai. The plain is broad in its northern part but narrows south of Gebel Hammam Faraon (about 80 km south of El-Shatt). The northern division has an average breadth of 30 km. It is very gently undulating and locally dotted with low hills of limestone. The surface of the plain is largely covered with drift sand which forms parallel crescentic dunes. In more southern parts, the coastal plain, which is traversed by several wadis, is of sandy marls and gypsum, covered in some parts with gravels.

The foreshore areas of the western coast of Sinai are subject to flooding with water of the Gulf of Suez. In the dry season these areas become covered with a thin mantle of white salts. The southern half of the coastal plain is crossed by well-defined ridges such as Gebel Hammam Faraon and Gebel Araba, the summits of which are over 500 m above sea level. Opposite these ridges the coastal plain is very much narrower.

The hills forming the eastern boundary of the foreshore plain become progressively higher southwards. The plateau of the southern subregion of Sinai drops by a number of steps to that plain. Along the Gulf of Aqaba the coastal plain is relatively narrow.

The central subregion of the Sinai Peninsula is called the El-Tih plateau (Fig. 5.2), but this is a misnomer because this plateau is only the southern part of central Sinai.

[1] El-Shatt is a city on the Sinai side of the Suez Canal that faces Suez.

The middle part of this central section is known as the Al-Ugma plateau and the northern part is the area of domes. These domes echo the alpine movements which occurred in the Eastern Mediterranean Region (Abu Al-Izz, 1971). The important peaks of central Sinai are:

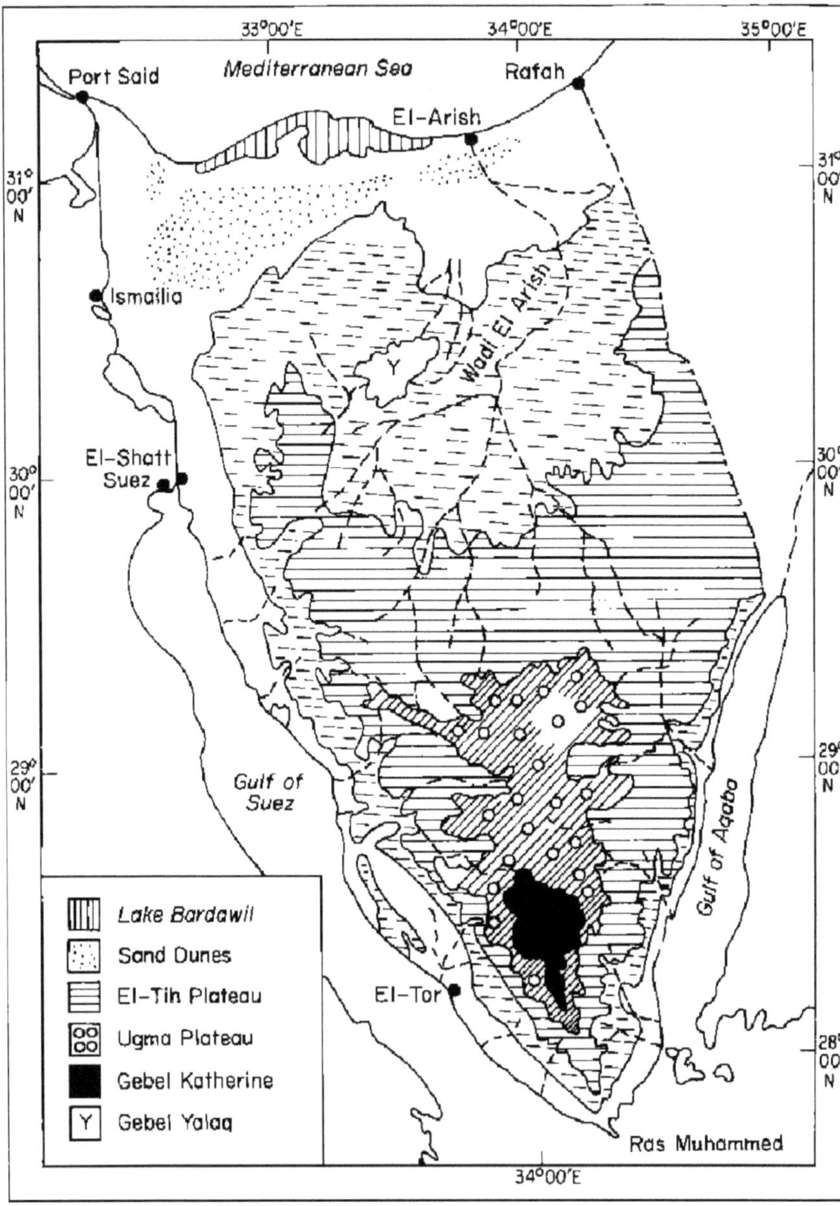

Fig. 5.2 The main geographical features of the Sinai Peninsula. Wadis are shown by *broken lines*; increased elevation by *heavier shading*

Gebel El-Halal (890m), an anticline with its axis running from northeast to southwest, parallel to the axis of the domes. For 7km, Wadi El-Arish crosses this mountain with a narrow course.

1. Gebel El-Maghara (500–700m), which includes several secondary domes, is an asymmetrical fold covering an area of about 300km^2. It is the largest Jurassic exposure in Egypt.
2. Gebel Yalaq (Yellag), about 1100m, is a great asymmetrical anticline with its southern side steeper and with abundant faults.

The El-Tih plateau slopes from the above-mentioned heights down to the north. It is dissected by drainage channels, most of which flow to the Mediterranean Sea. These channels are generally much shallower and more open than the wadis of the southern mountainous subregion.

The northern subregion of the Sinai Peninsula is about 8000km^2, or 13% of the area of the peninsula. It is bordered by the region of the folds in the south and the Mediterranean Sea in the north and extends to the Suez Canal in the west (Figs. 5.1 and 5.2). This region consists of a wide plain sloping gradually northward. It narrows in the east because of the presence of Gebel El-Maghara. There are sand dunes of 80–100m which extend for several kilometers landward in this wide plain, forming a continuous series parallel to the sea. This flat northern belt of Sinai is an extension of the Mediterranean coastal area of Egypt-the eastern (Sinai) Mediterranean coast. The northern sand dunes of this belt are elongated while at its southern edge the dunes are of crescentic type (Migahid et al., 1959). Aeolian deposition has thus played a great role in this section of Sinai, forming dunes which parallel the northwesterly winds. The dunes near Gebel El-Maghara extend from southwest to northeast, perhaps because of the influence of the mountains.

The northern Sinai dunes absorb and store rain water, the low lands between them being a permanent source of fresh water that can be tapped by digging shallow wells. The best quality and highest volume of water is in the delta of Wadi El-Arish (Hume, 1925). The water supply of the wells varies and its quantity depends upon rainfall. The depth of the wells can be as little as 3m or as much as 60m (Abu Al-Izz, 1971).

The northern subregion of the Sinai Peninsula is characterized by two important morphological units – Wadi El-Arish and Lake Bardawil.

5.1.1 Wadi El-Arish

Wadi El-Arish is one of the most important geographical features of northern Sinai. Its basin is about 20,000km^2 of Sinai in the Mediterranean drainage system. Its length is about 250km, narrow in its upper reaches through cutting across the El-Tih plateau (where the old pilgrimage road used to be). The wadi is joined there by two tributaries, one from the west, Wadi Al-Burak, and an eastern tributary, Wadi Al-Aqaba. Downstream other tributaries enter from both the east and west.

Wadi El-Arish can be divided into three sections: upper, central and lower (coastal). The upper section is about 100 km with a gradient of 6:1000. In the central section the wadi descends from 400 m to 150 m in about 100 km, i.e. with a gradient of 2.5:1000. The coastal section covers the last 50 km where the wadi has a gradient of 3:1000. Along the wadi, fluvial deposits form three terraces, having, at the town of El-Arish, elevations of 35, 22 and 13 m above sea level.

5.1.2 Lake Bardawil

North Sinai has a straight coastline bordered by sandy bars. Between these bars is the shallow Lake Bardawil (Fig. 5.1), that extends eastward of Al-Muhammadiya (45 km east of Port Said) for 98 km. It is elliptical and its total area is about 69,035 km^2. This large area is not continuously covered by water, but becomes separate ponds and lakes during summer. The lake is moreover affected by moving sand; so its area is decreasing, and it is taking the form of a playa rather than a typical lake.

Joined to Lake Bardawil are a number of bays and ponds that cover about 14% of its area. The islands in this lake are formed of old sand bars. There is no silt since this lake is far from the old branches of the delta.

Lake Bardawil is very closely connected with the Mediterranean Sea. The low sand bar which divides it from the sea is often covered by sea water. Lake Bardawil is, thus, the most saline of the northern Egyptian lakes (Chapter 6), for it is connected only with the sea.

5.2 Climate

Being a part of Egypt at the extreme northeast of Africa, the Sinai Peninsula belongs, climatically, to the dry province. According to Ayyad and Ghabbour (1986), the Sinai Peninsula can be divided into two main climatic zones: arid and hyperarid. The arid zone includes the northern subregion: summer is hot, winter is mild and rainfall usually occurs in winter. It is distinguished into two provinces (UNESCO/FAO Map, 1963): (1) the coastal belt province under the maritime influence of the Mediterranean Sea with a relatively shorter dry period (attenuated) and annual rainfall ranging between 100–200 mm, and (2) the inland province with a relatively longer dry period (accentuated) and an annual rainfall of 20–100 mm. The hyperarid zone, however, covers the central and southern subregions of the peninsula. It is also distinguished into two provinces: (1) hyperarid province with hot summer, mild winter and winter rainfall which is considered an extension of the Eastern Desert and includes central Sinai or the El-Tih plateau together with the western and eastern coasts of the Gulfs of Aqaba and Suez; (2) hyperarid province with cool winter and hot summer located around the summits of the Sinai mountains.

During the winter months some areas of Sinai experience short periods of brief but heavy rainfall that may cause the wadis to overflow.

Air temperature in Sinai is subject to large variations, both seasonally and spatially. Minimum winter temperature ranges from 19 °C at Sharm El-Sheikh to 15 °C at El-Tor, 14 °C at El-Arish, 9 °C at Nakl to 0 °C at St Katherine (–4 °C during January 1987). Maximum summer temperature also shows a large variation, and ranges from near 20 °C at St Katherine, with its high elevation the coolest in the peninsula, to more than 50 °C at El-Kuntilla. During summer the Gulf of Suez region is much warmer (35 °C) than the northern Mediterranean region (30 °C). The amount of rainfall in Sinai decreases from the northeast to the southwest. The relatively highest amount of rain is in Rafah (304 mm/year) followed by that of El-Arish (99.7 mm/year), but only about 10.4 mm/year occurs in the southwest. Rainfall decreases in the plateau region to about 23.3 mm/year, but then increases in the southern mountainous region to about 62 mm/year in St Katherine where precipitation may occur as snow that may last for 4 weeks (Migahid et al., 1959). In some years more than one snowfall may occur whereas in others snow may be absent. Precipitation may occur as hail on the high peaks. Water derived from melting snow or hail is usually insufficient to infiltrate the desert soil at the foot of the mountains appreciably.

Rain decreases in southerly (towards Ras Muhammed), easterly (towards the Gulf of Aqaba) and westerly (towards the Gulf of Suez) directions and averages only 12 mm/year in the Gulf of Suez (23 mm/year in Suez, 22 mm/year in Abu Redis and 9.3 mm/year in El-Tor).

Rainfall occurs in Sinai mainly during the winter season (November-March) and during spring or autumn. It decreases markedly or completely lacking from May to October. However, summer rain resulting from the influence of the Red Sea depressions causes floods. Tropical plants of Sudanian origin may germinate in the moist wadi beds following summer rain. Also, summer fogs and dew are frequent in Sinai north of El-Tih but absent in southern Sinai. In some years dew provides more moisture than does rainfall. Lichens are particularly effective in capturing moisture from fog and dew. Overall the annual average rainfall for the entire Sinai Peninsula is 40 mm, of which 27 mm is estimated to come from individual storms of 10 m or more.

The mean annual maxima and minima of relative humidity at El-Arish, El-Tor and Suez are: 79% and 56%, 70% and 50%, 73% and 30% respectively, and the mean values of annual evaporation of the three areas are: 4.3 mm/day, 10.2 mm/day and 9.4 mm/day (all Pier *Climatic Normals of Egypt*, Anonymous, 1960).

Winds in Sinai usually blow in winter from the west or southwest but in summer are mostly from the northeast to northwest. The speed of these winds generally do not exceed 10 knots and gentle wind frequently occur on the northern coast and sometimes along the Gulf of Suez. Strong winds with speeds higher than 34 knots do not usually occur more than 1 day every 3 years. In the northern area Khamsin winds with gusts with velocities reaching 64 knots are common during February-March. These winds, which are extremely hot and dry, blow over Sinai mostly from the southwest to east, causing sharp reductions in the relative humidity (to be less than 15%) and a sharp increases in air temperature (up to 45 °C).

5.3 Water Resources

In the Sinai Peninsula there are three indigenous water sources: rainfall, surface and ground water. The availability of Nile water for use in Sinai may also be considered (Anonymous, 1985).

Along the Mediterranean coastal area, rain water is stored in the sand and can be raised to the surface by digging shallow wells. In the desert country of the central subregion of Sinai the rain is scanty and the water supply is less abundant and more saline than in the Mediterranean coast.

In the southern mountainous country, rainfall, though scanty, is decidedly effective. There are broad catchment areas. Water falling on the mountains runs over the slopes and collects in the narrow deep wadis where it forms perpetual steams and rivulets. This water is normally fresh and evaporation is lower in the mountains. Rain storms sometimes burst upon the mountains causing torrents and terrible floods. On rare occasions, such storms change a dry wadi into a mighty river for some time. The excess water percolates and becomes stored underground in rock crevices. It can be obtained by digging wells or it appears at the surface as springs or streams of fresh water. The largest number of fresh-water wells and streams is near the monasteries of St Katherine and Feiran. Rain water, collecting in rock clefts, may sometimes form small pools of fresh water.

Another source of water supply in the montane country of Sinai is the snow which covers the summit of the high mountains in winter. The snow mantle may reach a depth of one metre or more in some places where it may long persist during the winter. As it melts with the advent of warm weather, water runs down the mountain slopes and adds to the resources of the wadis. Zohary (1935) believes that some of the higher mountains enjoy an annual precipitation of not less than 300 mm. This large resource of fresh water has made possible the growth of rich wild vegetation in southern Sinai as well as the development of oasis-like areas and *Tamarix* scrub in many wadis. It has been also possible to cultivate cereals and fruit (grapes, figs, olives, etc.), especially in areas of human settlements such as the monasteries.

In Sinai there were seven dams; three are masonry ones e.g. Rawafa Dam, Wadi Gharandal Dam and Wadi Shellal Dam. The most interesting is Rawafa which is an arched masonry dam on Wadi El-Arish, about 52 km south of El-Arish town. It was built in 1946 and reportedly had an initial capacity of about 3 million m^3, but, like the others, it is silted up.

Apart from dams, several methods have traditionally been employed in Sinai to store or conserve run-off water for drinking and for agriculture. In many areas cisterns are being employed to collect and store water for livestock and domestic use.

The bedouins in Sinai make use of flood water for agriculture. In many areas, the land in the wadi bed is ploughed and cultivated after the first rains of a season. This is the practice, for example, in Wadi El-Arish northeast of Nakl where the bedouins cultivate barley, corn (maize), tomato, sesame, grapes, pomegranates, olives and watermelons.

In many areas the bedouins have constructed spreader dykes in the wadi bed and cultivation of beans, wheat and other cereals is carried out just upstream of the dykes (Anonymous, 1985).

5.4 The Vegetation

5.4.1 General Features

Sinai is of special ecological interest because of its variable environment, beautiful landscape, distinctive flora and above all its uniqueness and contrasts. Sinai is the meeting point of two continents: Africa and Asia. This union is reflected in the physiography, climate and plant cover of Sinai. Although Sinai is surrounded on three-fifths of its perimeter by water, the climate is dry and the peninsula contains large tracts of desert and high rugged mountains. No permanent rivers flow from Sinai, yet during winter storms torrents of water rush down the usually dry wadis and can wash away concrete bridges and roadbeds.

Because of its unique character, the Sinai Peninsula has been the subject of several studies in different fields. Its flora has attracted the attention of many explorers and botanists since the seventeenth century and even earlier. Täckholm (1932) and Zohary (1935) give accounts of important contributions published on the flora of Sinai. Batanouny (1985) gave a full report describing the activities of explorers and botanists in the peninsula from 1761 to the present.

Among those making major contributions to the botany of the area are Delile (1809–1812) who visited Egypt in 1778 under the command of Napoleon I, and Fresenius (1834) whose *Beitrage zur Flora von Aegypten und Arabien*, based on the work of the German Wilhelm Ruepell, can be considered the first account of the flora of Sinai. This publication enumerated 38 families and 142 species, including many previously undescribed. Decaisne (1834), however, listed no fewer than 233 species in *Florula Sinaica*.

Phytogeographically, the Sinai Peninsula stands in a middle position between three well-defined phytogeographical regions of the world – Saharo-Scindian (African-Indian Desert region of Good, 1947), Irano-Turanian (west and central Asiatic region of Good, 1947) and the Mediterranean. Accordingly, the flora of Sinai combines the elements of these three regions (El-Hadidi, 1969). Zohary (1935) recognized the presence in Sinai of 942 species belonging to different elements or connected elements. The most important of these are: (1) Saharo-Scindian, represented by 299 species; (2) Irano-Turanian, represented by 98 species; (3) Mediterranean, represented by 118 species; and (4) Sudano-Deccanian, represented by 41 species. In addition, species of other biregional connection groups constitute about 40% of the total. Important connection groups are: (a) Mediterranean-Irano-Turanian and (b) Saharo-Scindian-Irano-Turanian. There are also some plants of foreign origin isolated in certain depressions of the El-Tih Plateau and on the high elevations of the southern mountains.

According to Hassib (1951), the total number of species in the flora of the Sinai Peninsula was 532, as follows: 38 nanophanerophytes, one stem succulent, 95 chamaephytes, 142 hemicryptophytes, 27 geophytes, 10 hydrophytes and helophytes, 216 therophytes and 3 parasites.

El-Hadidi (1969) states that in the Sinai Peninsula there are about 36 endemic species, most of which are confined to the mountain region and belong to the Irano-Turanian element. Only a few endemics belong to the Saharo-Scindian element. Species of the Mediterranean climate are characteristic of the central and northern sections of the peninsula.

According to Täckholm (1974), the flora of Egypt comprises 2080 species whereas the checklist of Boulos (1995) comprise 2121 species and 153 infraspecific epithets (subspecies, variety and forma) of native and naturalized vascular plants. Cultivated plants are excluded with the exception of 27 culivated grasses. "This leaves us with 2094 native and naturalized species, a very close figure to the 2080 species on Täckholm (1974)", Boulos (1995) stated. These species belong to 121 families comprising 742 genera. Out of these, 63 species are endemic distributed in different phytogeographical regions of Egypt. More than 65% of these endemic species (41) occur in Sinai: 25 species in Sinai only and 16 species in Sinai and other regions of Egypt (Boulos, 1995, 1999, 2000, 2002, 2005). Most of the endemic species of Sinai (>70%) are recorded from the Southern mountain.

5.4.2 *The Mediterranean Coastal Area*

(a) Coastal Habitats

The northern Mediterranean coast of Sinai (eastern section of the Egyptian Mediterranean coast) extends from Port Said eastwards to Rafah for about 240 km. This coastal area includes Lake Bardawil (Fig. 5.1).

The natural vegetation of the coastal area is very sparse; three main habitats can be recognized: sabkhas, sand dunes and open sand plains. Sabkhas are present in the northern part near the coast of Lake Bardawil, sand dunes occupy most of the southern part of the tract while the sand plains are between these two areas.

Sabkhas (salt-affected lands) may be distinguished into four basic types according to the distribution of plants: (1) salt-encrusted sabkhas; (2) wet sabkhas; (3) dry sabkhas; and (4) drift sand-covered sabkhas. Salt-encrusted sabkhas have almost no vegetation owing to their extremely high salinities. Only about 2–5% of the area in wet sabkhas is vegetated, varying with the amount of soil moisture. Of the three halophytes which occur, *Halocnemum strobilaceum* is the dominant, its cover being 70–95%. The other two halophytes are *Arthrocnemum glaucum* and *Suaeda vera*.

In the dry sabkhas, vegetation cover is about 5–10%; *H. strobilaceum* is also the dominant, associated with *A. glaucum, Cressa cretica, Juncus rigidus, Limoniastrum monopetalum, Phragmites australis, Suaeda vera* and S. *vermiculata*. In sabkhas covered with a sheet of drift sand, the plant cover varies from <5% to 15%.

These areas are vegetated by two communities dominated by *Zygophyllum album* and *Anabasis articulata*, with cover of 5–10% and 10–15% respectively. Other associates are *Cressa cretica, Cyperus laevigatus* and *Salsola kali.*

The northern Mediterranean coastal area of Sinai is of sand dunes-littoral, plain and inland. The littoral dunes are mostly in two parallel lines with lows (pans) between. These lows act as drainage basins where halophytes dominate. The vegetation of the littoral dunes is largely of patches of *Ammophila arenaria.* Although the cover may reach 50% within the patches, not more than 5% of the total area of the dunes is plant-covered (Kassas, 1955). Beside the dominant grass, the flora of littoral dunes commonly includes *Eremobium aegyptiacum, Lotus arabicus, Moltkiopsis ciliata, Polygonum equisetiforme* and *Salsola kali.* Among other associates are *Atriplex leucoclada, Cressa cretica, Cyperus laevigatus* and *Juncus acutus* (as halophytes in the lows). Species occasionally present are *Artemisia monosperma, Astragalus tomentosus, Cyperus capitatus, Echinops spinosissimus, Elymus farctus, Euphorbia paralias, Pancratium maritimum, Silene succulenta* and *Thymelaea hirsuta.*

Plain dunes are sand drifts that are lower and less mobile than the littoral ones. In these dunes *Artemisia monosperma* is the dominant. This community is subject to intense human interference by cutting and grazing. In places far from human settlements, the cover may reach 70% or more. The contrast between the vegetation inside a barbed-wire fence and outside is very striking (Kassas, 1955). The flora of an *A. monosperma* community includes several characteristic species, e.g. *Haplophyllum tuberculatum, Neurada procumbens, Panicum turgidum, Pituranthos tortuosus* and *Urginea maritima.* The last species is subject to selective cutting for its medicinal value. *Cynodon dactylon* is a common species everywhere. Stabilized mounds covered by *Lycium europaeum* are local. In one locality *Lagonychium farctum* is very abundant, growing on the leeward side of these dunes. Other associates are *Alhagi maurorum, Astragalus spinosus, A. tomentosus, Atractylis prolifera, Bassia muricata, Chrozophora verbascifolia, Citrullus colocynthis, Cleome arabica, Echinops galalensis, Eremobium aegyptiacum, Euphorbia terracina, Heliotropium luteum, Ifloga spicata, Launaea glomerata, Lotus creticus, Mentha* sp., *Stipagrostis plumosa, Tamarix aphylla* and *Ziziphus spina-christi.*

The vegetation of inland sand dunes depends on their history which influences the composition of the plant cover. Dunes formed on sabkhas are dominated by *Zygophyllum album* at one stage of development and may contain, at an advanced stage, some other halophytes, e.g. *Nitraria retusa.* The stabilized inland dunes of this type are, in general, dominated by *Panicum turgidum* with cover up to 60%. Associate species are *Anabasis articulata, Artemisia monosperma, Convolvulus lanatus, Cornulaca monacantha, Echiochilon fruticosum, Eremobium aegyptiacum, Fagonia arabica, Moltkiopsis ciliata, Noaea mucronata, Stipagrostis scoparia* (abundant) and *Thymelaea hirsuta* (abundant).

The second type of inland sand dune of the Mediterranean coastal area of Sinai may be formed by the accumulation of sand on desert mountains (Kassas, 1955). At the final stage there are huge sand dunes with rocky centres. In this type *Panicum turgidum* is the dominant on the lower dunes whereas *Artemisia monosperma* dominates on the higher ones. *Anabasis articulata, Convolvulus lanatus, Noaea mucronata* and

Thymelaea hirsuta are common associate species. Other associates include *Aerva javanica, Asparagus stipularis, Asthenatherum forsskaolii, Echiochilon fruticosum, Fagonia arabica, Haplophyllum tuberculatum, Pituranthos tortuosus, Stipagrostis plumosa* and *Teucrium polium*.

The third type of inland sand dunes is those formed by the accumulation of sand on desert plains. In these dunes three communities have been recognized: one dominated by *Panicum turgidum*, one by *Stipagrostis scoparia* and one co-dominated by *P. turgidum* and S. *scoparia*. *Panicum* is a good fodder plant and is, thus, subject to heavy grazing. S. *scoparia* is less grazed. As it occurs on the relatively higher dunes, S. *scoparia* is partly protected: the sites are not in easy reach of animals. The average cover of the *P. turgidum* community is 40% and with *S. scoparia* is 50% (Kassas, 1955). Common plants are *Artemisia monosperma, Convolvulus lanatus, Fagonia arabica, Gymnocarpos decander* and *Thymelaea hirsuta*. Apart from the species in the other two communities of inland sand dunes, the flora of the third one also contains *Citrullus colocynthis, Hyoscyamus muticus, Reaumuria hirtella* and *Salsola volkensii*.

(b) Sample Areas

The landward successive communities of the Mediterranean coastal land have been studied by M.A. Zahran, M.A. El-Demerdash and AA. Sharaf (1987, unpublished) in three line transects that extend for several kilometres from the shore-line southwards,

(i) Transect 1

This transect was in the most eastern part of the Sinai Mediterranean coast at Rafah (Figs. 2.1 and 5.1). In the first zone of the transect, the sand dunes are vegetated by two communities dominated by *Ammophila arenaria* and *Silene succulenta*. The cover of *Ammophila* stands is thin (5–10%); associates are *Pancratium arabicum, Silene succculenta* and *Tamarix nilotica*. In stands dominated by *S. succulenta*, the cover is 20–30%. *Acacia saligna* (semi-wild), *Ammophila arenaria, Pancratium arabicum* and *Tamarix nilotica* are associate species. In the second zone of the transect (1 km landward) are huge sand dunes richly vegetated by the semi-wild shrub *Acacia saligna*. The ground layer is almost covered (about 75%) with the succulent xerophytic-halophyte *Mesembryanthemum forsskaolii*. Other associates are *Casuarina stricta* (cultivated), *Cyperus capitatus, Nicotiana glauca. Phoenix dactylifera* (semi-wild). *Polygonum bellardii, Silene succulenta* and *Xanthium pungens*.

The third zone of the transect (6 km landward) comprises sand dunes dominated by *Artemisia monosperma*. This vegetation extends landward for a considerable distance (about 22 km: from 6 km to 28 km south of the shore-line). In the northern stands, the cover is high (up to 80%), contributed mainly by the dominant xerophyte (about 50%) and partly by the abundant associate species: *Senecio desfontainei* (about 20–25%) and *Neurada procumbens* (about 5%). Other associates, *Astragalus alexandrinus, Cyperus capitatus, Onopordum alexandrinum* and *Urginea maritime* have negligible cover. The cover of the A *monosperma* community decreases gradually southwards. At 25 km south of the coast, the cover is about 70% contributed by the dominant (40–45%), *Stipagrostis scoparia* (about 20%) and *Neurada procumbens* (about 5%). *Senecio desfontainei*, abundant in the northern stand, is absent from the southern ones where *Silene succulenta* is commonly present. At 28 km landward, the cover of A. *monosperma* on the dunes is reduced to only 5%. *Eremobium aegyptiacum* is abundant. Other associates are *Ononis serrata, Onopordum ambiguum* and *Stipagrostis scoparia.* In this zone *Thymelaea hirsuta* dominates in scattered patches in the low areas between the dunes.

(ii) Transect 2

This transect was in the coast of Sheikh Zuwayid (about 20 km west of Rafah) and extended for about 22 km from sea landward. Along the whole stretch of the transect are extensive sand dunes. In the beach zone the dunes are dominated by *Ammophila arenaria*, with cover ranging between 30% and 50%. *Salsola kali* and *Silene succulenta* are the only associates. The dominance by A. *arenaria* continues landward for about 3 km. The cover decreases in the landward stands to 20–25% and the number of the associates increases to four (*Acacia saligna, Artemisia monosperma, Calligonum comosum* and *Silene succulenta*). Gradually, *A. monosperma* increases landwards and becomes dominant, associated with an increase in the number of xerophytes. The cover of the A. *monosperma* community ranges between 25 and 35%, contributed mainly by the dominant. Associate species include *Cynodon dactylon, Haplophyllum tuberculatum, Lycium shawii, Lygos raetam, Onopordum ambiguum, Panicum turgidum, Tamarix nilotica* and *Thymelaea hirsuta.* In the inland sections of the transect, *Thymelaea hirsuta* is abundant in low areas between the dunes.

(iii) Transect 3

This transect was in the downstream section of Wadi El-Arish (about 45 km west of Rafah) extending for about 24 km from the sea landward. In the first zone the semi-wild *Phoenix dactylifera* vegetation has a cover of 10–20%. Common plants are *Echinops spinosissimus, Mesembryanthemum crystallinum, Pseudorlaya pumila* and *Silene longipetala.* The dune formation after about 1 km landward is vegetated with *Ammophila arenaria* (cover <5%). Associates include *Artemisia monosperma, Cutandia dichotoma, Echinops spinosissimus, Erodium bryoniaefolium, Lygos raetam* and *Moltkiopsis ciliata.* The dunes in the third landward zone of the transect are dominated by *Tamarix nilotica* with cover up to 40%. Asssociates are *Artemisia monosperma, Cornulaca monacantha, Lobularia libyca, Ononis serrata, Silene succulenta, Stipagrostis ciliata* and *Thymelaea hirsuta.* As in the other two transects in the southern section, the dunes are dominated by *Artemisia monosperma.* The cover of this community thins gradually from about 40% in the northern stands to <20% in the southern ones. The number of the associate species of the A *monosperma* community is relatively high and includes *Asparagus stipularis, Carthamus tenuis, Cornulaca monacantha, Cyperus capitatus, Echinops spinosissimus, Erodium bryoniaefolium, Heliotropium digynum, Hernieria hemistemon, Hyoscyamus muticus, Ifloga spicata, Lycium shawii, Neurada procumbens, Panicum turgidum* and *Thymelaea hirsuta*. The dominance of A. *monosperma* on the dunes continues southward but the cover is thinner. The areas between the dunes support a community dominated by *Thymelaea hirsuta* with 35–40% cover contributed mainly by the dominant shrub and partly (about 5%) by the common associate *Hammada elegans*. Other associates include *Artemisia monosperma, Cornulaca monacantha, Fagonia arabica, Hyoscyamus muticus, Panicum turgidum, Peganum harmala, Zilla spinosa* and *Zygophyllum album*. In some patches of the wadi bed, *Fagonia arabica* dominates, with thin cover (<5%). *Anabasis articulata, Cleome africana, Farsetia aegyptia* and *Pancratium sickenbergeri* are the associate xerophytes of this community.

(c) Lake Bardawil

Lake Bardawil, on the northern Mediterranean shore of Sinai, is considered a hypersaline lagoon bordering the sea. The narrow semicircular barrier-beach that forms the northern boundary of the lake separates it from the sea. Artificially maintained inlets connect the lake with the sea. The lake has no fresh-water supply and no major source of enrichment other than the Mediterranean. The influx of, water from the sea is very important to its ecology since this inflow and outflow maintain the salt concentration in the water at tolerable levels. The annual rainfall in the area of the lake is about 82 mm and the annual evaporation about 1600 mm. This excess of evaporation over precipitation has a marked effect on the salinity of the lake when the inlets are closed and circulation cut off. The salinity of the lake is normally about 45–55 parts/1000. Water temperature generally varies from 12 °C in January to 30.5 °C in June (Anonymous, 1982).

The vegetation of the shore-line in the wet salt marshes of Lake Bardawil is dominated by *Halocnemum strobilaceum* and *Arthrocnemum glaucum* and the elevated saline belt by *Zygophyllum album* which forms a pure stand. Within the benthic region of the lake, dense growth of *Ruppia maritima* occurs.

The Mediterranean coastal strip adjacent to Lake Bardawil is about 10 km wide and about 50 km long. This strip of land is poor in species. Depressed areas near the lake are subject to periodic inundation by salt water from the lake; the water evaporates to leave a saline residue. Only halophytes grow in these depressions, the vegetation being dominated by *Halocnemum strobilaceum* and *Arthrocnemum glaucum*. On the elevated parts of the area *Zygophyllum album* forms a more or less pure community and often occupies the whole area, except for a few microhabitats, up to the sand dune zone. The littoral strip also contains several localities with slightly elevated terraces, some 3–5 m above water level, dominated by *Nitraria retusa* with *Lycium europaeum* as a common associate. The elevated slopes of the dunes support

Artemisia monosperma, Lygos raetam, Moltkiopsis ciliata and *Thymelaea hirsuta*. Depressions between the dunes contain a characteristic salt marsh community of *Juncus subulatus* (dominant) and *Cynodon dactylon, Lycium europaeum, Nitraria retusa* and *Phragmites australis* as associates. In the slightly saline depressions are *Artemisia monosperma* and *Lygos raetam*. At the foot of these dunes are semi-wild date palms (*Phoenix dactylifera*).

Recent studies by Khedr and Gazzar (2006), El-Bana (2003) & Khalil and Shaltout (2006) recognized six main habitats in Lake Bardawil and its surroungings, namely: 1. open water, 2. wet salt marshes, 3. saline sand flats and hummock (nebkas), 4. stabilized sand dunes, 5. interdune depressions and 6. Mobile sand dunes. In the open water, only three flowering plants (*Cymodoceae nodosa, Ruppia cirrhosa* and *Halodule uninervis*) are growing. The first two species are widely distributed in the lake whereas *H. uninervis* has not been reported from the Mediterranean water before and seems to have been escaped from the Red Sea through the Suez Canal (Täckholm, 1974; Boulos, 1995). The wet salt marshes of Lake Bardawil are characterized by 16 halophytes and helophytes. *Halocnemum strobilaceum* is the dominant species and the associates include: *Arthrocnemum macrostachyum, Frankenia pulverulenta, Phragmites australis, Suaeda maritima* and *Tamarix nilotica*.

A total of 34 species were recorded from the saline flats and hummoks (nabkas), *Zygophyllum album* dominates the saline flats whereas *Nitraria retusa* predominates the nabkas. The associate species include: *Bassia muricata, Cutandia dichotoma, Lotus halophylus, Mesembryanthemum creystallinum, Salsola tetragona, Spergularia marina* and *Schismus barbatus*.

The stabilized sand duens are characterized by two groups of islets: eastern and western, with floristic assemblage of about 78 species. The characteristic species are *Retama raetam* and *Stipagrostis plumosa* in the eastern islets and *Asparagus stipularis, Deverra tortusa* and *Echium angustifolium* in the western islets. Associate species include: *Anabasis articulata, Bupleurum semicompositum, Daucas litoralis, Echiochilon fruticosum, Launaea capitata, Lycium shawii* and *Schismus barbatus*.

The interdune depression is an adverse habitat being divided into saline and non-saline facies due to the level of ground water. In non-saline areas, *Artemisia monosperma, Panicum turgidum* and *Thymelaea hirsuta* are the characteristic species. The associates include: *Atractylis carduus, Cynodon dactylon, Ifloga spicata, Pancratium maritimum* and *Salvia lanigera*. Halophytes, e.g. *Arthrocnemum macrostachyum* and *Halocnemum strobilaceum* predominate in the saline areas.

Total of 49 species have been recorded on the mobile sand dune, *Stipagrostis scoparia* and *Calligonum comosum* are the characteristic species. *Artemisia monosperma, Cornulaca monacantha, Eremobium aegyptiacum, Heliotrpium digynneum. Lotus halophulus, Malva parviflora, Pancratium sickenbergeri, Retama raetam* etc. are the associated species.

The flora of all habitats of Lake Bardawil contains a total of 136 species (109 genera and 42 families) representing 30% of the recorded species of the Mediterranean coastal plain of Sinai (Gibali, 1988). Gramineae is represented by 12.5% followed by Chenopodiaceae (11%), Compositae (9.6%), Leguminosae, (8.1%),

Caryophyllacaeae (6.6%), Cruciferae (3.7%) and Cyperaceae (1.5%). Most of the plant diversity of Lake Bardawil are therophytes (44.1%) followed by chamaephytes (25%), geophytes (11.8%), phanerophytes (5.1%), parasites (2.2%) and hydrophytes (0.7%).

Following the IUCN Red Data Book (El-Hadidi and Hosny, 2002), 6 threatened species are recorded in Lake Bardawil namely: *Astragalus camelorum, Bellevalia salah-eidii, Biorum oliveri, Iris mariae, Lobularia arabica* and *Salsola tetragona*. The first four species are endemic or near endemic (Khalil and Shaltout, 2006).

5.4.3 The Gulf of Suez Coast

The western coast of the Sinai Peninsula (eastern coast of Suez Gulf) is bounded by the Gulf of Suez in the west and the limits of the coastal desert and wadis that drain into it in the east. It extends from El-Shatt (Lat. 30°N) in the north to Ras Muhammed (Lat. 27°40 N) in the south for about 340 km (Fig. 5.2). The width of this coastal area depends mainly upon the geomorphology of the area and reaches its maximum (1–2 km) south of El-Tor while at Ras Bakr (about 90 km south of El-Shatt) the hills extend to the shore-line.

Four types of habitat have been recognized in the western coast of Sinai: mangrove swamps, littoral salt marshes, oases and coastal desert.

(a) Mangrove Swamps

Mangrove swamps are absent from the whole stretch of the eastern coast of the Gulf of Suez (as in its western coast). However, at the cap of the Sinai Peninsula where the Suez Gulf meets the Aqaba Gulf (Ras Muhammed, Fig. 5.1), there is a shallow and narrow lagoon extending from the Gulf of Suez landward. This lagoon provides a suitable site for the growth of mangal vegetation. Thickets of a pure community dominated by *Avicennia marina* are present in this lagoon which has a muddy substratum (Ferrar, 1914; Zahran, 1965, 1967, 1977, 2007a). Shrubs of this mangrove grow near to the banks of the lagoon as well as in its shallow water channel.

(b) Littoral Salt Marshes

This is the salt-affected land that runs parallel to the Gulf of Suez being strongly influenced by the saline water of the Gulf. Its width depends on the maximum reach of the Gulf's water as well as on the topography of the coast.

In the halophytic vegetation of these littoral salt marshes ten communities with different dominants have been recognized, namely: *Halocnemum strobilaceum, Arthrocnemum glaucum, Aeluropus massauensis, Zygophyllum album, Nitraria retusa, Tamarix nilotica* (common communities), *Halopeplis perfoliata, Limonium pruinosum, Cressa cretica* and *Juncus rigidus* (local communities).

1. *Halocnemum strobilaceum* community and 2. *Arthrocnemum glaucum* community. *H. strobilaceum* and *A. glaucum* are usually closely associated, *A. glaucum* being an abundant associate in stands of *H. strobilaceum* which is abundant in stands dominated by *A. glaucum*. Many stands are co-dominated by these two succulent halophytes. The vegetation of the stands of these two communities (cover ranges between 40% and 90%) includes *Aeluropus massauensis, Atriplex leucoclada, Cressa cretica, Schanginia hortensis* and *Suaeda vermiculata* as less common species and *Nitraria retusa* and *Zygophyllum album* as common associates.
2. *Aeluropus massauensis* community. This species and the community which it dominates are largely restricted to a stretch of about 70 km (between 11 and 81 km south of El-Tor). It grows in two forms, the normal mat form covering sheets of moist salty soil and a peculiar cone-like form, noted only in one stand at 81 km south of El-Tor. The two forms have also been observed in the Red Sea coastal land (Chapter 4). The cover of this community is high: 60–90% contributed by the dominant grass; associates (*Cressa cretica, Nitraria retusa* and *Zygophyllum album*) have negligible cover.
3. *Zygophyllum album* community. Along the Sinai coast of the Suez Gulf, *Z. album* is abundant and recorded as an associate species in almost all the communities in addition to the one which it dominates. It is dominant in two types of habitat: dry salt-affected areas in the zone inland to that dominated by *A. massauensis* and also the areas having phytogenic sand mounds along the coast. The cover of the stands is thin (5–15%); however, the number of the associates is relatively high (15) and includes a mixture of xerophytes and halophytes. These are *Nitraria retusa* (abundant), *Hammada elegans* (common), *Suaeda vermiculata* and *Tamarix nilotica* (occasional) and species rarely present which include *Alhagi maurorum, Anabasis articulata, Atriplex leucoclada, Halocnemum strobilaceum, Hyoscyamus muticus, Neurada procumbens, Pergularia tomentosa, Phoenix dactylifera* (semi-wild), *Salsola tetrandra, Zilla spinosa* and *Zygophyllum coccineum*.
4. *Nitraria retusa* community. *N. retusa* is a widely distributed species along the whole stretch of the western coast of Sinai. It builds sand mounds and hillocks of considerable size. The community dominated by this shrub occupies the most landward zone of the salt marsh ecosystem separating it from the non-saline desert ecosystem. The vegetation of the *N. retusa* community contains a considerable number of associates (18) – xerophytes and halophytes. These are *Zygophyllum album* (abundant), *Atriplex leucoclada* (common), *Aeluropus massauensis, Alhagi maurorum, Salsola tetrandra, Schanginia hortensis, Suaeda vermiculata* and *Tamarix nilotica* (occasional), *Arthrocnemum glaucum, Cressa cretica, Halocnemum strobilaceum, Hammada elegans, Juncus rigidus, Limonium pruinosum, Lygos raetam. Phoenix dactylifera, Zygophyllum coccineum* and (rare) *Z. simplex* (Zahran, 1967). *N. retusa* is present along the whole eastern coast of the Suez Gulf but its dominance starts at 100 km south of El-Shatt and continues to 276 km. The cover of this community ranges between 5% and 50%.

In one part of this coastal area, extending between 257 and 263 km from El-Shatt mainly dominated by *N. retusa*, the plants build huge hillocks that cover more than 50% of the area.

5. *Tamarix nilotica* community. *T. nilotica* is a widely distributed bush (or tree) in the western coastal area of Sinai. Its community is the most elaborately organized of the salt marsh ecosystem, growing in a variety of habitat conditions and showing varied physiognomy. *T. nilotica* bushes are sand binders, capable of building hillocks. *T. nilotica* is usually dominant in the deltas of the large wadis, e.g. Wadis Sudr, Gharandal, Sidri and El-Tor. This community may also occupy the most landward zone of the salt marsh ecosystem in areas where *Nitraria retusa* is absent. The flora of the *T. nilotica* community includes a large number of associates (22): 16 are xerophytes and 6 halophytes. *Launaea spinosa* is interesting as it dominates a community in the northern section of the western coast of the Suez Gulf (Chapter 4) but is rare on the eastern coast. In the delta of Wadi Sidri is an open salt marsh scrub dominated by *T. nilotica.* The water channel of the wadi is blocked by the hillocks built by *T. nilotica.* The silty terraces that bound the wadi channel are also dominated by its growth but hillocks are not built here. *Nitraria retusa* and *Zygophyllum album* are the abundant associates, while *Z. coccineum, Hammada elegans* and *Lygos raetam* are common. Four species, namely, *Hyoscyamus muticus, Phoenix dactylifera, Reaumuria hirtella* and *Salsola tetrandra* are occasional. Rare species are *Achillea fragrantissima, Alhagi maurorum, Atriplex leucoclada, Diplotaxis acris, Fagonia glutinosa, Gymnocarpos decander, Halogeton alopecuroides, Mesembryanthemum forsskaolii, Polycarpaea repens, Schanginia hortensis* and *Zilla spinosa.*
6. *Halopeplis perfoliata* community. This community is recorded in only one locality – the coast of Abu Zenima (124 km south of El-Shatt); elsewhere, *H. perfoliata* is not even an associate species. Associates of this community are: *Arthrocnemum glaucum, Halocnemum strobilaceum* and *Zygophyllum album* (common) and *Nitraria retusa* (rare). Pure stands of *H. perfoliata* are also present.
7. *Limonium pruinosum* community. Like the preceding community, the presence of the *L. pruinosum* community is confined to a limited stretch (about 0.5 km) in the Ras Bakr area (91 km south of El-Shatt) which is a high beach lacking a definite salt marsh, the hills being close to the rocky shore, and covered with rock detritus. The cover of this community is thin (2–5%); associates are *Hammada elegans, Nitraria retusa, Salsola tetrandra, Suaeda vermiculata* and *Tamarix nilotica.*
8. *Cressa cretica* community. *C. cretica* is an associate species in several localities of the eastern coast of the Suez Gulf. Dominance of this mat-forming halophyte is restricted to two sites in the south part of the coast – at 1 and 17 km south of El-Tor. C. *cretica* usually grows in pure stands on the moist sandy shore-line.
9. *Juncus rigidus* community. *J. rigidus* is not a common halophyte in the eastern coast of the Suez Gulf, being recorded as an associate species in the *Nitraria retusa* community only. It is dominant in four localities: 25 km south of El-Shatt where it is associated with *Nitraria retusa* and *Zygophyllum album*, and south of

El-Tor where there are three localities in which it occurs either in pure stands or associated with *Aeluropus massauensis, Cressa cretica* and *Nitraria retusa*. The cover of the stands of this community is up to 100%.

(c) The Oases

The coastal eastern stretch of the Suez Gulf is characterized by oasis-like depressions, e.g. Ayon Musa (Musa springs) and Hammam Musa (Musa bath). These two oases are at 20 and 240 km south of El-Shatt, respectively. Scrubland of, *Tamarix nilotica* typifies these oases, with abundant growth of *Alhagi maurorum, Cressa cretica, Desmostachya bipinnata, Juncus rigidus, Nitraria retusa* and *Zygophyllum album*. In areas where water is exposed, species such as *Phragmites australis* are present. These oases are bounded by the desert country supporting xerophytes which include *Asteriscus graveolens, Chenopodium murale, Diplotaxis acris, Fagonia glutinosa, Gymnocarpos decander, Halogeton alopecuroides, Hyoscyamus muticus, Peganum harmala, Plantago amplexicaulis, Pulicaria undulata, Reaumuria hirtella, Zilla spinosa* and *Zygophyllum coccineum*. Date palm (*Phoenix dactylifera*) occurs in groves as well as individual trees. The presence of these trees is an indicator of a fresh-water zone among the underground water layers of the oases (Abdel Rahman et al., 1965b).

(d) Coastal Desert

The eastern coastal desert of the Gulf of Suez is characterized by several wadis that run from the mountains of southern and central Sinai and flow into the Gulf. An account of the vegetation of these wadis and of the El-Qaa plain in the south of this coastal area is given.

(i) Wadi Sudr

This is one of the most developed wadis of the northern section of the western coast of Sinai. It is bounded by Gebels Raha (c. 600 m) in the north and Sinn Bishr (c. 618 m) in the south. The main trunk of the wadi extends roughly in a NE-SW direction for about 55 km and flows into the Suez Gulf at Ras Sudr (c. 55 km south of El-Shatt).

The vegetation of Wadi Sudr includes a variety of communities and species with wide tracts covered by plants. In the main channel of its downstream part there is an open scrub of *Tamarix aphylla* with frequent *T. nilotica*. The course of the wadi here reaches its greatest width and receives maximum water revenue and surface deposits. This vegetation is present in the downstream 17 km, but further upstream *T. aphylla* is replaced by *T. nilotica, Lygos raetam* and *Hammada elegans* being important elements of the vegetation.

The courses of the upstream tributaries of Wadi Sudr are more defined, narrow and with the central part (water course) devoid of vegetation and fine sediments, being occasionally swept by torrents The vegetation in these tributaries is confined to side terraces and dominated by *Tamarix nilotica* and/or *Lygos raetam*. In some of these tributaries where the mountains of the El-Tih plateau bounding the course are high, the channel is devoid of side terraces and the sediments are compact. In these tributaries there is an open growth of *Acacia raddiana*. *Capparis cartilaginea* is occasional in the crevices and fractures of the cliffs and the rocky sides of the wadi.

The vegetation of the large affluents and runnels draining the gravel formations is a *Panicum turgidum* grassland. In the finer affluents and runnels the vegetation is dominated by *Artemisia judaica* and *Jasonia montana* or *Zygophyllum decumbens*.

In Wadi Sudr there is a fresh-water well – Bir Sudr – in the flood channel of a main tributary of the wadi. Its fresh water is utilized for domestic purposes. The vegetation of Bir Sudr is, however, halophytic dominated by *Juncus rigidus* with abundant *Tamarix nilotica, Nitrara retusa* and groves of date palm.

(ii) Wadi Gharandal

Wadi Gharandal extends east-west for about 80 km. It originates to the north of Gebel Pharaon and drains into the Gulf of Suez at Hammam Pharaon 80 km south of El-Shatt. The inland portion of the wadi is in the El-Tih plateau (Eocene) whereas the outlet section runs through the Pleistocene and Recent coastal belt of the Gulf of Suez (Said, 1962).

The main channel in the downstream portion of the wadi is covered with thickets of *Tamarix nilotica* where the course is well defined and bounded by low hills. Where the water course is broad, sand deposits are frequent and the underground water is deep. *Zygophyllum albun* is dominant in these parts and *Hammada elegans* is abundant. This downstream portion of Wadi Gharandal, which extends eastwards for about 14 km, is especially rich in water springs that form localized swamps. Water runs in the lower parts of the main channel creating swamps and salt marsh habitats dominated by *Phragmites australis* and *Typha domingensis*. These swamps are surrounded by lawns of *Cyperus laevigatus* and of *Juncus rigidus*. *Alhagi maurorum* am *Desmostachya bipinnata* form localized patches in the higher parts *Tamarix nilotica* thickets occur where the sand deposits are deep. *Nitraria retusa*, however, forms discontinuous narrow patches on the terraces bordering the main channel. This downstream portion of Wadi Gharandal may be considered as one of the most outstanding agricultural settlements of Sinai where horticultural crops are prosperously cultivated. Palm groves are dense in this area.

The vegetation of the upstream affluents of Wadi Gharandal is dominated by *Hammada elegans* and *Zilla spinosa*. In the finer runnels *Achillea fragrantissima, Artemisia judaica, Jasonia montana* and *Zygophyllum decumbens* are dominants. *Capparis cartilaginea* makes dense growth on the side hills bordering Wadi Umm Lasseifa (an affluent of Wadi Gharandal). This dense growth is unique to that wadi as it contrasts with the usual sporadic distribution of C. *cartilaginea*. In the delta of Wadi Lasseifa is a well of good water quality, used for domestic purposes by the bedouins.

Ain Hegiya is a spring in Wadi Hegiya (another affluent of Wadi Gharandal), associated with swamp and salt marsh vegetation: *Typha domingensis, Phragmites australis, Cyperus laevigatas, Juncus rigidus* and *Tamarix nilotica*. In the water creeks *Veronica beccabunga* is common.

(iii) Wadis Taiyba, Matulla and Nukhul

These three wadis are grouped together because of their common characteristics. They all cut through limestone (Cretaceous-Eocene) formations forming well defined, much ramified courses and have very deep sides. Their channels are short: 22 km, 10 km and 15 km respectively. They have no terraces and their catchment areas are not extensive. The three wadis drain into the Gulf of Suez at Abu Zenima (about 120 km south of El-Shatt).

The downstream parts of these wadis are covered with shallow deposits that are occasionally swept by torrents. The vegetation is dominated by *Zygophyllum coccineum* but the extent of this community varies in the three wadis – from 3 km in Wadis Nukhul and Taiyba to 2 km in Wadi Matulla. The extensive growth of *Z. coccineum* in Wadi Matulla may be attributed to the relatively short and narrow course of the wadi and the shallowness of its surface deposits.

Few springs are present in the downstream parts of Wadis Nukhul and Taiyba and these are associated with swamp and salt marsh vegetation formed by *Phragmites australis, Tamarix nilotica, Juncus rigidus* and *Alhagi maurorum*.

The midstream portions of these wadis contain friable sediments of coarse sand mixed with rock detritus. The deposits in these parts are deeper than in the downstream ones. The vegetation is dominated by *Hammada elegans. Tamarix* grows on the sides of Wadis Nukhul and Taiyba. *Capparis cartilaginea* and C. *aegyptia* are present on the sides of the wadis at different heights. Whereas *C. aegyptia* may grow on the rocky bed of the wadi, *C. cartilaginea* does not. *C. aegyptia* is usually confined to the limestone formations whereas *C. cartilaginea* has a wider range.

The upstream parts of the channels of these wadis are often narrow, cutting across Nubian sandstone formations, and covered with coarse sand intermixed with dark pebbles and gravels. The vegetation here is an open scrub *of Acacia raddiana* with *Hammada elegans* forming most of the undergrowth.

(iv) Wadi Baba

Wadi Baba is one of the important drainage lines of Sinai. The water course near the downstream portion is broad and bounded by moderately high hills. In some parts of the wadi, the course is flanked by rugged mountains of over 300 m (Ball, 1916). The surface deposits are mostly sand with some dark gravels, rock fragments and boulders of different origin. A number of springs and wells have been reported in the main channel and side tributaries of the wadi by Ball (1916) who noted the existence of extensive palm groves and rushes in several sites of the wadi.

Wadi Baba drains into the Suez Gulf at Abu Rudis (about 130 km south of El-Shatt). The vegetation of its downstream section is an open scrub dominated by *Acacia raddiana* with undergrowth mainly of *Hammada elegans*. The presence of A. *raddiana* scrub in the downstream part is different from the other wadis in which this scrub is usually confined to upstream parts.

Wadi Baba is characterized by narrow gorges where limited patches of *Juncus rigidus* and a few plants of *Zygophyllum coccineum* grow. In the fractures of the slopes of the igneous mountains bounding the gorges *Acacia raddiana* may be present on high areas while *Capparis cartilaginea, Lycium shawii* and *Ochradenus baccatus* occur at the foot and a little up the slopes.

(v) Wadi Sidri

The main channel of Wadi Sidri runs NE-SW for about 80 km and receives a number of tributaries and feeders. During its course Wadi Sidri cuts across rocks of different origin: Eocene-Cretaceous limestone in the downstream part and Nubian sandstone and igneous and metamorphic rocks for most of its length. The wadi flows into the Gulf of Suez at about 150 km south of El-Shatt. In the downstream part the surface deposits are deep sand and the bed is covered with boulders. The vegetation is dominated by *Hammada elegans*. Limited patches of *Tamarix aphylla* scrub occur on side terraces of the main channel and at the confluence of side tributaries.

The midstream part of the wadi cuts into Eocene-Cretaceous limestone in the west for about 20 km and then makes its course in igneous and metamorphic rocks with scattered patches of Cretaceous Nubian sandstone for another 20 km. The vegetation of this part of the main channel is dominated by *Hammada elegans* with abundant *Acacia raddiana* in the eastern part and *Artemisia judaica* in the west. *Capparis cartilaginea* grows regularly at different heights on the flanking slopes and cliffs of the wadi whereas *C. aegyptia* is sporadic.

The difference in the origin of the substrata of the tributaries and runnels of the eastern and western parts of the wadi affects the type of vegetation. The tributaries and runnels of the west, draining limestone formations, have very coarse deposits of whitish gravels and boulders. The large tributaries are dominated by *Hammada elegans* whereas the smaller affluents are dominated by *Artemisia judaica* or co-dominated by *A. judaica, H. elegans, Cleome droserifolia* and other calcicolous species such as *Fagonia mollis, Iphiona mucronata* and *Reaumuria hirtella* occur in these affluents. On the other hand, the tributaries of the eastern parts cut mainly across basement complex formations with occasional patches of Nubian sandstone. These tributaries are relatively narrow, bordered by high mountains. The surface deposits are sandy with dark rock fragments on the surface. The vegetation here is an open scrub of *Acacia raddiana* with *Hammada elegans* dominating the undergrowth.

The upstream tributaries of Wadi Sidri drain the northeastern fringes of the southern mountainous area formed of Nubian sandstone. The surface deposits are of deep loose reddish sand, mostly with no gravels or boulders. The dominant is *Hammada elegans. Lygos raetam* and *Panicum turgidum* are abundant and *Lycium shawii* is occasional. On the sides of the course is sporadic growth of *Acacia raddiana*, which is largely replaced by *Lygos raetam* in the Nubian sandstone tributaries (Girgis and Ahmed, 1985).

(vi) Wadi Feiran

Wadi Feiran is the longest and broadest wadi of southern Sinai. It rises from the high mountains surrounding the Monastery of St Katherine at 2500 m or so above sea level. It descends steeply to the north, then turns to the west until it terminates in the Suez Gulf about 165 km south of El-Shatt.

The downstream part of Wadi Feiran extends for about 20 km, covered by sediments of rock boulders and fragments in a sandy-clay matrix. *Hammada elegans* dominates in this habitat, growing in distantly spaced patches forming huge hummocks. In addition trees of *Acacia raddiana* are widely spaced on gullies and rocky slopes. Common associates include *Anthemis pseudocotula, Artemisia judaica, Cleome arabica, Diplotaxis acris, Fagonia arabica, Farsetia aegyptia, Lygos raetam, Mentha longifolia* ssp. *typhoides, Pituranthos tortuosus, Zilla spinosa* and *Zygophyllum simplex. Moricandia sinaica* is rare in this xeric habitat. About 22 km east of the wadi mouth, fine sandy-clay soil constituents increase. Here, *Aerva javanica* v. *bovei* appears in addition to the above-mentioned common associates.

Feiran Oasis is about 43 km east of the mouth of Wadi Feiran and appears as a deep, fertile extension of the wadi surrounded by high red mountains crowded with trees (*Acacia raddiana, Phoenix dactylifera* and *Tamarix aphylla*). The oasis extends over a distance of 10 km. Abundant ground water and deep sandy-clay deposits (wadi terraces), as well as the natural protection of the locality against wind, favour the utilization of the oasis as a productive area, e.g. to cultivate fruit trees.

(vii) El-Qaa Plain

The El-Qaa plain is a depression of about 1125 km^2 on the southern section of the eastern coast of the Gulf of Suez desert (Fig. 5.1). It lies between sea level and 200 m, sloping gently towards the southeast and stretches in a NW-SE direction for some 120 km, reaching an average width of 20 km. This plain is bounded on the east by the western outskirts of the rugged montane area of south Sinai and on the northwest by several isolated blocks (Ahmed, 1983). The surface of the plain is slightly undulating and covered with outwash deposits originating from the neighbouring highlands. It is dissected by shallow drainage lines originating mostly from the eastern and western high lands, which flow towards the central channels, running more or less parallel to the coastal highways crossing the plain.

Wadi Thegheda is a secondary channel in the northeast of the El-Qaa plain. The flood channel of the wadi has an average width of about 60 m. The bed forms gradually elevated terraces running along fault planes striking in a NE-SW direction. The upstream part of the wadi does not extend much into the montane area. In the uppermost site, relicts of the old wadi-terraces are extensively eroded to considerable depths by the action of torrential floods, giving a network of pathways for the small volume of seeping water. Seepage and eventual evaporation of this brackish water create swampy and wet salt marsh habitats. The swampy habitat is dominated by *Phragmites australis* with abundant *Typha domingensis* whereas *Juncus subulatus* dominates the wet salt marsh. *Phoenix dactylifera* groves are also present.

The slightly elevated rocky parts of the wadi covered with boulders and stones are dominated by pure stands of *Zygophyllum coccineum*, with scattered plants growing well in cracks and concavities filled with transported sand and silt. In the shallow ill-defined water course, where rock fragments are covered with coarse sand and fine sediments, *Alhagi maurorum* dominates, associated with *Zygophyllum coccineum. A. maurorum* is an indicator of underground water. In stands where soil is saline, *Cressa cretica* replaces *Z. coccineum.*

Hammada elegans dominates in areas of Wadi Thegheda where the substratum is rocky, with surface deposits containing a substantial fraction of sandy calcareous materials intermixed with rocks of various sizes. Associate species include *Fagonia arabica, F. schimperi, Iphiona mucronata, Zilla spinosa* and *Zygophyllum coccineum.*

Wadi Ratama extends for some 55 km to the north of El-Tor where its trunk has a width ranging between 40 and 110 m. The broad, moderately developed course of the wadi is eroded under the influence of many side runnels that drain the surrounding highlands as well as the elevated boulder-strewn original surface of the plain. *Zygophyllum coccineum* is the dominant in the low terraces of the sides of the water course of the wadi. Patches of *Ochradenus baccatus* are scattered along these terraces. Associate species are *Ephedra alata, Fagonia glutinosa, Lygos raetam, Panicum turgidum* and *Zilla spinosa* In other parts of this wadi the outwash plains have deep soft alluvial deposits of sandy calcareous materials free from rocks. *Ephedra alata* is the dominant, associated with *Zygophyllum coccineum* (abundant) and other xerophytes such as *Fagonia glutinosa, F. sinaica, Hammada elegans* and *Zilla spinosa.* Another landform that occupies the extreme sides of the wadi, particularly in its broad part, is the outwash plain. The surface of this plain is dissected by small water courses to which the perennials are confined. The transported water-borne deposits are of sandy calcareous materials covered and intermixed with rock detritus. *Hammada elegans* dominates this habitat, *Fagonia sinaica* and *Zygophyllum coccineum* being rare.

Throughout Wadi Ratama is a series of intermediate terraces built by the saxicolous, soil-binding xerophyte *Ephedra alata.* These terraces are confined to the wide, relatively shallow parts of the course where they are 10–20 m wide and 40–120 cm high. Other plants of this habitat are *Fagonia glutinosa, F. sinaica, Hammada elegans, Lygos raetam, Zilla spinosa* and *Zygophyllum coccineum.* The midstream part of this wadi is relatively deep and narrow and is devoid of intermediate terraces, though scattered small hummocks are formed here and there. This habitat supports two communities, one dominated by *Lygos raetam* and one by *Zilla spinosa*. In the *Lygos* community, *Zilla spinosa* is abundant, and other associates include *Artemisia inculta, Fagonia glutinosa, Gymnocarpos decander, Panicum turgidum* and *Zygophyllum coccineum*. In the stands of *Zilla spinosa, Lygos raetam* is an abundant associate with other associates mentioned in the *Lygos* community.

Wadi Hebran is one of the longest well-defined drainage lines initiated along faults dissecting the western outskirts of the southern montane area of Sinai. It extends in a nearly NE-SW direction for more than 20 km and reaches Wadí Feiran through mountainous passes accessible to camels. The entrance of this wadi from the El-Qaa plain lies 26 km north of El-Tor city. Throughout the downstream part of this wadi fresh-water springs issue at the surface as well as at the fractured bounding cliffs. This is the source of drinking water for the bedouins.

The vegetation of Wadi Hebran is similar to that of other wadis of the area. However, with the plentiful fresh water supplies, the vegetation is dense and growth vigorous. In addition to the scattered dense groves of *Phoenix dactylifera* and *Acacia raddiana*, the following species occur in the areas of the springs: *Citrullus colocynthis, Cyperus laevigatus, Francoeuria crispa, Hyoscyamus muticus, Juncus rigidus, Mentha longifolia* ssp. *typhoides, Zygophyllum coccineum* and *Z. simplex*. Moreover, scattered individuals of *Capparis aegyptia* are recorded hanging on inaccessible heights of the bordering rocky cliffs. *Verbascum fruticulosum* is occasionally found and rarely *Cleome droserifolia, Lavandula pubescens* and *Solanum nigrum* in the fractured basement rocks in parts adjacent to the constructed water ditches.

The runnels of the deltaic part of Wadi Hebran have small ill-defined courses. Although their beds are almost completely covered with boulders of various sizes, a feature attributed to the infrequent washing, examination of the subsurface deposits indicates the abundance of soft alluvial materials intermixed with rock detritus. In this: habitat, *Pergularia tomentosa* is the dominant and *Panicum turgidum*, and *Zilla spinosa* are abundant. Other associates are *Acacia raddiana, Fagonia arabica, Iphiona mucronata, Lygos raetam, Ochradenus baccatus* and *Zygophyllum coccineum*.

Wadi El-Tor is one of the prominent features of the western coast of Sinai. It is marked particularly by a huge, steeply curved, meander with its apex facing east in its central region some 15 km northwest of El-Tor city. This part of Wadi El-Tor cuts across deep alluvial deposits that form wadi terraces, 2–5 m high, dissected by affluents draining the surrounding water-collecting areas. At some 4 km from the entrance to the central part, the flood course bends gradually southward and then to the southwest. The downstream part extends southwest to the sand beach, which represents the coastal fan of the wadi on the Suez Gulf.

In the flood channel (wadi bed) of the upstream part of Wadi El-Tor there is a community co-dominated by *Artemisia judaica* and *Zilla spinosa*. Associates are *Anthemis pseudocotula, Fagonia kahirina, F. mollis, Hammada elegans, Lygos raetam, Matthiola elliptica* and *Panicum turgidum*. In this part the wadi terraces, 40–60 cm high, are dominated by *Ephedra alata* associated with *Aerva javanica, Iphiona mucronata, Ochradenus baccatus, Pituranthos tortuosus, Zygophyllum coccineum* and *Z. simplex* (Anonymous, 1981). In the rocky run-off habitat, where boulders and gravels cover the substratum, *Ephedra alata, Gymnocarpos decander* and *Zygophyllum simplex* are abundant.

In the downstream part of Wadi El-Tor communities dominated by *Zygophyllum coccineum* and by *Hammada elegans* occur. The common associates of the *Z. coccineum* community include *Fagonia glutinosa, Francoeuria crispa, Iphiona mucronata* and *Ochradenus baccatus* together with *Zilla spinosa* (abundant). The stands of *H. elegans* are either pure or *Z. coccineum* is the only associate. In the saline areas of this part of the wadi *Tamarix nilotica* dominates; associates include *Cressa cretica, Cynodon dactylon, Cyperus laevigatus, Juncus acutus* and *Zygophyllum album*. Groves *of Phoenix dactylifera* are also present.

5.4.4 The Gulf of Aqaba Coast

The western coast of the Aqaba Gulf (the eastern coast of the Sinai Peninsula) extends for about 235 km from Aqaba southwards to Ras Muhammed where it meets the southern part of the eastern coast of the Suez Gulf (Fig. 5.2). The eastern foothills of Sinai descend sharply towards the Gulf of Aqaba and the width of the coastal plain is greatly reduced. However, the southern part of the coastal plain is broader than the northern. Alluvial fans derived from magmatic and metamorphic rock cover most of this plain. In the southern section a large area near the beach is of fossil coral reef.

The vegetation of the western coast of the Gulf of Aqaba may be divided into: mangal, littoral salt marsh and coastal desert types. The mangal vegetation is represented by limited shore-line swamps dominated by *Avicennia marina* in the coastal area of Nabq (about 50 km north of Ras Muhammed, Fig. 5.1). The coral reefs of the shoreline in the southern section of Aqaba Gulf as well as the warm temperature (mean annual temperature 26 °C) enable the mangrove plants to dominate. North of Nabq no mangroves have been recorded. The absence of these plants from the northern section of the Aqaba Gulf western coast may be attributed to the following factors.

1. The mean temperature of the coldest month is below the necessary requirement for their successful growth. Mangroves, in general, are a tropical formation and a high temperature in the coastal areas is a prerequisite for their presence (Chapman, 1975). These plants require a mean temperature of not less than 15 °C in the coldest month of the year. In the southern section of the Aqaba Gulf the mean temperature of the coolest month is 18.2 °C, within the range of the mangrove tolerance. In the northern section, however, temperature appears to be lower.
2. The steep cliffs of the coastal hills in the northern section prevent the development of a suitable shore-line for the growth of mangrove.

The littoral salt marsh vegetation of the Gulf of Aqaba western coast is in zones roughly parallel to the coast. The first zone, close to the shore-line, dominated by *Limonium axillare*, is subject to periodic flooding with sea water. In the Nabq area a few shrubs *of Avicennia* may grow as associates in the *L. axillare* community. The second landward zone is co-dominated by *Nitraria retusa* and *Zygophyllum album*.

The soil is highly saline and the water-table shallow. The third zone is dominated by the xerophytic shrub *Salvadora persica* which grows in a prostrate form and builds sand mounds.

The coastal desert and wadis of the western coast of the Aqaba Gulf bear xerophytic vegetation of the following species: *Abutilon fruticosum, Aerva javanica, Artemisia judaica, Blepharis edulis, Capparis cartilaginea, Cleome chrysantha, Crotalaria aegyptiaca, Cymbopogon schoenanthus, Cyperus jeminicus,*[2] (= *C. conglomeratus*), *Eremopogon foveolatus, Gymnocarpos decander, Hammada elegans, Heliotropium arbainense, Hibiscus micranthus, Lasiurus hirsutus, Launaea spinosa, Lindenbergia sinaica, Otostegia fruticosa* ssp. *schimperi, Panicum turgidum, Pterogaillonia calycoptera,*[3] *Salsola cyclophylla,*[4] *S. schweinfurthii, Seidlitzia rosmarinus, Solenostemma oleifolium, Taverniera aegyptiaca* and *Zilla spinosa*. In the large wadis there are xerophytic trees and shrubs, *e.g. Acacia raddiana, A. tortilis, Calotropis procera, Capparis decidua, Leptadenia pyrotechnica, Moringa peregrina, Salvadora persica, Tamarix aphylla* and *T. nilotica*. Groves of *Hyphaene thebaica* are also present (Danin, 1983).

Acacia raddiana usually dominates in the alluvial fans at the foot of the coastal hills. In these alluvial fans and after rain storms a dense vegetation of *Pulicaria desertorum* appears and persists as long as there is enough water. In other alluvial fans much *Schouwia thebaica* develops also after strong storms and lives even longer than *P. desertorum*.

The fossil coral reefs of the Aqaba Gulf western coast support a vegetation dominated by *Pulicaria desertorum* and *Schouwia thebaica*. This habitat receives sufficient run-off to support long-living shrubs and trees. In the coral reef wadis *Capparis decidua* and *Leptadenia pyrotechnica* are dominants (Danin, 1983).

5.4.5 The Montane Country

(a) Physiography

The montane country of Sinai is of triangular shape, the apex of the triangle being near the cape of the peninsula (Ras Muhammed). It is encompassed by the Gulf of Suez in the west and the Gulf of Aqaba, in the east. In the north this montane country is bounded by the calcareous plateau of El-Tih which slopes down northward into a wide coastal plain with sand dunes (Fig. 5.2).

The ranges of mountains that form the montane country of Sinai are in the southern and central subregions of the peninsula and comprise Gebel St Katherine (the highest peak in Egypt, 2641 m) as well as Gebels Musa, El-Tih, Halal, El-Ugma, Yalaq, El-Maghara and others. On the eastern side of the peninsula, the mountains

---

[2] Species not recorded by Täckholm (1974).

[3] Genus not recorded by Täckholm (1974).

[4] See Footnote 2.

are so close to the Gulf of Aqaba that there is almost no coastal land. On the other hand, the mountains on the western side of Sinai are relatively far from the Suez Gulf and there is a wide coastal belt along the whole stretch.

The montane country of Sinai is dissected by narrow wadis with deep slopes and is characterized by the presence of springs around which there are oases and human settlements.

Rain water falling on the mountains runs over the slopes and into the narrow deep wadis where it forms perpetual streams or pools. Some of this water percolates into the substratum and is stored in rock crevices. It can be obtained by digging wells or it may appear at the surface as springs or streams of fresh water. Snow is another source of water in the Sinai montane country as it covers the summits of mountains higher than 1000 m during winter. When it melts with the advance of warm weather, water runs down the mountain slopes and into the wadis. Because of its altitude the total amount of rainfall of this montane country is about 60 mm/year, mostly orographic rain. If snowfall is added to rainfall, the water supply in this montane area is enormous in comparison with that of other desert regions of Egypt. The climate of the Sinai montane country is determined primarily by the altitude, the effect of which masks that of latitude (El-Hadidi et al., 1970). There is a wide difference in temperature between summer and winter. August is the hottest month (mean temperature 24.5 °C) and January the coldest (mean temperature 8.7 °C). Throughout the winter, the mean monthly temperature is below 10 °C. Absolute minima of less than 0 °C (−4 °C during the winter of 1987 and −6 °C during the winter of 1966) are of frequent occurrence between November and March at the highest altitudes. Similarly summer temperature is relatively lower than in any of the inner and coastal deserts of Egypt. The diurnal range of temperature is very wide, varying from 16 °C in winter to 20 °C in summer. With regard to relative humidity, this montane country is the driest part of Egypt in all seasons, as the relative humidity is less than 40%.

(b) Vegetation

The flora and vegetation of the montane country of Sinai proper have been studied by many workers, e.g. Täckholm (1932, 1956, 1974), Zohary (1935, 1944), Zohray (1966), (1972) & (1978), Shabetai (1940), Murray (1953), Migahid et al. (1959), Ahmed (1983), Danin (1981, 1983), Danin et al. (1985), El Hadidi (1991) and Ali (2004).

A large number of plants grow in the different habitats of the montane country of Sinai, most of which are chasmophytes. In areas of high water resources there are oases and cultivated gardens.

The characteristic habitats of this montane country include small, narrow, blind rocky wadis, upstream parts of large wadis (that originate within the hills and run eastwards to flow into the Gulf of Aqaba, westwards to flow into the Gulf of Suez and northwards to flow into the Mediterranean Sea), gullies, terraces, rock crevices and slopes of mountains at different levels. A rocky substratum is the general feature of these habitats. Sediments are coarse with or without fine particles.

The rock habitat is unfavorable to the growth of plants because of the high resistance to root penetration, a thin depth of soil and deficient water content. For these reasons only certain plants, chasmophytes, can tolerate the adverse conditions of the habitat. Some of the rock plants of Sinai are firmly attached to the smooth surface of the rock by means of hook-like roots, e.g. *Galium sinaicum* and *Origanum syriacum*. Other plants grow in rock crevices, which, though narrow, are sometimes very deep. Soil and plant litter accumulate in these crevices, retaining water and forming a fertile substratum through which plant roots penetrate. Deep crevices support several species of shrubs and trees, e.g. *Capparis cartilaginea, Cupressus sempervirens. Ephedra alata, Ficus pseudosycomorus* and *Moringa peregrina* (Migahid et al., 1959). Rock plants may also be found in surface notches and depressions in which soil and decaying matter are retained. Another rocky medium is terraces and flat plateaux on the surface of which a small depth of soil is deposited.

The mountainous district of Sinai is the coolest owing to its high elevation. The flora is diverse and includes Irano-Turanian, Mediterranean and Sudanian species that are isolated from their main areas of distribution. The nearest regions for several of the isolated species in these mountains are in Iran or in Mount Hermon (Anti-Lebanon). *Crataegus sinaica* and *Scrophularia libanotica* occur both in Sinai and on Mount Hermon (Danin, 1983). *Primula boveana* is a rare endemic which has been isolated in Sinai since the Tertiary (Wendelbo, 1961). Its nearest relatives are in eastern Africa, Yemen and the Zagros mountains of Iran. There are several other endemics, most of which are restricted to smooth-faced rock outcrops.

The flora of the Sinai mountains is dominated by Irano-Turanian species and the most common plant is *Artemisia inculta*. It is accompanied by *Gymnocarpos decander* in fissured rocks at lower elevations and by *Zilla spinosa* and *Fagonia mollis* in stony alluvium. *Anabasis setifera, Atraphaxis spinosa* v. *sinaica* and *Halogeton alopecuroides* are associates in soil derived from dark volcanic rocks. *Stachys aegyptiaca* and *Pyrethrum santolinoides* accompany *Artemisia* at the foot of smooth-faced rock outcrops at low elevations and on stony slopes at higher sites.

Rock vegetation of these mountains is rich in semishrubs, shrubs and trees (Danin, 1983). Characteristic species are *Cotoneaster orbicularis, Crataegus sinaica, Ficus pseudosycomorus, Pistacia khinjuk, Rhamnus disperma, Rhus tripartita* and *Sageretia brand-rethiana*. The common annuals include *Boissiera squarrosa, Eremopoa persica, Gypsophila viscosa, Lappula sinaica* and *Paracaryum intermedium*.

Wadi El-Raha is a short, broad wadi ending blindly in a granitic mass. It is very close to St Katherine Monastery.

The rocky slopes of the mouth of Wadi El-Raha support two communities, one dominated by *Alkanna orientalis* and one by *Varthemia montana*. Associate species of the first community are *Achillea fragrantissima, Stachys aegyptiaca, Varthemia montana* and *Zilla spinosa*. In the stands of the V. *montana* community, associates are *Achillea santolina, Alkanna orientalis, Cynodon dactylon, Lavandula stricta, Stachys aegyptiaca* and *Stipa capensis*. The species grow at heights of 4–10 m from the bed level of the wadi. They are rooted in crevices of the granite rock and are widely spaced. *Zilla spinosa* tends to decrease progressively with height up the slopes whereas *Varthemia* and *Stachys* increase. In the wadi bed *Z. spinosa* dominates, with cover of 20–30%. Associates include *Achillea fragrantissima, Artemisia inculta, Diplotaxis harra, Fagonia mollis, Francoeuria crispa, Gomphocarpos sinaicus, Peganum harmala* and *Reseda pruinosa*.

A gully at a higher level on the northern side of Wadi El-Raha (about 4–5 m broad) is typical of other gullies of the mountains. The surface of the gully bed is covered with boulders to which *Galium sinaicum* sticks firmly. On the flat soil of the gully bed, between the boulders, are *Ephedra alata, Fagonia mollis, Gomphocarpos sinaicus, Lavandula stricta, Teucrium polium* and *Zilla spinosa*. The following grow in fissures of the boulders: *Alkanna orientalis, Artemisia inculta, A.judaica, Ballota undulata, Capparis spinosa, Fagonia mollis, Parietaria alsinifolia, Scirpus holoschoenus, Stachys aegyptiaca, Teucrium polium* and *Varthemia montana*.

The rocky run-off slopes of the El-Raha plain support vegetation, at different levels relative to ground water availability, comprising the following: *Ajuga iva, Centaurea aegyptiaca, Delphinium* sp., *Echinops spinosissimus, Farsetia aegyptia, Gomphocarpos sinaicus, Hyoscyamus muticus, Lotus* sp. and *Stachys aegyptiaca* (Ahmed, 1983).

At the head of Wadi El-Raha is an oasis supporting a number of species of wild and cultivated fruit trees. In the middle of this oasis is a well containing fresh water 3 m below the soil surface. During rain periods, the water surface rises to within only 0.5 m of the surface. The cultivated fruit trees and shrubs include, besides the date palm, carob, pomegranate, peach, almond and apricot.

Wadi El-Arbaeen is another narrow steep wadi, the mouth of which lies opposite that of Wadi El-Raha. On its bed are scattered boulders and large stones. There are successive broad terraces reducing to a deep narrow channel flooded by spring water. Round these springs are hygrophytic shade plants as well as aquatic and salt marsh species, e.g. *Adiantum capillus-veneris, Equisetum ramosissimum, Mentha longifolia* ssp. *typhoides* and *Origanum syriacum. Ficus pseudosy-comorus* is rooted in crevices near this vegetation. Where the water supply is abundant it allows the development of oases with cultivated gardens where palm trees, pomegranate, almonds, plums, grapes, apples, pears, peaches and *Cupressus sempervirens* are cultivated. The herbs *Solanum nigrum* and *Verbascum schimperianum*[5] grow abundantly in these oases.

The bed of Wadi El-Arbaeen has a rich flora which includes *Ammi majus, Anchusa aegyptiaca, Brachypodium distachyum, Carduus arabicus. Euphorbia peplus, Lactuca orientalis, Onopordum ambiguum, Plantago ciliata, Pulicaria arabica, Sisymbrium irio* and *Sonchus oleraceus*. Also, *Asperugo procumbens, Hypericum sinaicum* and *Verbascum schimperianum* are recorded in the southern mountains. In the mouth of Wadi El-Arbaeen the vegetation is thin (cover about 10%), dominated by *Peganum harmala* associated with *Zilla spinosa* (abundant), *Achillea fragrantissima* and *Stachys aegyptiaca* (common). *Aikanna orientalis, Artemisia judaica, Ballota undulata, Origanum syriacum* v. *aegyptiacum, Phlomis aurea* and *Teucrium polium* are rare.

[5] *V. schimperianum* is endemic to the montane country of Sinai (Täckholm, 1974).

St Katherine Monastery lies at the bottom of a narrow flat depression surrounded by steep high mountains on all sides except the west leading to the prophet Aaron's tomb at the meeting point of Wadi El-Raha and Wadi El-Arbaeen. In the mountains surrounding the monastery the flora includes, e.g. *Achillea fragrantissima, Alkanna orientalis, Andrachne aspera, Echinops glaberrimus, Fagonia arabica, Ficus carica* v. *rupestris* (semi-wild), *F. pseudosycomorus, Gomphocarpos sinaicus, Heliotropium arbainense, Hyparrhenia hirta, Iphiona mucronata, Launaea spinosa, Orobanche muteli* v. *sinaica, Peganum harmala, Pituranthos tortuosus, Scrophularia libanotica, Solanum nigrum* and *Varthemia montana*. These plants occur at different levels on the mountain slopes, the more xerophytic tending to be more abundant in the lower zones.

Gebel Musa is located southeast of St Katherine mountain. The northern, windward, slope of this mountain is richer in vegetation than the southern slope of the opposite mountain on the other side of the monastery. On the northern slope there are many shrubs on the. rocks, e.g. *Cupressus sempervirens, Ephedra alata, Ficus carica* v. *rupestris* and *F. pseudosycomorus*. The following species are also common: *Artemisia inculta, Astragalus fresenii, Atraphaxis spinosa* var *sinaica, Bromus tectorum, Callipeltis aperta, Crataegus sinaica, Echinops glaberrimus, Isatis microcarpa, Lactuca orientalis, Nepeta septemcrenata, Origanum syriacum, Oryzopsis miliacea, Phagnalon sinaicum, Phlomis aurea, Pituranthos triradiatus, Plantago ciliata, Pyrethrum santolinoides, Scandix stellata* and *Silene leucophylla* (Danin et al., 1985).

About midway to the summit of Gebel Musa is a broad flat area named Farsh El-Gebel, in which springs and fresh-water streams are present. At one side is a runnel sloping down steeply towards the Farsh El-Gebel and having a thick layer of silt at the surface superimposed on the rocky substratum. On the alluvial soil is a community of *Thymus decussatus* (endemic) and *Artemisia inculta*. The plant cover is high (70%) in the densest upper part, but decreasing to 50% towards the foot of the slope and to 40% in the lowermost flat part at Farsh El-Gebel. The flora also includes: *Phlomis aurea* (abundant), *Pyrethrum santolinoides* (common), *Scirpus holoschoenus, Stipa parviflora, Teucrium polium* and *Varthemia montana* (rare).

In the flat part of Farsh El-Gebel the vegetation is thinner than in the runnels, cover not exceeding 40%. This may be related to the more deficient water supply, rain water being evenly distributed over the whole area but accumulating by run-off in the runnels. Here, *Aristida coerulescens* v. *arabica* dominates, *Artemisia inculta* is abundant whereas *Phlomis aurea* and *Pyrethrum santolinoides* are common.

Around the fresh-water spring of Farsh El-Gebel is a dense vegetation dominated by *Scirpus holoschoenus*. Associates include *Stipa capensis* (abundant), *Anagallis arvensis, Bromus rubens, Galium sinaicum, Juncus bufonius, Phlomis aurea, Polypogon monspeliensis* and *Veronica anagallis-aquatica* (common). Algae form a green scum alongside the spring.

The rocky slopes of the mountains on the two sides of the upstream section of Wadi Feiran are characterized by a number of trees and shrubs, e.g. *Acacia* spp., *Ficus pseudosycomorus, Moringa peregrina, Tamarix nilotica, Capparis cartilaginea* and *Ephedra alata*. Except for *Ephedra* and *Ficus*, these species are absent in the cooler, less arid, granite mountains of southern Sinai. They grow abundantly at lower levels of the slopes but their cover decreases on higher ground.

Gebel Ugma is one of the mountains of the central subregion of Sinai. Its vegetation on the slopes changes with elevation. In the wadis at the foot of the mountain, *Artemisia inculta* is dominant on the gravelly terraces where the chalky material is leached. *Hammada elegans* is abundant on the sandy terraces. In the chalky substratum of the small wadis, species such as *Salsola delileana, S. tetrandra, Halogeton alopecuroides, Krascheninnikovia ceratoides* and *Reaumuria hirtella* occur. The large wadis are vegetated by *Achillea fragrantissma, Atriplex halimus, Lygos raetam* and *Zilla spinosa*.

Slopes up to 600 m are bare except for *Salsola tetrandra*, most plants being dead. At 600–1500 m the number of shrubs of S. *tetrandra* increases, perhaps bacause microhabitats at higher elevations have improved moisture regimes (Danin, 1983). In wet years *Atriplex leucoclada* is abundant, particularly on the lower slopes. *Artemisia inculta* and *Halogeton alopecuroides* are common at high elevations of north- and south-facing slopes respectively. The highest belt of vegetation on Gebel Ugma includes the *Chenolea arabica – Atriplex glauca* community on hard chalk (Danin, 1983; Danin et al., 1985). In wet years this belt is covered with annuals – *Anthemis melampodina* and *Leontice leontopetalum*. Scanty shrubs of *Tamarix* sp. grow in this high belt.

Gebel El-Tih is in the central part of the El-Tih plateau. It has slightly inclined strata. Small outcrops of smooth-faced limestone occur in the flanks at high elevations. Many springs, including Ain Sudr, Moyet El-Gulat, Ain Shallal and Ain Abu Ntegina, occur in canyons draining the mountain. Annual rainfall is 50–100 mm and mean annual temperature 16°–20 °C.

The smaller wadis in the area of the El-Tih plateau mountain are dominated by *Anabasis articulata, Artemisia inculta, Hammada scoparia, Gymnocarpos decander, Salsola tetrandra* and *Zygophyllum dumosum*. However, the larger wadis are co-dominated by *Lygos raetam* and *Achillea fragrantissima*.

The inclination of the strata influences weathering patterns, water regime and vegetation. At higher altitudes, the *Z. dumosum* community is replaced by *A. inculta. H. scoparia* is dominant on marl outcrops having a salt regime.

The steep escarpment of Gebel El-Tih supports a pioneer community dominated by *Anabasis setifera* and *Halogeton alopecuroides*.

The vegetation of Gebel El-Tih varies with the type of rock. *Artemisia inculta* or *Zygophyllum dumosum* dominates those slopes consisting of hard rock containing little marl. Small horizontally bedded hard strata are dominated by *Hammada scoparia*. Softer rocks support *Z. dumosum* along with *Salsola cyclophylla* and other xerohalophytes (Danin, 1983; Danin et al., 1985). This habitat also supports a *Halogeton alopecuroides-Salsola schweinfurthii* community on the slopes of the mountain and *Hammada scoparia* in runnels at the summit of Ithe plateau. Outcrops of smooth-faced limestone are restricted to the dip-slopes of inclined rock strata. *Halogeton poore* grows in this habitat.

The other species of Gebel El-Tih area include *Pistacia atlantica*[6] (on which grows the woody *Loranthus acaciae*, (a very common parasite of *Acacia* in the

[6]Not recorded neither by Täckholm (1974), nor Boulos (1995, 1999).

Egyptian Eastern Desert), *Anabasis articulata, A. setifera, Gymnocarpos decander, Noaea mucronata, Reaumuria hirtella, R. negevensis* and *Varthemia iphionoides*[7].

The district of Gebel Halal (892 m) and Gebel El-Maghara (750 m) includes several folds of Cenomanian-Turanian age with limestone, chalk, dolomite and marl outcrops. Extensive erosion in Gebel El-Maghara has exposed a sequence of 2000 m of Jurassic limestone, shales and sandstone. Large outcrops of smooth-faced limestone and dolomite occur at Gebel Halal. The wadis are filled with sand-covered alluvium.

Rainfall in this district is 50–100 mm/year distributed during January-February; the rest of the year is almost rainless. Mean annual temperature is in the range 16°–20 °C; the highest in June-July and the lowest in January-February. Average temperature rarely exceeds 30 °C and rarely goes below 10 °C. Extremes of up to 40 °C, however, are recorded (Boulos, 1960; Danin, 1983).

The large wadis of the northern limestone district of the El-Tih plateau where Gebels Halal and El-Maghara are situated are dominated *by Acacia raddiana, A. gerrardii* ssp. *negevensis, Tamarix aphylla* and *T. nilotica.*

Gebel Halal is characterized by an erosion crater. The old strata at the bottom of the crater are covered with alluvium derived from the weathering of adjacent ridges. Areas of the crater floor, used for bedouin encampments, are dominated by *Anabasis syriaca*, a xerophyte that can tolerate high concentrations of nitrogen (Danin, 1983). The slopes inside the crater are covered with stony alluvium which, with the favourable water regime, supports sparse, semi-shrub vegetation. The southern slopes of some alluvial hills bear branches of *Caralluma sinaica.* Most of the semi-shrubs on the slope are xerophytes, e.g. *Anabasis setifera, Atriplex leucoclada, Halogeton alopecuroides, Reaumuria hirtella* and *Suaeda palaestina.*

In the northwestern flanks of Gebel Halal bedded limestone, chalk, marl and limestone and hard dolomite are exposed. Wadis cutting through these flanks produce slopes facing in various directions. Limestone on south-facing slopes and marl and chalk on all slopes support mixed or monospecific communities dominated by *Atriplex glauca, Halogeton alopecuroides, Reaumuria hirtella, R. nevegensis, Salsola schweinfurthii, S. tetrandra* and *Suaeda palaestina.* The bedded limestone on north-facing slopes above 800 m mostly supports communities dominated by *Artemisia inculta* and *Noaea mucronata.* In the spring of rainy years the vegetation is accompanied by the geophytes *Anemone coronaria, Ranunculus asiaticus* and *Tulipa polychroma* and many annuals.

The most interesting species of Gebel Halal is *Juniperus phoenicea*, which grows in the crevices of the smooth-faced outcrops of the hard limestone and dolomites of the northwest slopes as well as in the wadis. Some *Juniperus* trees may reach 10–12 m and individuals of 4–8 m are common. According to Täckholm (1956, 1974), Boulos (I960), El-Hadidi (1969) and Danin (1983), *J. phoenicea* is absent from all regions of Egypt except this area of Sinai. It occurs throughout the Mediterranean coastal region except for Libya and Egypt. Hundreds of trees of *J. phoenicea* are present in Gebel El-Maghara, dozens are in Gebel Yalaq and thousands are in Gebel Halal. "Vines" *of Ephedra aphylla* cover many trees of *J. phoenicea.* Another

[7] See Footnote 2.

species accompanying *J. phoenicea* in rock habitats is *Origanum isthmicum* which is, according to Danin (1969), endemic to Gebel Halal of Sinai. He reported that the entire world population of 1000–2000 individuals of *O. isthmicum* occurs within an area approximately 5×2 km on the northwest flanks of Gebel Halal. Other notable associates are *Astoma seselifolium, Ephedra campylopoda, Rubia tenuifolia* and *Sternbergia clusiana*. All of these plants are absent from other regions of Egypt.

The lowest parts of the northwest flanks of Gebel Halal bear a community dominated by *Zygophyllum dumosum*. The wadis at the foot-hills of Gebel Halal support the growth of *Acacia gerrardii* and *A. raddiana*.

Gebel El-Maghara (750 m), some 110 km southwest of El-Arish, consists of Jurassic rocks surrounded by marine Lower Cretaceous exposures which form a conspicuous topographic low separating it from the outer slopes which are occupied by Upper Cretaceous formations (Shata, 1956). The vegetation of Gebel El-Maghara has been studied by Zohary (1935, 1944), Boulos (1960), Shmida and Orshan *l* (1977) and others. The following are the common species collected by Boulos (1960) from the Gebel Maghara area.

(i) Common plants of rocky habitat

Achillea fragrantissima. Allium artemisietorum, Anabasis setifera, Anastatica hierochuntica, Anthemis melampodina, Asparagus stipularis, Ballota undulata, Callipeltis cucullaria, Caralluma sinaica, Centaurea eryngioides, Cocculus pendulus, Colutea haleppica, Cornulaca monacantha, Ephedra alata, Eryngium glomeratum, Euphorbia erinacea, Globularia arabica, Gomphocarpos sinaicus, or *Gymnocarpos decander Helianthemum lippii, H. ventosum, Juniperus phoenicea* (on high altitudes), *Linaria floribunda, Lycium europaeum, Matthiola livida, Micromeria sinaica, Muscari racemosum, Noaea mucronata, Notholaena vellea, Ochradenus baccatus, Oryzopsis miliacea, Parietaria alsinifolia. Paronychia sinaica, Pennisetum elatum, Rorippa integrifolia* (endemic, Täckholm, 1974), *Salsola tetrandra. Salvia aegyptiaca, Scrophularia xanthoglossa, Silene setacea, Stachys aegyptiaca, Stipagrostis ciliata, Tetrapogon villosus, Tricholaena teneriffae, Urginea maritima, Varthemia montana* and *Zosima absinthifolia*.

(ii) Common plants on sandy habitat

Adonis cupaniana, Aizoon canariense, Anabasis articulata, Anchusa milleri, Andrachne telephioides, Arnebia decumbens, Astragalus sinaicus, Atriplex leucoclada, Avena alba, Bromus fasciculatus, Calendula micrantha, Carduus getulus, Carthamus glaucus, Citrullus colocynthis, Cleome arabica, Convolvulus elarishensis, C. oleifolius, Cucumis prophetarum, Diplotaxis acris, Echiochilon fruticosum, Erodium laciniatum, Erucaria uncata, Fagonia mollis, Farsetia aegyptia, Francoeuria crispa, Frankenia pulverulenta (in moist sandy soil), *Hammada elegans, Heliotropium undulatum, Herniaria hemistemon, Hippocrepis unisiliquosa, Ifloga spicata, Iris sisyrinchium, Koniga arabica, Lappula spinocarpos, Launaea angustifolia, Linaria tenuis, Lotus glinoides, Lygos raetam, Malva parviflora, Moltkiopsis ciliata, Neurada procumbens, Ononis reclinata, Panicum turgidum, Papaver hybridum, Pergularia tomentosa, Phagnalon barbeyanum, Polycarpon succulentum, Pteranthus dichotomus, Pterocephalus papposus, Pulicaria undulata, Reseda decursiva, Savignya parviflora, Schismus barbatus, Senecio desfontainei, Silene villosa, Sonchus oleraceus, Telephium sphaerospermum, Teucrium polium, Thymelaea hirsuta* and *Urospermum picroides. Cuscuta brevistyla* is a common parasite in Wadi El-Maghara on a variety of species, e.g. *Artemisia* J *inculta, Centaurea sinaica, Gymnocarpos decander, Helianthemum kahiricum, Lycium shawii, Salvia aegyptiaca* and *Stipa capensis. Sedum viguieri* grows in dense tufts on high altitudes and on slopes, in protected moistened areas. *Juncus rigidus* and *Lamarckia aurea* are common in salt ground near the wells of Wadi El-Maghara. *Orobanche cernua* and *O. ramosa* are common root parasites on a variety of plants.

During the last four decades and since Boulos (1960), the flora of Gebel Maghara area was, and is still, under various environmental stresses including: climatic aridity, mining pollution, unmanaged quarrying activities, over explotiation of woody and medicinal plants, over grazing and over collection of the palatable plants for livestock as well as denudation of wide areas of the natural vegetation to be replaced by the cultiavation of crop and fruit plants and establishment of new settlements. Such threats are not only badly affecting the biodiversity of the area but also are causing fragmentation of the different habitats. The consequence is the considerable decline in the capacity of the ecosystem to be unable to provide certain services to the human well-being.

Through a Millenium Ecosystem Assessment Project, the flora of Gebel Maghara area has been studied (Zahran, 2007b). Compairing these recent results with those of Boulos (1960), i.e., after 47 years, one could determine considerable changes in the floristic composition of the area. These are summarized in the following points:

1. Total numbers of species recorded by Boulos (1960) was 192 declined to 122 on 2007.
2. Most of the medicinal, woody and palatable species recorded in 1960 are absent from the list of 2007.
3. Many of the still growing species which are considered medicinal e.g., *Juniperus phoenicea* are endangered.

Accordingly, for the goods and services of the local inhabitants (the Bedouins) of Gebel Maghara area and even for all areas in Sinai and other Egyptian deserts, conservation and management of the natural wealth of the flora are a must. Local inhabitants should be invited to shair such activates to guarantee the success and sustainabilities of the planned programmes.

5.4.6 Features of the Flora of Sinai

The vegatation of Sinai, being a bridge between Africa and Asia, reflects the influence of three phytogeographical regions that meet and overlap in the peninsula. Most of the species of Sinai, a desert of the Saharan type, originate from the

Saharo-lowlands. Near the Gulfs of Suez and Aqaba many Sudanese species thrive. Irano-Turanian species, typical of North African and Iranian steppes, grow on the high mountains and plateaux where the climate is wetter than in low land; the Mediterranean flora is not widespread.

The estimated number of species in the different habitats of the three subregions of Sinai ranges between 942 (Zohary, 1935), 1247 (Täckholm, 1974), 687 (Abdual-lah et al., 1984), 886 (Danin, 1972) (Danin et al., 1985), 984 (El-Hadidi et al., 1991) and 1262 (Boulos, 1995). Notably, the high mountains in the southern subregion and hills of the central subregion support a richer flora than the northern subregion, particularly the rock types. Higher rainfall, greater cloudiness and larger amounts of dew and mist result in a milder climate in the highlands than that of the lowlands. Relatively poor in species are the boulder-strewn deserts where the vegetation is concentrated in wadis, sand areas and salinas or salt springs.

The total number of endemic species in the three subregions of Sinai is 41 species most of these endemics (about 65%) are present in the southern montane country. The other two subregions contain about 35% of the endemics of Sinai: 32% in the central subregion and 3% in the northern subregion. However, Boulos (1995) stated that species endemic to Sinai Peninsula are 33, another 4 endemics are known from Sinai and other regions in mainland of Egypt. He added: "60.7% of the endemics to Egypt are known from Sinai, of which 54% are restricted to Sinai".

Some species recorded in Sinai belong to damper regions of the world than Sinai. These species grow in the rocky mountains of the peninsula. Many of these are believed to be relicts from periods when a more humid climate prevailed in the Middle East (Danin, 1983). *Juniperus phoenicea* which dominates in Gebels Halal, El-Maghara and Yalaq, where rainfall is about 100 mm/year, is a well-known tree of the Mediterranean area as it grows in Greece where there is 500–550 mm of rainfall annually as well as in southern France and the Pyrenees where the annual rainfall is 700–1000 mm. Prehistoric excavation and palaeobotanic records in conjunction with carbon dating have shown that *J. phoenicea* dates from 4000 to 34,000 years ago in Gebel El-Maghara (Danin, 1983, after Haas, 1977). This finding indicates that *J. phoenicea* was widespread in Sinai during that period, rainfall being higher than today. As the climate of Sinai became drier, shrubs and trees of *J. phoenicea* grew in pockets of soil in outcrops of smooth rocks or in wadis with such rocks in their catchment areas. A few species of Mediterranean woodland climbers, e.g. *Ephedra campylopoda* and *Rubia tenuifolia*, grow together with the old Juniper trees and shrubs.

Origanum isthmicum is very rare in Gebel Halal (Täckholm, 1974) being endemic, as already noted, in about 5×2 km of this Gebel (Danin, 1969); it is recorded as absent from Gebel El-Maghara (Boulos, 1960). Danin (1983) states "*O. dayi* is endemic to Irano-Turanian territory of NE Negev and the southern Judean Desert and *O. ramonense* is endemic to the Central Negev highland. The closest relatives of these species are in Europe."

In the subalpine mountains of the southern Sinai subregion are relicts of species which are found in Lebanon. These include *Arenaria deflexa* and A *deflexa* v. *glabrata, Campanula dulcis* and *Scrophularia libanotica*.

The smooth rocks support endemics and very rare species such as *Cotoneaster orbicularis, Hypericum sinaicum, Micromeria serbaliana* and *Silene schimperiana.* The rarest endemic of Sinai, *Primula boveana,* is considered a relict of the Tertiary when the wet and cooler climate enabled it to reach Sinai. When the climate changed, a few plants of P. *boveana* survived, growing near small springs of the high mountains (above 1700 m) where the climate is cool. These springs flow throughout the year and are found mostly on red granitic rocks. Four closely related species *of Primula* are also local endemics; one in Yemen and the Horn of Africa, one in Turkey and two in the Himalayas (Danin, 1983). These five endemic species of *Primula* are derived from one ancestral species that was extensively distributed in Asia and Africa during a cool wet period about 6 million years ago (Wendelbo, 1961). Though the southwest mountains of Saudi Arabia are midway between Sinai and the Yemen mountains, no species of *Primula* has been recorded in the Hegaz mountains (Migahid, 1978).

Varthemia spp. are plants of smooth-faced rock outcrops and cliffs, and are Mediterranean elements that extend into the Irano-Turanian and the Saharo-Arabian assemblages. In Sinai two *Varthemia* spp. have been recorded: *V. montana* which grows in southern Sinai and V. *iphionoides* which typically grows in the crevices and fissures of smooth-faced limestone and dolomite outcrops and in wadis in which the water channel is in hard limestone. *V. montana,* however, grows mainly in sandstone, magmatic and metamorphic rock. It is used by bedouins for making tea, but V. *iphionoides* is said to be better for this purpose (Danin, 1983).

Rhus tripartita grows in the Sinai desert in areas with cliffs and canyons. Its typical habitats are deep-fissured limestone, dolomite, magmatic and metamorphic rock, all of which weather into large blocks. It is likely that *R. tripartita* was more widely distributed when Sinai had a wet climate (Danin, 1983).

Chapter 6
The Nile Region

6.1 Geomorphology

The River Nile extends from Lake Tanganyika in Tanzania (Lat. 3 °S) to the shore of the Mediterranean Sea (Lat. 31°15'N) for a length of about 6625 km. In this long course (Fig. 6.1), the river flows essentially a south to north path; both its source in Equatorial Africa and its mouth in the Mediterranean Sea lie within 1° of the same meridian of longitude (31 °E). It drains an area of about 3 million km^2 and connects regions which differ from each other in relief, climate, geological structure and soils. It derives its waters from the southern lake area of the Sudan basin and from the Ethiopian highlands which form part of the East African coalescing series of plateaux traversed by the great African rift system. In its northward passage, the Nile drains the major interior Sudan basin across the high Nubian area into Egypt and the Mediterranean by way of a series of cataracts.

The River Nile basin, modestly estimated as over 2,849,000 km^2 (Hurst, 1952), is a part of a larger area that is associated with this great natural system, including substantial parts of Egypt, Sudan, Ethiopia, Kenya, Uganda, Tanzania, Rwanda, Congo (Kinshasa), Central Africa and Chad. The territories embraced by this definition include catchment areas of tributaries that no longer contribute to the water of the principal channel, and include upstream rivulets of other river systems. The Nile is not merely a river flowing across some 34° of latitude, but also a complex system including various forms of water bodies (lakes, marshes, streams, canals, drains, etc.) and landforms (highlands, plains, valleys, etc.). These territories represent a great variety of climate, vegetation and land-use and also a number of biogeographical regions.

A series of barrages and dams has been built across the River Nile and its principal tributaries. The first were the Delta Barrages (1843–1861) and the most recent is the Aswan High Dam (1960–1968) which brings the downstream section of the River Nile under full control. Further projects include the construction of dams on the equatorial lakes and on Lake Tana in Ethiopia and the Junglei diversion canal of Sudan. Associated with river control works are considerable changes in land-use

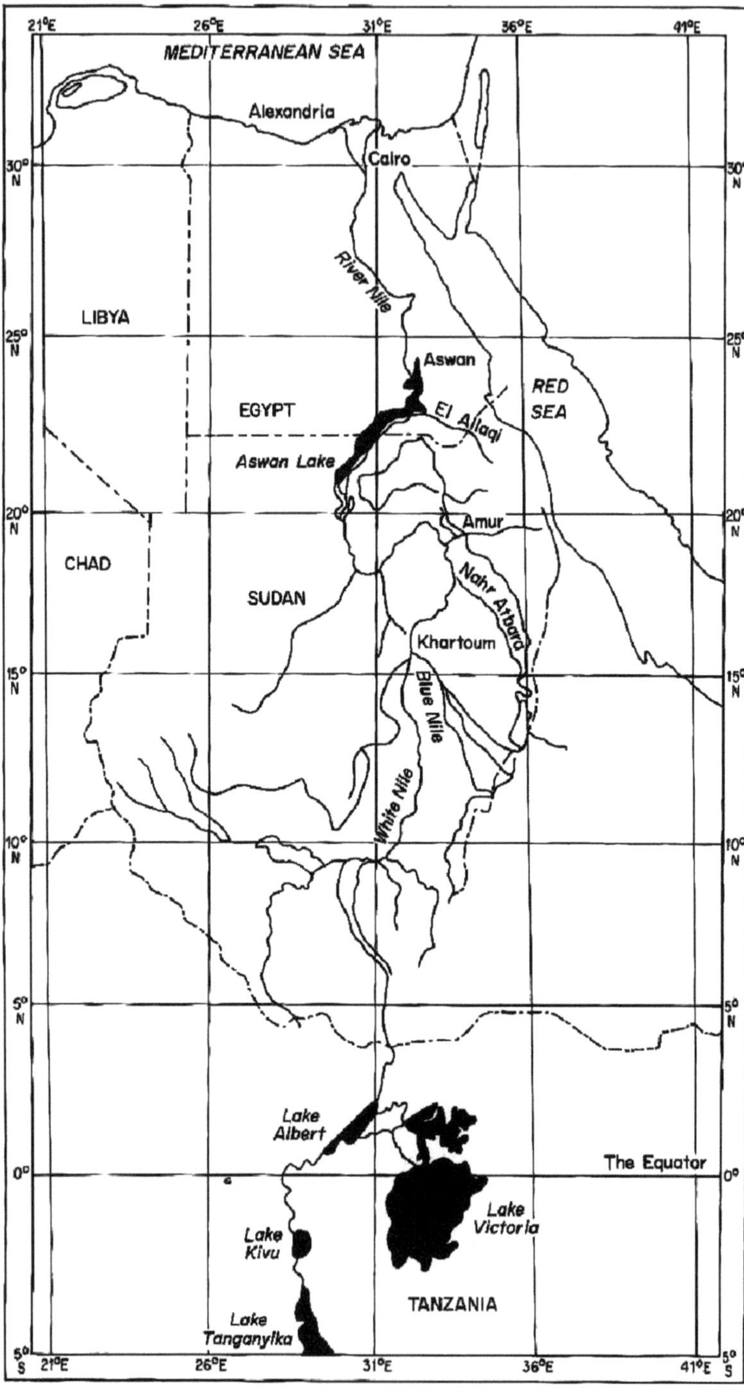

Fig. 6.1 The Nile Basin

pattern in Egypt and Sudan, in the biogeography of the river fauna and flora and in the development of the Nile Delta and the biota of its shores.

The Nile obtained most of its prominent features during the Pleistocene period and three phases can be distinguished: (a) Protonile, about 600,000 years BP with a more westerly course than the present river; (b) Prenile, about 500,000–125,000 years BP when the course of the Nile was still west of its present one, but east of the Protonile; and (c) Neonile, about 30,000 years BP during which most of the Nile sediments in Nubia were deposited and the Nile Valley attained its present form.

Of the total course of the River Nile only the terminal 1530 km lie within the borders of Egypt. Throughout this part of its course, and except for the dry wadis of the Eastern Desert (Chapter 4), the River Nile receives not a single tributary. The annual discharge in the main Nile reaching Egypt is 94 billion m^3, of which 58 billion m^3 are contributed by the Blue Nile, 24 billion m^3 by the White Nile and 12 billion m^3 by the Atbara River in the Sudan (Hurst, 1952).

After entering Egypt from the Sudan at Wadi Halfa, the Nile flows for more than 300 km in a narrow valley bordered by abrupt cliffs of sandstone and granite before reaching the First Cataract which starts about 7 km upstream of Aswan. Narrow strips of alluvial land could, until recently, be cultivated on both banks of the Nile in many parts of the Wadi Halfa-Aswan reach, but these, together with other features below the 180 m contour, are now drowned, being located in the area of the High Dam Lake. Downstream of the cataract, the valley begins to broaden, and flat strips of cultivated land between the river and the cliffs gradually increase in width northward. The total area of Upper Egypt (the Nile Valley) is about 12,000 km^2, stretching over a distance of more than 1000 km (Fig. 2.1).

Lower Egypt (the Nile Delta) is twice the area of Upper Egypt. Beside the apex it spreads in a plain studded with an intricate network of canals and drains; the former lie along the higher tongues of land, the latter in the hollows. According to Said (1981), seven major branches of the delta are mentioned by various historical documents and in ancient maps. These branches are: Canopic Branch (the present Rosetta Branch), Bolbitinic Branch, Sebennitic Branch, Fatmetic ranch (the present Damietta Branch), Mendisy Branch, Tanitic branch and the Pelusiac Branch (Fig. 6.2). Five of these branches generated and silted up in the course of time whereas two branches; Rosetta (about 239 km) and Damietta (about 245 km), are still running. The whole mesh loses itself in a coastal marsh belt of wastelands (Berari), punctuated with a number of coastal and inland lagoons.

The delta of the Nile appears as a triangle broader at the base than the sides. The length of the delta from south (20 km north of Cairo at the Delta Barrages) to north (the Mediterranean Sea) is 170 km and from east to west its breadth is 220 km. The area of the delta is about 22,000 km^2 and thus it comprises about 63% of Egypt's fertile land.

The northern coast of the Nile Delta close to the Mediterranean Sea is characterized by three shallow lakes: Manzala (in the east), Burullus (in the middle) and Idku (in the west) (Fig. 6.3). These lakes receive the main bulk of the drainage water from the Nile Delta land.

Formerly all the lands of the Nile Valley and Nile Delta were watered by inundation of the basin system. The silt of the water, before the construction of

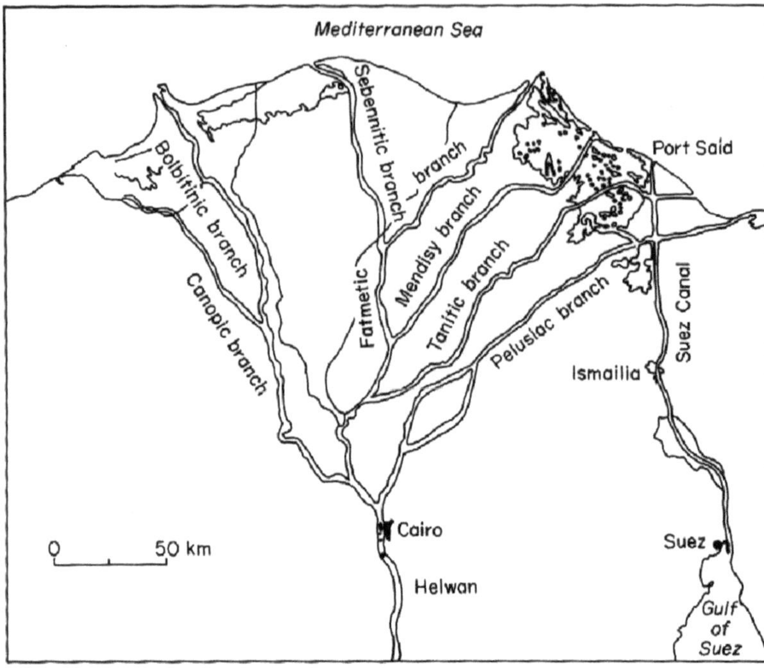

Fig. 6.2 The Nile Delta showing the seven ancient branches of the Nile

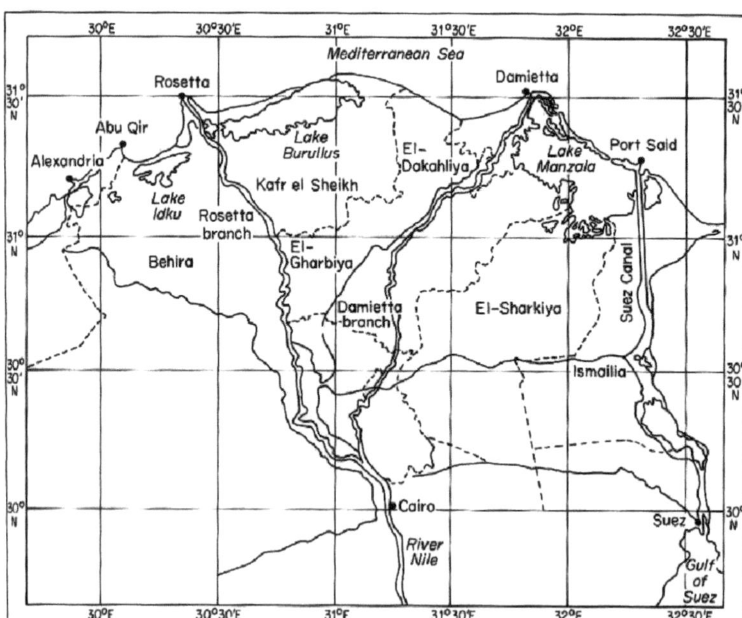

Fig. 6.3 The chief features of the Nile Delta

Aswan High Dam, gave an annual increase in sediment. There was an increment in the thickness of mud of 1 m every 1000 years. The silts carried by the Nile water from Ethiopia, through the Blue Nile, formed the fine fertile land of black or reddish colour in layers with sand between. Nowadays, no silt reaches the Egyptian lands.

6.2 Climate

The climate of the Nile region of Egypt is shown in Table 6.1, which includes climatic records for six meteorological stations. Aswan and Qena stations represent the Nile Valley; the Delta Barrages station (20 km north of Cairo at which the River Nile diverges into the Damietta and Rosetta branches) represents the northern end of the Nile Valley and the beginning of the Nile Delta. Tanta station is in the centre of the Delta while the deltaic Mediterranean coastal area is represented by the Rosetta and Damietta stations.

An extremely arid climate prevails in the Nile Valley: high temperature, low relative humidity, high evaporation and negligible rainfall (1.4 mm–5.3 mm/year). Climatic aridity gradually decreases northwards. At the Delta Barrages and Tanta the annual rainfall is 20.8 mm and 45.5 mm respectively. The climate of the deltaic coastal belt of Egypt is an extension of that of the western Mediterranean coast. The annual rainfall is about 160 mm in Rosetta and 102 mm in Damietta. Winds are generally light but violent dust storms and sand pillars are not rare. El-Khamsin winds blow occasionally for about 50 days during spring and summer.

The El-Fayium Depression is in the arid part of Egypt with annual rainfall of about 14 mm, mean annual maximum and minimum temperatures are 29.5°C and 14.5°C respectively, mean annual evaporation is 6.9 mm/day (Piche) and mean annual relative humidity is 66%, 32% and 51% at 6 a.m., noon and 6 p.m. respectively.

Fayium Province is one of the depressions of the Western Desert of Egypt. Being the nearest to the Nile Valley (Fig. 3.4) and after being connected with the River Nile by a large irrigation canal (Bahr Yusuf), the Fayium Depression is considered as a part of the Nile Region. The lowest part of the depression is occupied by a shallow saline lake – Qarun Lake – which is about 4.5 m below sea level and about 200 km^2 in area (Fig. 3.4). The depression has a total area of about 1700 km^2. Its floor just above the lake level is about 23 m above sea level (Ball, 1939).

6.3 Vegetation Types

The Nile Region of Egypt may be ecologically divided into two main subregions:

1. The deltaic Mediterranean coast;
2. The Nile system.

Table 6.1 Climatic data for six meteorological stations in the Nile region of Egypt (means of 1931–1960) (*Climatic Normals of Egypt*, Anonymous, 1960)

	Aswan					*Qena*					*Delta Barrages*				
	Temp. (°C)					*Temp. (°C)*					*Temp. (°C)*				
Month	*Mx*	*Mn*	*Rh*	*Ev*	*Rf*	*Mx*	*Mn*	*Rh*	*Ev*	*Rf*	*Mx*	*Mn*	*Rh*	*Ev*	*Rf*
Jan	24.2	9.5	52	8.6	0.1	22.7	6.7	66	3.4	0.2	19.8	6.3	78	2.6	3.3
Feb	26.5	10.6	45	10.0	Tr	25.3	7.6	59	4.3	1.0	21.3	6.9	75	3.4	5.0
Mar	30.7	14.2	36	13.9	0.1	30.3	11.1	47	6.6	0.1	23.9	8.6	69	4.0	2.5
Apr	35.7	18.6	30	17.0	0.3	35.4	15.9	35	9.3	Tr	28.4	11.2	66	5.4	0.4
May	40.3	23.5	26	17.2	0.6	39.0	20.7	31	11.0	0.3	32.4	14.9	62	7.2	0.7
Jun	42.0	25.1	28	21.8	Tr	40.9	22.9	35	12.6	Tr	34.5	17.6	66	7.9	0.0
Jul	41.9	26.1	29	20.1	0.0	40.8	23.7	38	11.6	0.0	35.6	19.3	72	7.7	0.0
Aug	42.0	26.4	32	20.0	0.0	40.8	24.1	39	11.8	0.0	34.6	19.9	73	6.0	0.0
Sep	40.0	24.0	37	19.0	Tr	38.1	22.2	52	9.0	0.0	32.0	18.4	76	4.7	0.0
Oct	37.5	21.7	41	16.9	0.2	35.1	18.9	57	8.1	0.6	30.3	16.1	76	4.1	1.5
Nov	31.4	16.5	49	11.8	0.1	29.8	13.6	61	4.5	2.2	25.9	12.5	79	3.0	0.8
Dec	26.5	13.2	53	9.0	Tr	24.3	8.9	66	3.3	0.9	21.5	8.3	76	2.4	6.6
M. An.	34.9	19.1	38	15.4	–	33.5	16.4	49	8.0	–	28.4	13.3	72	4.9	–
Total	–	–	–	–	1.4	–	–	–	–	5.3	–	–	–	–	20.8

Table 6.1 (Continued)

Month	Tanta					Rosetta					Damietta				
	Temp. (°C)					Temp. (°C)					Temp. (°C)				
	Mx	*Mn*	*Rh*	*Ev*	*Rf*	*Mx*	*Mn*	*Rh*	*Ev*	*Rf*	*Mx*	*Mn*	*Rh*	*Ev*	*Rf*
Jan	19.7	6.0	83	2.1	9.0	18.1	11.3	76	3.7	44.4	18.4	8.2	83	2.9	23
Feb	21.0	6.4	82	2.8	7.5	18.5	11.1	76	3.9	31.7	18.7	8.8	80	3.4	17.9
Mar	23.7	8.1	77	3.7	4.2	20.1	12.6	73	4.2	8.2	20.4	11.0	75	4.1	10.3
Apr	27.6	10.7	69	5.4	1.9	22.5	14.5	72	4.6	4.5	23.2	13.7	71	4.7	2.6
May	31.8	14.5	63	7.0	4.2	25.6	17.8	73	4.5	3.1	26.6	17.0	69	5.3	1.9
Jun	33.8	17.2	65	7.6	Tr	27.6	21.2	74	4.6	0.0	29.1	19.8	69	5.4	0.2
Jul	34.5	19.1	73	6.6	Tr	29.1	23.2	76	4.6	0.0	31.0	21.2	74	5.1	Tr
Aug	34.6	19.3	78	5.7	0.0	30.2	23.9	74	4.8	0.0	31.3	21.5	75	4.9	Tr
Sep	32.5	17.4	79	4.5	0.2	29.3	22.8	71	5.0	1.0	29.4	20.2	72	4.7	0.7
Oct	30.1	15.4	80	3.6	4.2	27.5	20.4	72	4.8	8.9	27.6	18.7	72	4.4	4.9
Nov	25.9	12.2	83	2.6	4.6	24.1	17.2	73	4.1	19.8	24.0	15.4	78	3.6	16.1
Dec	21.3	8.1	83	2.0	9.7	20.4	13.1	74	3.6	38.2	20.0	10.6	84	2.8	24.7
M. An.	28	12.9	76	4.5	–	24.4	17.4	74	4.4	–	24.0	15.5	75	4.3	–
Total	–	–	–	–	45.5	–	–	–	–	160	–	–	–	–	102.3

Rh, relative humidity (%); Ev, evaporation (mm/day Piche); Rf, rainfall (mm); M. An., mean annual; Tr, trace; Temp, temperature (°C); Mx, maximum; Mn, minimum.

The deltaic Mediterranean coast is a narrow belt influenced greatly by the sea. It is the coastal area between Abu Qir eastward to Port Said: 180 km from west to east and about 15 km from sea landward (Fig. 6.3). The Nile system however, encompasses the lands affected mainly by the water of the River Nile in Egypt and this includes: the Nile Valley between Aswan in the south northward to the Delta Barrages, the man-made lakes south of Aswan to the Sudano-Egyptian border, the Nile Delta between the Delta Barrages northwards to the inland border of the Mediterranean coastal belt, and the Nile Fayium at about 60 km southwest of Cairo in the Western Desert. The vegetation types of the different habitats of these two sub-regions are described.

6.3.1 The Deltaic Mediterranean Coast

(a) General Features

The deltaic Mediterranean coast of Egypt (the middle section of the Egyptian Mediterranean coast) is characterized by villages and summer resorts such as Rosetta, Baltim, Gamasah and Ras El-Bar (Figs. 6.3 and 6.4). The three shallow lakes of this coastal area receive the main bulk of the drainage water from the Nile Delta and are connected with the sea by outlets through strips of land fringing them. Along the shore-line of the delta are sand dunes associated with the eastern banks of the pres-

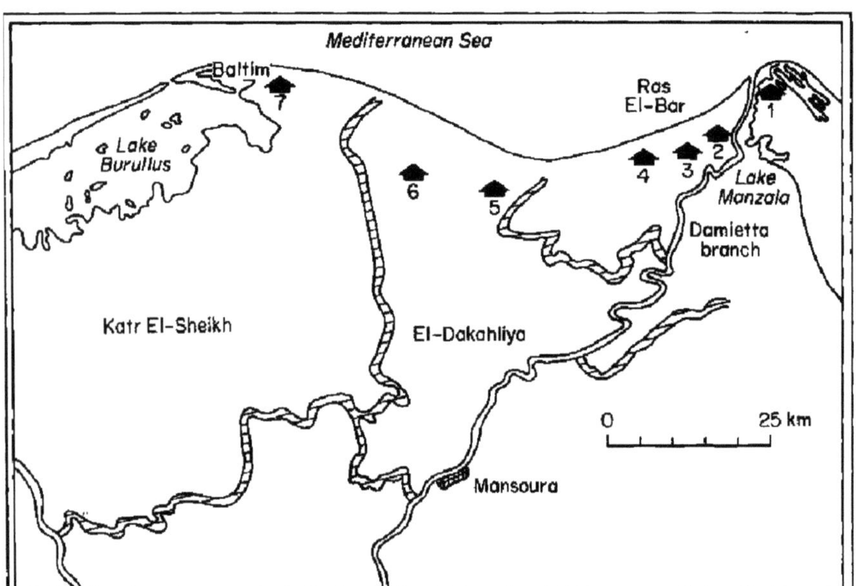

Fig. 6.4 The position of the seven studied sites (1–7) in the coastal region of the Nile delta. *Shaded lines* show boundaries of provinces

ent and past branches of the Nile. Two promontories are associated with the mouths of the Damietta and Rosetta branches of the Nile. The land between the two mouths extends into the sea to Lat. 31°36'N, i.e. 12 and 5 seconds further than the tips of the Rosetta and Damietta promontories respectively. This middle part is the site of the mouth of an old branch of the Nile – the Sebennitic.

The coastal land to the east of the exit of Lake Burullus is covered by sand dunes that extend for about 15 km east of the outlet. The strip of land to the west of this outlet has no sand dunes comparable to those on the east. This pattern is repeated at the sites of all the several branches of the Nile that emptied directly into the Mediterranean until the ninth century. The sand was obviously brought to the shore-line with the Nile sediments discharged at the mouth. Softer silts and clays travelled further into the sea or were transported along the shore-line for long distances by littoral currents. Sand was deposited at the shore-line and pushed to the eastern side of the mouth by the eastward currents that prevail throughout the main part of the year (Ball, 1939; Kassas, 1972a).

Though the climate of the deltaic section does not vary greatly from climates of the western and eastern Mediterranean sections, its vegetation is not the same. Unlike the western and eastern sections, the middle one is not only affected by sea water but it is also affected by the water of the northern lakes and the Damietta and Rosetta branches.

During both ancient and recent times, the western Mediterranean coast of Egypt has been extensively studied (Chapter 3). Fortunately, after the success of the Egyptians to get the Sinai Peninsula returned to its motherland from the Israeli occupation (1981), the Egyptian Government is actively promoting scientific studies and investment projects in Sinai, including the northern Mediterranean coast. However, the ecological studies carried out in the middle section of the Mediterranean coast of Egypt (which was previously considered part of the eastern section[1]) are few. Apart from the works of Montasir (1937) on Lake Manzala and Kassas (1952b,c) on *Alhagi maurorum*, there appears to be no published work about the vegetation of this part of Egypt before the papers of Zahran (1984), Zahran et al. (1985a, 1988, 1990) and El-Demerdash et al. (1990).

(b) Plant cover

The vegetation of the deltaic Mediterranean coastal land can be divided into zones landward that vary in dominance, composition and extent depending upon (a) landform and (b) distance from the sea, lakes and cultivated lands. The communities of this coastal vegetation are described in seven representative line transects in selected sites along the coast. These transects extend from the shore-line landward to the border of the cultivated land (Fig. 6.4).

[1] The Mediterranean coast of Egypt was divided into two sections: Western and Eastern (Hassib, 1951).

(i) Transects

Site 1. Ezbit El-Burg

This is a small village near Damietta City and Ras El-Bar summer resort. The vegetation occupies four habitat types progressively landwards: sand formation, salt marsh, swamps and potentially cultivated land. The sand formation is of low and narrow bars close to the sea (zone 1). It is dominated by *Zygophyllum aegyptium* (Hosny, 1977) which forms a pure stand with thin plant cover (<5%). The salt marsh occupies a wide area to the south of this zone and is dominated by *Arthrocnemum macrostachyum. Halocnemum strobilaceum* is the abundant associate and it may codominate with *Arthrocnemum* in certain stands. Other associates in this zone include *Cressa cretica, Frankenia pulverulenta, Halimione portulacoides, Inula crithmoides, Juncus acutus* and *J. rigidus*. The inland low areas of the transect catch the water seeping from both the sea and Lake Manzala. This swampy habitat is dominated by *Typha domingensis* with abundant *Phragmites australis. Cyperus laevigatus* is also abundant in the wet saline fringes of the swamps. In the arable land limiting the inland extent of this transect, *Cynanchum acutum*, which is absent from the northern zones, is very abundant.

Site 2. El-Senaniah

This site is within the borders of the new Damietta sea-port where interference by man is substantial. The area was previously planted with thousands of date palms but most of these trees have been removed to build new establishments of the port. Also, the natural vegetation has been removed and, accordingly, the recognized zones are restricted to salt affected lands dominated by *Arthrocnemum macrostachyum*. Associate species are *Halimione portulacoides, Inula crithmoides* and *Tamarix tetragyna. Cynanchum acutum* is abundant in the inland arable land. *Alternanthera sessilis, Ceratophyllum demersum, Echinochloa stagnina, Eclipta alba* and *Eichhornia crassipes* are abundant in the water canals and on their moist banks.

Site 3. Kafr El-Batikh

Kafr El-Batikh is a small village of the deltaic Mediterranean coast about 25 km southwest of Damietta. The transect of this site extends from sea landward for about 1 km and nine zones may be distinguished. The first is a narrow sand-line (about 100 m width) which is almost barren except for the remains of dead algae and sea-grasses (seaweeds). The second zone is also narrow (150 m wide), with a sandy substratum and is characterized by undershrubs of *Zygophyllum aegyptium* that build low sand mounds forming pure stands with thin cover (<5%). The third zone represents the beginning of the salt marsh habitat which extends throughout six zones (from zone 3 to zone 8). *Arthrocnemum macrostachyum* is the dominant of zone 3 (about 200 m wide) where cover is thin (<10%) contributed mainly by the dominant. *Halocnemum strobilaceum* is the only associate. The fourth zone is a wide strip of salt land (about 600 m wide), almost without vegetation except for widely spaced dead plants of *A. macrostachyum* on small scattered hummocks. The fifth zone is narrow (about 100 m wide), formed of low sand embankments co-dominated by *A macrostachyum* and *Inula crithmoides*. Of the 18 associate species, 14 are perennials and four annuals. Most of these species are salt tolerant. *Zygophyllum aegyptium* and *Halocnemum strobilaceum* are abundant in this zone whereas *Lippia nodiflora* and *Sporobolus virginicus* are locally abundant. *Juncus rigidus* and *Phragmites australis* are frequent here, *Cyperus laevigatus, J. acutus* and *Phoenix dactylifera* occasional. *Halimione portulacoides, Saccharum* sp. and *Tamarix tetragyna* are local. Other associates are *Cynodon dactylon* and *Spergularia* sp. The associate annuals include *Cakile maritima* and *Senecio desfontainei* (occasional) and *Lotus* sp. and *Melilotus indica* (rare).

The fifth zone can be considered as transitional between that dominated by *A. macrostachyum* and the sixth zone dominated by *Inula crithmoides*. The latter zone is a dry salt marsh area about 250 m wide. The associates are 13 perennials and six annuals. *A macrostachyum* is abundant and *H. strobilaceum* frequent. *Aeluropus massauensis* is locally abundant and *Calligonum comosum* local but less frequent. *Alhagi maurorum, Juncus rigidus* and *Phragmites australis* are occasional, whereas *Cynodon dactylon, Lippia nodiflora, Phoenix dactylifera* and *Zygophyllum aegyptium* are rare. Associate annuals include *Launaea angustifolia, Lotus* sp., *Melilotus indica, Rumex pictus* and *Salsola kali*.

The seventh zone is dominated by the succulent halophyte *Suaeda vera*. The relatively high plant cover (about 50%) is contributed mainly by the dominant and partly by the abundant associates (*Senecio desfontainei* and *Cakile maritima*). Locally abundant associates are *H. strobilaceum, Rumex pictus* and *Z. aegyptium* and locally common associates are the perennials *A. macrostachyum, Cynodon dactylon* and *Phoenix dactylifera* and the annual *Cutandia memphitica*. *Erodium hirtum* and *Reichardia orientalis* are occasional.

South of zone 7, the level of the land drops and soil moisture content increases, zone 8 being a wet salt marsh dominated by *Juncus rigidus*. The cover of this zone is about 60–70%, mostly of *Juncus* tussocks. *Inula crithmoides* is abundant, *J. acutus* and *Phragmites australis* are frequent and *A. macrostachyum* and *Z. aegyptium* are local. Occasional species are *Cynodon dactylon* and *Tamarix tetragyna*. Rare associates include *Cakile maritima, Carex distans, Melilotus indica* and *Polygonum equisetiforme*.

The most landward (ninth) zone of this transect is a swampy habitat dominated by *Typha domingensis*. These swamps are formed by water seeping from Manzala Lake as well as from the cultivated land. The high cover (up to 90%) is contributed mainly by the dominant. *Phragmites australis* is frequent. *Juncus acutus* and *Scirpus tuberosus* are common, growing on the very wet banks. The relatively dry banks are characterized by the growth of halophytes such as *Halimione portulacoides, Inula crithmoides, Salsola kali* and *Tamarix tetragyna*.

Site 4. Umm Reda

The sequence of zones of the vegetation of the transect at this site is similar to that of site 3. The first narrow shore-line zone is barren, followed by another narrow zone dominated by *Zygophyllum aegyptium* which builds low sand mounds. The third zone is a salt marsh dominated by *Arthrocnemum macrostachyum* associated with *Halocnemum strobilaceum, Inula crithmoides, Juncus acutus* and *Tamarix tetragyna*. The fourth zone is a wide barren black of alkali land followed by a vast new reclaimed uncultivated area (zone 5) where *Cakile maritima* is very abundant. The sixth zone contains three habitats: (a) sand dunes dominated by *Zygophyllum aegyptium*, (b) depressed areas where seeped water accumulates, creating a swampy habitat co-dominated by *Typha domingensis* and *Phragmites australis* and (c) barren saline areas. The landward zone is of arable land with many weeds, e.g. *Cakile maritima* and *Launaea angustifolia*.

Site 5. Zaiaan

In the transect at this site the first four zones are similar to those of the preceding transect. However, the fifth is a wide salt-affected land on which there are two types of sand dunes: mobile and stabilized. On the mobile dunes, which are the bigger, there are scattered plants of *Elymus farctus, Stipagrostis ciliata, S. scoparia* and *Zygophyllum aegyptium*. These are pioneer plants in the psammosere succession. The stabilized sand dunes, on the other hand, are characterized by the growth of *Echinops spinosissimus* and *Astragalus tomentosus*. Other associates may also be present, e.g. *Erodium hirtum, Lycium europaeum* and *Moltkiopsis ciliata*. The salt-affected land between the sand dunes is usually wet and sometimes waterlogged during the winter season. During summer high evaporation rate increases water loss from the soil, and salt crusts can be seen on the surface layers of the soil. These saline lands are almost without vegetation except for a few widely spaced plants of *Arthrocnemum macrostachyum, Inula crithmoides* and *Tamarix tetragyna* in areas of relatively low salt content. The landward zone is arable land covered with sand sheets where some vegetables, e.g. *Citrullus* spp. and tomato, have been cultivated. *Heliotropium kassasi* is very common in this zone.

Site 6. Abu Madi (Qalabshu)

This is the area of huge coastal mobile sand dunes (>60 m high) which are barren except for a few scattered plants of *Phragmites australis*. The presence of this reed on these coastal mobile sand dunes may indicate that the area was swampy before being covered with the chains of dunes that run east-west parallel to the shoreline. Between dunes are large water runnels with saline conditions where *Arthrocnemum macrostachyum* and *Halocnemum strobilaceum* co-dominate with *Zygophyllum aegyptium* as a common associate. The presence of this saline habitat may further suggest the previous presence of a swampy habitat. Parts of these runnels are covered with a sand sheet, a factor that decreases the salt content of their substrata. In these patches some psammophytes, e.g. *Elymus farctus* and xerophytes e.g. *Calligonum comosum*, may grow. These barren coastal sand dunes are formed of loose coarse sand, poor in organic matter and with low salt content.

The vegetation of the salt marsh habitat starts at the foot of these huge dunes in the south; cover is up to 60%, contributed mainly by the dominants *A. macrostachyum* and *H. strobilaceum. Frankenia revoluta* is abundant; *Limoniastrum monopetalum* is frequent and *Cressa cretica* is occasional. Three species, *Limonium pruinosum, Polygonum equisetiforme* and *Suaeda vera*, are local. Annual associates include *Bassia muricata, Mesembryanthemum crystallinum, Paronychia arabica* and *Salsola kali.*

The second zone of this transect occupies an area characterized by fixed small dunes dominated by *Thymelaea hirsuta. Asparagus stipularis*, a co-dominant in zone 5 of this transect, is frequently present on these dunes. *Lygos raetam* is locally abundant. *Lycium europaeum*, dominant in zone 4 and co-dominant in zone 5, is here a rare associate. Other associates are *Echinops spinosissimus, Pancratium arabicum, Silene succulenta* and *Stipagrostis ciliata.* The ground of this zone is a natural extension of the salt marsh area, and is characterized by the growth of halophytes in areas between the fixed dunes. *Limoniastrum monopetalum* and *Tamarix tetragyna* are abundant. *Frankenia revoluta* and *Zygophyllum aegyptium* are commonly present. Other associates are *Aeluropus massauensis, Limonium pruinosum, Lotus halophilus, Moltkiopsis ciliata, Paronychia arabica, Polypogon monspeliensis, Schismus barbatus, Sporobolus spicatus* and S. *virginicus.*

The third zone is a barren salt-affected land that extends landward for about 300 m. It separates the zone of the fixed dunes dominated by *T. hirsuta* (zone 2) from that of the small fixed dunes (sand hummocks) dominated by *Lycium europaeum* (zone 4) where there are 20 associates (10 perennials and 10 annuals). Abundant associates of the *L. europaeum* zone are *Asparagus stipularis* and *Moltkiopsis ciliata. Limoniastrum monopetalum* is frequently present. Eight species are local; three are perennials (*Limonium pruinosum, Frankenia revoluta* and *Suaeda vera*) and five are annuals – *Mesembryanthemum crytallinum, Ononis serrata, Paronychia arabica, Plantago indica* and *Rumex pictus.* Three perennials – *Lygos raetam, Tamarix tetragyna* and *Thymelaea hirsuta* – and two annuals, *Ifloga spicata* and *Polypogon monspeliensis*, are occasional. *Centaurea glomerata, Echinops spinosissimus* and *Launaea angustifolia* are rare.

The fifth zone may be considered as an extension of the fourth, being also characterized by fixed small dunes co-dominated by *Asparagus stipularis* and *Lycium europaeum*, both of which contribute equally to the cover (up to 60%). Eighteen associates have been recorded in this community: 14 perennials and 4 annuals. Frequently present associates are *Limoniastrum monopetalum* and *Moltkiopsis ciliata* (perennials) and *Rumex pictus* (annual). Two annuals (*Ifloga spicata* and *Plantago indica*) and three perennials (*Echinops spinosissimus, Erodium hirtum, Limonium pruinosum*) are occasional. *Frankenia revoluta, Suaeda vera, Tamarix tetragyna* and *Thymelaea hirsuta* are rare in this zone.

The last landward zone, which separates the cultivated fields from the natural vegetation of the coastal area at Abu Madi village, is a saline land dissected by salt water creeks. This area is periodically inundated with drainage water of the nearby fields and thus it is very suitable for the growth of rushes, sedges and halophytes. The dominant species here is *Juncus acutus. J. rigidus* is abundant and *J. subulatus* is frequent. Two sedges, *Cyperus laevigatus* and *Scirpus tuberosus*, are occasional. Halophytes occasionally present *are Aeluropus massauensis, Cressa cretica, Halimione portulacoides, Inula crithmoides, Polygonum equisetiforme* and *Tamarix tetragyna. Arthrocnemum macrostachyum* is rare and *Halocnemum strobilaceum* is absent. *Typha domingensis*, the dominant "reed" in other parts of this coastal area, is rare here. Annual associates usually occur in the dry parts and include *Bassia muricata, Parapholis marginata, Polypogon monspeliensis* and *Salsola kali.*

Site 7. Baltim

Baltim is a summer resort of the deltaic Mediterranean coastal area. The line transect in this site shows six successive zones: beach zone, sand sheet zone, Gebel El-Nargis zone, salt marsh zone, palm trees – sand dunes zone and the swamps.

The beach is a narrow strip of sand except for some dry remains of "seaweeds" (sea-grasses), mainly *Posidonia oceanica* and *Cymodocea major* and marine algae. The second zone landward is also narrow with a sandy substratum dominated by *Silene succulenta* associated with 29 species: 13 perennials and 16 annuals. *Cakile maritima* (annual) is a co-dominant during winter and spring. Associate perennials, mostly rare, include *Acacia saligna, Alhagi maurorum, Cynodon dactylon, Cyperus capitatus, Erodium hirtum* and *Polygonum equisetiforme*. Other occasional perennials are *Dactyloctenium aegyptium, Lippia nodiflora* and *Paspalidium geminatum. Ricinus communis* (semi-wild) is locally abundant. The natives of Baltim village usually cultivate grapes (*Vitis vinifera*), figs (*Ficus carica*) and water melon (*Citrullus vulgaris*) in this sandy-sheet zone. Apart from *Cakile maritima*, other annuals include *Mesembryanthemum crystallinum* and *Senecio desfontainei* (abundant), *Emex spinosus, Launaea angustifolia* and *Polypogon maritimus* (occasional) and *Cutandia memphitica, Ifloga spicata, Lotus halophilus, Malva parviflora, Melilotus indica* and *Salsola kali* (rare).

The third zone of this transect is of the large sandy dunes called Gebel El-Nargis dunes (Hills of Lilies) that reach more than 10 m high. *Silene succulenta* is the dominant on both north- and south-facing slopes of the dunes but the cover varies. On north-facing slopes it is about 30% whereas on the south-facing slopes it is about 60%. These differences in cover may be because the south-facing slopes are not subject to the direct effect of cold, strong north winds which lower the temperature and may uproot many seedlings on the north-facing slopes. On both slopes the dominant, *Silene succulenta,* contributes the main bulk of the cover. However, on the south-facing slopes some associates, e.g. *Elymus farctus,* rarely recorded on north-facing ones, are abundant and co-dominate certain stands, with considerable cover (up to 15%) during the wet season. Although the number of associate species (29: 14 perennials and 15 annuals) on north-facing slopes is similar to that of the south-facing slopes (32: 14 perennials and 18 annuals), the plant cover of these associates on the north-facing slopes is negligible. Associate perennials of both slopes are *Alhagi maurorum, Desmostachya bipinnata, Erodium hirtum, Imperata cylindrica, Ipomaea stolonifera, Lycopersicum esculentum. Pancratium arabicum, Polygonum equisetiforme* and *Stipagrostis ciliata. Cyperus capitatus* has been recorded as a frequent associate on north-facing slopes whereas *Echinops spinosissimus* is known only on south-facing ones. The annual associates are *Bromus rubens, Cakile maritima, Cutandia memphitica, Ifloga spicata,*

Lotus halophilus, Melilotus indica, Polypogon maritimus, Salsola kali and *Senecio desfontainei* (on both slopes), *Parapholis marginata, Plantago ciliata* and *Rumex pictus* (on north-facing slopes) and *Carthamus glaucus, Daucus bicolor, Malva parviflora, Ononis serrata, Polypogon monspeliensis* and *Pseudorlaya pumila* (on south-facing slopes). *Citrullus vulgaris* is cultivated on both slopes.

The fourth zone of the Baltim transect is a salt marsh dominated by *Arthrocnemum macrostachyum* with *Halocnemum strobilaceum* as an abundant associate and in some stands a co-dominant. Other associates include *Aethiorhiza bulbosa, Cressa cretica, Cyperus conglomeratus, Frankenia revoluta, Limoniastrum monopetalum, Limonium pruinosum, Moltkiopsis ciliata, Sporobolus virginicus* and the annuals *Mesembryanthemum crystallinum* and *Reichardia tingitana*. Though the number of associates of this community is relatively low (16), plant cover is high (60–70%), contributed mainly by the dominant halophyte.

The fifth zone is another zone of huge sand dunes with a community dominated by semi-wild palm trees (*Phoenix dactylifera*). Within this community are some stands dominated by other species. In the low areas in these dunes where water is exposed forming a local swampy habitat *Typha domingensis* dominates. In the saline patches of the runnels within the sand dunes there are areas dominated by *Arthrocnemum macrostachyum, Imperata cylindrica, Schoenus nigricans, Sporobolus virginicus* and *Zygophyllum aegyptium*. Perennial associates of this zone include *Cynodon dactylon, Moltkiopsis ciliata* and *Pancratium arabicum* (abundant), *Aetheorhiza bulbosa, Alhagi maurorum, Cressa cretica, Cyperus capitatus, Dactyloctenium aegyptium, Desmostachya bipinnata, Echinops spinosissimus, Frankenia revoluta, Juncus acutus* and *Silene succulenta* (occasional or rare). Among the annuals *Mesembryanthemum crystallinum, Ononis serrata, Pseudorlaya pumila, Rumex pictus* and *Salsola kali* occur in scattered patches in this zone. Other annual associates include *Adonis dentata, Carthamus glaucus, Cutandia memphitica, Launaea angustifolia, Lobularia libyca, Malva parviflora, Reichardia tingitana, Senecio desfontainei* and *Sinapis arvensis*.

The sixth innermost zone is a depression which receives drainage water from the nearby cultivated land of Baltim village and the water seeping from Lake Burullus. The dominant of this swampy habitat is *Typha domingensis* which covers about 80–90% of the zone. The widespread reed in Egypt, *Phragmites australis*, is here occasional. On saturated banks of these swamps, which are periodically inundated, *Cyperus conglomerates* and *Juncus rigidus* occur. The relatively dry banks have a saline substratum supporting some halophytes, e.g. *Cressa cretica, Halimione portulacoides, Inula crithmoides, Polygonum equisetiforme, Suaeda vera* and *Tamarix tetragyna* (perennials) and *Mesembryanthemum crystallinum* (annual).

(ii) Main habitats

Four main types of habitat can be recognized in the deltaic Mediterranean coast of Egypt: sand formations, salt marshes, swamps and potentially cultivated lands. The sand formations comprise the low and small sandy mounds that form low sand bars along the shore-line. In this habitat *Zygophyllum aegyptium* predominates with no associate species and with very thin cover. Also, there are the huge mobile sand dunes of Abu Madi (60 m high) which are barren except for a very few plants of *Phragmites australis* which may be an indicator of a previously swampy habitat. The partially stabilized sand dunes are dominated by the pioneer psammophytes *Stipagrostis ciliata* and *Elymus farctus* with abundant presence of *Alhagi maurorum* and *Echinops spinosissimus* etc. Stabilized dunes are dominated by *Asparagus stipularis, Echinops spinosissimus, Lycium europaeum, Silene succulenta* and *Thymelaea hirsuta*. The semi-wild *Phoenix dactylifera* is also a characteristic feature of the sand dune vegetation of this coastal belt.

The vegetation of the salt marsh ha'bitat is of communities dominated or co-dominated by *Arthrocnemum macrostachyum* (widespread), *Halimione portulacoides, Halocnemum strobilaceum, Inula crithmoides, Juncus acutus, J. rigidus, Limoniastrum monopetalum, Suaeda vera, Tamarix tetragyna* and *Zygophyllum aegyptium.*

The frequent swampy habitats are usually in low areas of the landward zones where the water seeping from the lakes and/or drained from cultivated lands accumulates. *Typha domingensis* dominates in these swamps with the common presence of *Phragmites australis*. In the saline-saturated fringes of these swamps, rushes, e.g. *Juncus acutus* and *J. rigidus*, sedges, e.g. *Carex extensa, Cyperus laevigatus* and *Scirpus tuberosus*, grow.

The potentially cultivated land occupies the most landward areas of this coastal belt. Being less saline, this habitat supports the growth of annual weeds. The most abundant weed is *Cakile maritima*; common ones include *Amaranthus ascendens, Launaea angustifolia* and *Senecio desfontainei.*

6.3.2 The Nile System

The plants of the Nile system of Egypt (about 553 species[2]) represent about 29.9% of the total flora of Egypt (Hassib, 1951). About 126 species are not recorded elsewhere in Egypt. Some 149 species are present in the Nile valley, 291 occur in the Nile Delta and 64 characterize the Nile Fayium. Therophytes represent 59.4%, hydrophytes and helophytes 9.8%, hemicryptophytes 8.0%, chamaephytes 7.6%, geophytes 6.4%, nanophanerophytes 2.9%, micronanophanerophytes 1.8% and parasites 2.6%. Apart from *Opuntia ficus-indica* (Cactaceae), usually cultivated as a fence plant and for its edible fruits, no stem succulents are present in the Nile system flora. Megaphanerophytes, mesophanerophytes and epiphytes are also absent. The percentage of hydrophytes and helophytes (9.8%) is, however, higher than in other regions of Egypt.

The Nile system of Egypt includes a number of habitats formed and/or greatly influenced by the water of the River Nile. These are:

1. The aquatic habitat
2. The swampy habitat
3. The canal bank habitat

[2] The number now known is higher and may reach >600 (Täckholm, 1974).

4. The cultivated lands
5. The northern lakes
6. The artificial lakes
7. The Nile Islands

(a) Historical

Early man penetrated the Nile Valley in Egypt during the lower Palaeolithic period some 250,000 years ago. The linear pattern of the Nile and the concentration of resources along its main course promoted the establishment of many settlements, of varied size, character and density during the various periods and cultures. Most of these settlements used the resources of the Nile; their proximity to the major desert wadis also indicates the utilization of the wadi fauna and flora (El-Hadidi, 1985).

According to Roubet and El-Hadidi (1981), the Late Pleistocene of Egypt extended from 20,000 years BP (or earlier) to 12,000 years BP and included several cultures of the Middle and Late Palaeolithic periods. Very few plant remains are recorded from the sites of this period. These include fragments or charred remains of *Acacia, Salsola* and *Tamarix* which are mainly sources of firewood and charcoal. In these early times, man must have practised some selection of plants for their food characteristics. Many of these food plants would have been annual herbs which are now recorded as common weeds of cultivation. Such species are likely to be found on the flood plains exposed following the regression of the Nile water at the end of each season's flood[3] (El-Hadidi, 1985). Thus, early each spring, the green foliage of numerous annuals would include such species of Compositae as *Cichorium pumilum, Lactuca* spp. and *Sonchus oleraceus* whereas the Cruciferae provide wild *Raphanus sativus, Brassica nigra* and *Eruca sativa*. It is suggested that other species eaten would include *Beta vulgaris, Cynodon dactylon, Corchorus olitorius* and *Rumex dentatus*. Some of these wild forms later became the progenitors of the present cultivars of radish and other useful plants. A wild form *of Lactuca*, endemic to Upper Egypt, is believed to be the progenitor of the Cos cultivar which has been known in Egypt since Pharaonic times and was early associated with Min, the God of Vegetation and Procreation (El-Hadidi, 1985, after Derby et al., 1977).

The end of spring would be the season for harvesting seeds and grain. Numerous wild annual legumes would have provided ripe seeds of high nutritive value, probably of *Astragalus, Cicer, Lathyrus, Lens, Lotus, Lupinus, Pisum, Trigonella, Vicia* and *Vigna*. Wild grasses as a source of grain were presumably utilized, although to a lesser extent. Species of Panicoid genera (*Pennisetum* and *Sorghum*) were more likely than those of Festucoid genera such as *Triticum* and *Hordeum*. In late spring there would have been the collection and consumption of non-leguminous seeds, such as those of oil and spice plants. Of these, the following are thought to have been known to early Egyptian man: *Anethum graveolens, Carthamus tinctorius, Coriandrum sativum, Lepidium sativum, Sinapis alba*.

[3] With the establishment of the Aswan High Dam (1965), no floods now occur in Egypt.

The summer floods and the winter months were periods when collection activities would have decreased because of lack of available species and important food sources would be the ripening fruits of trees and shrubs. Among these were the palms: *Phoenix dactylifera, Hyphaene thebaica* and *Medemia argun*. Other species with edible fruits might include *Balanites aegyptica, Ficus sycomorus* and *Ziziphus spina-christi*.

Over-wintering storage organs can also provide food when other sources are unavailable. Bulbs of wild onions and garlic (*Allium* spp.) as well as the tubers of *Cyperus esculentus* are known food sources from the Neolithic period (Täckholm and Drar, 1950) and apparently earlier. The rhizomes of the water lilies *Nymphaea coerulea* and *N. lotus* were valued for their high nutritive content.

(b) The aquatic habitat

In Egypt, where a warm climate prevails most of the year, the hydrophytes of the River Nile and its irrigation and drainage systems are greatly developed. The establishment of the Aswan High Dam in the most extreme south of Egypt (Fig. 2.1) controls to great extent the flow of water in the Nile and its Damietta and Rosetta branches. This control has led to numerous ecological changes in the Nile system, the effect of damming on downstream reaches being marked. Changes due to damming include silt-free water running downstream which results in the extensive use of fertilizers to compensate for the lack of the silt. Side effects also include: changes in the chemical and physical characteristics of irrigation water; the presence of water in the canals all the year around; the level of water in the Nile system, particularly in Lower Egypt, being noticeably lower and the current being of decreased velocity. The absence of silt in the Nile below the High Dam has made it no longer necessary to dredge the canals. Dredging removes large quantities of seeds and perennating organs of water plants; such factors are causing a noticeable and considerable increase in the growth rate and densities of the fresh-water hydrophytes of the Nile system. Also, new water-weeds, e.g., *Myriophyllum spicatum*, started to appear after the establishment of Aswan High Dam (El-Kholy, 1989).

Before the establishment of the Aswan High Dam, the water-weeds of the Nile system were not especially troublesome. But now this type of vegetation presents an acute problem; plants are causing so much trouble that the Egyptian Government spends millions of Egyptian pounds (and hard currency) to control these weeds chemically and mechanically.

The aquatic weeds of the Nile system of Egypt are some 35 species of 19 genera of 15 families (Täckholm, 1974). The plants are either entirely submerged, free-floating or their roots may penetrate the soil at the bottom of the stream. Some of these bottom-rooting plants have floating leaves. Recently, Zahran and Willis (2003) stated that the aquatic vegetation of the River Nile system of Egypt comprises 36 communities dominated by 8 submerged, 9 floating and 19 emerged species. An account of these aquatic weeds follows.

(i) Family Araceae

Pistia stratiotes is a free-floating, stemless, stoloniferous herb, consisting of a leaf-rosette and a tuft of fibrous roots beneath. Its presence in Egypt is recorded in a limited area of the northern section of the Nile Delta, in calm and stagnant canals of Faraskur (about 20 km south of Damietta).

P. stratiotes is a troublesome aquatic weed where it grows, but, unlike *Eichhornia crassipes* which is widespread in Egypt, as it is of very limited occurrence its impact is not a major one.

(ii) Family Ceratophyllaceae

Three *Ceratophyllum* spp. have been recorded in the Nile system of Egypt; two are rare (C. *submersum* and *C. muricatum*) and the third is common (C. *demersum*). *C. demersum* grows in both shallow and deep waters. It usually forms a pure community, but is sometimes associated with *Utricularia inflexa* (Hassib, 1951). It is an undesirable submerged weed of canals and drains. Reproduction is by seeds and also by broken pieces of the brittle branched stem which float away and rhizomes then anchor them to the mud (Simpson, 1932). In Nubia it was recorded for the first time by Springuel (1981) in the shallow water of the Nile near Aswan.

(iii) Family Haloragidaceae

Myriophyllum spicatum has been fairly recently recorded invading the River Nile and its system. This troublesome submerged weed was never seen in Egypt before the establishment of the Aswan High Dam (Simpson, 1932; Täckholm, 1956, 1974; Tawadrous, 1981). Fayed (1985) reported that in 1970–1980, *M. spicatum* seemed to be vigorously invading the Nile system of the southern provinces of Egypt: Qena, Sohag, Assiut and Minya. In 1972, the southern border of the distribution of this plant in Egypt was in the Idfu area. In April 1986 it was recorded by Springuel (1987) in the shallow water near the west Nile bank north of Aswan. According to El-Kholy (1989), *M. spicatum* is invading the River Nile from Aswan to Giza; its presence in the Nile Delta has not been observed. Zahran and Willis (2003) reported it as invasive weed in Lake Nasser (High Dam Lake). Khedr and El-Demerdash (1995) observed that *M. spicatum* is very abundant in some irrigation and drainage canals of the Nile Delta. Khedr and Zahran (1999) showed its presence in Lake Manzala but absent from Lake Burullus. Also, Shaltout and Khalil (2005) did not record it in Lake Burullus.

(iv) Family Hydrocharitaceae

Three submerged perennial fresh-water weeds of this family – *Elodea canadensis, Ottelia alismoides* and *Vallisneria spiralis* – are all rare in Egypt. *Ottelia* normally grows in rice fields. *Elodea* is usually recorded as a casual in the suburbs of Cairo and *Vallisneria* was formerly recorded only near Aswan (Täckholm, 1974). *Vallisneria* is now, however, a common weed in all Nubian water bodies including High Dam Lake, the Aswan Reservoir and the River Nile north of Aswan (Springuel, 1987).

(v) Family Lemnaceae

This family comprises a group of very small free-floating water plants without well-developed stems and leaves but with tiny leaf-like fronds forming green masses on the surface of the water ponds and stagnant pools.

In the Nile system, six species of three genera – *Spirodela, Lemna* and *Wolffia* – are present. *Spirodela punctata* is common in the Nile Delta whereas S. *polyrrhiza* is common in the stagnant water of the Nile system. *Lemna gibba* is the most common and widely present species of this family. *L. perpusilla* and *L. minor* are rare in the Nile Delta. *Wolffia hyalina* is common in the stagnant fresh and brackish waters.

(vi) Family Lentibulariaceae

In Egypt only one genus, *Utricularia*, of this family occurs. The plant is floating with finely dissected leaves with bladders in which small animals are caught. *U. inflexa* is a rare species, usually of rice fields in the Nile Delta.

(vii) Family Marsileaceae

Marsilea is the only genus of this family present in Egypt. This aquatic fern has erect petioles and creeping rhizomes. The fronds (leaves) consist of four digitate segments (leaflets) with forked veins radiating from the base. *M. aegyptiaca* is common in all waters of the Nile system while *M. capensis* is rare, being present only in the Nile Delta.

(viii) Family Najadaceae

Najas spp. are submerged weeds in the fresh and brackish waters but not in the saline ones. They spread like a mat along the bottom of shallow water.

N. pectinata, N. minor and *N. graminea* are rare in the Nile Delta, and absent from other parts of the Nile system. *N. armata*, however, is a common annual weed of the channels of the Nile Delta and Fayium (Täckholm, 1974). In 1973 *N. armata* was recorded in the High Dam Lake by El-Hadidi (1976). At the same time its presence in the northern section of the lake was reported by Entz (1976). Now, according to Springuel (1987), *N. armata* is a common weed in all the Nubian water bodies and its vigorous growth was reported during the period 1980–1986. Before the establishment of the Aswan High Dam, this species was not known in the Aswan area.

(ix) Family Nymphaeaceae

Species of *Nymphaea* are aquatic herbs of shallow water with solitary scapose large flowers and rounded floating leaves on long petioles, produced from perennial rhizomes. Two species have been recorded in the Nile system: *N. coerulea* and *N. lotus*. Both are common in the Nile Delta but rare in or absent from the Nile Valley. *N. coerulea* is the Sacred Lotus of ancient Egypt. Its tuberous roots are edible and the flowers are showy and fragrant. The large leaves, which spread out flat on the surface of the water, exclude all light for submerged plants which consequently die out (Simpson, 1932). *N. coerulea* is the blue water lily while *N. lotus* is white.

(x) Family Onagraceae

Jussiaea repens is a floating perennial herb rooting at the nodes. It produces clusters of white to pink inflated spongy fusiform roots of unusual appearance. It is a rare hydrophyte in the Nile Delta and absent from other areas of the Nile system.

(xi) Family Pontederiaceae

The genus *Eichhornia* includes two species of free-floating plants that occur in Egypt: *E. crassipes* and *E. azurea* (Täckholm and Drar, 1950). *E. azurea* is occasionally cultivated in gardens in Cairo, flowering in June–August. Its leaves are not in a rosette and their petioles are not, or hardly, swollen. It is not a troublesome weed. *E. crassipes*, however, is very common and widely spread in Egypt. Its leaves are in rosettes with inflated bladder-like petioles. It flowers in May-September and sometimes the flowering period extends to December. Both species reproduce vegetatively and by seeds.

E. crassipes is the worst aquatic weed in Egypt. It is called the Nile Lily, Ward El-Nil or Water Hyacinth. It has a conspicuous 5 cm pale violet flower, with the upper lobe larger than the rest, having a patch of blue with a yellow centre.

E. crassipes is a native of South America (Brazil) and is now naturalized in many other warm countries. It flourishes in fresh and in brackish waters but the salt content of sea water does not favour its growth. It has a fast growth rate at warm temperatures (optimum is 27.6°C); growth rate decreases substantially at lower temperatures (Bock, 1969). This may explain its absence from the waters of cold countries. The standing crop of *E. crassipes* in the deltaic region of the Nile is generally highest in autumn and early winter, a dry weight of over 500 g/m^2 being recorded in the Rosetta branch of the Nile in January 1988 (Serag, 1991). On average, however, fresh weights are about 2500 g/m^2 and dry weights 230 g/m^2, the highest production being in nutrient-rich (eutrophicated) waters.

E. crassipes was introduced to Egypt as an ornamental plant during the rule of Khedive Tawfiq (1879–1892); hence, it has, for many years, been grown to a limited extent in certain public and private gardens of Cairo and Alexandria (Zahran, 1976). Percheron (1903) warned against its danger if it spread in the Egyptian canals without control. No one could, at that time, understand his warnings. Thirty years later Simpson (1932) reported that *E. crassipes* was widely distributed in the freshwater canals of the Nile Delta and near Cairo and Alexandria and in the brackish water of the northern lakes, Manzala, Burullus, Idku and Mariut. He showed that the Egyptian environment was suitable for its successful growth and warned that "every year delay makes difficult to clear the water from it". During the last decades and after the establishment of the Aswan High Dam (1965), the growth and distribution of *E. crassipes* and other water weeds are becoming very serious in Egypt. It is difficult to find a canal, stream or drain not infested by *E. crassipes*. In the Nile Delta and Fayium the growth of *E. crassipes* is so dense that the plants are interwoven such that "man can walk on it crossing the threatened canal as on a bridge" (Zahran, 1976).

Apart from the Nile system which is badly infested by water hyacinth, the possibility of the infestation of the High Dam Lake is expected. Kassas (1972a) states "Invasion of High Dam reservoir by water weeds is subject only to their migration efficiency and local conditions of water depth". Also, Tawadrous (1981) noted that it is probable that *E. crassipes* will appear in High Dam Lake as has *Potamogeton pectinatus* which has been recorded from its seasonally inundated shores.

(xii) Family Potamogetonaceae

Potamogeton is the only genus of this family in Egypt and is represented by a group of submerged plants (pondweeds) which grow in fresh and/or brackish waters. Ascherson and Schweinfurth (1889a,b) recorded five *Potamogeton* species and one variety known to occur in Egypt: *P. natans, P. natans* v. *serotinum, P. lucens, P. crispus, P. pusillus* and *P. pectinatus*. Muschler (1912) also noted five species but did not record the variety. Täckholm, Tackholm and Drar (1941) included five species of *Potamogeton* in the *Flora of Egypt: P. nodosus, P. crispus, P. schweinfurthii*, (= *P. lucens*, Boulos, 1999), *P. pectinatus* and *P. panormitanus*. Täckholm (1956) added a further species, *P. perfoliatus*, to the list of 1941. El-Hadidi (1965) recorded *P. trichoides* and gave an explanation of its recent introduction to Egypt. Täckholm (1974 and Boulos, 1995) listed seven *Potamogeton* spp. in Egypt (the list of 1956 together with *P. trichoides*). All are perennial submerged hydrophytes. Except for *P. panormitanus* which is very rare, present only in the oasis region of Egypt, all of the other *Potamogeton* species are present in the Nile system. *P. perfoliatus, P. trichoides* and *P. schweinfurthii* are rare. *P. crispus* is widespread, always choking the canals that feed the fields. It is one of the worst submerged weeds of Egypt. *P. pectinatus* is also common. Both *P. pectinatus* and *P. crispus* can occur in fresh and brackish streams and their dense growth usually takes place during the summer season. *P. nodosus* is the most serious submerged weed in Egypt; its stems are strong enough to impede navigation.

P. crispus is spread chiefly by turions (perennating buds) or rhizomes. There are no records of reproduction by seeds. New growth of this plant usually starts in November after a resting stage, beginning in September. Full vegetative growth occurs from April to August when flowering and fruiting may also be seen. *P. crispus* is very common in the Nile system north of Aswan. It is gradually invading the water bodies south of Aswan, including the High Dam Lake (Springuel, 1987).

The phenology of *P. pectinatus* is similar to that of *P. crispus*. It starts growth in November, full growth being made in April to July. The peak flowering is during April and May. *P. pectinatus* reproduces by tubers and creeping rhizomes, but development from seeds has not been seen. Resting and perennating organs are present from May to November. *P. pectinatus* is common in Egypt throughout the Nile system. It is widely distributed in Nubia and in the northern part of the High Dam Lake (Boulos, 1966b; El-Hadidi, 1976; Springuel, 1987).

New growth of *P. nodosus* is made from November to June. Full growth occurs from April to September. Flowers and fruits are usually seen from May to September. Seeds are produced in substantial numbers, but germinating seeds have not been seen. "For all the three common species of *Potamogeton* seeds are not the means of propagation" (Tawadrous, 1981). *P. nodosus* is common north of Aswan but was collected only once near Kom Ombo and in High Dam Lake. Later surveys (Springuel, 1987) have failed to find it in this lake.

P. perfoliatus was recorded in the Nile system at Nubia even before the establishment of Aswan High Dam and now grows in shallow water along the Nile banks north of the Aswan Dam (Springuel, 1987).

(xiii) Family Ranunculaceae

Four annual hydrophytes of the genus *Ranunculus* have been identified in the Nile system of Egypt: *R. rionii, R. saniculifolius, R. sphaerospermus* and *R. trichophyllus*. All of these species are either rare in the Nile Delta or absent from Upper Egypt. Perennial forms of *R. saniculifolius* and *R. trichophyllus* may occur (Täckholm, 1974). Boulos (1995) recorded 10 species belong to *Ranunculus*, 6 of these occur in the Nile Region of Egypt. These are: *R. peltatus* subsp. *sphaerospermus, R. rionii, R. trichophyllus, R. bulosus, R. sceleratus, R. marginatus* and *R. arvensis*.

(xiv) Family Ruppiaceae

The single genus *Ruppia* is represented in Egypt by one perennial submerged species, *R. maritima*, which occurs in two varieties: v. *spiralis* and v. *rostrata*. The first is very rare in the salt water of the Mediterranean coastal area whereas the second is very common in all types of water in Egypt.

(xv) Family Zannichelliaceae

Zannichellia is a genus of submerged perennial hydrophytes that grows in fresh, brackish and salt water in Egypt. *Z. palustris* is a very common weed, present in three varieties: v. *genuina*, v. *pedicellata* and v. *major. Z. palustris* is widespread in the Nile system, including all water bodies of Nubia and High Dam Lake (Springuel, 1981, 1985a,b, 1987). However, Boulos (1995) recorded *Zannichellia* with only one variety (*Z. palustris* var. *pedicellata*).

(c) The swampy habitat

In the swampy parts of Egypt reeds grow near water usually with their root system and lower parts of their shoot system below the water level. In every case the weeds invade the water either from the banks or from shallow water. Some of these plants are rooted at the normal water level and their stems and branches spread out over the water; they usually cannot spread by shoots arising from creeping stems below the water unless the water is still, very slow or very shallow, e.g. *Agrostis semi-verticillata, Alternanthera achyranthoides, Diplachne fusca, Echinochloa stagnina* and *Jussiaea repens*. Others spread from the bank in a similar manner but their strong stems may develop from submerged creeping systems, some only from rather shallow water, e.g. *Cyperus alopecuroides* and *Typha domingensis*. A third type of swamp plant grows in relatively deep water, e.g. *Cyperus articulatus* and *Phragmites australis*. Simpson (1932) noted that the seeds of probably nearly all swamp plants have to germinate in moist soil or mud, or in very shallow water, but vegetative parts of the plant may grow and found new colonies when entirely submerged.

Phragmites australis is the most serious reed in the Egyptian swamp habitats. It is deep-rooting, of strong growth and spreads rapidly. A community of this weed may overcome *Typha domingensis* which is usually its major competitor in a water channel. *Phragmites* can withstand a higher salt concentration in soil or water than *Typha. P. australis* is a tall perennial grass varying considerably in height; on sandy banks it may be only 50 cm tall whereas in an undisturbed swampy habitat it may be some 4 m above the water. It has an extensive rhizome system rooted in the substratum, but occasionally rhizome-like structures are at the surface and may extend long distances (sometimes up to about 18 m floating on slow-running canals).

P. australis has a wide range of habitat. It grows in fallow land where it gives a low cover and also in brackish swamps where it forms a thick vegetation. The most luxuriant growth of P. *australis* in Egypt is in the drain banks and northern lakes of Egypt: Manzala, Burullus, Idku and Maruit (Shaltout et al., 2004).

Echinochloa stagnina is one of the most troublesome invaders in *Egypt*, both in canals and drains. This perennial grass is sometimes 2 m high when growing within a *Phragmites* community. Creeping rhizomes bear some stems which are prostrate and the plant is entirely glabrous. This grass makes rapid growth and cutting makes it develop even more luxuriantly. The standing crop is generally highest in January, with fresh weights of about 1400 g/m^2 and dry weights of 280 g/m^2 (Serag, 1991).

In some parts of the northern Delta, e.g. in Damietta, *E. stagnina* is cultivated to be used as fodder for livestock as it is reported to have a high sugar content (Simpson, 1932). Owing to the danger of its forming sudd in the irrigation channels, its cultivation, if undertaken at all, should be carefully controlled.

Echinochloa crus-galli grows in fields of rice, maize, vegetables and root crops, i.e. both dry and moist fields. It is a native of central and east Asia but is now of world-wide occurrence. In Egypt, *E. crus-galli* is common in the moist swampy areas and occasional in dry places of the Nile region. It is a noxious weed as it acts as a source of a "yellowish infection" of rice. Its growth reduces yields of maize by up to 80% and it may promote diseases and insect attack in maize and other crops (Abu Ziada, 1986).

Diplachne fusca is a tall (1–1.5 m) glabrous perennial grass with long leaves and branched culms, rooting below. This weed is present everywhere in the canals and drains.

Typha domingensis is a tall marsh herb often of more than 2–2.5 m high, with stout creeping rhizomes. It is a very serious weed which completely chokes many canals. *Typha* (Typhaceae) usually forms a pure stand and is less salt tolerant than *Phragmites*.

Jussiaea repens (Onagraceae) is a creeping floating herb rooting at the nodes. It is a serious invader of waterways in Egypt. Though *Jussiaea* is said to be poisonous to cattle its leaves are eaten by moorhens and snails (Simpson, 1932). It can be spread from the bank for a considerable distance on account of the "floats" (inflated roots) which support the plant. It is very easily distributed by its numerous seeds and by small pieces of the plant that can root in the mud when broken from the stem. *J. repens* is common in the Nile Delta.

Polygonum salicifolium (Polygonaceae) is common everywhere in the Nile region. It is a serious pest in many places but more so in the smaller runnels. It invades the larger ones but fills up the smaller slow-running channels. This slender perennial weed forms dense masses near the water level up to 80 cm tall. Its rootstock is black, woody and creeping; the stems are prostrate at the base, rooting at the nodes but much branched above.

Polygonum senegalense is a common serious weed in the water channels of the Nile system. It is a tall (often 1.5 m), stout, glabrous plant with a very thick swollen stem and broadly lanceolate leaves. It appears to be spreading rapidly. It increases both by seeds and pieces broken from the stem which root in the mud. With its creeping rootstock large clumps are formed. Although a troublesome invader, *P. senegalense* is not found in deep swamps. It completely blocks small channels and restricts streams near the banks encouraging other plants; pieces of weed become entangled in its branches and these may take root and grow. Simpson (1932) states "P. *senegalense* would be a very dangerous weed in the future". This prediction has now become a fact, both in the Nile Delta and Upper Egypt (Zahran and Willis, 2003).

Paspalidium geminatum is an undesirable invading grass in channels, swamps, rice fields and river-shores of the Nile system and is common everywhere. However, *P. obtusifolium* is a rare grass in the Nile Delta where it grows in ditches and rice felds.

Polypogon semiverticillatus (= *P. viridis*, Boulos, 1995) is an annual creeping grass that is a nuisance in irrigating channels in all areas of the Nile system.

Alternanthera repens (Amaranthaceae) is a rather troublesome invader which roots in the banks and spreads some 50–100cm into the water but is rare in the Nile Delta.

Veronica anagallis-aquatica (Scrophulariaceae) is a very common perennial swamp plant in all areas of the Nile system.

Besides the species mentioned above, other plants of the swampy habitats of Egypt include *Alisma gramineum, A. plantago-aquatica, Bacopa monnieri, Carex divisa, Cyperus articulatus, C. difformis, C. longus, C. mundtii, C. polystachyos, Damasonium alisma, Eleocharis caribaea, E. palustris, Fuirena ciliaris, Juncus acutus, J. subulatus, Limosella aquatica, Scirpus litoralis* and S. *supinus*.

(i) Papyrus and lotus: their history and occurrence

Papyrus (*Cyperus papyrus*) was used by ancient Egyptians in paper sheet making whereas lotus (*Nymphaea lotus*) was their sacred flower (Täckholm, 1951). Täckholm and Drar (1950) reported that papyrus became almost extinct in Egypt more than 150 years earlier. The last traveller to notice it was Baroness v. Minutoli at Damietta and the banks of Manzala Lake. After her time, no further record was made and the plant was considered extinct. The few specimens cultivated in the gardens of Cairo and Alexandria are of recent introduction, brought to Egypt from Paris in 1872 (El-Hadidi, 1971). Lotus is confined to the canals and drains of the Nile Delta and Fayium but absent from the Nile Valley (Täckholm, 1956). According to El-Hadidi (1971), the disappearance of papyrus from Egypt and the restriction of lotus to the channels of the Nile Delta may be related to changed conditions which have become unfavourable for their natural growth. During the last 100 years, a permanent "perennial" irrigation system has been introduced and established to replace the classic basin irrigation system known in Egypt for thousands of years. Also, the construction of dams and barrages and the development of drainage systems have resulted in the drying up and shrinkage of numerous ponds and swamps that existed along the Nile and that were associated with the ancient basin irrigation system. Such ponds and swamps were the natural habitat of papyrus and lotus of ancient Egypt.

Not very many years ago (1963–1966) a few stands of lotus (*Nymp-haea lotus* v. *aegyptiaca*) were discovered in the Nile Valley region (Beni Suef, about 100km south of Cairo, Fig. 2.1) growing in small drains by El-Hadidi (1971). Other species recorded in these stands include *Arundo donax, Ceratophyllum demersum, Cyperus articulatus, Potamogeton crispus* and *Typha domingensis*. Also, in July 1968, the same author discovered a stand of about 20 plants *of Cyperus papyrus* among other reeds in a fresh-water marsh close to Um Risha Lake of the Wadi El-Natrun Depression. *C. papyrus* was growing safely and well protected among the "reeds" of *Typha elephantina, T. domingensis* and *Phragmites australis*. Other plants of this stand are *Berula erecta, Fuirena pubescens, Panicum repens* and *Scirpus litoralis* v. *subulatus*. This locality of papyrus seems to be the only one known in Egypt. However, dense stands dominated by *C. papyrus* have been discoverd by both Serag (2000) and Hussein (2000) in the downstream section of the Damietta Branch as well as in two islands of the River Nile in the area of Cairo.

(d) Canal bank habitat

The plants of the canal bank habitat may be categorized into three types: (a) bank retainers; (b) "aggressive" species (smotherers); and (c) sand controllers.

The bank retainers are plants having bankholding qualities. In addition to the fact that their roots bind the soil they shade out other species which may be harmful. Plants in this category are cultivated trees and shrubs: *Acacia nilotica, Ficus sycomorus, Melia azedarach, Morus alba, M. nigra, Parkinsonian aculeata, Salix safsaf, Tamarix arborea* and *Ziziphus spina-christi*, and undershrubs, herbs and grasses e.g. *Alhagi maurorum, Arthrocnemum glaucum, Arundo donax, Chenopodium ambrosioides, Conyza dioscoridis, Cynodon dactylon, Desmostachya bipinnata, Imperata cylindrica, Panicum maximum* and *Suaeda vermiculata*. The manner in which these plants retain the soil differs in the various species. Some have an intricate root system forming an intertwining network that holds the soil together, being effective because the dense clumps prevent disturbance of the surface soil.

"Aggressive" species (smotherers) are those which make such rapid and robust growth that they prevent many smaller and more slow-growing plants, including weed species, from establishment. Quick growing species which form mats, either perennials, e.g. *Cyperus laevigatus* and *Lippia nodiflora*, or annuals, e.g. *Trifolium resupinatum*, soon cover patches of bare soil at water level and stifle any seedlings that may germinate. Many potential invaders succumb at the most vulnerable seedling stage and any surviving are eliminated by the rapidly growing dense species. Competition operates through the large root system and by shading of the leaves so that they produce inadequate photosynthates. Apart from the previously mentioned species, *Canna indica, Inula crithmoides, Saccharum spontaneum* v. *aegyptiacum, Silybum marianum* and *Sphaeranthus suaveolens* are also vigorous and dense growing plants on salt soils.

The sand controllers are plants that can tolerate and at least partly stabilize drift sand. The efficient windbreak trees, shrubs and other perennials include *Arundo donax, Casuarina equisetifolia, Dalbergia sisso. Eucalyptus citriodora, E. rostrata, Parkinsonia aculeata, Ricinus communis, Salix babylonica, S. safsaf* and *Tamarix aphylla*. Some plants of this group spread when once established, either by seeds, e.g. *Dalbergia sisso* and *Ricinus communis*, or by a creeping underground system, e.g. *Arundo donax*. Some species are especially useful in trapping sand in salt marsh areas, e.g. *Arthrocnemum glaucum, Halimione portulacoides, Inula crithmoides, Juncus rigidus, Nitraria retusa* and *Sporobolus spicatus. Opuntia ficus-indica*, a fence plant, is also a sand controller.

The flora of the canal banks may also include *Ambrosia maritima, Andropogon annulatus, Coronopus niloticus, Eclipta alba, Ethulia conyzoides, Glinus lotoides, Gnaphalium pulvinatum, Potentilla supina, Urospermum picroides* and *Verbena supina*.

Kochia indica (= *Bassia indica*) is a richly branched herb which is one of the very common canal bank species, especially in the Nile Delta and Fayium. It is extending southwards in the Nile Valley. *K. indica* is salt tolerant and drought resistant; as it is of rich nutritive value, it can be used as green or dry fodder for livestock (Zahran, 1986; Zahran, Muhammed, and El-Dingawi, 1992).

(i) Vegetation succession of the canal banks

The banks of the canals and drains in Egypt are usually cleared of weeds once or twice a year. Soon after clearance the weeds start to appear again. The first to show are usually those with deep creeping, underground parts, as many of these escape the clearing operation. These are usually grasses such as *Cynodon dactylon, Desmostachya bipinnata, Imperata cylindrica, Phragmites australis* and *Saccharum spontaneum*. New growth soon appears from the stumps of *Conyza dioscoridis, Salix safsaf* and *Tamarix aphylla*. Seedlings also appear either from seeds carried by the wind or by currents. Among species with windborne seeds are the following: *Conyza dioscoridis, Imperata cylindrica, Inula crithmoides, Phragmites australis* and *Typha domingensis*.

As soon as the vegetation reaches the water level, pieces of weed are caught up in it and may take root in the bare soil. Such weeds as *Echinochloa stagnina, Panicum repens* and *Paspalidium geminatum* may get a hold in this way.

The vegetation on the banks arises from the banks themselves, from windborne seeds and from the water. The first source cannot be controlled but the invasion from windborne seed can be reduced by clearing areas of the seed sources. There is competition between the plants already on the banks and those reaching the banks. High on the bank *Cynodon dactylon* is outcompeted by *Arundo donax* which, in turn, may be replaced by *Imperata cylindrica* and *Desmostachya bipinnata* or perhaps by *Saccharum spontaneum* which will usually occupy all zones unless it comes into competition with well established bushes of *Tamarix aphylla. Arundo donax*, if no *Saccharum* is present, may be replaced at the top of the bank by *Imperata* and/or *Desmostachya* and occur lower on the bank. In the lower zone it may be in competition with *Tamarix* and *Conyza* which remain successful but it may dominate between the bushes. If *Sphaeranthus suaveolens* is already established, *Arundo* may die out. Weeds in the lower zones, being near the water level, will get a strong hold unless some of the more vigorously growing plants are there to check them. Many of the weeds are quicker growing than the useful species, especially if such plants as *Conyza* and *Tamarix* have to develop from seedlings. Thus, the advantage of having a "smotherer" plant there, such as *Sphaeranthus*, is obvious. Cutting the weeds or pulling them will promote the "smotherer" plants enormously in these early stages and will substantially reduce the work to be done later. The "smotherer" plants and bank retainers are then likely to become quickly dominant. If weed species dominate at an early stage in the lowest zone nothing can be done except to cut them at regular intervals. If, however, the "smotherer" plants get a hold before the marginal weeds, the aquatics have less chance of being caught up near the bank. Furthermore, if the bank retainers get an early start, the danger from bankslip is reduced and the floating pieces of weed have less chance to gain a hold.

Bankslip should be very carefully avoided for this is the most difficult zone for weeds to invade. *Conyza* and *Tamarix* help little here but *Sphaeranthus* is one of the few plants which is of use.

The vegetation of the banks is subject to frequent changes as *Tamarix* is cut as fuel, *Diplachne* is cut for fodder, animals will graze on some weeds and others will be taken for making mats or shelters.

(e) The cultivated lands

(i) General survey

The cultivated lands of Egypt, formed since prehistoric times, belong phytogeographically and edaphically to more or less distinct regions. The largest part of these lands is in the Nile Region[4] which owes its existence to the alluvial deposits of the Nile. This region extends from the south of the Nile Valley (Nv subregion) and northwards as the Nile Delta (Nd subregion), the Mediterranean element influencing the vegetation of the latter subregion. The western expansion of the Nile Region, the Fayium as Nile Fayium (Nf subregion) was originally a marshy depression in the Western Desert and had been cultivated for a very long time. The Nf subregion still contains marsh vegetation. Irrigation of the cultivated lands of the Nile Region ($<25{,}000\,km^2$) is by the Nile water using the perennial irrigation system.

Two crops are grown in a seasonal sequence: a summer and a winter crop. Among each set of crops, there is at least a cereal and a leguminous or an oil crop. The weeds of these lands are generally annual of either summer or winter growth.

It follows that a crop rotation is accompanied by a weed rotation (El-Hadidi and Kosinova, 1971).

Other parts of the cultivated lands of Egypt depend on rain or underground water resources. Rain-fed agriculture in Egypt is restricted to the winter and spring months and is confined to a narrow strip of land (5–25 km wide) that runs parallel to the Mediterranean coast which includes the Mareoticus, deltaic and Sinai sections. The underground water, however, provides a permanent supply for irrigation in the oases and depressions of the Western Desert. The flora of the cultivated lands of these oases and depressions contains, apart from their characteristic elements, certain weeds and halophytes introduced with new crops (El-Hadidi and Kosinova, 1971).

Rainfall and/or underground water are the source of irrigation in a group of small oases among the mountainous blocks of the Sinai Peninsula. The flora of the fields of these oases includes several endemics, together with native species and aliens (El-Hadidi et al., 1970).

The high rate of population increase in Egypt required the expansion of the cultivated land. Reclamation of desert plains took place along the Nile Region in the

[4] Accordingly this section is included in the Nile Region in this book.

southern section and on both sides of the Nile Delta. Land reclamation is taking place also in the oases and depressions of the Western Desert.

(ii) Weed vegetation

The weeds of the cultivated lands of Egypt are mainly short-lived (ephemerals, annuals and biennials) herbs. Perennial herbs, under-shrubs and shrubs may also be present in certain neglected areas where the soil is affected by salt (with halophytes), is swampy (with helophytes) or is dry (with xerophytes).

The most common crop in permanently irrigated land during winter is wheat (*Triticum vulgare*). Other important crops include broad beans (*Vicia faba*) and Egyptian clover (*Trifolium alexandrinum*). The coastal land along the western Mediterranean coast is characterized by the rain-fed cultivation of barley (*Hordeum vulgare*). During summer the main local cereal crop is maize (*Zea mays*). This is replaced in the southern provinces of Upper Egypt and the oases by the warmth-requiring millet (*Sorghum durra*). It is customary to have an associate crop in the same field, namely cow-pea (*Vigna sinensis*).

Cotton (*Gossypium barbadense*) is cultivated on a larger scale in the delta and in the northern provinces of the Nile Valley. It is also cultivated in some newly reclaimed areas of Tehrir Province (west of the Delta), New Nubia (Kom Ombo area) and New Valley (Kharga and Dakhla Oases) (El-Hadidi and Kosinova, 1971).

Rice (*Oryza sativa*) is cultivated on a large scale in Lower Egypt whereas sugar cane (*Saccharum officinarum*) is widely cultivated in Upper Egypt for the production of sugar. S. *officinarum* remains on the land for several years. In the Nile Delta, Fayium and northern provinces of the Nile Valley, the varieties of sugar cane cultivated are only for juice production.

Weeds of very common occurrence in the cultivated lands of winter crops in all regions of Egypt are *Anagallis arvensis, Brassica nigra, Chenopodium album, C. murale, Convolvulus arvensis, Cynodon dactylon, Melilotus indica, Polypogon monspeliensis, Sonchus oleraceus* and *Trifolium resupinatum*. Most of these species are natives of the Mediterranean region, whereas *Cynodon, Sonchus* and *Chenopodium album* are often found in the warm temperate regions of the world. These three weeds tend to continue their growth during the summer. New seedlings of *Sonchus* and *Chenopodium* appear during early summer whereas *Cynodon* persists throughout the whole year.

The very common weeds in summer crops include *Amaranthus angustifolius, A. ascendens, Convolvulus arvensis, Corchorus olitorius, Cynodon dactylon, Portulaca oleracea, Solanum nigrum* and *Sonchus oleraceus. Cynodon, Convolvulus, Solanum* and *Sonchus* may also be present in the winter months, although less common. The others are obligate summer weeds.

Less common weeds of winter crops are *Emex spinosus, Malva parviflora, Polygonum equisetiforme, Solanum nigrum* and *Vicia calcarata*. Less common weeds of summer crops include *Cichorium pumilum, Dactyloctenium aegyptium, Eragrostis pilosa, Gynandropsis gynandra, Urochloa reptans* and *Xanthium brasilicum*.

Among the, winter weeds, *Silene rubella* is rare, but there are more rare summer weeds, e.g. *Amaranthus chlorostachys, Ammi majus, Avena fatua, Beta vulgaris, Brassica nigra, Lolium perenne, Lotus corniculatus, Malva parviflora, Panicum repens, Plantago lagopus, Polypogon monspeliensis, Reichardia orientalis, Rumex dentatus* and *Xanthium spinosum*. Three of these weeds, *Amaranthus, Panicum* and *Xanthium*, are obligate summer weeds.

During winter, a group of plants may be recognized as characteristic of the barley cultivation of the Western Mediterranean coastal area; these include *Achillea santolina, Adonis dentata, Anacyclus glomerata, Bupleurum subovatum, Carduus getulus, Centaurea alexandrina, Cutandia dichotoma, Echium sericeum, Eryngium creticum, Filago spathulata, Herniaria hemistemon, Koeleria phleoides, Lathyrus pseudocicera, Linaria haelava, Onobrychis crista-galli, Peganum harmala, Plantago albicans, Reseda decursiva, Roemeria hybrida, Salvia lanigera, Scorzonera alexandrina* and *Trigonella maritima*. During the summer, a few stands of the western Mediterranean newly reclaimed and cultivated areas were investigated by El-Hadidi and Kosinova (1971) where permanent irrigation was introduced, and where other crops (maize and cotton) were also tried. New species started to appear, e.g. *Astragalus annularis, Enarthrocarpus strangulatus, Erucaria microcarpa, Lotus creticus, Matricaria aurea* and *Sphenopus divaricatus*.

The extensive increase in recent years of the North American *Parthenium hysterophorus* is a very striking example of the important impact of the introduction of aliens, the present population of this plant in Egypt apparently arising from a single sowing in 1960 of impure grass seed imported from Texas (Boulos and El-Hadidi, 1984).

The weeds of the cultivated lands of the western Mediterranean coastal belt include a group of halophytes and helophytes which also occur in similar habitats within other regions, e.g. Nile Fayium and Oases. These plants include *Aeluropus lagopoides, Centaurium pulchellum, Cressa cretica, Cyperus laevigatus, Diplachne malabarica, Imperata cylindrica, Kochia indica, Phragmites australis, Schang-inia baccata, Scirpus tuberosus, Spergularia marina* and *Typha domingensis*.

The Nile Delta is the site of a few species not known in other parts of Egypt. *Melilotus siculus, Senecio aegyptius*, and *Setaria verticillata* are recorded for the winter.

The rice fields, mainly in the Nile Delta, are characterized by a weed flora many of which are helophytes, e.g. *Carex divisa, Cyperus alopecuroides* (*C. dives*), *C. articulatus, C. difformis, C. longus, Diplachne fusca, Echinochloa colona, E. crus-galli, E. stagnina, Foeniculum vulgare, Lythrum junceum, Paspalidium obtusifolium, Polygonum lanigerum, Scirpus fistulosus* and *S. supinus*. In the newly reclaimed cultivated lands of the Nile Delta, other species have been recorded, e.g. *Andrachne racemosa, Physalis angulata, Striga asiatica* and *Tagetes minuta*.

The weed flora of the cultivated lands in the oases is a mixture of xerophytes, mesophytes, halophytes and helophytes, e.g. *Aeluropus lagopoides, Asphodelus tenuifolius, Bassia muricata, Boerhavia diandra, Citrullus colocynthis, Corchorus tridens, Cressa cretica, Fagonia indica, Haplophyllum longifolium, Heliotropium supinum, Hyoscyamus muticus, Kochia indica, Lagonychium farctum, Launaea cassiniana, Linaria aegyptiaca, Pulicaria arabiea, Salsola baryosma, Sckouwia thebaica, Scirpus tuberosus, Sporobolus spicatus, Tribulus longipetalus, Typha domingensis* and *Withania somnifera.*

Weeds of general occurrence in all phytogeographical regions of Egypt must have spread to the remote oases through the continuous introduction of new crops; such weeds as *Lathyrus hirsutus, Thesium humile* and *Trachynia distachya* are common to both the Mediterranean coastal area and the oases. This may be attributed to the introduction of barley from the Mariut district to the oases, e.g. Siwa Oasis. This group includes also *Astragalus corrugatus, Erodium malacoides, Phalaris paradoxa, Plantago pumila, Silene nocturna, S. villosa* and *Vaccaria pyramidata.*

The results of cultivation in the newly reclaimed areas of the New Valley show several interesting features in respect of invasion by weeds. In the recently cultivated areas, which were completely weed-free, few weeds, e.g. *Amaranthus paniculatus, Brassica nigra, Cynodon dactylon, Euphorbia aegyptiaca* and *Medicago hispida*, were recorded. This may indicate that *Cynodon* is one of the early invaders of cultivated land. *Brassica* and *Medicago* are winter weeds introduced to this area during its first cultivation. On the other hand, the record of *Amaranthus paniculatus* is the first in the whole oasis region of Egypt (El-Hadidi and Kosinova, 1971; Täckholm, 1956, 1974). In addition, the newly reclaimed locality of Dakhla Oasis supports some xerophytes, e.g. *Bassia muricata, Salsola baryosma, Schouwia thebaica* and *Tephrosia apollinea*, common desert plants and likely to occur here. The root parasite *Striga hermonthica* was recorded for the first time in the oases of Egypt in a maize field in a recently reclaimed area at Wadi El-Natrun. Broad bean seeds which were cultivated in the oases region were subjected to special treatment to destroy any accompanying *Orobanche* seeds.

(f) The northern lakes

The northern lakes of the Nile Delta, namely Lake Manzala, Lake Burullus and Lake Idku (Fig. 6.3) are very close to the Mediterranean Sea. They are separated from it by strips of land that are very narrow in several places and are connected with the sea through narrow outlets (straits). These straits are either remnants of the mouths of old deltaic branches or merely gaps in weak sections of the bars known as tidal inlets (Abu Al-Izz, 1971). The areas of these northern lakes are affected by several factors, such as the continued degradation and deposition, the accumulation of the remains of vegetation, the blowing of sand and man-made desiccation, e.g. closing of some irrigation canals and construction of levees. All these factors and others have caused a decrease in the areas of these lakes which receive the major part of the agricultural drainage water of the Nile Delta. According to Al-Kholy (1972), the areas of these lakes have decreased since 1799 as shown in Table 6.2. However, further and extensive reduction in these lakes is serious affecting their abiotic and biotic composition. For example, according to Shaltout and Khalil (2005), the area the Lake Burullus decreased from 1092 km^2 in 1801 to 55.65 km^2 in 1913 (i.e. 49% reduction), to 502.7 km^2 in 1972 (54% reduction) to 401 km^2 in 1984 (59.7% reduction) and to 410 km^2 in 1997 (62.5% reduction).

The ecology of one of these lakes (Lake Manzala) is described.

Table 6.2 Areas of lakes of the Nile Delta (After Al-Kholy, 1972)

	Area in feddans (= 4200 m^2)				
Year	1799	1889	1912	1956	1970
Lake Manzala	470,000	460,000	410,000	326,000	300,000
Lake Burullus	270,000	180,000	140,000	136,000	130,000
Lake Idku	80,000	80,000	45,000	33,000	17,000

(i) Lake Manzala

Environmental Characteristics

Lake Manzala is the largest of the northern deltaic lakes of Egypt. It is in the northern quadrant of the delta between the Mediterranean Sea to the north, the Suez Canal to the east, the Damietta Branch of the Nile and the provinces of Sharkiya and Dakahliya to the west. Thus, Lake Manzala serves five provinces, namely: Damietta, Port Said, Ismailia, Sharkiya and Dakahliya (Fig. 6.3). According to Abu Al-Izz (1971) and Al-Kholy (1972), after Andreossy (1799), one of Napoleon's senior Generals during the Egyptian Campaign, the longest dimension of Lake Manzala was that joining Damietta on its west to the Pelusiac Branch of the Nile on its east, a distance of about 84 km. Its shortest dimension was that joining Matariya on the south and the Mediterranean Sea on the north, a distance of about 17 km. The area of Laka Manzala during that time was 470,000 feddans. At present Lake Manzala is about 47 km long and 30 km wide. It narrows in the middle to only 17 km and its area is about 300,000 feddans. Lake Manzala, like the other deltaic lakes, is shallow, the depth ranging between 0.7 m and 1.5 m. The depth increases considerably in the navigable Manzala Canal. The coasts are irregular and indented.

Lake Manzala is not of maritime origin, having no relation to the Mediterranean Sea in its formation. It is believed to have been formed as a result of the accumulation of the Nilotic water in the low area of the northeastern delta. Earthquakes caused this land to subside at the end of the sixteenth century and the sea crossed to the sandy barriers (Abu Al-Izz, 1971). Nile water thus has been mixed with the sea water which rushes to the lake through the action of the north-east winds which prevail in this locality. This is affirmed by the presence of silt mixed with sand at the bottom of the lake and by the water being slightly brackish.

Up to the seventh century, Lake Manzala (formerly known as Lake Tanis) was traversed by the Pelusiac, Tanitic and Mendisy branches of the Nile Delta. Herodotus mentioned that the Fatmetic Branch (now the Damietta Branch) of the Nile had increased in width at the expense of the Pelusiac and Tanitic. These two branches became blocked at their mouths through the accumulation of silt and sand by the action of the north-east winds which forced their waters back into the Nile proper. In that way their water reinforced that of the neighbouring branch, the Fatmetic.

Originally Lake Manzala was connected to the Mediterranean Sea by two main openings which represented the mouth of the Mendisy and Tanitic branches of the

Nile. Between these mouths, there had been a smaller opening which Herodotus described as a false one. Now Lake Manzala is joined to the Mediterranean Sea by Strait Ashtum El-Gamil, which was the outlet of the Tanitic Branch. In addition to this strait there are other openings through which the lake reaches the sea.

The southern coast of Lake Manzala has many inlets through which water drains into it. “The flow of drainage water into the lake diminishes the salinity which ranges between 0.8% and 1%” (Abu Al-Izz, 1971). The salinity of the Mediterranean Sea is between 3.3% and 3.9%.

Lake Manzala is characterized by a large number (1022) of islands. Some of these are clayey and extend from NE to SW. Others, oriented from NW to SE, are just bars of sand. Still others are composed of mollusc shells. Some of these islands have an area of 62–150 feddans; these are clayey ones, e.g. Ibn Salam, Kom El-Dahab and Gassa. Such islands were populated in former times and had an interesting history. The sandy islands are generally smaller than the clay ones and vary considerably in area, e.g. Aggag, Sameriat and Gamil. Those formed of mollusc shells are usually very small and irregular, e.g. Ghomein, Haggar, Hatab and Khara. However, a few of these islands are large, their areas reaching hundreds of feddans, e.g. Rami, Hammar and Sorgan (Montasir, 1937; Abu Al-Izz, 1971; Al-Kholy, 1972; Zahran et al., 1989).

As in other shallow lakes, rough waves do not occur in Lake Manzala. The waters are more or less calm except when winds blow and force the waters back, leaving the shores uncovered. The shores are not subject equally to this effect because such winds are mostly northeast and sometimes east; winds rarely blow from other directions. For this reason the differences in the density of vegetation on the different shores is easily observed. Some shores are densely covered with vegetation while others are more or less bare. High and low tides are not known to occur in Lake Manzala (Montasir, 1937).

The water of Lake Manzala was less saline than that of the sea and was used for drinking during times of flood. But now and after the establishment of the Aswan High Dam (1965) no flooding occurs and the salinity of the lake is increased (Al-Kholy, 1972). The temperature of the water of Lake Manzala varies from 25–30 °C in summer to 10–20°C in winter. It is almost uniform with depth owing to the shallowness of the lake (warm currents are not known).

The climate affecting Lake Manzala is generally arid. The absolute minimum temperature ranges between 10°C in January and 23.7°C in August whereas the absolute maximum temperature ranges between 19°C in January and 33.3°C in August. Relative humidity is very similar throughout the year, the air being humid all the year round; relative humidity is rarely less than 70% or more than 80%. The rainfall of Lake Manzala ranges between 47 mm/year and 88 mm/year; the rainy months are usually November, December, January and February. The total rainfall in any of these months varies between 5 mm and 20 mm and rarely amounts to 30 mm. Wind velo-city is almost uniform throughout the year; usually there is a gentle breeze. The wind velocity normally ranges between 10 and 20 km/h.

Vegetation

General Features

There are very few published studies on the vegetation of Lake Manzala. Muschler (1912) described some halophytes from the lake. Stocker (1928) recorded the water content and salinity of the soil of the halophytic vegetation near Port Said north-east of the lake. However, the ecology of the lake is reported by Montasir (1937), who states that the vegetation of Lake Manzala includes halophytic and helophytic species growing mainly on shores and islands of the lake. Included are 27 species as follows: *Arthrocnemum glaucum (A. macrostachyum), Arundo donax, Atriplex farinosa, Cistanche phelypaea, Cressa cretica, Cyperus laevigatus, Halimione portulacoides, Halocnemum strobilaceum, Halopeplis perfoliata. Inula crithmoides, Juncus acutus, J. rigidus, Limoniastrum monopetalum, Limonium delicatulum, Phragmites australis, Salicornia fruticosa, S. herbacea, Salsola kali, S. longifolia, Sporobolus spicatus, Suaeda pruinosa, S. salsa, S. vera, S. vermiculata, Tamarix aphylla, Typha domingensis* and *Zygophyllum aegyptium*. Ten of these species were dominants; these are *P. australis, A. macrostachyum, C. cretica, H. strobilaceum, H. portulacoides, I. crithmoides, J. acutus, J. rigidus, S. fruticosa* and Z. *aegyptium*. According to Zahran et al. (1989), i.e. more than 50 years after the records of Montasir (1937), seven of these species listed are still dominant. *I. crithmoides* and *C. cretica* are nowadays recorded as common associates but S. *fruticosa* is absent. Further, five species, *Atriplex farinosa, Halopeplis perfoliata, Limoniastrum monopetalum, Salicornia herbacea* and *Suaeda vermiculata*, which were included in the floristic list of Montasir (1937), are absent.

According to Montasir (1937), the hydrophytes commonly growing in the water of Lake Manzala include *Ceratophyllum demersum, Eichhornia crasssipes, Lemna* spp. and *Potamogeton crispus*. However Khedr (1989) states that apart from the dominant "reeds" (*Phragmites australis* and *Typha domingensis*), the water habitat of Lake Manzala is characterized by five dominant hydrophytes (*Eichhornia crassipes, Jussiaea repens, Najas armata, Potamogeton pectinatus* and *Ruppia maritima*) and four associates (*Ceratophyllum demersum, Lemna gibba, L. minor* and *Potamogeton crispus*). *Chara* sp. is also commonly present.

The Communities

According to Zahran et al. (1989), the islands of Lake Manzala are characterized by tracts of land which are converted into salinas as a result of the evaporation of water seeped from the lake. The level of the underground water and the soil properties seem to be the major limiting ecological factors. The vegetation of these islands, essentially halophytic, is described in seven communities dominated by *Phragmites australis, Juncus acutus, J. rigidus, Arthrocnemum macrostachyum, Atriplex portulacoides, Halocnemum strobilaceum* and *Zygophyllum aegyptium.*

Phragmites australis is widespread and its community occurs in all islands where it forms dense thickets along shore-lines with surface deposits of sand and silt. The plant cover ranges between 50% and 100%, contributed mainly by the dominant reed. *Inula crithmoides* and *Arthrocnemum macrostachyum* are common associates. Less common are *Atriplex portulacoides, Carex extensa, J. acutus* and *J. rigidus.*

Juncus acutus is common on all islands and its community usually occupies the first, low-level, zone of the sedge-meadow habitat, close to the reed zone, where the soil is wet, dark and slippery with thin salt crusts on the surface. The cover of this community is high: up to 100%. The common associates are *A. macrostachyum, I. crithmoides, J. rigidus* and *P. australis*. Other associates include *Atriplex portulacoides, Cressa cretica, Conyza dioscoridis, Cyperus laevigatus, Juncus bufonius, Paspalidium geminatum, Scirpus tuberosus, Suaeda pruinosa, S. salsa* and *Tamarix nilotica.*

Juncus rigidus dominates a widespread community in the second landward zone of the sedge-meadow with a relatively low moisture content. *Phragmites australis* and *Arthrocnemum macrostachyum* are the most common associates. Less common ones are *Atriplex portulacoides, Halocnemum strobilaceum, Inula crithmoides, Juncus acutus, Sporobolus virginicus* and *Suaeda pruinosa.*

Arthrocnemum macrostachyum is a common succulent halophyte and its community is a prominent feature of the vegetation of Lake Manzala Islands. It grows in patches covering the high sandy habitat amidst low areas where seeped water is accumulated. The zone dominated by *A macrostachyum* precedes upslope the zone *of Atriplex portulacoides* and follows that of *J. rigidus*. The plant cover of this community is generally open (50–70%), contributed mainly by the dominant. Common associates are *A. portulacoides, H. strobilaceum, I. crithmoides, J. rigidus* and *P. australis*. Other associates include *Cressa cretica, Cyperus laevigatus, Frankenia hirsuta, Juncus acutus, Scirpus tuberosus, Suaeda pruinosa, S. salsa* and *Tamarix nilotica*.

Atriplex portulacoides is widespread and its community occurs on the dry parts of the salt lands next to that of *Arthrocnemum* with cover ranging between 50% and 90%. *A. macrostachyum* is closely associated and may be considered as co-dominant in some stands. *P. australis* and *I. crithmoides* are common associates. This community is a mixed one as it includes halophytes, helophytes and psammophytes and this reflects the heterogeneity of the habitat conditions. The flora of this community includes also *Frankenia hirsuta, H. strobilaceum, Inula crithmoides, Kochia indica, Sporobolus virginicus, Tamarix nilotica* and *Zygophyllum aegyptium.*

Halocnemum strobilaceum produces the greatest part of the community cover, which is often dense (average 70%). It forms large patches with *A. macrostachyum* as the most common associate. *P. australis* and *A. portulacoides* are common. Less common associates include *Cressa cretica, Frankenia hirsuta, J. rigidus*, and *Z. aegyptium.*

Zygophyllum aegyptium is not common. It dominates the sandy islands, and may occur as a less common associate in habitats covered with sandy sheets in only two communities dominated by *A. portulacoides* and *H. strobilaceum*. It usually occupies the most landward zone in the littoral zonation. Higher up the land is barren. The plant cover of this community ranges between 30% and 60%, made up mostly by *Z. aegyptium. A. macrostachyum* and *H. strobilaceum* are common. Less common associates include A *portulacoides, Frankenia hirsuta, Mesembryanthemum nodiflorum, Phragmites australis*, and *Suaeda pruinosa*.

(g) The artificial lakes

In the most southern part of Egypt, south of Aswan, the Nile flows through a hilly desert region, the "Cataract Region" with a very dry, almost rainless, climate. The First Cataract starts 7 km upstream from Aswan. The cataract itself is a series of rapids resulting from the channel of the river being locally obstructed by a number of rocky islands (Said, 1981).

The valley upstream of the First Cataract has been converted into a reservoir by the construction of the first Aswan Dam (1902) across the Nile at the head of the cataract. The dam is a giant structure with a length of 2 km. It was designed to hold water to relative level 106 m. It has been subsequently raised twice (1912 and 1934); the level of the water during 1935 was at 121 m and the reservoir (artificial lake) was extended south for 25 km (Springuel, 1985a). The construction of the Aswan High Dam during the period 1959–1965 resulted in the formation of a great artificial lake and bounded the old reservoir between the two dams (Figs. 6.5 and 6.6). Accordingly, the River Nile in the most southern part of Egypt is characterized by two artificial lakes (reservoirs): (I) High Dam Lake and (II) Reservoir bounded by the two dams.

(I) High Dam Lake

A. General Characteristics

The High Dam Lake is within Lat. 22 °N and 23°58'N in Egypt. It extends southwards into the Sudan, nearly to 20 °N as Lake Nubia (Fig. 6.5). The lake as a whole is surrounded by rocky terrain, chiefly piedmonts and peneplains of sandstone (Raheja, 1973). The entire reservoir in Ègypt and Sudan has a gross capacity of 157,000 m^3. The river bed at which the High Dam has been constructed is 99 m above sea level. The mean depth of the lake at 180 m above sea level is expected to be 24.9 m and its mean width at the same level is about 19.9 km. The shore-line at the same level is expected to be 7875 km.

The ratio of the shore-line to the length of the entire reservoir is 18:1. The length of the eastern shore-line is almost double that of the western shore-line. Some of the arms (also called khors or lagoons) of the lake are over 50 km long, e.g. Khor Allaqi is about 80 km. There are 85 major khors, of which 48 are on the eastern shore and 37 on the western. The shore-line length of these khors is 969.9 km, of which 576.3 km are to the east and 393.6 km to the west (Raheja, 1973).

Shore-line morphology is primarily dependent on both erosion and deposition rates. The most important factor affecting the shore-line morphology is the sedimentation of the huge amounts of silt carried by the river during flood. Changes of shore-line morphology are expected, to occur at a faster rate during the filling stage of the reservoir than later, when the reservoir has reached its maximum holding capacity.

The water level of the High Dam Lake was expected to reach its maximum (180 m above sea level) in 1980 but this did not occur as since 1978 the level started gradually to decrease. During 1982 the level ranged between 170 m and 175 m above

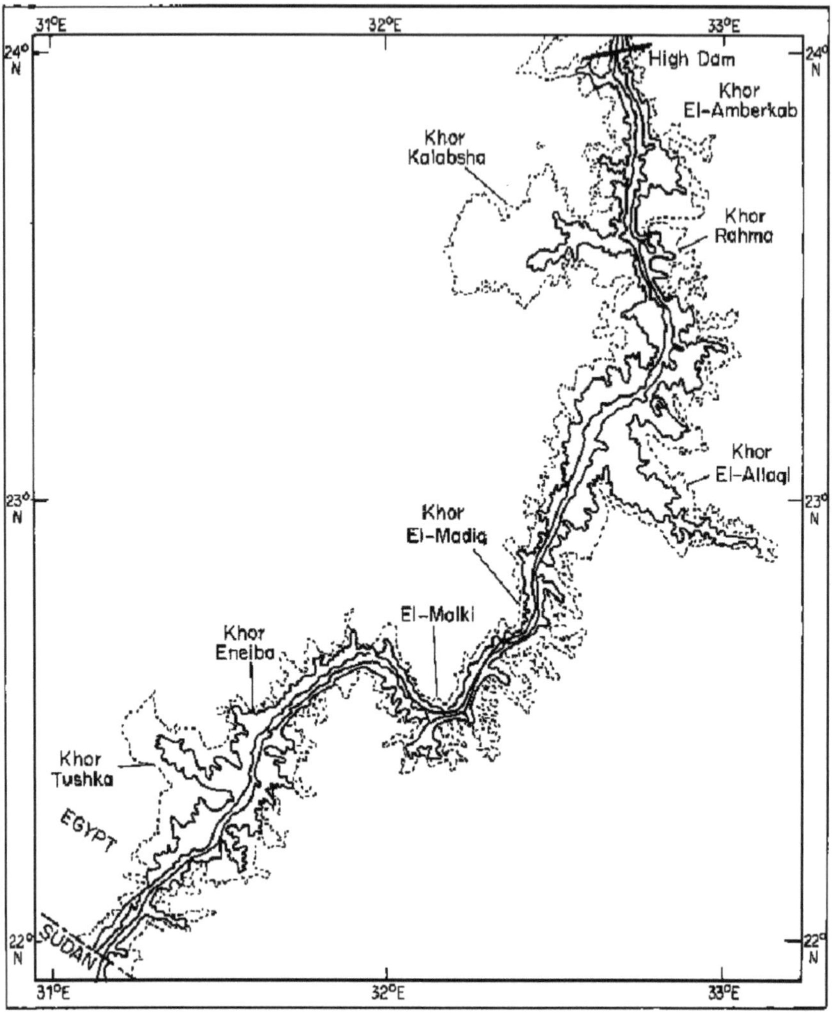

Fig. 6.5 The artificial lake south of Aswan High Dam on the Nile. *Continuous lines* show the margin of the lake; *dotted lines* the extent of the khore

sea level. Evaporation and mean annual discharge through floodgates and turbines result in an annual variation in water level of about 5 m (Springuel, 1985b).

The entire reservoir of Aswan High Dam lies in the extremely arid part of Egypt and the Sudan. Occasional showers may occur during any season of the year, an incident that may happen every few years.

B. Vegetation Types

The vegetation types associated with the High Dam Lake are presented here under two main topics: a. vegetation of the khors and b. shore-line vegetation.

a. Vegetation of the Khors

Khor El-Amberkab and Khor Rahma are on the eastern side of the lake, 40 km and 50 km south of Aswan High Dam respectively (Fig. 6.5). The winding nature of the shore-lines of the khors is reflected in the existence of many peninsulas. The elevated parts within the khors which are not inundated form numerous islands. Floristically, the shore-lines of both khors do not show a great diversity: only 16 species of terrestrial flowering plants and four hydrophytes are recorded. On the steep rocky slopes the perennial species frequently present are *Hyoscyamus muticus, Phragmites australis* and *Tamarix nilotica*. Less common and rare perennials include *Cynodon dactylon, Imperata cylindrica, Pulicaria undulata* and *Salsola baryosma. Glinus lotoides* is the abundant annual which, with some other annuals, grows in the crevices between the rocks and along the low banks near the water. These include *Amaranthus blitoides, Chenopodium album, Crypsis aculeata, Heliotropium ovalifolium, H. supinum* and *Rumex dentatus. Najas armata* and *N. minor* are abundant hydrophytes in the shallow water of these khors. Other aquatics are *Potamogeton nodosus* and *P. trichoides* (Springuel, 1985a).

Khor Kalabsha is the largest on the west side of the lake. Its shoreline is less rocky than shores of the eastern side. The silt and sand deposits form a fertile layer along the shore-line and could be used for cultivation. In this khor, 29 species are recorded: 22 terrestrial along the shore-line and seven aquatic in the shallow water. *Hyoscyamus muticus, Phragmites australis* and *Tamarix nilotica* are frequent perennials whereas *Glinus lotoides* and *Heliotropium supinus* are frequent annuals. *Calotropis procera, Crypsis alopecuroides* and *Leptadenia arborea* are recorded only from Khor Kalabsha. Other associates include *Astragalus vogelii, Chenopodium album, Crypsis aculeata, Echium rauwolfii, Imperata cylindrica, Portulaca oleracea, Rumex dentatus, R. vesicarius*, and *Salsola baryosma*. In the shallow water *Najas armata, N. minor* and *Potamogeton trichoides* are abundant. Four submerged hydrophytes (*Potamogeton crispus, P. nodosus, Vallisneria spiralis* and *Zannichellia palustris*) have scattered distribution.

Khor Allaqi, about 150 km south of Aswan High Dam, penetrates the Eastern Desert for about 80 km. It has a wide mouth and narrows towards its extremity. The flora of its shore-lines includes 28 species: 25 terrestrial and three aquatic growing in the shallow water near the banks. *Hyoscyamus muticus* and *Tamarix nilotica* are abundant perennials on the steep rock slopes. Abundant annuals – *Glinus lotoides, Heliotropium supinum, Portulaca oleracea* and *Rumex dentatus* – form a narrow belt on the low banks near the water. Less frequent species of Khor Allaqi are *Amaranthus lividus, Chenopodium album, Citrullus colocynthis, Crypsis aculeata, C. schoenoides* (absent from the other khors), *Cynodon dactylon, Cyperus michelianus, Diplotaxis harra, Fagonia arabica, Fimbristylis bis-umbellata, Heliotropium ovalifolium, Oligomeris linifolia, Rumex vesicarius, Salsola baryosma* and *Senecio aegyptius*. The aquatic plants are *Najas armata, N. minor* and *Potamogeton trichoides*.

Further south on the west side of the lake at Khors El-Madiq, El-Malki, Eneiba and Tushka (Fig. 6.5), the shore-line vegetation shows much the same floristic diversity. *Hyoscyamus muticus* and *Tamarix nilotica* are the abundant perennials and *Glinus lotoides* is the frequent annual. *Fagonia arabica* and *Salsola baryosma* are present in Khors El-Madiq and El-Malki, but are absent from Khors Eneiba and Tushka. Other plants recorded in these khors are *Citrullus colocynthis, Crypsis aculeata, Echium rauwolfii, Francoeuria crispa, Heliotropium ovalifolium, H. supinum, Phragmites australis, Portulaca oleracea, Rumex dentatus and R. vesicarius. Najas armata* and *N. minor* occur in the four khors whereas *Potamogeton trichoides* occurs only in Khors Eneiba and Tushka.

b. Shore-Line Vegetation

According to water supply, deposition processes and the pattern of plant cover, six habitats may be recognized in the shore-line of the Aswan High Dam Lake (Springuel, 1985b).

Habitat 1: Steep Banks of the Lake and Central Elevated Parts of the Islands

These areas are characterized by inadequate water supply. They receive water only during a short time in November and/or December of the high flood periods. Deposition processes are very slow and there is only a thin layer of sand and silt deposits in the rock crevices. This type of habitat provides suitable conditions for *Tamarix nilotica* whereas the dry remains of *Hyoscyamus* and *Phragmites* are conspicuous in the area. Annuals are infrequent and appear for only a short period after the annual high floods. The flora of this habitat includes *Citrullus colocynthis, Fagonia arabica, Francoeuria crispa* and *Salsola baryosma* (perennials) and *Echium rauwolfii, Glinus lotoides, Heliotropium supinum, Oligomeris linifolia, Portulaca oleracea* and *Rumex dentatus* (annuals).

Habitat 2: Areas at the Edge of Rocky Slopes and Low Banks

These areas are inundated for a longer period (November–January) than habitat 1. The soil is formed of sand and silt deposits. The deep accumulation of sand is the most suitable habitat for perennials such as *Hyoscyamus, Phragmites* and *Tamarix.* These three species form dense vegetation belts in certain localities. Annuals such as *Glinus* and *Rumex* are abundant in cracks between the rocks. The appearance of these annuals and their subsequent growth depend on annual fluctuations of water level. The vegetation of this habitat includes the species of habitat 1 in addition to *Calotropis procera, Morettia philaena, Pulicaria undulata* and *Senecio aegyptius.*

Habitat 3: The Low Banks Submerged During Certain Periods Between October and March

These areas are characterized by their deep silty soil with high moisture content when the water retreats. The vegetation grows in longitudinal rows each year corresponding to the retreat of the water level at regular intervals. Fringing vegetation of annual species occupies a narrow (1–3 m wide) strip of a previously inundated zone. The juvenile growth of *Phragmites* is very frequently found near the water. Dense carpets are formed, with 100% cover in some stands by *Glinus lotoides* and *Heliotropium supinum.* This is followed by a narrow belt of *Crypsis aculeata,*

C. alopecuroides and *C. schoenoides-Chenopodium album, Portulaca oleracea, Rumex dentatus* and *R. vesicarius* and other annuals seldom occur far from the water. Numerous seedlings of *Tamarix nilotica* grow on low banks but most of the seedlings do not survive the prolonged flooding. *Cynodon dactylon, Echium rauwolfii, Francoeuria crispa, Heliotropium ovalifolium, Morettia philaena, Oligomeris linifolia* and *Salsola baryosma* are occasional in this habitat.

Habitat 4: Embankments Covered by Gravel

This habitat also remains submerged under water for the same period as habitat III. During the summer, when the flood is over, the soil usually dries. Thus, this habitat is unfavourable for a great number of species and the cover is the lowest (2–20%).

Plants usually appear in the thin layers of sand deposits between the gravels. Perennials normally do not live beyond the seedling stage in this habitat (except *Hyoscyamus muticus*). Annuals such as *Glinus lotoides* and *Amaranthus* sp. are abundant; other plants present are *Crypsis* spp., *Heliotropium ovalifolium, H. supinum, Phragmites australis, Portulaca oleracea, Salsola baryosma* and *Tamarix nilotica.*

Habitat 5: Small Islets Close to the Water Level

These are subjected to temporary inundation. Such islets are numerous in all khors. They form the most favourable habitats for pioneer plants such as *Hyoscyamus muticus, Glinus lotoides* and *Rumex dentatus. Tamarix nilotica* is restricted to the central parts of these islets where softer soil accumulates and where some seedlings are able to survive the long inundation. The vegetation cover in these islets is usually sparse and only a few species are present. The flora of this habitat includes the species of habitat 4 as well as *Echium rauwolfii, Oligomeris linifolia* and *Rumex vesicarius.*

Habitat 6: Shallow Waters Along the Shore-Line and Islands

This is the habitat of aquatic vegetation where there is dense growth of submerged hydrophytes. These are *Potamogeton trichoides, Najas minor* and *N. armata* as abundant species with common presence of *Potamogeton nodosus* and *Zannichellia palustris.*

N. armata and *N. minor* have been reported by El-Hadidi (1976) from the southern part of the Nile system. During the last 10 years *Najas* spp. have become distributed everywhere along the High Dam Lake and form dense populations in its shallow water. Tawadrous (1981) notes that *Potamogeton pectinatus* was recorded from the seasonally inundated shores of the High Dam Lake.

Bishai et al. (2000) reported eleven submerged flowering and non-flowering species in the water of Lake Nasser. These are: *Najas marina* subsp. *armata, N. horrida, Vallisneria spiralis, Zannichellia palustris, Potamogeton lucens (= P. schweinfurthii), P. crispus, P. Pectinatus, P. trichoides, P. perfoliatus*, and *Myriophyllum spicatum* (flowering plants) in addition to the macro-algae *Nitella hyaline*. "*M. spicatum* is an invasive species that appeared in Lake Nasser in 1993, probably introduced into the lake through fishing nets" stated Ali (1998).

(II) Reservoir bounded by the two dams

A. General Characteristics

This reservoir is bounded on the north by the first Aswan Dam and on the south by the High Dam (Fig. 6.6). It is elongated with often dendritic side areas (khors), most of which are on the eastern side and a few small ones on the west. This reservoir is characterized by a number of granitic islands which are described in a separate section. The vegetation of this reservoir is mainly restricted to the shorelines and its different types of habitat are controlled by two main factors: soil formation and moisture content (Springuel, 1985a).

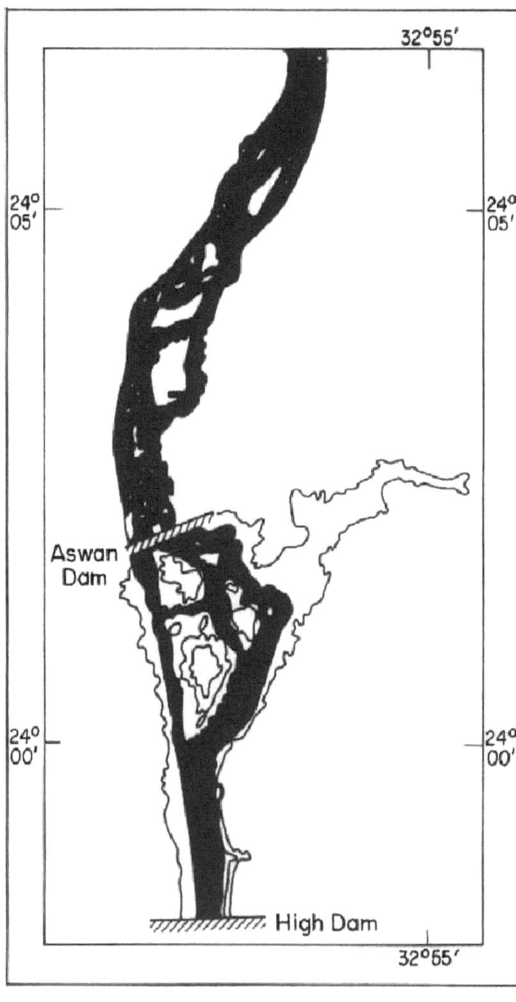

Fig. 6.6 The reservoir between Aswan Dam and High Dam in the Nile. Water is shown *black*

The process of soil formation along the shore-lines is largely confined to lower elevations, inlets and khors. The alluvium is covered gradually by an additional layer of silt. Higher reliefs and exposed sites are subjected to intensive erosion; they are devoid of soil cover and their surfaces are barren.

Soil moisture depends mainly on the diurnal and monthly water fluctuations of the reservoir. During 1965, after the construction of the High Dam, fluctuation of water level in the reservoir was apparently high: 105.1 m in July and 113.4 m in November as compared with levels during 1980 which were 106.9 m and 107.9 m during November and August respectively. During 1983–1984, daily fluctuations were apparently high as observed by Springuel (1985a) who states Fresh remains of *Ceratophyllum demersum* and *Potamogeton crispus* were often found on the branches of *Sesbania sesban* shrubs at 3–4 m above the water level. These daily water fluctuations result from the operation of both the Aswan (old) Dam and the High Dam which have significant effects on the soil moisture of the shore-lines of the reservoir.

B. Habitat and Vgetation Types

Seven types of habitat have been recognized along the shore-lines of the reservoir. These habitats differ in intensity of alluvium deposition and water supply and are distinguished within two generalized ecological series. Four habitats (aquatic, lower lever of the terrace, higher level of the terrace and transitional zone) are associated with the gentle slopes of the banks and the other three (lower level, middle level and upper level of rocky slopes) are associated with steep rocky areas.

Habitat 1: The Aquatic Habitat

This habitat occupies the narrow zone of shallow water along the banks of the reservoir (and around the islands). The depth of water where plants grow varies from 10 to 400 cm, depending upon fluctuation. This habitat is characterized by submerged hydrophytes dominated by *Ceratophyllum demersum* with *Potamogeton crispus* as an abundant associate (or co-dominant in certain areas). Very often both species form separate belts parallel to the shores of the reservoir, and are rarely mixed. Other species are *Vallisneria spiralis*, a summer hydrophyte of relatively deeper waters, *Potamogeton pectinatus* and *Najas armata*.

Habitat 2: Lower Level of the Terrace

This zone, close to the water level, varies in height from 0 to 1 m, being subjected to a longer period of inundation than the higher terrace. In this habitat the soil formation processes are slow and the water-table is near the surface. This zone is characterized by its hydromesophytic meadows in which two units may be recognized. First, there is a mixed herb community with a mosaic pattern, different species being dominant in different areas. *Phragmites australis* and *Fimbristylis bisumbellata* dominate in small depressions while elevated parts of the silty substratum are covered by the growth of *Cynodon dactylon*. Other species in this habitat include *Apium leptophyllum, Aster squamatus, Conyza dioscoridis, Coronopus didymus, Cyperus laevigatus, C. rotundus, Dichanthium annulatum, Eragrostis pilosa, Francoeuria crispa, Glinus lotoides, Lotus arabicus, Polygonum equisetiforme, Portulaca oleracea* and *Sesbania sesban*. Second, there is an *Aster squamatus* community which may be considered as a pioneer community on the lower zones. It is associated with the above-mentioned species as well as with, for example, *Cyperus rotundus, Echinochloa colona, Euphorbia hirta, Phragmites australis* and *Sonchus oleraceus*.

Habitat 3: Higher Level of the Terrace

This habitat lies 1–3 m above the water level. It is submerged by water for short periods, mainly as a result of daily water fluctuations. The soil is deep silt, and with high humus content. The vegetation here is of xeromorphic river-bed scrub dominated by *Sesbania sesban* (5–6 m high) associated with herbaceous mesophytes: *Cynodon dactylon, Cyperus rotundus* and *Polygonum equisetiforme*. Other associates *are Apium leptophyllum, Aster squamatus, Conyza dioscoridis, Coronopus didymus, Cyperus laevigatus, Dichanthium annulatum, Echinochloa colona, Eragrostis pilosa, Fimbristylis bisumbellata* and *Portulaca oleracea*.

Habitat 4: Transitional Zone

This habitat lies 3–6 m above the water level and links the terrace with the steep slope of the desert cliff. It may, occasionally, be inundated with water. The ground surface is often studded with rock fragments covered by shallow silt deposits.

Four communities dominated by *Alhagi maurorum, Calotropis procera, Dichanthium annulatum* and *Imperata cylindrica* have been recognized in this habitat. Other species include all those species recorded in habitat III as well *as Acacia nilotica, Chenopodium ambrosoides, Euphorbia hirta, Plantago lagopus, Solanum nigrum* and *Trigonella laciniata*.

Habitat 5: Lower Level of the Rocky Slopes

This habitat, being about 1 m above water level, is subject to long periods of inundation. The vegetation is an open grassland dominated by *Dichanthium annulatum* with *Sesbania sesban* as a co-dominant in some stands. Common associates are *Aster squamatus, Calotropis procera, Cynodon dactylon, Francoeuria crispa* and *Leptadenia heterophylla*. Less frequent are *Euphorbia prostrata, Imperata cylindrica, Lotus arabicus, Sorghum halepense* and *Tamarix nilotica*.

Habitat 6: Middle Level of the Rocky Slopes

This part is about 3 m above the water level and is subject to daily water fluctuations. Silt deposits and humus accumulation result in a thin soil. As in habitat 5 the vegetation is co-dominated by *Dichanthium annulatum-Sesbania sesban* associated with e.g. *Calotropis procera, Conyza dioscoridis, Cynodon dactylon, Echinochloa colona, Eragrostis pilosa, Euphorbia hirta, Leptadenia heterophylla* and *Trichodesma africanum*.

Habitat 7: Upper Level of the Rocky Slopes

This zone, which is about 7 m above the water level, borders the desert plain and is occasionally inundated. Coarse sediments and sand from the desert accumulate between the rocks and form a shallow soil layer poor in humus.

The vegetation here is a mesoxerophytic scrubland co-dominated by *Alhagi maurorum-Calotropis procera*. Plant cover is low (5–40%). Associate species are *Acacia nilotica, Amaranthus graecizans, Ambrosia maritima, Argemone mexicana, Citrullus colocynthis, Dichanthium annulatum, Euphorbia granulata, Francoeuria crispa, Glinus lotoides, Leptadenia heterophylla, Lotus arabicus, Polypogon monspeliensis, Ricinus communis, Solanum nigrum, Sonchus oleraceus* and *Trichodesma africanum*.

The vegetation of the shore-line of this reservoir follows an ecological series of decreasing soil moisture content, from the completely submerged habitat with hydrophytic plants (*Ceratophyllum demersum* community) followed by a meadow of hydromesophytes dominated by *Phragmites australis, Fimbristylis bisumbellata* and *Aster squamatus* and then a river-bed scrub of xeromesophytes dominated by *Sesbania sesban*, a habitat of xeromesophytic grassland dominated by *Dichanthium annulatum* and *Imperata cylindrica* and finally scrub of meso-xerophytes dominated by *Alhagi maurorum-Calotropis procera* (Springuel, 1985a).

(h) The Nile islands

The main streams of the River Nile in Egypt and its Damietta and Rosette branches are embracing more than 300 islands mainly covered with silt. The number and areas of these islands increase northwards. A short account on the habitat types and plant life of islands representing the most southern section (Group I, Aswan area) and the middle section (Group II, Minia area) of the main stream of the River Nile is given below:

Group I: Aswan area

(A) Ecological characteristics

The Nile at Aswan, north of the High Dam, is interrupted by about 30 islands of varied size and structure. Most of these are small uninhabited granite islets which support a natural vegetation believed to be the only remains of the original plant cover of the Nile Valley that survived after the establishment of the Aswan Dam (1902–1930) and the High Dam (1960–1965). The vegetation of these islands represents a relict of woodland growth that became scarce (El-Hadidi and Springuel, 1978).

North of Kalabsha Gorge (60 km south of Aswan) the Nile cuts its course through sands for 20 km after which igneous and metamorphic rock appear (for more than 35 km). The northern part of the latter section is known as the First Cataract which is about 7 km south of Aswan with a length of 12 km. The First Cataract forms the least obstacle in the course of the river on its way north (Abu Al-Izz, 1971). The area of the First Cataract begins at the southern tip of Haysa Island, the largest igneous island in this vicinity. It is followed by a group of smaller islands where the river is deflected westwards and the current becomes stronger, indicating the beginning of the cataract. Here the old Dam was built across four granitic islands which divided the now northward-deflected river into five channels. The river runs north of Aswan Dam in a strong torrent; its water flows through the westernmost channel, where a navigation canal, 2 km long, has been dug. Beyond Suhayl Island the speed of the river increases; this is where the First Cataract ends, north of which small islands in the Nile divide it into sectors. The most famous of these are Suluga Island and Aswan Island. The latter was known during the Roman era as Elephantine Island (Springuel, 1981). The other islands include Biga, Gezel, Awad, Burbur and Duns. Most of these igneous islands are of coarse-grained granite. There is also fine-grained granite evident on the surface, e.g. Awad and Haysa Islands south of old dam and in isolated spots on the Island of Suluga north of the dam. The coarse and fine granites (basement complex) have been subjected to weathering and appear in a variety of shapes, sometimes spheroidal masses or parallel epipeds when deflated.

The Aswan area represents one of the extremely dry areas of the world, almost rainless (average 1.0 mm/year). Temperature is consistently high during most of the year with a daily average of about 34 °C between April and October. The building of Aswan High Dam and the establishment of its Lake had an effect on the climate of the Aswan area. The air temperature has decreased in maximum and increased in minimum values by about 2 °C in recent years. Evaporation seems to have increased, being higher in summer than in winter. The total rainfall, however, shows no change.

The vegetation of three representative islands – Duns Island, Burbur Island and Gezel Island – has been studied quantitatively by Springuel (1981) using the transect method. The vegetation of Duns Island (about 800 m long and 200 m broad) was surveyed along 16 transects of 100–180 m. The vegetation of Burbur Island (800 m long and 250 m broad) was studied along a 250 m transect and the vegetation of the Island Gezel (900 m long and 50–100 m broad) was surveyed by eight transects of 100–200 m. Selected transects are presented as illustrative examples. All transects were east-west through the whole length of each island.

In Duns Island, aquatic plants, e.g. *Ceratophyllum demersum* and *Potamogeton crispus*, usually occupy the shallow water near the bank, followed by a zone of recent shallow sand deposits to a depth of a metre occupied by mesophytes such as *Eragrostis* sp., *Vahlia dichotoma* and *Cynodon dactylon* (open cover). *Acacia seyal* forms dense thickets. The partly submerged land is dominated by *Polygonum senegalense. Francoeuria crispa* grows in cracks of the rocks. In mixed salt-sand terraces, 2 m above the water level, is a dense growth of *Imperata cylindrica* with spaced trees of *Acacia seyal*. The next lowest level, of recent sand deposits (1–1.5 m above water level), supports scattered shrubs of *Tamarix*. A very narrow belt of perennial plants (*Cynodon, Cyperus, Panicum* etc.) marks the ecotone between the land and water. In shallow water or partly inundated areas, *Polygonum senegalense* and *Phragmites australis* form floating mats. Aquatic species in the permanently submerged land are *Ceratophyllum demersum* and *Potamogeton crispus*. Perennial grasses and sedges include *Panicum repens* and *Cyperus* spp., in the seasonally submerged land, whereas in the sand terraces of recent deposits are *Tamarix nilotica* and *Acacia albida*. The east and west slopes of the hills of this island are characterized by a dense vegetation mainly *of Acacia seyal* and *Tamarix nilotica*.

In Burbur Island, the aquatics *Ceratophyllum demersum* and *Potamogeton crispus* are present in the submerged land, while in partly submerged areas *Polygonum senegalense* and *Phragmites australis* dominate, with some other swamp plants, e.g. *Panicum repens* and *Cyperus* spp., which may dominate low lands close to this swampy habitat. *Acacia seyal* forms thick vegetation on the west slopes of the hills and on the steep west slopes is *Imperata cylindrica*.

In Gezel Island, the aquatics are similar to those of the other islands. On partly submerged land *Phragmites australis* dominates and *Salix* sp. and *Mimosa pigra* are associates. The high rocky habitat is almost barren except for a few plants of *Francoeuria crispa, Imperata cylindrica* and *Tamarix nilotica* in the rock crevices.

B. Habitat and vegetation types

Five main habitats have been recognized in the rocky islands of the River Nile in the Aswan area: submerged land, partly submerged land, seasonally submerged land, occasionally submerged land and dry land (Springuel, 1981).

Habitat 1: Submerged Land

This habitat includes shallow areas along the banks of the Nile, around the islands and areas between the islands where the water current is slow. The ground is mostly sandy and occasionally muddy. In areas where the current is strong the bottom is mostly covered by gravel and aquatic plants are absent.

The aquatic vegetation of this habitat is co-dominated by *Ceratophyllum demersum* and *Potamogeton crispus* with a cover from 10% to 100%. The water depth ranges from 80 cm to 250 cm for *C. demersum* and *P. crispus* (dominants), *P. perfoliatus* and *Zannichellia palustris* (perennial associates) and *Najas armata* and *N. pectinata* (annual associates). Rarely *Typha domingensis* has been recorded in the shallow parts of the fringes of this habitat.

The aquatic vegetation of these islands is less subject to the influence of human activity than that of other habitats. An exception is *P. crispus* which is sometimes collected and dried for use as fodder for livestock. Fishermen also fish in the zone of *C. demersum* as it is a very important spawning medium for fish.

Habitat 2: Partly Submerged Land

This habitat is dominated by *Polygonum senegalense* which is an aquatic/terrestrial, tall glabrous plant with thick, swollen horizontal stolons, just above the soil surface, up to 3 m long and with vertical stems reaching 1.5 m.

In areas of the rocky islands, *P. senegalense* may form, with its associate species, continuous belts surrounding the islands. It is restricted to slow-running water, ranging in depth from 10 cm to 150 cm. This depth is subject to annual fluctuation of the river water. The roots of *Polygonum* are anchored in the firm soil of moist places and where the water is shallow and the plant sends long shoots into deeper water, forming floating mats. In strong flowing water and where gravel accumulates on the bottom *P. senegalense* is absent.

The vegetation of this habitat includes 17 species in addition to the dominant. *Phragmites australis* is abundant. Common are *Cyperus longus, Mimosa pigra, Oxystelma esculentum, Panicum repens, Salix* spp., *Tamarix nilotica* and *Typha domingensis*. Other associates are *Conyza dioscoridis, Cynodon dactylon, Cyperus mundtii, C. rotundus, Leptadenia heterophylla, Ricinus communis, Sesbania sesban* and, *Veronica anagallis-aquatica* (perennials) and *Avena fatua* (annual).

The *Polygonum* zone is inundated for the greater part of the year, and even in dry periods, when the level of the river water is low, a belt of a few metres from the banks of the islands is still partly submerged. The cover ranges from 50% to 100%. The diversity of the vegetation changes with depth of water and distance from land. In the optimum depth of water, there is a population of *Polygonum* with two associate "reed" plants: *Phragmites australis* and *Typha domingensis*. With decreasing depth of water approaching land, species such as *Cyperus longus, Oxystelma esculentum, Salix* sp. and *Tamarix nilotica* appear. On the wet borders between water and land the diversity increases with the growth of such plants as *Cynodon, Cyperus* and *Panicum*. The soil near the banks of the islands bears a well-developed turf reaching 20 cm in depth. This turf is formed from the partly decaying remains of plants and the accumulation of soft silt and clay between the roots.

Human interference in the reed swamp habitat of these islands is considerable and leads to destruction of the vegetation. Young shoots of *Polygonum, Phragmites* and *Typha* are usually cut for cattle, especially buffaloes. *Phragmites* and *Typha* are used by the local inhabitants for making mats. In dry periods this vegetation is partly burned.

Habitat 3: Seasonally Submerged Land

a. Low Bank Meadow

The main characteristic of the meadow grasses is their capacity for vegetative spread which tends to exclude colonization by new species and provides a sufficiently large population to withstand flooding. This habitat is generally dominated by *Cynodon dactylon* and *Panicum repens* which form a narrow belt (2–5 m) on the low banks of the islands. The zone of the meadow grass is from 10 to 100 cm above the river water, so its moisture content is high. It is flooded with water during a short period of the year. The soil is a recent alluvial deposit.

The phytocoenosis of the *Cyperus-Panicum* meadow (cover 70–100%) of these islands shows considerable floristic richness. A total of 46 species has been recorded: 11 trees and shrubs, 16 perennial herbs and 19 annuals. The rich flora may be related to the intermediate topographic position of this zone between water and land. During flooding, such species as *Polygonum senegalense* and *Cyperus longus* extend quickly throughout this zone. When the dry period starts these species die back to the original boundary near the water and species from the dry areas, such as *Francoeuria crispa, Imperata cylindrica, Leptadenia heterophylla* and *Vahlia dichotoma*, start to colonize this zone. In this dry period many *Acacia* seedlings appear. When the cycle repeats itself and the area is flooded, *Acacia* disappears and only plants which have as strong a resistance to floods as to drought can form the established vegetation in this zone.

The flora of this meadow includes *Cyperus mundtii* and *Lotus arabicus* (abundant), and *Cajanus cajan, Cyperus rotundus* and *Sesbania sesban* (common). Less common associates are *Cyperus longus, Leptadenia heterophylla, Polygonum senegalense, Senecio aegyptius* and *Tamarix nilotica*. Other associates include *Acacia nilotica, A. raddiana, A. seyal, Avena fatua, Brassica arabica, Chenopodium album, Conyza dioscoridis, Crypsis schoenoides, Digitaria nodosa, Eclipta alba, Eremopogon foveolatus, Francoeuria crispa, Heliotropium ovalifolium. H. supinum, Hemarthria altissima, Imperata cylindrica, Medicago sativa, Mimosa pigra, Oxystelma esculentum, Ricinus communis, Saccharum spontaneum, Salix subserrata, Solanum nigrum, Sonchus oleraceus*, and *Veronica anagallis-aquatica*.

The vegetation of the meadow grass area is intensively used by the Nubians and especially the dominants. These plants are usually cut as fodder for cattle. The Nubians also use this zone for cultivation of *Cajanus cajan*, but as they do this in a very primitive way by simply scattering the seeds without preliminary preparation of the soil, this activity does not disturb the natural vegetation of this zone (Springuel, 1981).

b. Low Rocky Habitat

This habitat is also seasonally submerged and is characterized by the presence of a large number of species, with different pioneer plants on the low rocky areas, randomly distributed along the banks of the islands. In addition to summer flood, the habitat is occasionally inundated for brief periods in the year. Heterogeneity in the microrelief and physical make-up of recent deposits between the rocks are main reasons for the differing soil moisture throughout the area. The deep depressions between the rocks containing stagnant water are characterized by very wet soil, whereas the higher elements of the micro-relief with deep sand deposits between the rocks are dry. The soil is formed from the accumulation of deposits brought by flood water and sand brought by wind; differences in the particle-size fractions of these soils depend on the position in relief, on water current and on the wind velocity.

This rocky habitat supports a rich flora (49 species): 15 trees and shrubs, 17 perennial herbs and 17 annuals. In spite of such high floristic diversity, the vegetation is fairly open with a low cover (but up to 70%). With the great range of species with different ecological characteristics and the presence of a large number of common plants, there is no clear dominant. The most frequent shrubs (*Leptadenia heterophylla* and *Tamarix nilotica*) are those with very low cover values and hence do not much influence the physiognomy of the vegetation. The plants with high cover are *Cynodon dactylon, Francoeuria crispa* and *Panicum repens* (perennials) and the annual *Lotus arabicus*. Common plants with low cover include *Conyza dioscoridis, Mimosa pigra, Oxystelma esculentum* and *Sesbania sesban* (shrubs), *Cajanus cajan, Cyperus longus, C. mundtii, Imperata cylindrica* and *Polygonum senegalense* (perennials) and *Conyza aurita, Heliotropium* spp., *Senecio aegyptius* and *Trigonella hamosa* (annuals).

The deep depressions between the rocks containing stagnant water are occupied by helophytes, e.g. *Cyperus longus, Mimosa pigra, Phragmites australis* and *Typha domingensis*.

The vegetation of this area is not greatly disturbed by man. The large number of stones and the abundance of the poisonous annual legume *Lotus arabicus* make this area dangerous for grazing.

Habitat 4: Occasionally Submerged Land

This habitat is characterized by recent, mobile and deep sand deposits. It consists of flat terraces at the water level from 80 to 150 cm high which are occasionally flooded (once every few years). The bare surface of the sand is heated by the sun during the day to very high temperatures, thus increasing evaporation. Such factors make the habitat less favourable for plant growth, especially at the seedling stage. On the other hand, the large difference in day and night temperatures promotes the condensation of water at night which is sufficient for some annual species.

The vegetation of the occasionally submerged land is an open type dominated by two species with broad ecological tolerance to these conditions of moisture and soil – *Tamarix nilotica* and *Leptadenia heterophylla*.[5]

The phytocoenosis of this habitat shows low diversity. Eighteen species are recorded: 6 trees and shrubs, 9 perennial herbs and 3 annuals. The plant cover is heterogeneous, ranging from 5% to 80%, but is usually less than 40%, the main contributors to the cover being the canopy of *Acacia albida, Leptadenia heterophylla* and *Tamarix nilotica*. Species such as *Acacia raddiana, A. seyal, Cynodon dactylon, Eragrostis aegyptiaca, Francoeuria crispa* and *Vahlia dichotoma*, though common, make little contribution to the cover. Other species in this habitat include *Acacia nilotica, Conyza aurita, Cyperus mundtii, C. rotundas, Fimbristylis bisumbellato, Imperata cylindrica, Lotus arabicus, Panicum repens, Saccharum spontaneum* and *Tephrosia apollinea*.

Habitat 5: Dry Land

The dry land of the Nile islands at Aswan may be divided into (a) silty habitats and (b) high rocky habitats.

a. Silty Habitats

The silt deposits on the dry land are present in the central parts of the islands with heights from 3 m to 6 m above the river water level. The main attribute affecting the

[5] Täckholm (1956, 1974), records two species of *Leptadenia* in Egypt: *L. pyrotechnica*, very common in the Egyptian deserts and Nile Valley, and *L. heterophylla*, a very rare species in the Aswan area.

existence of plants in areas where the source of moisture is the subsurface water at a depth of 3–6 m is the presence of a well-developed root system, capable of penetrating deposits down to the subsurface water. Trees and shrubs best satisfy such a requirement.

A characteristic feature of this habitat is that the vegetation has four layers. The upper layer is formed by *Acacia* trees, the second is a shrub layer of *Tamarix nilotica*, the third is of undershrubs, and the fourth is the ground layer formed by some perennials, e.g. *Cynodon*, and annuals, e.g. *Sonchus*.

In Egypt there are 13 species of *Acacia* (Täckholm, 1974), six of these – *A. albida, A. arabica, A, laeta*, A *nilotica*, A *raddiana* and A *seyal* – being present in the silt habitat of Aswan Island. The silt deposit of this island may be low (<5 m above the river water level) or deep (>5 m above water level). The vegetation in the low deposit area is clearly layered. The uppermost, tree layer (3–5 m high) is represented by the common trees *Acacia nilotica* and *A. seyal* as well as the rare A *arabica, A. laeta* and A. *raddiana*. The shrub layer (150–300 cm high) comprises *Tamarix* and A *albida* which also grows as shrubs. The undershrub layer includes *Cajanus cajan, Desmostachya bipinnata, Hyoscyamus muticus, Imperata cylindrica, Francoeuria crispa* and *Tephrosia apollinea*. The ground layer includes *Cynodon dactylon* and *Cyperus mundtii, C. rotundus* and *Panicum repens* (perennials), *Conyza aurita, Lotus arabicus, Schoenefeldia gracilis* and *Senecio aegyptius* (annuals).

The vegetation cover of the low silt deposits area ranges between 10% and 100%. The plants grow in groups differing in size and diversity. Between these groups is usually a thin shrub layer of *Tamarix nilotica* (1–1.5 m high) associated with *Francoeuria crispa* and with *Leptadenia heterophylla* which twines around *Tamarix* and covers *Francoeuria* or grows on the bare surface of the silt. *Acacia albida, A nilotica* and *A. seyal* usually form thickets. Some stands of *A albida* are very dense and impassable thickets interwoven with *Leptadenia*. In these thickets, *A. seyal* is present either in tree form with an umbrella-shaped crown or as a shrub if the main trunk has been cut. *A. laeta* and *A. raddiana* usually grow on the slopes of the silt terraces and form groups with open canopies together with other *Acacia* trees. *T. nilotica* grows between the groups of trees and usually forms a shrub layer not more than 1.5 m high, with an open canopy.

The vegetation of the deep silt deposits is an open one with maximum cover of 50%. It is of low diversity. The most common species are *Acacia nilotica* (tree), *Leptadenia* (twiner) and *Tamarix* (shrub). Less common ones include *A. albida, A. seyal, Calotropis procera, Desmostachya bipinnata, Francoeuria crispa, Imperata cylindrica* and *Ziziphus spina-christi*.

The silty habitat of the river islands at Aswan has been more influenced by human activity than any of the other habitats. This is due, primarily, to the presence of very useful cultivable land, of deep deposits with a considerable amount of silt and clay. Some of these areas are now under cultivation, most cultivations being restricted to the upper parts of the relief where the land is wide enough. Another aspect of human interference is the cutting of trees and shrubs of *Acacia* and *Tamarix*. When the separate trees in the thicket are cut down, new coppices soon appear and the removal of a few trees does not much influence the growth of the thickets, except for the entry of some new species to the open area such as *Francoeuria* and *Tamarix*. The ecological regime of the thicket as a whole does not change (Boulos, 1966b). On the other hand, burning of trees of *Acacia* causes much damage. If the area has been burned, it is vegetated by *Desmostachya bipinnata, Imperata cylindrica* and *Tamarix nilotica*.

b. High Rocky Habitat

The high rocky habitat of the dry land of the Nile Islands at Aswan is of granitic rocks which are different in size and height and represent the base for deposits of soil-forming material brought by the river water around them. The biggest rocks may be up to 30 m; usually rocks higher than 20 m lack plant cover. A very thin layer of wind-borne soil accumulates in the cracks between the rocks where water also accumulates. However, the water supply of this habitat is very poor, there being no contact with the river water owing to the granitic bedrocks. The only water reaching the soil is the negligible amount of rain (rainfall rarely occurring), and the condensation of water resulting from the large fluctuation in day and night temperatures.

The best adapted plants for such dry conditions are *Francoeuria crispa* and *Leptadenia heterophylla*. Both have a high tolerance to drought and undemanding soil requirements. The plant cover ranges from 5% to 20%. Common associates are *Acacia raddiana. A. seyal* and *Tamarix nilotica*. Less common species include A. *albida.A. laeta, Calotropis procera, Desmostachya bipinnata, Imperata cylindrica, Ziziphus spina-christi* and the annual *Ceruana pratensis*.

Group II Minia Area

The plant life of 43 islands of the middle section of the main stream of the River Nile in Minia Area was studied by Mohamed and Hassan (1998). Most of these islands (34) are cultivated and the rest (9) are uncultivated, their areas ranged between 0.005 km^2 and 1.046 km^2.

Four habitat types are recognized in these islands, namely: wet, canal bank, cultivated lands and inhabited areas. The hydrophytes and helophytes inhabiting the wet habitat include: *Ceratophyllum demersum, Cyperus articulatus, Ludwigia stolonifera, Persicaria lapathifolia, P. salicifolia, Phragmites australis, Potamogeton nodosus*, and *Typha domingensis*.

The canal bank habitat is organized into 2 terraces. The first (lower) terrace is subjected sometimes to water flooding due to the fluctuation of water level of the Nile. The floristic assemblage of this terrace comprises hydrophytes, e.g. *Cyperus* spp., *Phragmites australis, Saccharum spontanum* and *Typha domingensis* and mesophytes, e.g. *Cynodon dactylon, Aster squamatus, Fibristylis bisumbellata, Alternanthera sessilis, Conyza bonariensis, Pluchea dioscoridis, Sencio aegyptius* and *Potentilla supine*. The second higher levelled terrace and its slope are characterized by the growth of *Sesbania sesban, Acacia nilotica, Cynanchum acutum, Ricinus communis, Imperata cylindrica, Desomstachya bipinnata, Cynodon dactylon,*

Alhagi graecorum, Tamarix nilotica and *Cyperus* spp. A group of therophytes were recorded in this terrace, e.g. *Polypogon monspeliensis, Solanum nigrum, Phyla nodiflora, Glinus lotoides, Eclipta alba, Chenopodium album, C. murale, Gnaphalium luteo-album, Ammi majus, Amaranthus lividus* and *Senecio aegyptius.*

The cultivated areas of these islands are characterized by weed plants associated with field crops and vegetables. The main aggresstal weeds (inside the fields) are: *Solanum nigrum, Malva parviflora, Melilotus indicus, Orobanche ramosa, Euphorbia peplus, Trifolium ruspinatum, Trigonella glabra, Brassica tournefortii, Spergularia marina, Beta vulgaris, Chenopodium album, C. mural, Cyperus rotundus, Convolvulus arvensis, Sonchus oleraceaus, Ammi majus, Cynodon dactylon, Echinochloa colona, Polypogon monspeliensis, Leptochloa fusca, Setaria verticillata, Rumex dentatus, Protulaca oleraceae, Cuscuta pedicclata* and *Amaranthus lividus*. The radial weeds (outside the cultivated plots) include: *Cyperus articulatus, C. alopecaroides, Desmostachya bipinnata, Juncus hypridus, Sphaeranthus suaveolens, Pluchea dioscoridis* and *Tamarix nilotica.*

The areas inhabited by the farmers contain some of the shade trees and shrubs e.g. *Ficus retusa, Eucalyptus rostrata, Ricinus communis, Delbergia sisso* and *Phoenix dactylifera.*

Chapter 7
The History of the Vegetation: Its Salient Features and Future Study

7.1 The History of the Vegetation

In previous chapters descriptions are given of the different types of vegetation of Egypt represented at present. But what did the vegetation of Egypt look like in the past? To what extent was it similar to that today and what changes over what intervals of time have taken place? The study of fossil pollen grains in sedimentary environments provides a fruitful approach to the elucidation of the vegetation of ancient times which, in turn, is necessary to interpret the vegetation of the present day.

The use of pollen data for the reconstruction of past vegetation clearly requires an understanding of the relationships between modern vegetation types and pollen rain. Wright, McAndrews and van Zeist (1967) have pointed out that the reconstruction of the former vegetation of an area by pollen analysis is little more than speculation unless the fossil pollen assemblages can be related to a vegetation of known structure and composition. A study of this kind has been carried in Egypt by Ayyad (1988), who has investigated three landward transects in the coastal area of the Nile delta. These transects were chosen along the Mediterranean coast of the delta, since this area shows distinctive zonation patterns in the different habitats represented. The zones were differentiated by subjectively defined communities. The cover, abundance and floristic composition of each of these types of community were described. The recent pollen grains in the surface samples beneath the vegetation of the different zones of the transects were quantitatively analysed.

The application of detrended correspondence analysis (DECORANA) to the pollen data collected has proved appropriate for indicating the similarities and contrasts between the different types of vegetation and their pollen in the soil. This technique has proved successful in comparing the fossil pollen assemblages with the current pollen rain. The results show that the surface pollen samples are separated according to habitats (Fig. 7.1). Along axis 1 from left to right a wet to dry environmental gradient is represented. The results show that certain communities can be easily distinguished on the basis of their pollen assemblages, e.g. swamps and some sand dunes, whereas

M.A. Zahran, A.J. Willis, *The Vegetation of Egypt*,

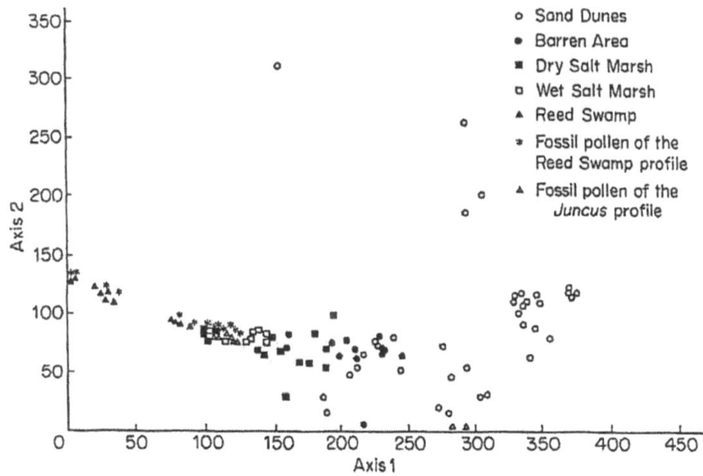

Fig. 7.1 Detrended correspondence analysis ordination of surface pollen samples of the Nile delta

others overlap and are not easily distinguished on this basis, e.g. salt marsh and barren areas. Moore (1976) pointed out that reconstruction of past vegetation can be considerably aided by information on the dispersal potentiality of different species. Distances moved by pollen grains depend on their density and aerodynamic properties, which vary from species to species. Local surroundings of the pollen-producing plants also have an important influence on both the amount of pollen produced and the distance travelled. Ayyad (1988) found pollen grains of such Mediterranean taxa as *Alnus* and *Corylus*, not native to Egypt, in the surface soil of the coastal area of the Nile delta. Presumably, such pollen had been transported long distances by wind from southern Europe. Ritchie (1985) has also reported scattered occurrences of low frequencies of Mediterranean trees such as alder, pine and oak, not native to Egypt, in analyses of the modern pollen spectra from Dakhla Oasis, Western Egyptian desert.

The prevalent wind direction at flowering times can profoundly influence the proportion of long-distance transported pollen in studied samples. Since the predominant wind direction in the Egyptian coastal area is from the north and northwest, the winds will have crossed many hundreds of kilometres of water or treeless vegetation before reaching the coast; this no doubt has diminished the longdistance pollen input into the area.

Moore and Stevenson (1982) noted that geographic locality, aspect and topography, and low local pollen production can all play a part in the presence of high tree pollen rain in an unforested area. They also stressed that a thorough study of the current pollen rain in a desert area is essential for palaeoecological interpretation of local fossil pollen profiles.

In the study area of the Nile delta, Ayyad (1988) found tree pollen input was low, even less than that of tundra sites. The lack of trees in the pollen catchment of the Egyptian sites investigated means that tree pollen input is at a very low level. Of great importance is the geography and topography of the sites. In treeless environments,

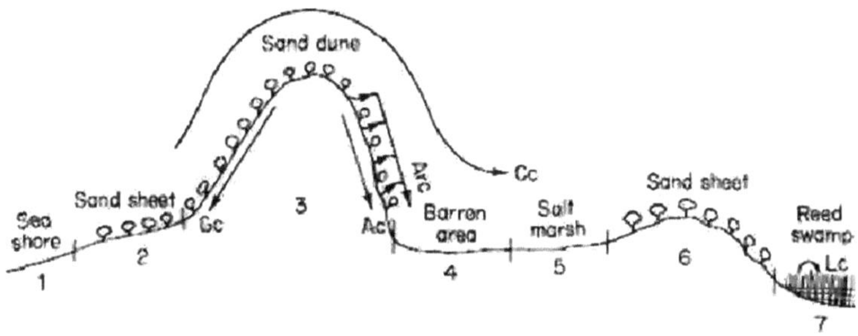

Fig. 7.2 A model of pollen input into different habitats (1–7) of the Nile delta (adapted from Tauber, 1965). Ldc, long distance transport pollen; Cc, canopy air-borne; Ac, alluvial input; Gc, colluvial input; Arc, aerial resuspension input; Lc, local input

such as semi-arid and tundra sites, the pollen transport models of Tauber (1965), applicable to temperate woodlands, have to be modified. As proposed by Ayyad (1988), pollen transport in semi-arid environments (Fig. 7.2) involves:

1. A long-distance component (Ldc), comparable to that of Tauber, the importance of this component depending on local pollen productivity, prevalent wind direction at flowering times and storm frequency and movement.
2. A canopy component (Cc) as given by Tauber. Some pollen produced within the canopy, or escaping from below, will be carried by air currents above the canopy itself and some may be trapped by eddies in the surface of the canopy and slowed down so that it sinks through the canopy. The canopy is unlikely to be dense and is often composed simply of low shrubs, many of them entomophilous.
3. A water-fed movement of pollen at ground level in the form of sheet flooding, the alluvial component (Ac).
4. A gravity-fed movement of pollen, the colluvial component (Gc).
5. An aerial resuspension component (Arc), depending on the importance of sand storms.
6. A local pollen component (Lc), from the local vegetation.

This model differs from that of Tauber mainly because of the structure of the vegetation. The importance of each of the components varies according to the biological and geographical factors at each site; in general, however, the sum of the long-distance and canopy components is greater than the sum of the rest.

7.1.1 Fossil Pollen in Soil Profiles

The interpretation of the type of vegetation from examination of fossil pollen assemblages may be complicated by considerable variation in the pollen spectrum of sites within the same vegetation type. A further complication is that some of the most important species (which may be dominant in the sites) may not be well represented in the pollen rain, e.g. *Juncus, Phoenix* and *Zygophyllum* (Fig. 7.3). Nevertheless, some striking changes in vegetation can be recognized from pollen studies. To elucidate the history of the vegetation, Ayyad (1988) analysed two profiles in the Nile delta, one of a reed swamp and one of a *Juncus* site (Fig. 7.1). The reed swamp profile, 35 cm deep, sampled every 5 cm, was from an area dominated by *Typha domingensis*. The high percentages, up to 75%, of pollen of the Chenopodiaceae in the lowermost layer and the low values, up to 6%, of *Typha domingensis* pollen in this layer suggest that the site

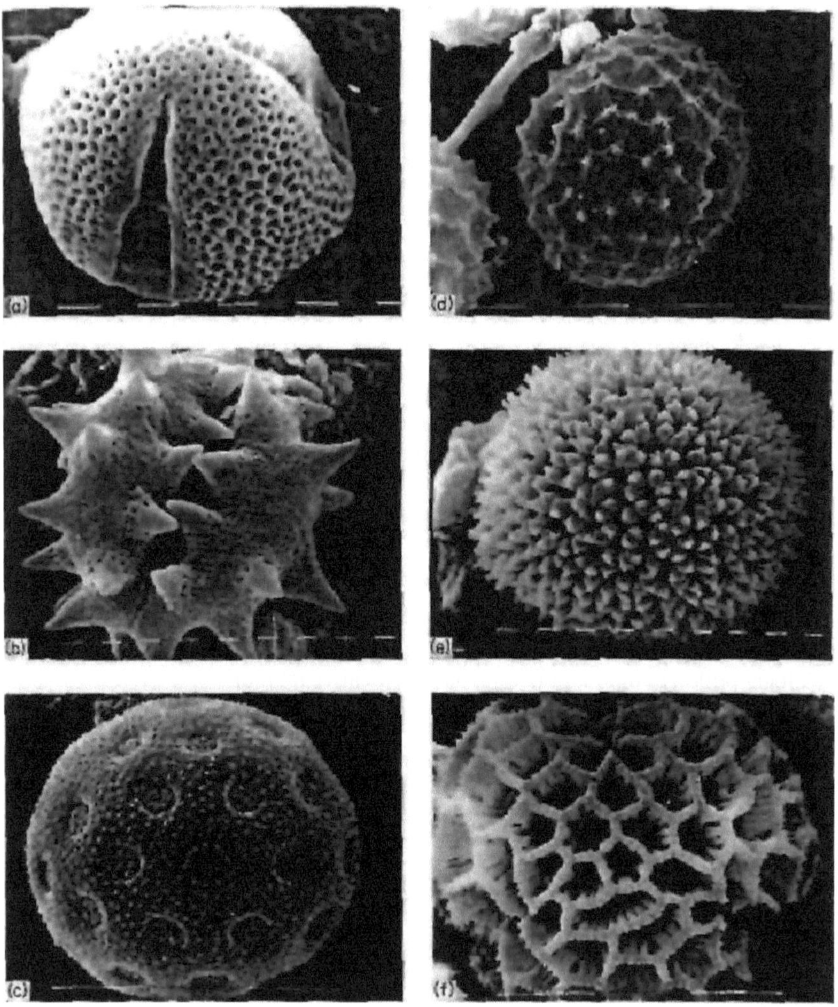

Fig. 7.3 Electron micrographs of pollen of species of the modern Egyptian flora: **(a)** *Zygopkyllum aegyptium,* 16 Jim diameter grains; **(b)** *Conyza dioscoridis,* 36 |im; **(c)** *Silene succulenta,* 60 fim; **(d)** *Carthamus glaucus,* 61 um; **(e)** *Thymelaea hirsuta,* 35 nm; **(f)** *Limonium pruinosum,* 46 |im

was once a salt marsh. Gradual decrease in pollen of the Chenopodiaceae to 6% and increase in *Typha* to 88% in the uppermost layer reflect a change from saline to fresh conditions. Members of the Cyperaceae formed a successional stage in the development of the vegetation. The second profile, 20 cm deep, was in an area dominated by *Juncus*. Pollen of *Juncus* does not survive well in soil and could not be retrieved from the samples. Pollen of the Chenopodiaceae was dominant throughout the profile, this suggesting a salt marsh area. In only one layer (at 10–15 cm) did pollen of *Lotus* dominate, with 75%. This could indicate a very local presence of the genus.

Ayyad (1988) has applied detrended correspondence analysis to the surface pollen and fossil data. Samples rich in fossil pollen of *Typha domingensis* were found to be at the left of the ordination, representing the reed swamp habitat, and samples with the fossil pollen of *Lotus* were at the right, indicating the drier conditions of a sand-dune habitat (Fig. 7.1). She has suggested that this numerical technique has a potential value in unravelling the history of the vegetation, especially in tracing successional developments, and further work should be done on Egyptian environmental history by analysing deeper cores.

7.1.2 Information on the History of Agricultural Activities and of the Vegetation from Archaeological Sites

The history of agricultural activities may be traced by the introduction of palynological methods into Egyptian archaeology. Ayyad et al. (1992a) have presented for the first time a palynological investigation of mudbrick as a potential bearer of information relating to agriculture. Palynological studies have been made of mudbrick samples taken from the Giza Pyramid area and further investigations have been conducted on various archaeological sites of Egypt (Fig. 7.4). In the Tel El-Roba area (the ancient capital of the 16th district of Lower Egypt, 3000 BC), for example, pollen samples were analysed from an excavated site 3 m deep (Ayyad and Krzywinski, 1992). Only one kind of well-preserved pollen {*Vicia faba* type) was found. The highest amounts were at 215 cm depth, indicating the probable ground level at the time concerned and that this legume was cultivated as a food plant in Lower Egypt in ancient times.

Pollen studies of the mudbrick samples have yielded well-preserved pollen grains, mainly of cereal type, and the rest of the pollen assemblage can be considered as consisting mainly of common weeds growing within the crop. The fact that clusters of pollen grains of Gramineae and whole segments of grass stems and leaves were present in the samples indicates that the plant specimens in the bricks were derived from human activity rather than from random natural processes such as transportation by wind and water.

The materials used in making bricks in Ancient Egypt were Nile mud, chopped straw and sand. These were mixed in varying quantities to produce bricks with different characteristics. The process of brickmaking in Pharaonic Egypt was similar to that used today.

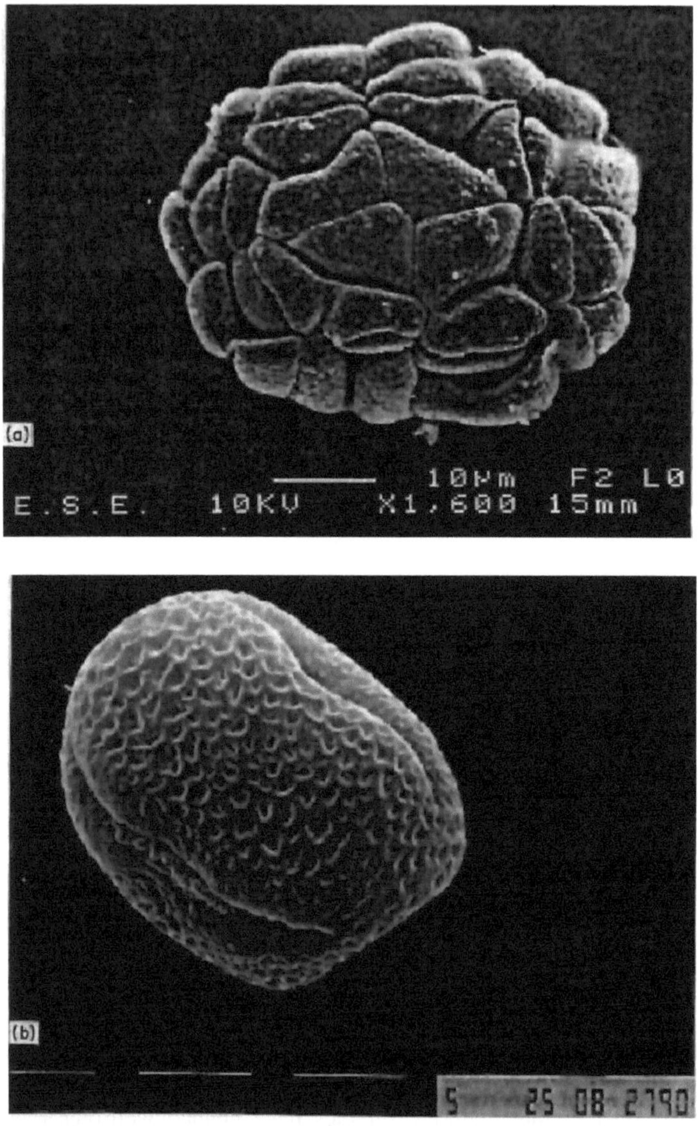

Fig. 7.4 Egyptian Fossil pollen: **(a)** *Acacia albida,* 48 μm, from mudbrick from a template, Giza Pyramids, 2485–2457 BC; **(b)** *Vicia faba,* 26 μm, from the archaeological site Tel El-Roba, Northern Nile Delta, 268–256 BC

This type of study demonstrates that material obtained from mud-brick can be a valuable source of environmental information. One pollen grain of the tree *Celtis integrifolia* and another grain of *Erica* sp. have been found in the mudbrick samples. Presumably these plants (not now known in the Egyptian flora) once grew in Egypt and disappeared with the change of climate, or possibly the grains had been transported from the south with the alluvial sediments of the Nile. Arkell (1949) reported that the population which inhabited Khartoum in Holocene times gathered the fruit of wild *Celtis integrifolia*. Ritchie (1987) found pollen grains of C. *integrifolia* along a core (2.3 m) from a lacustrine sediment in the Saharan Desert of Northern Sudan, radio-carbon dated c. 9500–4500 BC.

The results of this study show that mudbricks can, indeed, serve as cultural time-capsules, particularly concerning the state of agriculture when and where they were made. The ubiquitous distribution of mudbricks in Egypt throughout its history suggests that they have a great potential for providing geographically and chronologically specific data concerning agriculture and vegetational history.

The food plants of prehistoric and Predynastic Egypt have been described by El-Hadidi (1985); some details of these are given in Section 6.3.2(a). Although very few plant remains are known from the Late Palaeolithic, in the Late Pleistocene a considerable range of plants was probably collected for food over the different seasons, and this exploitation may have pre-adapted the early Egyptians to an agricultural mode of life. Changed climatic conditions after 10,000 BP, with decreases in native grasses, may have stimulated the development of agriculture, seed being sown on the banks of the Nile after the recession of the flood waters. Carbonized barley and emmer wheat found in settlements at Fayium and elsewhere in Egypt, dating from about 6000 BP, provide some of the earliest well-established evidence of agriculture in Africa (El-Hadidi, 1985).

From the dawn of recorded history Egypt was a major source of agricultural produce, particularly of grain. In the later phases of its ancient history it was also a major exporter of grain (Garnsey, 1988). Herodotus, the Greek historian, who travelled widely in the eastern Mediterranean and Near East in the middle of the fifth century BC, stated that the delta was the easiest land to work in the known world (Butzer, 1976).

In recent years research into the emergence of incipient farming in the Nile Valley and on its margins has focused on Upper Egypt and the southern part of the Western Desert (Hobler and Hester, 1969). Data acquired during the period of intense excavation in Lower Nubia during the 1960s and 1970s indicated that people there may have made considerable use of ground wild grains as early as 12,000 BP (Wendorf and Schild, 1984). Subsequent finds of naked barley, hulled six-row barley, and hulled barley at Nabta Playa in Upper Egypt indicate that "domesticated" cereals were grown there between 8000 and 7000 BP.

It is paradoxical that, although Egypt is generally recognized as among the first to establish an economy based mainly on large-scale agriculture, Egyptologists have not carried out direct palaeo-environmental or archaeo-ethnobotanical investigations to any great extent; in more recent cultures, however, where agriculture was less important, such investigations have been extensive. There have, nevertheless, been some attempts to reconstruct the history of vegetation in Egypt. Butzer (1959a) was able to build up a knowledge of the natural vegetation of the floodplain in former times (up to some 4000 years ago), by using macro-remains from geological deposits and above all archaeological finds in tombs and from excavations. The botanical material present in the tomb of Tutankhamun has been particularly thoroughly investigated (Hepper, 1990). These studies indicate much about the occurrence of plants, especially those used in garlands, perfumes and for fibre or food, in Pharaonic times when there were dense groves of papyrus (*Cyperus papyrus*) in the Nile delta. The picture of the vegetation which Butzer (1959) constructed was aided by the study of the collection of plants and tree remains from Predynastic and earlier Dynastic times, supplemented by tomb reliefs. Butzer (1959) stated that the characteristic indigenous trees of Egypt were beyond doubt the Nile acacia (*Acacia nilotica* v. *nilotica*) and tamarisk (*Tamarix nilotica, T. articulata*). The Mediterranean species sycamore (*Ficus sycomorus*) was probably also indigenous, as well as for example, the Egyptian willow (*Salix subserrata*). *Balanites aegyptiaca, Ceratonia siliqua, Mimusops schimperi* and *Ziziphus spina-christi*. The wild date palm *Phoenix sylvestris* has been found in the Upper Pleistocene deposits of the Kharga Oasis. Among the characteristic elements of marsh vegetation were *Nymphaea lotus, N. caerulea* and *Cyperus papyrus*. However, a systematic review of the endemic plants did not give a picture of the successive natural vegetation. As Butzer (1959) stated, each element has to be grouped into associations forming individual types of vegetation. With pollen analysis it is possible only to provide analogies derived from the principles of plant geography and plant ecology.

Saad and Sami (1967) studied the fossil pollen content of 22 samples in the delta, at Berenbal, near Rosetta, at depths down to 30 m (lower depths, of marine sand, were devoid of pollen). They described the changes which were believed to have occurred in this site during its past history. This region of the delta was once part of the sea, but after marine regression the Nile water began to reach the area. Xerophytes and halophytes appeared and limited lagoons were formed. Nearly 20,000 years ago, the Nile was blocked, high rainfall converting the area into swamps, cold periods probably following as shown by the pollen of *Betula* and *Ulmus*. More recently rainfall decreased, the swamps diminished and land plants began to flourish. The most recent deposits reflect continued siltation and the stabilization of cultivated land, mycorrhiza being abundant.

Mehringer et al. (1979) analysed pollen of sediments from two cores dating from about 1650 and 1920 AD, and elucidated the history of Birket Qarun and Fayium over the last 325 years. From about 1650 the shallow east arm of the lake was first occasionally dry and then continuously shallow. Maize pollen was first observed in sediments deposited toward the end of this shallow phase which lasted through the 1700s. The deluge of 1817–1818 is represented by redeposited marine hystrichospheres, whereas perennial irrigation, in 1874, is reflected by an increase in pollen of cat's-tail (*Typha*). The recent abundance of pollen of olive and date, accompanied by an increase in cereal pollen, resulted from accelerated agricultural development following World War I. Recently introduced exotic trees were represented by pollen of *Casuarina* and *Eucalyptus*.

El-Shenbary (1985) made palynological studies of 20 samples collected from two caves at different depths in the Mariut ridge at Burg El-Arab. She recognized four different palynological zones in one of the caves. The lower zone (120 cm deep) is dated to Graeco-Roman times, with the presence of pollen grains *of Olea* and *Vitis*. The second zone (100–80 cm deep), characterized by the presence of believed non-native pollen types such as bisaccate pollen (including *Pinus* and

Cedrus) and Pteridophyte spores, might be associated with the ancient Canopic Branch of the Nile. In the third zone (60 cm deep), such pollen grains were absent and plants tolerant of a dry and warm climate became abundant. In the upper zone (40 and 20 cm deep), pollen grains of the Boraginaceae appeared along with those of other plants such as *Ephedra* and *Thymelaea*, indicating a very arid climate. In the other cave, she found a concentricyst form (at a depth of 90 cm), indicating that the sediments were of marine origin and suggesting a marine transgression during that time. The increase of pollen of the Chenopodiaceae in the following bed supports this assumption, this family indicating saline habitats. The uppermost layers included pollen grains of plants that could tolerate an arid and hot climate.

Ritchie (1985), in a study of the modern pollen spectra from Dakhla Oasis, Western Egyptian desert (the hyperarid Eastern Sahara), found small frequencies (5%) of Saharo-Arabian taxa of indicator value and large proportions (30% each) of Cheno-Amaranth and Gramineae pollen. There was also a small proportion of naturalized anemophilous trees (*Casuarina* and *Eucalyptus*), and scattered occurrences in low frequencies of such non-native Mediterranean trees as alder, birch, pine and oak.

Ayyad et al. (1992b) studied the pollen grains of the deltaic Mediterranean coast of Egypt. Surface soil samples collected from a range of vegetation types along sea-landward transects at different points along the coast were palynologically analyzed. Five main groups of vegetation have been recognized (i) beaches and sand sheets, (ii) mobile and fixed sand dunes, (iii) salt marshes, (vi) highly saline areas and salt flats and (v) reed swamps.

The resulting data has been ordinated using multivariate techniques (DECO RANA). The pollen groupings that emerged from this analysis are generally closely related to the vegetation groups from which they were derived, although it proved impossible to separate saline barren areas from the vegetated salt marsh sites on the bases of their pollen grains. In order to test the results against fossi material, two short profiles of alluvial sediment were execrated and their soil samples were analyzed for pollen. Ordination of the data from these soil sections provided information on the past history of the sites, mainly related to local successional processes in these maritime, less arid, environment of Egypt. The results, also, illustrate the potential of surface studies as a bases for environmental reconstruction in the coastal region of the Nile Delta of Egypt.

The eco-palynological studies in the extreme arid part of Egypt's deserts represented by two wadis in the Eastern Desert (Wadis Bir Al-Ain and Qere) and Kharga Oasis of the Western Desert, show that the vegetation of the study area is categorized under 13 communities as follow: 4 communities in Wadi Bir Al-Ain dominated by *Fagonia bruguieri, Francoeuria crispa, Zilla spinosa* and *Zygophyllum coccineum*, 4 communities in Wadi Qena dominated by *Haloxylon salicornicum, Zilla spinosa, Zygophyllum coccineum* and *Copela cinerea* and 5 communities in Kharga Oasis dominated by *Aeluropus lagopoides, Alhagi graecorum, Imperata cylindrical, Tamarix nilotica* and *Lagonychinum farctum* (Zahran, Ayyad and El-Khatib, 1995). Pollen grains analysis included the preparation of type slides from fresh and erbarium material and extraction of fossil pollens from soil sediments. Three profiles were dug in Wadi Bir Al-Ain, Wadi Qena and Kharga Oases at depths 240 cm, 300 cm and 300 cm, respectively. Soil samples were collected every 20 cm (in Wadi Bir Al-Ain) and 100 cm (in Wadis Qena and Kharga Oasis). The results clearly elucidate that pollen grains of members of 33 families were recorded in the current and old vegetation types of the studied wadis and Oasis in addition to *Acacia* spp. These are: 14 families and *Acacia* spp. occur in both the current and ancient vegetation types mamely: Asclepiadaceae, Boraginaceae, Capparaceae, Chenopodiaceae, Compositae, Cruciferae, Cyperaceae, Euphobiaceae, Gramineae, Leguminosaea, Palmae, Resdaceae, Tamaricaceae and Zygophyllacea, 13 families were recorded in the current vegetation only namely: Aizoaceae, Cleomaceae, Convolvulaceae, Cucurbitaceae, Cynomoriaceae, Frankeneaceae, Gentiaraceae, Juncaceae, polygonaceae, Ranunculaceae, Scintalaceae, Scrophulariaceae and Solanaceae. On the other hand, pollen grains of members of 6 families (Amaranthaceae, Casuarinaceae, Liliaceae, Malvaceae, Myrtaceae and Typhaceae) have been recorded in the ancient vegetation only.

The current and past vegetation types of Kraman Islands, River Nile, Sohag Governorate, Upper Egypt is described by El-Khatib (1997). Two main habitat types are recognized wet and dry. In these habitats the distribution of the plant communities of the current vegetation is controlled by edaphic and hydrological factors. Four communities dominated and co-dominated by six species, namely: *Eichhornia crassipes, Ceratophyllum demersam, Phragmites australis, Typha domingensis, Lotus arabicus* and *Ranunculus sceleratus* inhabit the wet habitat. The dry habitat is occupied by *Francoeuria crispa, Gnaphalium luteo-album, Tamarix nilotica, Desmostachya bipinnata* and *Imperata cylindrical* communities. Twenty two associate species had been identified including, for example: *Myriophyllum spicatum, Cyperus rotundus, Juncus bufonus, Cyperus articulatus*, and *Typha domingensis* (wet habitat) and *Solanum nigrum, Plantago major, Ambrosia maritima, Polypogon monspeliensis* and *Alhagi graecorum* (dry habitat).

Past vegetation of the Kraman Island has been determined through investigation of the pollen grains in soil samples collected at 5 cm downward distances from a 60 cm profile dug in a vegetationless area. Cluster analysis of pollen grains percentages at different levels of the profile shows that pollen assemblages are separated into three distinct groups. Group A comprises families of aquatic habitat namely: Haloragaceae, Pontederiaceae, Potamogetonaceae, and Ceratophyllaceae. These families of the hydrophytes were not represented in the upper strata (0–5 cm and 5–10 cm) but have considerable percentages in the middle ones (15–40 cm), in the lower strata (45–60 cm), their pollen assemblages attain their highest values: 22%, 34%, 23% and 21%, respectively. Group B includes the pollen assemblages of familes of the swampy habitat (helophytes) namely: Cyperaceae, Typhaceae, Polygonaceae and Poaceae. The dominance of these assemblages is the main character of the middle level of the profile. Their pollen percentages ranged between 10%–20%, 10%–22%, 3%–13% and 17%–31%, respectively.

Group C is represented by the only land families, mamely: Tamaricaceae, Fabaceae, and Brassicaceae. The pollen assemblages of these families increased upward the profile reaching in the upper statum 20%, 25%, 10%, and 8%, respectively. In the lower layer their pollen grains disappeared. On the other hand, the deposit pollen grains of families Ranunculaceae, Solanaceae, Convolulaceae, Chenopodiaceae, Amaranthaceae, Boraginaceae and Asclepiadaceae, contributed only to a small part of the pollen assemblages especially in the upper strata of the profile. This reflects change from hydric to mesic habitats.

On the basis of pollen assemblages, one may conclude that the studied island was under different changes in the habitat conditions, strong fluctuation and change in over flooding by the Nile water within the last years. Thus, in the largest part of the island, plants of medium water requirement (mesophytes) replaced the hydrophytes of the past vegetation.

These results demonstrate that there have been considerable changes in the vegetation of Egypt over the past 20,000 years (since the last glacial advance in higher latitudes). However, these studies are still at a very early stage and much remains to be discovered about the vegetation and flora of Egypt in former times.

7.2 Future Study of Phytosociology and Plant Ecology

The accounts of the plant communities and ecological features of the major species previously described here may be considered to provide the scientific basis of further studies on the vegetation of Egypt which are needed. The chief ecological factors controlling the distribution and success of species need to be quantitatively investigated. Also, the Egyptian flora provides considerable scope for research, for example, in plant population dynamics, population genetics and physiological ecology. Environmental conditions range from extremes of aridity and salinity to those of fresh water. Much remains to be elucidated regarding the morphological, phenological, physiological and biochemical characteristics which influence the occurrence and relative abundance of species in the diverse habitats represented in Egypt. Phytosociological units have frequently in the past been recognized on a subjective basis, taking into account the general appearance (physiognomy) of the vegetation, the dominant species and their associates. In this way communities have been characterized based on the English tradition developed by A.G. Tansley (see, e.g. Tansley, 1939), in which the importance of features of the habitat in influencing communities is reflected as well as the succession of vegetation, culminating in a climax usually determined by climatic, edaphic or biotic factors. An alternative approach involves the phytosociological units of the Zurich-Montpellier school which have been progressively adopted. In this continental phytosociology constancy and "character-plants" are used as a basis for characterization and the need is stressed to select uniform stands for study (for relevées). Now objective methodology and quantitative procedures are being increasingly used, vegetational units being recognized on the basis of techniques involving classsification and ordination.

As yet there are relatively few investigations in Egypt which are based on an objective and quantitative approach and there is much scope for such studies. One investigation of this type has been made by Abdel-Razik et al. (1984) in a 20 km transect, perpendicular to the shore 80 km west of Alexandria. This transect passed through sand dunes, salt marshes, ridges and saline and non-saline depressions. Multivariate analysis (involving indicator species analysis and reciprocal averaging) applied to vegetational variation in these different habitats showed the importance of salinity and fertility gradients as well as of soil texture. The species *Asphodelus microcarpus, Echiochilon fruticosum* and *Plantago albicans* correlated with high sand percentage (and occur on loose sandy soil) whereas the group of *Helianthemum stipulatum, Limonium pruinosum, Pituranthos tortuosus* and *Scorzonera alexandrina* correlated with pH, organic matter and silt percentage on the environmental gradients, in agreement with the occurrence of these species on compact fertile soils. The study by Serag (1991) on the canal bank vegetation of the deltaic region of the Nile involving classification and ordination showed the presence of 16 distinct communities, the major edaphic influences being carbonate and organic carbon content, pH, salinity and water-holding capacity. Another objective investigation, by Dargie and El Demerdash (1991) was of two contrasted environments, the Sinai coastal plain and part of the Eastern Desert, by use of quantitative procedures involving ordination and "integrated interpretation". In the Sinai coastal plain, moisture content and water quality were shown to be the major controls of the vegetation; in the Eastern Desert moisture status was also of much importance but so too was disturbance from grazing and the collection of wood for fuel. "Dummy" variables were found to be useful for factors difficult to quantify on limited field visits; it is suggested that future work on desert vegetation should place less emphasis on sampling surface soil and concentrate on assessing critical factors such as depth to water-table, water quality and grazing pressure.

Although allelopathic effects (growth inhibition of its neighbouring plants by a plant releasing chemical substances) are known for many desert plants, including *Imperata cylindrica*, these have been little studied in Egypt. *Artemisia herba-alba* has been shown to suppress annnuals in the Negev desert of Israel (Friedman et al., 1977) and *Tribulus terrestris to* suppress annuals in an abandoned field in the desert of Kuwait (El-Ghareeb, 1991); phenolics of the leachate from shoots of *T. terrestris* were found to inhibit germination and radicle extension of associated annuals. There is scope for investigations of this kind in Egypt. For example, Eisa (2007) stated that rhizomes of *Imperata cylendrica* excude allelopathic substances which inhibit the growth of other plants. This grass usually grows in pure stands.

Only few studies have been made of mycorrhizal relationships in spite of their established importance in vegetation in many parts of the world. However, vesicular-arbuscular mycorrhizas are reported to be widespread in major crops in Egypt (Ishac and Moustafa, 1991). The composition of VA-mycorrhizas is found to vary with host plants, but spores are mostly of the genus *Glomus*. Mycorrhizal inoculation is known to promote crop yield and to reduce the incidence of disease in Egypt, but much further research on mycorrhizal relationships is needed, especially of native species.

Root nodules of members of the Egyptian flora have also been little studied, but many species of the Leguminosae have roots with nodules containing nitrogen-fixing *Rhizobia*. Although a number of species *of Acacia* bear active bacterial root nodules, in some species, e.g. *Acacia tortilis*, they are rare or absent and their ecological importance needs investigation.

7.3 The Main Types of Vegetation and Its Features: Synopsis

Although in an arid region of the world, Egypt has a rich natural flora which may be classified under seven major types of vegetation to provide a summary statement of the salient features of the plant communities represented:

1. Desert vegetation
2. Salt marsh vegetation
3. Mountain vegetation
4. Sand dune vegetation
5. Reed swamp vegetation
6. Fresh-water vegetation
7. Saline water vegetation

The desert vegetation is by far the most important and characteristic type of the natural plant life of Egypt. It covers vast areas and is formed mainly of xerophytic shrubs and undershrubs, e.g. *Anabasis, Hammada, Leptadenia, Lycium, Thymelaea* and *Zilla*, robust grasses, e.g. *Panicum*, and a few trees of varied height and vigour, e.g. *Acacia* and *Balanites*, usually growing in the large wadis of the desert. In rainy periods short-lived plants (ephemerals, annuals and biennials) may change the yellow desert into a green carpet. The germination characteristics of many of these plants are distinctive and merit further study.

Salt marsh vegetation is the second most important type of vegetation in Egypt, occurring in the extensive salt-affected areas along the coast but also in the inland oases and depressions. This vegetation is also present in the Nile region, particularly in neglected areas. Succulents, e.g. *Arthrocnemum, Halocnemum, Salicornia* and *Zygophyllum*, make up a large component of the vegetation; halophytes are chiefly excretive, e.g. *Aeluropus, Limoniastrum, Sporobolus* and *Tamarix*, but a few are cumulative, e.g. *Juncus*.

The mountain vegetation is restricted to the high lands: chains of mountains along the Red Sea coast, mountains of southern Sinai and the Uweinat mountains in the most southwesterly area of the Western Desert. Although these mountains are in extremely dry deserts, rainfall is relatively high, with a more favourable climate for plants than in other parts of the country. The flora includes *Caralluma, Cocculus* (liana), *Dodonaea, Dracaena, Moringa* and *Rhus*, typical of the mountains. In the high parts of the Sinai Mountains, air temperatures are lower than elsewhere in Egypt, being usually below freezing point in winter; some plants of fairly cold regions of the world occur here, e.g. *Juniperus*.

Sand dune vegetation is both coastal and inland. Along the Mediterranean coast psammophytes such as *Ammophila arenaria, Euphorbia paralias* and *Silene succulenta* are frequent in the lines of dunes. Smaller dunes on the Red Sea coast, especially in the southern part, are dominated by *Halopyrum mucronatum.* This grass builds dunes in very limited coastal areas in Egypt, and also in Saudi Arabia and Eritrea. Fixation of the inland dunes of the oases and depressions of the Western Desert of Egypt is a pressing need; major dune builders here are the grasses *Aristida* and *Stipagrostis*; also *Populus euphratica* (present in the Siwa Oasis) has potential in this respect. Further study is desirable on the autecology of these and other dune-fixing plants.

Reed swamp and fresh-water vegetation is widespread in the Nile region. Many of the species are troublesome weeds which need to be controlled and further studied. *Phragmites australis* and *Typha domingensis* are the most common "reeds", but *T. elephantina* is restricted in Egypt to Wadi El-Natrun Depression. *Eichhornia crassipes* is the most troublesome floating hydrophyte, causing problems in the Nile and its irrigation and drainage systems. Submerged hydrophytes include *Potamogeton* spp. and *Ceratophyllum* spp. which also need to be controlled. On the other hand, vegetation of saline water is much limited. However, submerged marine vascular plants (seaweeds or sea-grasses) such as *Cymodocea, Halophila* and *Posidonia* are a notable feature of the coastal waters of Egypt. Mangrove vegetation is absent from the Mediterranean and the northern parts of the Gulfs of Suez and Aqaba, being present only along the Red Sea coast of Egypt from Hurghada southwards as well as at Ras Muhammed at the southern extremity of the Sinai Peninsula. *Avicennia marina* is the dominant mangrove but *Rhizophora mucronata* occurs in a few stands in the most southerly part of the coast. A national project aimed at establishing mangrove trees along the Egyptian coast should be considered; coastal areas might then be changed into green forests, with the woody plants being of economic importance. Fortunately, this has been realized through a project entitled "Assessmant and Management of Mangrove Forest in Egypt for Sustainable Utilization and Development" funded by ITTO, Japan and supervised by MALR, MSEA and EEAA, Cairo during September 2003–December 2007 (Anonymous, 2006).

Chapter 8
Remote Sensing and Vegetation Map of Egypt

B.B. Salem,[1] Andersen, G.L.[2] and Zahran, M.A.

8.1 Introduction

Remote sensing is an extensive science drawing from many areas for support and development. It is an interesting and exploratory, as it provides images of areas in a fast and cost-efficient manner, and attempts to demonstrate what is happening right now in a study area. Satellite and digital imagery play an important role in remote sensing; providing information about land studies. Remote sensing (RS) can be defined as "the science and art of obtaining information about an object, area or phenomenon through the analysis of data acquired by device that is not in contact with the object, area or phenomenon under investigation" (Lillesand and Kiefer, 1994). It (RS) was born to serve military purposes. Aerial photography was the first stage and in the past world war period it was adapted to civilian use.

Remote sensing provides important coverage, mapping and classification of land cover features, such as vegetation, soil and water. The benefits of remote sensing continue to arise. It can be used to assess hard to reach areas for field work and provides a more detailed permanent and objective survey that offers a different perspective. Regarding vegetation distribution on the ground and assessing its changes with time, remote sensing mapping proved to have many advantages over the other traditional methods (Lewis et al., 2001; Salem, 2003, 2007). One of the primary applications of remote sensing is to identify patterns of vegetation distribution on the ground and to assess changes in vegetation over time. Remote sensing has several advantages over traditional methods of vegetation mapping, but also some limitations. The focus of vegetation detection by remote sensing has shifted to the interpretation of satellite imagery. Images in digital format allow for numerical

[1] Departmenet of Environmental Science, Faculty of Science, Alexandria University, Alexandria, Egypt.

[2] Department of Biiology, University of Bergen, Bergen, Norway.

M.A. Zahran, A.J. Willis, *The Vegetation of Egypt*,

processing and analysis and the application of multivariate classification methods to the data.

Spatial analysis has become the most rapidly growing field in ecology (Fortin and Dale, 2005). Wiens (1989) stated that, landscape, ecosystems and habitats are changing rapidly and thus, the importance of time, space and scale has become clearer. However, our ability to actually incorporate and study spatial and temporal processes is largely connected to the recent development in computer and satellite technology. According to Goodchild (1994), Pettorelli et al. (2006) & Kerr and Ostrovsky (2003), Remote Sensing (RS), Geographical Information System (GIS) and Global Positioning System (GPS) are presently important tools that have facilitated studies of spatial and temporal processes within ecology and vegetation sciences.

Since the evolution of civilian satellite based remote sensing in the 1970s, RS imagery has played an essential role in the worldwide focus on ecological and vegetational changes in arid lands. The NOAA-AVHRR, primarily a weather satellite, with low spatial but high temporal resolution was used to monitor changes in African dry lands (Sahel) on continental scale. At that time, the first civilian earth observation satellite, Landst MSS had a comparatively high spatial resolution but a low temporal resolution and was designed to monitor changes on landscape scale. In recent years, MODIS has started to follow up the NOAA-AVHRR monitoring of changes across African dry lands reflecting the greening and disappearance of ephemeral vegetation within short-term rainfall fluctuations (Pickup et al. 1993; Anyamba and Tucker, 2005). Moreover, according to Andersen (2007), Landsat MSS and TM imagery are too coarse to detect scattered trees in hyper-arid areas or to separate trees from ephemeral vegetation. It is only in high resolution aerial photos or satellite images that individual trees can be detected. Because its high spatial resolution, CORONA KH-4A imagery has been a mean source of data permitting individual trees to be detected at the beginning of a period 1965–2003 which is long enough to give essential information on the long-term dynamics of these trees.

The ability to monitor vegetation using optical sensors is related primarily to the interference of leaves with sunlight.[3] F Both the function (photosynthesis) and structure of the leaf generate a special spectral signature, characterized by very high absorption in the red wavelengths (620–700 nm) and very high reflectance in the near infrared region (740–1100 nm). According to these characteristics the different channels (bands) of optical sensors are defined. As well as red and near-infrared bands, optical sensors typically also have blue and green bands.

Classifying vegetation, using remote sensing is a valuable tool to determine vegetation occurrence and distribution. Also, the influence of the various environmental factors e.g. latitude, altitude, length of the growing season, solar radiation and other climatic factors, soil types, topographical aspects, pollutants etc. could be detected. Interpretation of satellite images of vegetation becomes easier if the

[3] Optical sensors are passive as they detect natural energym typically reflected sunlight. The other main type of remote sensor includes active ones, e.g. radar, that emit energy.

researcher understands what plants are native to the area and what influences their growth and distribution. There are several factors that influence the reflectance quality of vegetation on satellite and remote sensing images. These include brightness, greeness and moisture. Brightness is calculated as a weighted sum of all the bands and is defined in the direction of principal variation in soil reflectance. Greeness is orthogonal to brightness and is a contrast between the near-infrared and visible bands. It is related to the amount of green vegetation in the scene. Moisture in vegetation will reflect more energy than dry vegetation. Leaf properties that influence the leaf optical properties are the internal or external structure, age, water status, mineral stresses, and the health of the leaf. It is important to note that the reflectance of the optical properties of leaves are the same, regardless of the species. What may differ for each leaf, is the typical spectral features recorded for the three main optical spectral domains; leaf pigments, cell structure and water content. The Electromagnetic wavelengths also affect different parts of plants. These parts include leaves, stems, stalks and limbs. The wavelengths of each specific spectral band play a role in the amount of reflection that occurs. For example, tree leaves and crop canopies reflect more in the shorter radar wavelengths, while tree trunks and limbs reflect more in the longer wavelengths. The density of the tree or plant canopy will affect the scattering of the wavelengths. Within the electromagnetic spectrum, bands will produce different levels of reflectance rates. For example, in the visible bands (400–700 nm), a lower reflectance will occur as more light will be absorbed by the leaf pigments than reflected. The blue (450 nm) and red (670 nm) wavelengths include two main absorption bands that absorb two main leaf pigments. Field work (ground truthing) should still be undertaken to provide specific information that can not be obtained from images. Vegetation change ranges from the in-growth of a single tree to the entire deforestation by clear cut. Whether we can detect and monitor vegetation changes by remote sensing data depends on the spatial and temporal characteristics of the change and the type of remote sensor data to be used. Therefore, it is important for the plant ecologists to understand the nature of vegetation changes prior to analyzing remote sensing data (Salem, 2003, 2007).

8.2 Case Studies

Though hundreds of ecological studies had been and are still conducted on the natural vegetation of Egypt, yet the vegetation map of the country has not been yet prepared. Fortunately, during the last decades, plant ecologists and Remote Sensing specialists are actively working to produce the required vegetation map e.g. Salem (2003), Youssef et al. (2004, 2007), Salem and Wassem (2003) and Andersen (2007). The following pages throw light on three case studies about three vegetation types in three different habitats of Egypt, namely: mangrove vegetation, xerophytic scrublands and Moghra Oasis. In the three cases Remote Sensing technology and field studies (ground truthing) were together used.

8.2.1 Mangrove Forests

8.2.1.1 General Remarks

Mangrove forests are a characteristic feature of the shorelines of the tropical and subtropical seas and oceans, however, their optimum density, diversity and cover is in the wet tropics. Some mangrove swamps occur in the coastlines of the arid areas like those of the Red Sea and Arabian Peninsula's coastal belts (Zahran, 2007a). In Egypt, as mentioned before in Chapter 4, the mangrove forests comprise two main species, namely: *Avicennia marina* and *Rhizophora mucronata*. The distribution of these two mangrove species in Egypt's coastal belts is of ecological interest. Both are totally absent from the Mediterranean coast. On the other hand, *A. marina* grows and predominates along the whole stretch of the Red Sea Coast starting from Hurghada (Latitude 27° 12), 400 km south of Suez, southwards to Mersa Halaib (700 km south of Hurghada) in the Sudano-Egyptian borders (Latitude 22 °N). Though absent from the eastern and western coasts of the Gulf of Suez, *A. marina* occurs in some stands in the southern section of the Gulf of Aqaba as well as in the Ras Muhammed Cape of the Sinai Peninsula. The northernmost stand of *A. marina* of the Gulf of Aqaba in Al-Manquatta (Latitude 28°12 N) represents the northernmost *A. marina* forest in the world (Zahran, 2007a). *Rhizophora mucronata* occurs only in the most southern section of the Red Sea coast starting from Shalatein (Latitude 23°28 N) southward to Mersa Halaib (Latitude 22 °N). The presence of the mangrove forests is not only restricted to the main coastal belts of the Red Sea and Gulf of Aqaba but also in the shorelines of the offshore islands e.g. Abu Minqar, Safaga, Devil, Wadi El-Gemal, Sooma, etc …

According to Saenger (2003) and Anonymous (2006), the total area occupied by the mangrove vegetation in Egypt is about 525 ha (1250 feddans) in sites with different areas distributed along the shoreline of the Red Sea and its islands and Sinai Peninsula.

8.2.1.2 Mapping of Mangrove Forests in Egypt

With increasing population and developing techniques, people may be considered a main cause of environmental changes. The land cover, including natural vegetation, is altered primarily by direct and/or indirect human use. Youssef and Ghallab (2007) stated that, data of remote sensing satellite have become reliable for different aspects such as monitoring, mapping, infrastructure, and sustainable development.

In Egypt, mangrove forests are facing environmental threats e.g. oil pollution, overcutting, overgrazing, denudation etc … These factors are, unfortunately, causing great loss of wide areas of these valuable plants. For the conservation, management and sustainable utilization of the mangrove forests and their unique ecosystem, it is necessary to have comprehensive, accurate and up-to-date information about their present and past statuses.

Based on the relevant published works on the mangrove vegetation of Egypt, namely: Ferrar (1914), Zahran (1965, 1977) and Kassas and Zahran (1967), ground truthing studies were conducted through two projects: the first was funded by FAO (2002–2003) and the second was funded by the ITTO (2003–2006), Saenger (2003) and Anonymous (2006). Both projects were supervised by EEAA, Cairo.

Through the cooperation between NARS (National Authority for Remote Sensing and Space Sciences, Cairo) and EEAA, the mangrove forests in Egypt were mapped using remote sensing satellite images (Landsat MSS) in 2 years: 1975 and 2002. To prepare the required images, 10 major sites representing the mangrove forests in Egypt along the coasts of the Red Sea, two islands, Gulf of Aqaba and Ras Muhammad of the Sinai Peninsula were selected. These sites are:

Site 1: El-Ghargana (Nabq) swamps, Gulf of Aqaba coast, Sinai Peninsula.
Site 2: Ras Muhammad swamps, Head of Sinai Peninsula.
Site 3: Abu Minqar Island (facing Hurgada city), 400 km south of Suez, Red Sea coast.
Site 4: Safaga Island (facing Safaga city), 460 km south of Suez, Red Sea coast.
Site 5: South Safaga city, 478 km south of Suez, Red Sea coast.
Site 6: Qusseir (El-Quseir) site, 540 km south of Suez, Red Sea coast.
Site 7: Wadi Gemal (Mersa Alam) site, 730 km south of Suez, Red Sea coast.
Site 8: Hamata site, 780 km south of Suez, Red Sea coast.
Site 9: Mersa Shaab site, 960 km south of Suez, Red Sea coast.
Site 10: Mersa Abu Fissi (Halaieb) site, 1100 km south of Suez on the Sudano-Egyptian border.

Ten Remote Sensing Images for the above mentioned 10 sites were prepared. Each image includes information about the location, area in 1975 and area in 2002 of that particulate site. These areas are tabulated in Table 8.1and Fig. 8.1, the data of which clarify the following points:

Table 8.1 Areas of the mangrove forests in the ten sites surveyed by the Remote Sensing Technology along the Red Sea coast, Gulf of Aqaba, and Ras Muhammad coasts, South Sinai, Egypt in 1975 and 2002 (+ = area increased, − = area decreased).

Mangrove sites	Areas (m^2) in 1975	Areas (m^2) in 2002	Remarks
A. Sinai Peninsula coasts			
1. El-Ghargana (Nabq)	348	462	+14 m^2
2. Ras Muhammad	270	350	+80 m^2
B. Red Sea coast			
3. Abu Minqar Island (facing Hurgada)	333	276	−57 m^2
4. Safaga Island (facing Safaga)	275	269	−6 m^2
5. South Safaga	660	466	−94 m^2
6. El-Qusseir	89	81	−8 m^2
7. Wadi Gemal (Mersa Alam)	291	171	−120 m^2
8. Hamata	295	205	−90 m^2
9. Mersa Shaab	769	632	−137 m^2
10. Mersa Abu Fissi	675	580	−95 m^2

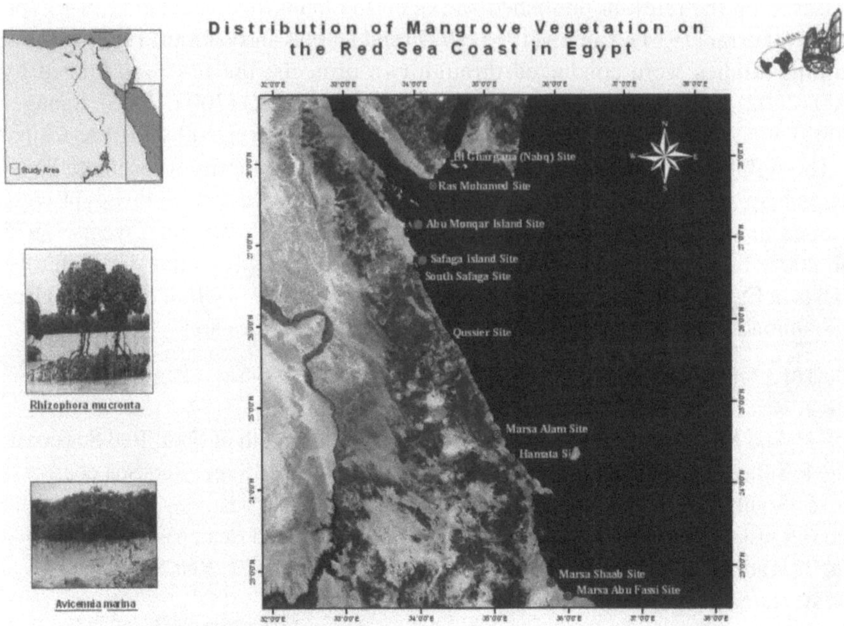

Fig. 8.1 Map showing the location of the selected 10 mangrove sites, Red Sea coast, Egypt

1. The area of the mangroves forests in the two sites of Sinai Peninsula (Nos. 1 & 2) increased from 348 m^2 to 462 m^2 (+14 m^2) in Nabq site and from 270 m^2 to 350 m^2 (+80 m^2) in Ras Muhammad Site.
2. Areas of all the eight sites (Nos. 3–10) of the mangrove forest along the Red Sea coast decreased. Reduction in areas varied.
3. The highest reduction has been detected in Mersa Shaab site (No. 9), being 137 m^2 (769 m^2 on 1975 and 632 m^2 on 2002), followed by that of Wadi Gemal site (No. 7), being 120 m^2 (291 m^2 on 1975 and 171 m^2 on 2002), then the area of Mersa Abu Fissi site (No. 10), being 95 m^2 (675 m^2 on 1975 and 580 m^2 on 2002), then the area of South Safaga site (No. 5), being 94 m^2 (660 m^2 on 1975 and 466 m^2 on 2002), then the area of Hamata site (No. 8), being 90 m^2 (295 m^2 on 1975 and 205 m^2 on 2002), then the area of Abu Minqar Island site (No. 3), being 57 m^2 (333 m^2 on 1975 and 276 m^2 on 2002). Low reductions in the areas of mangrove forests (−8 m^2 and −6 m^2) are noted in El-Qusseir and Safaga sites (Nos. 6 and 4) respectively.

The question that may be raised: Why did the areas of the mangrove forest of South Sinai coast increased while those of the Red Sea coast decreased?

Statistically, it is not possible to answer this question, but through the field observation of Prof. Zahran, ex-excecutive manager of the ITTO project, he recognized that man's uncontrolled interference and impacts are the major factors causing the deterioration of the mangrove forest along the Red Sea coast. The reverse in true for the mangrove forests in the South Sinai swamps where local people are few and are quite aware about the importance of these forests, they do not allow their livestock to graze the leaves of *A. marina* all the time. Also, they cut the wood of the mangrove trees in a very limited way. Apart from that, because the sites of the mangrove forest of South Sinai are among the attractive coastal places for the thousands of tourists visiting Sharm El-Sheikh resort, they are, to some extent, under the control of the rangers of the South Sinai Protectorate (EEAA). One of the responsibilities of these rangers is to keep the mangrove ecosystem away from any interference of the tourists. On the other hand, along the Red Sea coast, the situation is different; in addition to the uncontrolled wood cutting, camel trade is centered in the southern section of the Red Sea coast particularly in Shalateen area on the border between Egypt and Sudan. The normal route of the herds of camels traveling from Eritrea and Sudan to Egypt is the Red Sea coast. There is no alternative to feed these animals except the mangrove leaves. In addition, along the Red Sea coast, the establishment of new resorts for tourists is causing destruction of wide areas of the mangrove forest. Oil pollution is, in fact, another sever impact killing great number of not only mangrove trees but also other biota (fishes, etc …) living in the mangrove ecosystem.

To overcome this problem, we propose the following proposals:

1. More and extensive studies aiming at conservation, rehabilitation, management, and re-establishment of the mangrove forests in Egypt should be implemented.
2. Programs of public awareness for the local inhabitants should be considered.
3. Providing a feeding alternative for the local and transient animals (camels).
4. Prohibiting the establishment of new tourist resorts in the areas of mangrove forests.
5. Regular cleaning of the mangrove swamps from all types of pollutions e.g. oil, rubbish etc …
6. Strengthen the observation system of the rangers of the Red Sea and South Sinai protectorates to enable them to protect these valuable forests.
7. Adopting a scientific programme for sustainable development of the mangrove ecosystem in Egypt. Local and/or international authorities could be invited to support the implementation of the proposed program.

8.2.2 Xerophytic Scrublands

8.2.2.1 General Remarks

The major portion of the earth's land surface (about 41%) is drylands and deserts housing more than 2 billion people. In Egypt deserts cover more than 96% of its total area. But contrary to common belief deserts are not barren. Xerophytic trees inhabiting the wadis dissecting the Eastern Desert and Sinai Peninsula have for millennia been a primary natural resource for pastoral nomads, securing not only fodder and shade for their animals, but also wood and charcoal as fuel for domestic use (Krzywinski and Pierce 2001). During long dry periods trees are the only dependable vegetation resource. They provide microhabitats for other biota, are islands of soil fertility and can be considered as key species in the desert ecosystem.

High mortality and lack of recruitment of trees such as *Acacia tortilis* have been reported from arid regions elsewhere, and there are reasons to believe that the tree populations in the Eastern Desert too are in decline, mainly due to over-exploitation for non-traditional, commercial charcoal production. These essential resources need protection, but proper future monitoring and sustainable management strategies require better understanding of spatial and temporal trends among these tree populations. Such insight can only be obtained by analyzing variation and change between the present and the past; however, historical data with sufficient details arc few.

A valuable and underexploited source of historical data is imagery from the first generation US photo-reconnaissance satellite system CORONA. It operated between 1960 and 1972 and provided panchromatic photos with a best resolution of 1.8 m for the camera system KEYHOLE (KH) 4B. The CORONA image archive remained unknown to the civilians until it was declassified in 1995. Andersen (2006) and Andersen and Krzywinski (2007) present a comprehensive investigation of the potential and limitations of high resolution CORONA imagery for surveying and monitoring spatio-temporal changes in the xerophytic arboreal wadi vegetation of the Eastern Desert of Egypt.

8.2.2.2 The Eastern Desert

This study area is located in the coastal mountain range of the Eastern Desert (Fig. 8.2) between 24 °N and 25 °N parallels of latitude. In this hyper-arid region the coefficient of variation of rainfall reaches 200%. In addition to scattered showers and oreographic rain important water resources are dewfall, mist, fog and ground water (Kassas and Zahran 1971). The mountainous desert landscape with peaks reaching nearly 2000 m a.s.l. combined with its proximity to the sea results in a wide variation in moisture conditions due to the the topo-hydrology of the wadi system. These wadis house arboreal vegetation, e.g. *Acacia tortilis* and *Balanites aegyptiaca*; trees having long roots enabling them to reach deeply seated and permanently moist soil layers.

8.2.2.3 Tools to Study Population Dynamics and Change

Studies of change require detailed, comparable and accurately positioned observations from the past. The ability to detect trees from the panchromatic, high-resolution (2.7 m) CORONA KH 4A data was evaluated using present day (2003) field observations from 19 wadi sites as a reference (Fig. 8.2). A spatial overlay analysis compared tree maps derived from digitized, rectified and manually

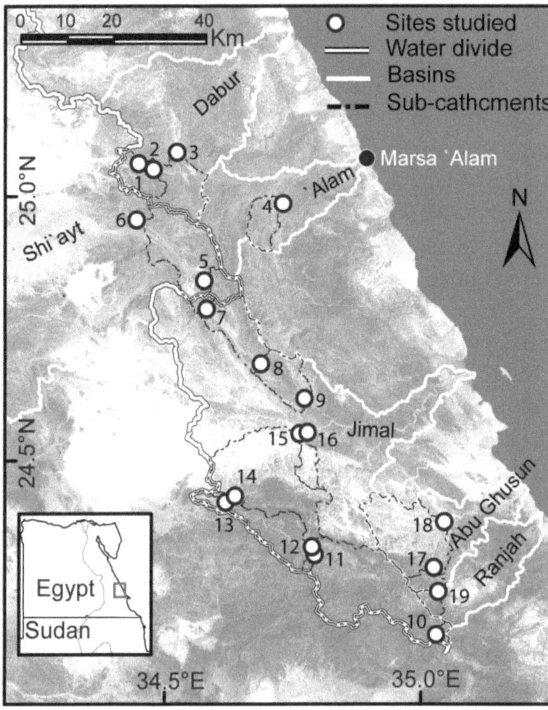

Fig. 8.2 Study area. Studied sites in relation to basins, sub-catchments and east-west water divide. A Landsat TM image is displayed in the background

interpreted CORONA imagery (1965) to present day reference data. Image interpretation was done independently of field data. Because of spatial errors in GPS (Global Positioning System) measurements, the identification of the same tree in both datasets was based on a buffer distance of 20 m. Individuals with a canopy area smaller than 6 m^2 were excluded from the analysis in accordance with image resolution (2.7 m at best). Recruitment and mortality estimates derived for all sites were interpreted in relation to topo-hydrological data extracted from the Shuttle Radar Topography Mission (SRTM) digital elevation model in order to detect possible effects of moisture conditions (Andersen and Krzywinski, 2007).

During field work in February and March 2003, individual trees of any size were mapped in detail at each site selected. Spatial positioning was based on GPS measurements from a handheld receiver (Garmin 12XL) with an external, extendable antenna that was used to receive GPS signals above the canopy. Several parameters describing the population structure were measured and each tree was photographically documented. Canopy area was estimated by pacing the greatest length of each canopy and the length orthogonal to it, and by assuming a mean circular shape. Pacing was later converted into meters. Traces of use such as browsing and pollarding were also registered for each tree.

8.2.2.4 Results

The spatial overlay shows that 382 (66%) of 576 individuals registered in the field with a canopy area larger than 6 m^2 are recognizable on the imagery. Of the 194 unrecognized trees, 85% are within a 20 m distances of landscape elements that reduce image contrast and thereby diminish an analyst's ability to interpret the imagery. Among sites, from 9 to 55% of the tree population was concealed by such elements, e.g. other trees, mountain sides, shadows etc. The total number of individuals remaining unrecognized (5%) was, however, reduced to 3% after a reinspection of images with field data overlaid. Some individuals were either not there or at least below minimum detectable size in 1965. However, for larger individuals in 2003 it is probably thin foliage or the effect of past pollarding, i.e. a reduced canopy area, that made them unrecognizable in 1965 imagery.

Visual inspection of imagery shows that relative differences in size among neighboring individuals in 2003 are similar to what they were in 1965, e.g. smaller trees in 2003 correspond to smaller or less distinct structures in 1965 imagery, while large trees in 2003 were also relatively large in 1965. The apparent small changes in size indicate that growth of trees is very slow. However, in the case of pollarding this pattern is disrupted; it was observed that distinct and larger structures in the 1965 imagery coincide spatially with individuals that had a small canopy area in 2003. In the opposite case, when pollarding of a mature tree prior to 1965 makes it undetectable in the 1965 imagery and it appears as a mature individual in 2003, a false impression of fast growth or even recruitment is given. It is therefore important in this type of change analysis to be aware of such possible misinterpretations.

In all, 766 individuals were identified on the imagery while only approximately half of them had survived into 2003, indicating an overall negative trend in population size in this part of the Eastern Desert. At some sites this negative trend is particularly alarming because the reduction in mature trees is substantial (more than 50% of the population was lost during the period) at the same time as recruitment is nearly absent, e.g. at Nuqrus Middle (8), Gaetri (13) and Jimal I (15; see Fig. 8.2). At a few sites there was a positive trend and better recruitment, e.g. at Dabur I and II (1 and 2), Kharrasha (6) and Hulus Upper (10; see Fig. 8.2). The effect of topo-hydrology (moisture conditions) on recruitment was tested in a generalized linear model, but no significant relationship was found. It has to be emphasized that the time period (ca. 40 years) is short in relation to sufficient recruitment rates in a system governed by remnant dynamics (Eriksson, 1996). For mortality, on the other hand, there are indications that hydrological variables and surface water in particular can explain some of the observed pattern in mortality. Still, the results indicate that direct human intervention, i.e. charcoal production, is the main cause of tree mortality in the Eastern Desert (Fig. 8.3).

8.2.2.5 Resume

The valuable paper of Andersen (2006) could be considered the first scientific report showing the potential of CORONA satellite imagery for studies of long-term changes in arboreal vegetation. Images offer a spatial and temporal dimension for

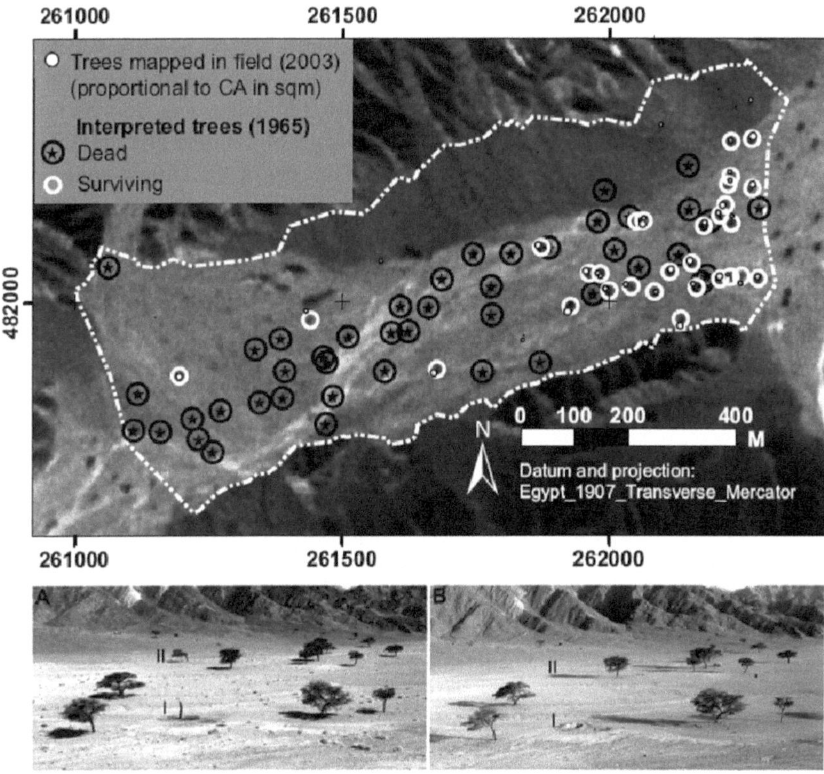

Fig. 8.3 A CORONA map shows the changes at the site in wadi Gaetri (13; cf. Fig. 8.1) between 1965 and 2003. In the *lower panel* photos from 1996 (**A**) and 2003 (**B**) indicates the utilisation of even root parts for charcoal prodcution (at I) and cutting of a green tree (at II).

ecological information which other data sources cannot furnish at a comparable cost, coverage, resolution or accessibility. A large percentage of trees (70%) are identifiable in the KH 4A imagery, and it should be noted that even better results should be achievable with the KH 4B data. The images give access to extremely valuable historical ecological information making it possible to assess ongoing changes and trends in mortality and recruitment in an environment where historical data are mainly lacking. Only with improved knowledge about these processes is it possible to lay a sound foundation for future management of desert tree populations in the Eastern Desert, which at some sites are in an alarming decline.

8.2.3 Moghra Oasis

8.2.3.1 Introduction

Moghra oasis occurs in the hinterland of the El-Omayed area in the north western coastal region of Egypt, where El-Omayed Biosphere Reserve (OBR) exists. It is considered as one of the most important Egyptian inland-water bodies. Its importance stems from dependence of the local inhabitants of El-Omayed area on it as an alternative rangeland in dry season, where the water resources support a dense vegetation cover and increase the grazing capacity of Moghra's rangeland. In time of natural forage shortage herds of El-Omayed region move to Moghra benefiting from its rangelands. This places the oasis under a growing pressure of overuse and exploitation of its resources, especially the vegetation cover. The aim of the current study is to study the habitats of Moghra oasis, plant composition and vegetation pattern by both field surveys and the highest spatial resolution satellite imagery possible. The process of evaluation will involve an estimate of their conservation values of the prevailing perennials. It is hoped that the results obtained from this study could be used to link Moghra oasis to El-Omayed Biosphere Reserve (OBR), which is situated at the northern coast and inland to Moghra oasis. Moghra oasis could serve as an additional core area linked to OBR by an ecological corridor, or a satellite protected area on its own.

8.2.3.2 Study Area

Moghra oasis is one of the two oases in the Qattara Depression (Moghra and Qara), which is the largest depressions in the Egyptian desert. Moghra oasis is a small uninhabited oasis (Lat. 30° 1<@1;> N, Long. 28° 5<@2;> E), situated on the north eastern edge of Qattara Depression and centered by a brackish-water lake (Fig. 2.1). The lake is about 4 km^2 including *Phragmites* swamp. On the boundaries salt marshes occur, encroached by wind blown sands in some areas. The lake comprises the lowest part in the oasis (−38 m) as displayed in the digital elevation model.

The shallow water table and the outward seepage of the lake's water accompanied by excessive evaporation create the wet salt marshes that outskirt the lake. Sand formations are dominant in the western and southern sides of the Moghra Lake. The deposits are in the form of dunes in areas adjacent to the lake or in the form of deep sand sheets in other places. Occasional silicified tree trunks are found in the gravel desert in the north eastern and eastern parts these seem to have been rafted, Said (1962).

8.2.3.3 Results

The satellite image used in this study is the product of the orbital Quick Bird satellite. QuickBird data produced from a mosaic of two QuickBird frames that cover Moghra oasis and its surrounding. Each of the two frames is composed of 16,913 columns and 8700 rows. The statistics of the digital numbers of the images were obtained, the images were preprocessed, processed using ERDAS (IMAGINE) 8.4 software. The processing involved enhancements for visual interpretation of the image and on-screen digitization of the major land/cover types guided by ground-truth data. The processing involved obtaining the false colour composite image of the extracted 1.41 km^2 of Moghra by mosaicing from the two QuickBird frames. The false colour composite FCC image of the resulted sub-scene was carefully examined visually for the preliminary visual interpretation and distinction of the different land cover classes present in Moghra Oasis. Confirmation on visual interpretation was made through field investigations that were carried out on several seasons for accurate identification of each type of the present vegetation. Besides, ground truth using base maps, reference points using a GPS for determining geographic locations and boundaries were also employed. The study area was visited 19 times; during which the area became well known. The features which were unclear and difficult to interpret in the FCC were identified and became more apparent and clear. The processing of the satellite image also involved classification to produce the most dominant and distinct land cover classes. Prior to classification procedure, the statistics of the digital image produced were used to determine from the spectral signature of the prevailing land cover and to assess the spectral pattern recognition of the prevailing features. The digital distribution of the subscene was examined through histograms and scatter plots and spectral signatures of each land cover class were extracted. The image was analyzed using the unsupervised classification, which involves algorithms that examine a large number of unknown pixels and divide them into a number of classes based on natural groupings present in the image values. The classes that result from the unsupervised classification are called spectral classes. Because they are based on the natural groupings in the image values, the identity of the spectral classes will not be initially known. The classified data must be compared with some form of the reference data. Classification accuracy measures were estimated for the resulted classification. Field surveys were conducted to verify the satellite image classification results (Lillisand and Kiefer, 2000). The false Colour Composite image (**FCC**; Fig. 8.4) was used for the preliminary identification of the prevailing land cover classes. From the visual interpretation of the FCC it was noticed that the lake is oriented along the NE-SW direction and appear in black patches. The salt marshes surround the lake from the western side taking the tones of grey. The reed swamp vegetation surrounds the lake from the eastern and southern side taking the tones red. It can also be investigated from the FCC image (Fig. 8.4) the sand dunes in the western and the southwestern parts skirting the salt marshes and mainly taking the yellow tone. All information extracted from the FCC was used as a base for the field surveys. The main areas of obvious cover change were visited for field checks, and collection of samples. Locations of stands for carrying out the field work were located on the FCC to indicate their relationship to each other and their prevailing land cover (Fig. 8.4). The image was analyzed using the unsupervised classification into 6 major land cover classes (Fig. 8.5), representing the most prevailing land cover types in Moghra Oasis. The total area of each class is presented in hectare in Table 8.2. As was observed from the classified image most of the study area is covered by dry salt marshes and sand dunes covering an area of about 565.5 ha accounting for about 40.1% of the total area represented in the classified image (Table 8.2).

The lake covered an area of about 156.8 ha (about 11.1% of the total area) surrounded from the eastern side by saline flats with reed swamp vegetation covering an area of about 247.3 ha (about 17.5% of the total area), and surrounded by wet salt marshes from the western side covering an area of 193.3 ha (about 13.7% of the total

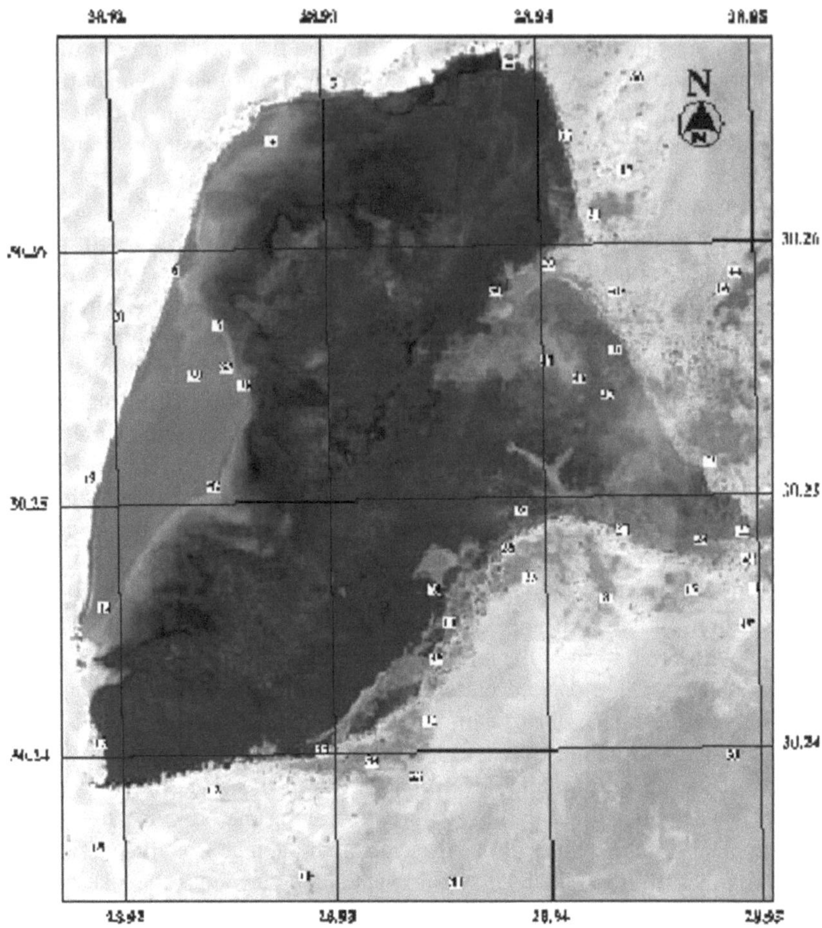

Fig. 8.4 False colour composite (FCC) of Moghra Oasis, with the location of some of the studied sampling stands

area). The gravel desert covers the southeastern corner and parts of the north eastern side of the study area with an area of 248.1 ha (about 17.1% of the total area).

Vegetation Characteristics: A total of 86 stands were selected and located on the FCC to represent major apparent variations in the physiognomy and vegetation and in edaphic features of the major types of the habitats present in the study area (Refer to Fig. 8.5). Through field studies it was possible to distinguish nine major habitats in which the plant communities were combined: reed swamp vegetation, salt marshes, salt marshes covered by sand accumulations, sand hummocks/salt marsh transition, sand hummocks, sand dunes slopes, sand dunes, sand plains and gravel desert. These habitats support the growth of about 31 species all of which are perennial species. Most of this species are of importance as grazing plants. It was noticed that sand plains support the largest number of species compared to the other

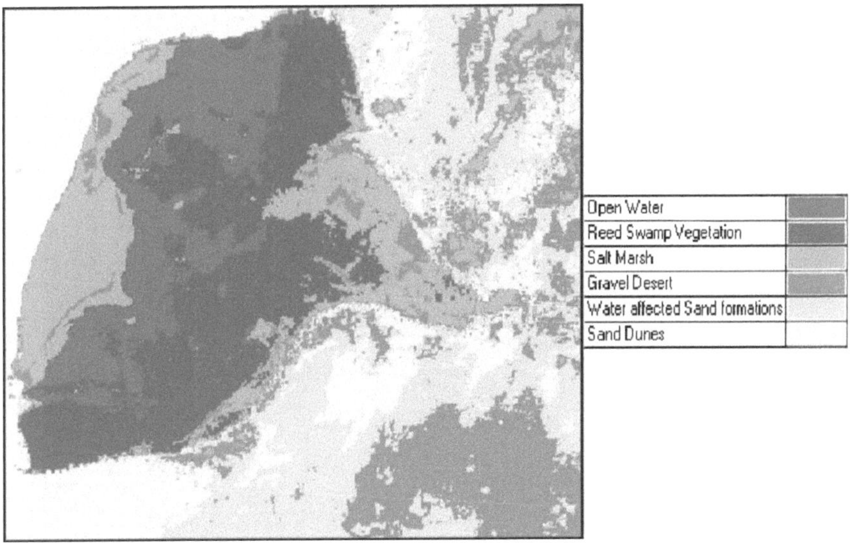

Fig. 8.5 Classified image of Moghra Oasis

Table 8.2 Area of the land cover classes in hectare (ha) as resulted from classification of the satellite image of Moghra Oasis

Land cover class	Area in ha	Percent of total area
Open Water	156.8	11.1
Saline Flats with Reed Swamp Vegetation	247.3	17.5
Wet Salt Marsh	193.9	13.7
	248.1	17.6
Salt Marsh covered by sand accumulations	303.1	21.5
	262.4	18.6
	1411.5	

habitats (16 species), followed by the salt marshes covered by sand accumulations which support the growth of 13 species. On contrary the sand hummocks, sand dunes and gravel desert habitats supports the lowest number of species (7, 8 and 9 respectively).

Chapter 9
Sustainable Development of Egypt's Deserts

This chapter is presented under 4 subtitles:

Definition, religious attitudes, ecological considerations, and renewable natural resources.

9.1 Definition

Sustainable development was defined as the development that meets the needs of the present without compromising the ability of future generations to meet their own needs. If needs are to be met on a sustainable basis, the Earth's natural resources must be well managed and enhanced. Also, it is defined as a development strategy that manages all assets, natural resources and human resources as well as financial and physical assets, for increasing long-term wealth and well-being (Anonymous, 1987a).

Development is a value-laden word implying change that is desirable. It may be considered a *vector* of desirable society objectives, the elements of which might include (Pearce et al., 1990):

1. increase in real income per capita,
2. access to resources, and
3. a "fairer" distribution of income.

Sustainable development may be conserved as a situation in which the development *vector* (D) does not decrease over time, and is consistent with justice to both nature and human generation.

Sustainable development has been described as a path toward the twin goals of social justice and environmental protection. It rejects policies and practices that support current living standards by depleting the productive base, including natural resources and that leaves future generations with poorer prospects and greater risks than our own. In its broadest sense, the strategy for sustainable development aims to promote harmony among human beings and between humanity and nature (Anonymous, 1987a).

M.A. Zahran, A.J. Willis, *The Vegetation of Egypt*,

In its general meaning, sustainable environmental development aims towards reducing income disparities and increasing access to health care, shelter, education, jobs and other essentials of life (Cunningham and Saigo, 1992). It requires management of all assets of natural and human resources to increase long-term wealth and well-being of all. Property claims are valid only when they leave as much good as they take.

Bases for sustainable development include: 1. reliable scientific information, 2. consensus on ethical principles, 3. hope for the future, and 4. consideration of personal interest and incentives. "We can manage our resources only if we know what we have and what we are doing to them. We need to agree on the reasons for preserving and distributing resources. We also need assurance that progress is possible and that we or our descendents will benefit from that progress" (Cunningham and Saigo, 1992) Also, Kassas (2004) stated that sustainable development could be realized through three main basis: 1. social equality, 2. economic efficiency and 3. environmental conservation. He added that it is the responsibility of the governmental institutions, in collaboration with the international and regional organizations, to implement programs aiming at the conservation of the natural resources, e.g. the biodiversity, and to protect them against deterioration.

9.2 Religious Attitudes

The universe, was created in due proportion and measure both qualitatively and quantitatively. God says: 1. "Verily, all things We have created by measure", 2. "Everything to Him is measured", and 3. "And produced therein all kinds of things in due balance" (The Holy Quran, Sura 54 Aya 49, Sura 13 Aya 8 and Sura 15 Aya 19, respectively).

In this universe with its main four spheres (atmosphere, hydrosphere, lithosphere and biosphere) and their various elements and processes, there is fulfilment of man's needs and evidence of the Creator's greatness. Sura 20, Aya 53 of the Quran states "He who has spread out the earth for you and enables you to go about therein by roads (and channels) and has sent down water from the sky, with it We have produced diverse pairs of plants each separate from the others, eat (for yourselves) and pasture your cattles, Verily, in this sign for men endowed with understanding".

Man is a part of this universe whose elements are complementary to one another in an integrated whole. However, man is a distinct part of the biosphere and has a special position among its other parts. According to Ba Kader et al. (1983), the relationships between man and the universe in general have been clarified in the Quran in five main rules:

1. A relationship of utilization, development and subjugation for man's benefit and for the fulfilment of his needs.
2. A relationship of meditation on and contemplation and exploration of the universe and what it contains.

3. Man is only a mere manager of the earth and not a proprietor, a beneficiary and not a disposer or ordained. He (man) has been granted inheritance to manage and utilize the earth for his benefit and for the fulfilment of his needs. Man, therefore, has to keep, maintain and preserve it honestly and has to act within the limits dedicated by honesty.
4. God has granted all people of this planet the inheritance of all sources, of life and resources of nature. Thus, the utilization and sustainable use of these resources is the right and privilege of all people. Hence, man should take every precaution to ensure the interests and rights of all others since they are equal partners on earth. Similarly, man should not regard such ownership as restricted to one generation. It is a joint ownership in which each generation uses and makes the best use of nature, according to its needs without disrupting the interests of future generations.
5. All religions (Islam, Christianity, Judaism, Buddhism etc.) consider that the environment is the source of life and the depot of resources of nature; accordingly we must protect, conserve and construct and develop its assets and prohibit their abuse and destruction. This is obvious in the idea of the revival and restoration or recovery of lands through agricultural activities. God says in the Quran Sura 11, Aya 61 "It is He who hath produced you from the earth and settled you therein". In his teachings, the Prophet Muhammed says "On Doomsday if anyone has in hand a sapling of a palm he should plant it". Such a positive attitude towards green plants is admirable. Thus, from the religious point of view, man has to take the correct measures to improve all aspects of his environment for his own benefit and for the betterment of life for future generations.

9.3 Ecological Considerations

Environmental stress has often been seen as the result of the growing demand on scarce resources (Anonymous, 1987a). Accordingly, conservation and sustainable development of natural resources should ideally be directed towards a common goal: the rational use of the earth's resources to achieve the highest quality of living for mankind. Dasmann et al. (1960) stated "In practice, economic development tends to place stronger emphasis on quantitative increases of production, aimed at enhancing the material well-being of people, whereas conservation, while concerned with sustaining quantitative yield, also emphasizes management of more qualitative aspects of human environment which can add depth and meaning to human life".

Natural ecosystems, terrestrial and aquatic, may be classified under two main types: 1. modified ecosystems and 2. unmodified ecosystems.

The sustainable economic development of both types remains subject to the ecological limitations which operate within natural systems. These limiting factors must be taken into account if the development is to succeed and be maintained.

Within any ecosystem each species of plants, animals and man, exists as a population, the growth or decline of which depends on the capacity of the system to provide its requirements. There must be limits to population growth of all plants and animals, including man. The environmental limits to growth determines the *Carrying Capacity* for any species which may be at: (a) subsistence level, (b) security level or (c) optimum level. The optimum level is, in fact, the normal objective for human societies, their domestic animals and their crops. These levels are controlled by the major factors affecting the ecosystem, namely: climate, soil, water and the complex of biotic factors including the transmission of energy and cycling of nutrients and the impact of parasites, diseases and other kinds of predation.

The use of ecology in the planning of the sustainable development has two aims: (a) enhancing the goals of development and (b) anticipating the effects of development activities on the natural resources and processes of the environment. Dasmann et al. (1960) state that in any particular area not yet opened to human use, e.g. Egyptian Deserts, wide ranges of options exist:

1. It can be left without any man's interference, i.e. in a completely natural state and reserved for scientific and educational uses, watershed protection and/or for its contribution to landscape stability.
2. It can be developed as a national park (protectorate) with the natural scene remaining largely undisturbed to serve as a setting for outdoor recreation and the attraction of tourism.
3. It can be used for limited harvest of its wild vegetation or animal life, but maintained for the most part in a wild state – serving to maintain landscape stability, support certain kinds of scientific or educational uses, provide for some recreation and tourism, and yield certain commodities from its wild populations.
4. It can be used for more intensive harvest of its biomass as in forest production, pasture for livestock, or intensive wildlife culling. In this case its value as a "wild" area for scientific study diminishes, its value for tourism and outdoor recreation may diminish but is not necessarily lost; its role in landscape and watershed stability is changed.
5. The wild vegetation and animal life having been removed in part, the area is intensively utilized for cultivation pasture or farming crops.
6. The wild vegetation and animal life having been almost completely removed, it can be used for intensive urban, industrial or transportation purposes.

The above mentioned options may be subjected to two questions:

1. Is it possible to carry out changes from one option to the other without substantial danger to the ecosystem?
2. What are the bases, upon which the decision makers depend, to select the option for the welfare of the people and the conservation of the environment?

The answer to the first question is that if one of the first three options is selected, the choice remains, to great extent, open to change from one of these uses to the other or to use the land for any of the latter three objectives. Selection of the fourth option reduces the possibility of restoring the land to any of the first three categories but does not eliminate restoration completely. On the other hand, if any of the fifth and sixth options was selected, any shift to the other options within a reasonable period of time would be costly and difficult.

The main considerations that lead to a correct selection of the option are ecological and economic. Some areas may have extremely high ecological values – they may represent unique ecosystems with rare or endangered species, or they may be essential to maintain soil stability and water yield in a river basin. Where alternative areas exist that are suitable for more intensive use, it would be unwise to use the environmentally unique area for any purpose likely to impare its ecological value. On the other hand, areas having high economic potential being, for example – 1. Unique source of valuable minerals, 2. provide the only logical site for a hydroelectric dam or reservoir, 3. support a high-yielding, high-value stand for commercial timber, or 4. have deep rich soil which could be put under intensive agriculture for long time- are suitable sites for the more far–reaching forms of development.

Under all conditions, considerations of both ecology and economy are required to maximize the total gains and to minimize the losses. In addition, it is wise to keep a range of resource use options available for future generations.

9.4 Renewable Natural Resources

The major renewable natural resources that have to be considered in the national and/or regional programmes for the sustainable environmental development of all deserts, e.g. Egyptian deserts, include: renewable energy, water resources, soil types, natural vegetation and animal life. Other biotic and abiotic elements, e.g. soil microorganisms, evaporation rate, topography, are also important. Man is the manager as well as the beneficiary of such developmental projects. The following is a short account of four major renewable natural resources of the desert ecosystem, namely: energies, water, soil and natural vegetation.

9.4.1 Energies

Non-renewable energy resources (fossil fuels and nuclear power) provide about 95% of the commercial energy used worldwide. The world will run out of affordable supplies of these non-renewable energies unless there is radical change in the consumption patterns and/or to find some major renewable energy resources found. According to Cunningham and Saigo (1992), the developed countries which have only 20% of the total population of the world consume about 66% of all commercial energy. In addition, energy consumption is rising rapidly in the developing countries and is expected to be four times in 2020 that of 1980.

In addition to the woody trees and shrubs (Anonymous, 1980), in the deserts, the renewable sources of energy are: hydroelectric, biomass methane, geothermal, wind and sun, the latter two sources are considered here.

9.4.1.1 Solar Energy

Sun energy is, actually, the ultimate source of all known energies (El-Qattamy, 1975). Prior to the age of fire, solar energy was humanity's sole source of heat, light and mechanical energy. Wood was, the principal fuel. This explains why Pharaohs and Babylonians worshipped the sun. Nowadays as reserves of oil and gas are dwindling, and as constraints on the use of coal are growing and the future of nuclear power is in doubt, solar energy is making a comeback to both new and familiar forms (Brown, 1981).

Apart from green plants which capture sun energy through the process of photosynthesis, new technologies permit solar energy to be harnessed in innumerable ways. It can be captured directly through such devices as rooftop collectors, photovoltaic cells and buildings incorporating solar architecture.

No ecosystem is without solar potential, but no two ecosystems (or countries) are precisely the same in this respect. The desert ecosystem does, in fact, receiving the greatest amount of sun energy compared to other ecosystems of the earth. The richest desert is the Empty Quarter of the Arabian Peninsula's Desert followed by that of Sahara of North Africa which includes the Egyptian Deserts (Table 9.1).

El-Qattamy (1975) stated that changing the total amount of sun energy (25,344,000 KW/h $\times 10^9$) reaching five deserts (Table 9.1) into electric energy produces about $130{,}000 \times 10^{12}$ KW/h, equivalent to 3000 (three thousands) times the amount of electric energy needed for the whole world.

This indicates the importance of the desert biome as a major receiver of sun energy. Comparison can be made with the clean space of the high levels of the atmosphere where there are neither absorbing nor reflecting materials for sun rays. In these clear and clean areas the strength of sun energy is about 1353 W/m^2 whereas in the desert the strength is about 800 W/m^2. In the areas where there are clouds and/or water vapour, the strength of sun energy decreases by about 30% due to the absorption of considerable sun radiation. In addition to clouds, dust, smokes and other air pollutants reflect sun radiation.

Latitudinally, the amount of sun radiation decreases gradually from the equator northwards and southwards. The best areas for the utilization of sun energy are situated

Table 9.1 Sun energies received by five deserts (After El-Saiegh, 1976)

Deserts	Area (km^2)	Heat Energy of Sun	
		KW/h/km^2 × 10^6	KW/h × 10^9
1. Sahara, Africa	7,770,000	2300	17,871,000
2. Arabian Peninsula	1,300,000	2500	3,250,000
3. Middle and Western Australia	1,550,000	2000	3,100,000
4. Kalahari, Africa	518,000	2000	1,036,000
5. Mogif, South California	35,000	2200	77,000
Total or mean rate	11,173,000	Rate = 2260	25,344,000

between latitudes 40° north and south of the Equator. The deserts located between these latitudes receive the highest amounts of sun energy (El-Saiegh, 1976).

The above discussion shows that the Egyptian desert represents a very suitable area for the efficient utilization of solar energy as a cheap, clean and renewable natural resource for the welfare of the Egyptians. Solar energy could also be used to obtain amounts of the badly needed fresh water by desalinating sea water. According to El-Saiegh (1976), 300 km^3 of sea water could provide $36 \times 10^7 m^3$ of fresh water/year for domestic purposes or into $50 \times 10^7 m^3$/year for agricultural and industrial purposes. Solar energy could also be used to drill for ground water. Securing adequate amounts of fresh water is the first requirement for any developmental programmes in the deserts upon which depends the following:

1. Agricultural schemes;
2. Establishment of new factories;
3. Establishment of new villages and settlements;
4. Availability of enough fresh water for domestic uses and attracting more people to inhabit the developed deserts.

In addition to desalination of sea water and pumping of ground water, solar energy could be used to produce electric power, operation of sun furnaces etc.

9.4.1.2 Wind Energy

About 2% of the sun's energy striking the earth ultimately results in winds which are formed in two major ways (Chiras, 1991). First because sunlight falls unevenly on the earth some areas are heated more than others. Warm air rises, and cooler air flows in from adjacent areas. The earth's principal circulation pattern develops as warm air near the equator rises, drawing cooler polar air towards the tropics. The earth's rotation then causes air to circulate clockwise in the Northern Hemisphere and anticlockwise south of the equator. The second major wind-flow pattern results from the unequal heating of land and water. Air over the oceans is not heated as much as air over the land. Therefore, cool oceanic air often flows landward and replaces warm rising air.

The potential of wind energy is enormous (Cunningham and Saigo, 1992; Miller, 1997). Tapping the globe's windiest spots could provide 13 times the electricity now produced worldwide and wind energy could supply 20–30% of the electricity of many countries. However, "Today, wind generated electricity accounts for only a tiny portion of the world's enormous energy needed" (Chiras, 1991).

Our question is: what are the advantages and disadvantages of the utilization of wind energy? Wind energy is a clean renewable source of electricity; harvesting takes limited areas of land and is safe to operate. Moreover, wind technologies do not preclude other land uses. Wind farms, for example, can be grazed by animals and also can be planted with various crops. "Wind power is an unlimited source of energy at favourable sites and large wind farms can be built in only 3–6 months. This system emits no carbon dioxide nor other air pollutants during operations.

They need no water for cooling and their manufacture and use produce little water pollution." (Miller, 1997). The same author stated that the costs of producing electricity with wind farms is about half that of a new nuclear plant and is competitive with coal. Expert opinion is that by the middle of the twenty-first century, wind power could supply more than 10% of the world's electricity. However, wind power can be used only in areas with sufficient winds, e.g. sea shores and high mountains, both occur in the Egyptian deserts. On the other hand, wind does not blow all of the time, so backup systems and storage are needed. Storage technologies seem to be one of the major weaknesses. In addition, wind generators may be noisy and may impair both television reception and telecommunications.

9.4.2 Water Resources

Water, which is the basis of life for man, animals, plants and microorganisms, is scarce in deserts. Precipitation (rainfall) is very low and is not reliable for developmental projects. Moreover, water is not only a matter of quantity; it is also a matter of bad quality water would be useless.

In view of the great scarcity of surface water supplies, or even their complete absence throughout much of the year, ground water assumes a vital importance in the deserts. Ground water is more reliable than rainfall in that it can be drawn upon throughout the year and is usually less affected by dry periods of successive years (Dixey, 1966). The Bedouins inhabiting the deserts, e.g. Egyptian deserts, are accustomed to sinking wells by simple equipment in suitable places to depths of as much as 100 m. However, greater amounts of water supplies may be obtained by means of bore-holes which can be constructed under difficult conditions of hard rocks or loose sand, and which can be drilled to greater depths. This, however is beyond the capacities of the local people.

Dixey (1966) warned that ground water could be misused by over pumping and/or by extracting water over successive years at high rate. The result is a gradual lowering of the water table in the vicinity of the bore-hole and the reduction and eventually exhaustion of the supply. In the parts of the Egyptian coastal deserts directly affected by sea water (littoral salt marshes), the ground water is in contact with salt water of the sea. In this case, over-pumping leads to a rise of salt water into the wells with consequent pollution of the supply, rendering it unfit for agriculture or other domestic use. In addition, due to the distribution of oil fields along parts of the Red Sea coastal deserts in Egypt, oil pollution may be expected and the water of the wells will be unfit for all uses.

Since ground water is replenished by rainfall which is very scarce in the Egyptian deserts, it might be expected that the volume stored is decreasing, a factor that decreases its economic potentiality as a permanent source of fresh water. Supplies of the required quantities of fresh water are essentially needed to initiate developmental programmes of the area. The most reliable way is the desalination of sea water which can be maintained along the whole stretch of the Egyptian coastal deserts. As mentioned earlier, solar energy is a cheap and effective option for water desalination. Accordingly, xerophytes and/or halophytes, proved to have agro-industrial potentialities, should be introduced as non-conventional crops for cultivation in the Egyptian deserts. Water requirements of these plants are very low an advantage that extends the economic use of ground water for hundreds of years.

9.4.3 Soil

Soils of the coastal and inland deserts accumulate soluble products of weathering in the upper part of the soil profile; these products are present as calcium carbonates and soluble salts (Jewitt, 1966). Soil texture varies widely from heavy clay to coarse sands, reflecting the influence of parent material, although most have one horizon of heavy texture. As mentioned before (Chapter 2), Dregne (1976) stated that the soils of arid lands (deserts) fall into two main orders: *Aridisols* or the essentially desert soils and *Entisols* or alluvial soils and soils of sandy and stony deserts. Aridisols are mineral soils distinguished by the presence of horizons formed under recent conditions of climate or those of the earlier periods. The surface stratum, the epipedon, is usually light-coloured and there may be a salty or clayey stratum near the surface. The salty layer is well represented in the littoral section directly affected by the sea water. "Most of the time when temperatures are favourable for plant growth, aridisols are dry or salty with consequent restrictions on plant growth". However, such growth restriction may not be applied to the salt tolerant plants (the halophytes) living in the Egyptian deserts. These plants withstand saline conditions and many of them are of economic value and can play a considerable role in the sustainable development. The Entisols, the most common type in the Egyptian deserts, are also mineral soils with little or no development of horizons.

The soils of the Egyptian deserts, like those of the other deserts, contain low levels of organic matter, are slightly acidic to alkaline in reaction at the surface, show accumulation of calcium carbonate within the topmost 1.5 m (5 ft) layer, have weak to moderate profile development, are of coarse to medium texture and have low biological activity. In the upstream parts of the wadis of the Egyptian deserts, soils show a thin surface layer of stones, pebbles or gravel that constitutes a desert pavement from which the fine particles have been removed by the action of wind and/or water.

Generally, the soil of the Egyptian deserts is suitable for different types of agriculture if fresh water is available.

9.4.4 Natural Vegetation

9.4.4.1 General Remarks

Many species of plants that supply 90% of the world's food, fodder, fiber, drugs etc. were domesticated from wild plants found in the tropics (Miller, 1997). Existing wild plant species, many of them still unclassified and unevaluated, remain interesting to plant ecologists, agronomists and genetic engineers to develop new crop strains; some of them may become important sources of food, livestock fodder, clothes, rubber, dyes, paper, drugs, perfumes etc.

Wild plants are vital components of ecosystems. Plants supply animals with food, recycle nutrients essential for agriculture and help to produce and maintain fertile soil. They also produce oxygen and carbon dioxide in the atmosphere, moderate earth's climate, help regulate water supplies and store solar energy as chemical energy. Moreover, green plants may help to remove poisonous substances, and make up a vast gene pool of biological diversity.

Ecologists and conservationists believe that hundreds of wild species are disappearing at an alarming rate. Preserving natural plants is a must, for their actual or potential usefulness as renewable natural resources for people. At the same time, wild species have an inherent right to exist; it is ethically wrong for us to cause the extinction of any plant species (Zahran, 2007b).

According to Miller (1997), extinction of plants (and animals) is a natural process. As the planet's surface and climate have changed over its 4.6 billion years of existence, species have disappeared and new ones have evolved in their places. However, since the dawn of agricultural activities, about 10,000 years ago, the rate of species extinction has increased sharply as human settlements have expanded worldwide. It was estimated that at least 4000 and possibly very many more species became extinct mostly because of man's activities and the number is rising. The greatest threat to most natural plants is destruction, fragmentation and degradation of their habitats especially on land.

Egypt is a part of the arid region of the world and its prime farmland and surface River Nile water are fully utilized. With the high population increase (20 millions on 1952 increased to 75 millions on 2007 to be about 90 millions on 2020) it is necessary to extend the agricultural land horizontally to its wide non-cultivated deserts westward and eastward and in the Sinai Peninsula. No dought the deserts are the only promising solution for the present and future high population increase.

Being the principal requirements for the initiation and continuation of life and without it no development could be achieved; the water is the limiting factor which should have priority. Accordingly, the planning of the sustainable developmental programmes in the Egyptian coastal and inland deserts should depend upon their available water resources which comprise the following kinds:

(a) Limited amount of the winter rainfall particularly along the Mediterranean coastal desert that extends from Sallum (in the west) to Rafah (in the East). Here, the annual precipitation ranges between >50 mm and <200 mm. In the inland as well as on the Red Sea coastal deserts rainfall is negligible, i.e. is not a reliable source of water.
(b) The non-renewable ground water, mostly brackish, stored in many parts of the Egyptian deserts.
(c) The Sea water along the coasts of the Red Sea, Mediterranean Sea and Gulfs of Aqaba and Suez of Sinai Peninsula.

9.4.4.2 Economic Potentialities of Desert Vegetation

Many publications e.g. Malcolm (1972), Anonymous (1980, 1987a,b, 2004, 2006, 2007), FAO (1980, 1994), Pessarakli (1993), Zahran (1993a, 2004), Ben Haider (1994), Abdel Razik (1994), Ahmed et al. (2000), Ashour et al. (2002), Hammouda (2005, 2006) etc... stated that salt tolerant and drought resistant plants may provide alternative resources for many developmental programs in the deserts under their extreme environmental conditions. In these deserts, fresh water from rain is too small, as well as brackish-ground water and sea water are too saline to irrigate conventional crops. However, the three kinds of water could be used to cultivate xerophytes and/or halophytes of economic values in the different habitats of the Egyptian deserts.

The natural wealth of the flora of the Egyptian deserts, the producers of their main ecosystems (mangrove, reed swamps, salt marshes, sand dunes, desert plains and wadis and mountains), comprises hundreds of annual and perennial species. These plants could be considered the backbone for the sustainable environmental development of these deserts. The annual species, e.g. *Zygophyllum simplex, Asphodelus tenuifolius*, usually cover a wide area between the perennial trees and shrubs during the wet seasons. Such seasonal green cover is usually used as natural range for the local domestic and wild animals. However, the permanent framework of the different vegetation types is formed mainly of perennial halophytes and xerophytes. Apart from being adapted to live under salinity and/or aridity stress and their water requirements are very low, these plants have various economic potentialities. Taking these advantages into consideration, these plants could be propagated in the Egyptian deserts depending only upon the available water. To highlight their economic value, the vegetative yields of the plants to be cultivated there, could be used as materials for certain industries, e.g. livestock fodder, paper, drugs, oil, etc...

Economically, the flora of the Egyptian deserts could be categorized under 5 main groups (Zahran, 2004):

1. Fodder producing plants,
2. Drug producing plants,
3. Fiber producing plants,
4. Oil and perfume producing plants,
5. Wood and fuel plants.

Many of these plants may have more than one usage. For example, the halophytic mangrove species, namely: *Avicennia marina* and *Rhizophora mucronata*, as well as the xerophytic trees *Balanites aegyptiaca* and *Acacia spp.* could be considered as food, fodder, drug and wood producing plants. Moreover, the desert natural vegetation contains a considerable number of psammophytes capable of living on and stabilizing sand formations.

9.4.4.3 Case Studies

In Egypt, different studies were carried out to make use of xerophytes and halophytes with agro-industrial potentialities. These include: afforestation of mangrove forests and establishment of fodder, fiber and drug producing plants. Also, plants used to stabilize sand dunes and palm trees are considered.

The following pages present a few case studies.

Afforestation of Mangrove Forests

Mangrove forests are tropical formation, representing one of the major ecosystems of the biosphere. They cover about 161,000 km^2 along 60–75% of the shorelines of the tropical seas and oceans (Anonymous, 2004). The goods and services of the mangrove ecosystems could be categorized under three groups: direct use values, indircct use values and non-use values (Saenger, 2001, 2003).

The direct use values i.e. providing several types of resources, including: wood, fisheries, food, energy, pharmaceutical, and fodder. The indirect services values include: shoreline protection, windbreak, storm protection, sediment regulations, nutrient retention, water quality maintenance, groundwater discharge and local microelements stabilization. Other non-use values include direct and indirect use values: biodiversity, uniqueness and heritage.

In Egypt, the mangrove forests cover about 525 km^2 distributed over 28 sites: 23 sites along the Red Sea coast including Abu Mingar and Safaga Islands, one site in Ras Muhammed, and 4 sites in the southern section of the Gulf of Aqaba. The northern most site of the Gulf of Aqaba (Lat. 28°12′N, long. 34° 25′E) represents the northern latitudinal limit of the Indo-Pacific-East African mangrove realm. As mentioned before, (Chapter 4), two mangrove species are recognized, namely: *Avicennia marina and Rhizophora mucronata. A. marina* is wide spread whereas *R. mucronata* occurs only in the most southern section of the Red Sea in the Sudano-Egyptian border.

Unfortunately, all sites of the mangrove forests along the Red Sea coast are threatened by: (i) human overuse (cutting, urbanization), (ii) overgrazing, and (iii) oil pollution. The areas of these sites are, thus, considerably decreased. The sites of the Sinai Peninsula's coasts are under complete protection being natural protectorates. This has been realized by Remote Sensing Mapping of 10 representative sites: 2 in Sinai Peninsula and 8 along the Red Sea coast using satellite images on 1975 and 2002 (Anonymous, 2006).

For their potentially high values and for the threats they are facing, management, conservation, silviculturing and rehabilitation of the mangrove forests in Egypt deserve special attention. Fortunately, through a cooperative project between the Ministry of Agriculture and Land Reclamation, the Ministry of Statc of Environmental Affairs, and the International Tropical Timber Organization (ITTO), Japan silviculturing and rehabilitation of mangrove plants (*A. marina* and *R. mucronata*) in representative sites of the shoreline of the Red Sea of Egypt have been successfully conducted (Anonymous, 2006). About 50 feddans have been cultivated with these two mangrove species. Rehabilitation of about 5 feddans have been also conducted.

Fodder Producing Plants

I. General Remarks

In Egypt, insuffiecient meat production is one of the major food problems which is often exacerbated by increasing demands. On the other hand, the annual production of the traditional green fodders, namely clover (*Trifolium alexandrinum*) and alfalfa (*Medicago sativa*) in the Nile Region as well as the natural ranges of the deserts are not enough to meet the increasing requirements of the meet producing animals. Fortunately, the natural flora of Egypt's desert comprises a considerable number of palatable species (perennials and annuals) belonging to different families. However, many of these species are endangerd due to overgrazing and uncontrolled cutting (El-Hadidi et al., 1991).

In Egypt, rainfall is negligible and consequently the natural ranges are not widespread but occur in limited areas of its deserts. Generally, the framework of these ranges is formed of perennial halophytes and xerophytes enriched during the short rainy season with annual vegetation that gets dry during the long dry period (>8 months). The, relatively, richest rangelands of Egypt are those distributed in the narrow belt of the northern Mediterranean coastal desert that extends from Sallum to Rafah (Fig. 2.1) where annual rainfall may reach up to 200 mm. Other parts of Egypt's deserts are practically rainless and, consequently, their natural ranges are not reliable. However, in the wadis associated with the montane countries of the Red Sea coastal desert (e.g. Gebel Elba) and Sinai Peninsula, Gebel Saint Catherine, orographic rain and run-off water enable range plants to grow and predominate.

The natural ranges of the northern Mediterranean coastal desert may supply the livestock animals with at least part of their requirements of green fodder all year around if correctly managed and sustainably utilized. Unfortunately, overgrazing and uncontrolled cutting are causing their deterioration, a factor that exacerbates desertification of these valuable range lands. To overcome these problems and maintain a permanent supply of feed from the renewable forage plants in Egypt's desert, several studies has been conducted aiming at the determination of their nutritive values and to estimate their biomass production as well. Propagation experiments have been also carried out to cultivate some of these plants and introduce them as non-conventional fodder producing plants under aridity and/or salinity stresses of Egypt's deserts. Here is a short account of some of these studies.

II. Representative Case Studies

A. Determination of the Biomass and Nutritive Values

Case study I: Ahmed and Nassar (1999) determined the nutritive values of 108 palatable plant species through phyochemical investigations. These plants were mixture of perennial halophytes and xeropytes as well as annuals belonging to 19 familes and growing in the different habitats of the Mediterranean coastal desert of Egypt. Leguminosae was represented by 38 species followed by Chenopodiaceae

(14 species) and Gramineae (12 species). Other families were represented as follow: 4 species for each of Cruciferae, Plantagraceae and Caryophyllaceae, 2 species for each of Scrophulariaceae and Umbelliferae and one species for each of Rubiaceae, Geraniaceae, Solanaceae, Polygonaceae, Cistacea, Labiatae, Plumbaginaceae, Tamaricaceae and Frankeniaceae. The results shown on Table 9.2 elucidate that:

1. The investigated plants, are all palatable, differ in their chemical constituents and, consequently, their nutritive values and palatability.
2. The relatively high mean values of crude proteins occur in the species of Leguminosae (15.5%) followed by Umbelliferea (14.6%), Rubiacea and Geraniaceae (14.1% each), Chenopodiaceae (13.2%) and Boraginaceae (12.7%). Gramineae plants (grasses) contain very low amount of crude protein (8.6%).
3. Mean values of crude fiber was the highest in the members of Labiatae (36.2%) followed by Cruciferae (27.1%), Gramineae (26.9%), Scrophalariaceae (25.6%) and Leguminoseae (22.3%).
4. Total carbohydrates were relatively high in the members of Gramineae (mean=27.5%) and Boraginaceae (mean=27.8%) followed by Frankeniaceae (24%) and Compositae (22.1%). Leguminosae contains 17.1%.
5. Mean values of the crude fats were high in the members of Cistaceae (9.0%), Cruciferae (7.3%), Solanaceae and Rubiaceae (6.5% each), Caryophyllaceae (5.9%), Scrophularieceae (5.3%), Geraniaceae (4.5%), Gramineae (4.48%), Leguminosae and Chenopodiaceae (4.22% each).
6. Succulence attained its high mean values in families Solanaceae, Tamaricaceae, and Chenopodiaceae (4.9%, 4.8% and 4.22%, respectively).
7. Chenopodiaceae contains a relatively high amount of total ash (mean=27.4%) followed by Tamaricaceae (mean=22%), Polygonaceae (20%) and Compositae (18.8%).
8. The amounts of minerals vary considerably between the species of the different families.

Case study II: The forage biomass of eleven halophytes of the Western Mediterranean Coastal Desert have been estiamated by Heneidy and Bidak (1996) for the following aims:

1. Exploring the contribution of these plants as renewable forage for the grazing animals, and
2. Assessing their grazing potentialities (nutritive values).

The studied halophytes were: *Arthrocnemum macrostachyum, Atriplex halimus, Halocnemum strobilaceum, Salsola tetrandra, S. tetragona* and *Suaeda pruinosa* (Chenopodiaceae), *Aeluropus lagopoides* and *Phragmites australis* (Gramineae), *Limoniastrum monopetalum* (Plumbaginaceae), *Frankenia revoluta* (Frankeniaceae) and *Juncus rigidus* (Juncaceae). Field observation showed that *A. macrostachyum, A. lagopoides*, and *S. tetragona* are very highly palatable whereas *A. halimus, J. rigidus* and *S. tetrandra* are highly palatable. The other five species have less palatability.

Table 9.2 Means of the chemical composition of 108 range plant species belonging to 19 families collected from the Western Mediterranean Coastal Deserts of Egypt (After Ahmed and Nassar, 1999) {S=succulence %, TA=Total Ash %, CP=Crude Protein, CFt=Crude Fat, TC=Total Carbohydrate, CFb=Crude Fiber,}

Families	%		Minerals (me/100 gm)							gm / 100 gm			
	S	TA	Na	K	Ca	Mg	P	Cl	SO⁴	CP	CFb	TC	CFt
1. Leguminosae	3.7	12.11	15.89	29.9	111.8	35.2	6.83	35.2	1.22	15.5	22.3	17.1	4.22
2. Gramineae	3.99	12.4	14.33	18.83	89.83	46.5	5.93	35.5	1.12	8.6	26.9	27.5	4.48
3. Chenopodiaceae	4.22	27.4	160.2	51.5	114.7	97.6	7.4	126.2	3.6	13.2	16.2	13.9	6.28
4. Compositae	3.23	18.8	23.8	59.5	96.6	36.9	5.3	41.8	1.89	11.4	18.3	22.1	4.22
5. Boraginaceae	3.6	15.0	24.0	30.3	113.6	36.5	5.8	31.4	2.29	12.7	19.3	27.8	2.0
6. Plantaginaceae	3.9	16.30	38.8	8.5	118.0	75.0	5.6	76.2	2.8	10.7	20.7	18.4	2.3
7. Caryophyllaceae	2.9	9.0	13.8	12.8	99.0	30.0	4.9	36.7	2.4	12.2	22.4	18.7	5.9
8. Umbelliferae	3.4	13.0	9.0	17.50	121.0	89.0	6.9	59.1	1.8	14.6	17.8	14.0	2.5
9. Scrophulariaceae	3.1	11.0	65.0	17.5	110.0	5.0	17.3	33.8	0.91	8.7	25.6	9.3	5.3
10. Rubiaceae	3.2	28.0	152.0	467.0	154.0	70.0	32.0	31.2	2.1	14.1	17.0	14.7	6.5
11. Geraniaceae	4.6	14.0	27.0	41.0	66.0	36.0	2.35	27.2	5.1	14.1	13.2	21.0	4.5
12. Solanaceae	4.9	12.0	31.0	66.0	110.0	100.0	13.0	38.0	2.69	9.6	13.2	9.9	6.5
13. Polygonaceae	3.9	20.0	25.0	87.0	88.0	38.0	13.3	23.4	1.7	6.0	20.0	9.2	3.6
14. Cistaceae	2.9	8.0	13.0	25.0	88.0	120.0	9.8	10.8	1.47	8.6	20.6	18.7	9.0
15. Labiatae	2.9	8.0	17.0	13.0	66.0	60.0	3.5	33.4	1.26	10.9	36.2	14.6	1.5
16. Plumbaginaceae	2.2	12.0	25.0	68.0	154.0	35.0	10.1	26.4	2.83	11.4	18.7	5.7	2.0
17. Tamaricaceae	4.8	22.0	355.0	62.0	100.0	5.0	2.9	20.4	7.14	12.3	21.8	19.4	0.5
18. Cruciferae	3.83	13.0	35.3	31.5	126.5	00.0	10.0	35.4	5.4	11.1	27.1	20.1	7.3
19. Frankeniaceae	3.4	6.0	49.0	7.0	66.0	60.0	7.8	66.7	4.5	1.4	14.7	24.0	1.0

Table 9.3 shows the values of the biomass and means of the chemical compositions of the eleven halophytes under consideration. Generally, the biomass of all species was lower in the wet period than in the dry one. *A. halimus*, other chenopods and *L. monopetalum* produced the relatively highest biomass during both the wet and dry seasons. These are followed by the biomasses of *J. rigidus* and *P. australis*. The lowest biomass productions were those of *A. lagopoides* and *F. revoluta*. The highest amount of NFE (38.6%) was recorded in *S. tetrandra* followed by *S. tetragona* (37.3%), the lowest value being that of *F. revoluta* (28.2%). The grass *A. lagopoides* contains the lowest amount of crude protein (7.2%), the highest is that of the other grass *P. australis* (16.9%). Crude fibers was the highest in the grazeable parts of *J. rigidus* (44.1%) and the lowest was that of *S. tetragona*. The reverse was true for the ash contents being 6.2% in *J. rigidus* and 36.6% in *S. tetragona*. Organic matter content was the highest (93.1%) in *J. rigidus* followed by *P. australis* (81.7%) and *H. strobilaceum* (76.2%), the lowest being that of *S. pruinosa*.

Among the grazing animals, Heneidy and Bidak (1996) observed that camel is the domestic animal, that uses halophytes more intensively. Also, goats feed on all of the above mentioned species except *L. monopetalum* whereas sheep do not eat *L. monopetalum, F. revolute* and *J. rigidus*. They concluded that, the halophytes could be considered important fodder reserve for livestock in Egypt's deserts. These salt tolerant plants may provide at least part of the requirements of the grazing amimals particularly in the dry season. Growth of halophytes may be a solution for shortage of green fodder supply in the marginal areas of the deserts particularly during the long dry period.

B. Propagation Experiments

Experiment I: Ahmed et al. (2002) tested the germination of 35 palatable plant species naturally growing in the Egyptian deserts. Out of these 29 species are perennials and 6 are annuals. The tested plants belong to 12 families, namely: Leguminosae, (20 species), Gramineae (4 species), Compositae (2 species), and one species belongs to each of Anacardaceae, Asclepiadaceae, Capparaceae, Chenopodiaceae, Convolvulaceae, Cruciferae, Labiatae, Resedaceae, and Rhamnaceae. Among the studied species five are endangered, namely: *Acacia millifera, Colutea istria, Delonix elata, Periploca angustifolia* and *Rhamnus lycioides*.

Results of this experiments revealed that pretreatment of seeds is required for seeds of three leguminous shrubs being with hard and resistant seed coats. These species are: *Acacia albida* (= *Faidherbia albida*), *A. raddiana* and *Lygos raetam*. For the germination of these seeds, soaking in H_2SO_4 for 1–3 h is necessary. On the other hand, seeds of the three leguminous shrubs, *A. milliffera, Indigofera articulata* and *Taverniera aegyptiaca* germinated without soaking. Acid scarification improved germination characteristics for seeds of six annual leguminous species, namely: *Hymenocarpos circinatus, Medicago polymorpha, Onobrychis crist-galli, Trifolium purpuream, Vicia cinerea*, and *V. monantha*. Members of gramineae, namely: *Chloris guyana, Lasiurus hirsutus, Panicum antidotale* and *P. turgidum* germinated directly. Meanwhile germination of *Capparis cartilaginea*

Table 9.3 Biomass (lowest and highest) productions and means of the chemical composition of eleven halophytes, western Med. Coastal desert, Egypt (after Heneidy and Bidak, 1996). (W=wet season, D=dry season, NFE=nitrogen free extract, CP=crude protein, CF=crude fiber, EE=ether extract, OM=organic matter)

Species	Biomass (kg dry wt ha)		Chemical composition					
	W	D	NFE	CP	CF	EE	Ash	OM
A. macrostachyum	2820–4234	3040–4524	32.5	15.6	14.8	8.9	26.7	72.1
A. halimus	4015–5822	4217–6035	34.4	9.8	21.2	8.1	25.8	73.4
A. lagopoides	210–415	107–325	30.2	7.2	22.6	3.2	34.8	65.6
H. strobilaceum	1907–3303	2010–3504	35.1	12.8	13.1	9.3	31.7	76.2
L. monopetalum	4410–4908	4615–5204	35.3	10.2	19.8	6.8	29.1	68.4
Frankenia revoluta	215–230	251–274	28.2	9.4	28.7	4.7	28.4	73.2
J. rigidus	1630–1680	1630–1800	33.4	8.5	44.1	6.9	6.2	93.1
S. tetrandra	2280–2430	2420–2620	38.6	10.4	12.9	6.7	35.9	56.6
S. tetragona	1752–2440	1844–2532	37.3	9.5	13.1	6.1	36.6	66.7
S. pruinosa	1150–1788	1222–1872	31.8	11.6	19.2	5.8	35.8	64.8
P. australis	585–1150	671–1315	32.6	16.9	21.3	10.9	18.3	81.7

and *Rhamnus lycioides* seeds required acid scarification because their seed coats are very hard.

The presence of growth regulators e.g. gibberellic acid seems to be necessary for the germination of some deserts seeds. For example, the germination of seeds of *Crotalaria aegyptiaca, Salvia aegyptiaca, Vicia cinerea* and *V. monantha* reached 100% when soaked in 200 ppm GA_3 for 24 h.

Dormancy of seeds may be broken by subjecting the moistened seeds to fluctuated temperature. The most suitable temperature for the best germination of most of the studied seeds ranged between 20–30 °C and 25–35 °C.

Experiment II: Growth and forage yield of the fodder shrub: *Atriplex canescens* had been experimented in the salt affected land of the delta of Wadi Sudr located at 60 km south of Suez (El-Shatt), east coast of the Gulf of Suez, Sinai Peninsula. The experiment focused on the effect of three different levels of soil amendments, namely: farm yard manure, town refuse and elemental sulfur on the growth forage yield and chemical constituents of *Atriplex canescens* (El-Housini et al., 2004). The soil of the experiment was sandy-clay in texture, calcareous (56% $CaCO_3$) and its EC = 8.5 mmhos/cm and pH = 7.9. The sole source of irrigation was the underground brackish-saline water containing about 4500 ppm dissolved salts. Seedlings of *A. canescens* were raised from seeds sown in polyethylene bags containing mixture of sand-clay soil and peat moss in 1:1:1 ratio. Seven month old healthy seedlings were transplanted and cultivated in rows of 2 m apart and spaced in 2 m between plants within the row. The experiment was irrigated monthly and at 45 days intervals in hot (summer) and cold (winter) season, respectively. A complete block design was used for applying the three soil amendments. The experiment continued for 2 years. Measurements of growth parameters was conducted twice/year in dry and wet seasons. Plant samples were analysed chemically. The results elucidate that:

1. *A. canescens* could be propagated as forage halophyte in the salt affected land.
2. The biomass production (fresh and dry yields) increased by increasing the levels of the manures in both wet and dry seasons.
3. Total amounts of carbohydrates and protein content of *A. canescens* increased with increasing the levels of manures. On the other hand, crude fiber content decreased significantly with increasing the levels of amendment. Such reduction may be attributed to the indirect effect of the increasing amount of protein. Ash content showed no clear trend in its response to soil amendment.

Experiment III: The growth of the fodder halophyte kallar grass (*Leptochloa fusca*) in both coastal and inland salt affected lands of Egypt's desert was experimented by Ashour et al. (2002).

The field trials were carried out during two successive years 1998/1999 and 1999/2000 using the available local water (diluted sea water and brackish drainage water) for irrigation of *L. fusca*. The biomass production and the chemical composition of the grass were evaluated.

(i) The first trial was conducted in the salt affected land associated with the eastern coast of the Gulf of Suez (Sinai Coast). Four dilutions of sea water were used for irrigation: 5000 ppm, 10,000 ppm, 15,000 ppm and 20,000 ppm. Four plots (4 m2 each) were transplanted with root stumps of *L. fusca*. The plots were irrigated weekly in autumn and winter and two times weekly in spring and summer. Four harvests were made at 2 months intervals.

The results of this experiment revealed that the biomass of *L. fusca* was the highest (39.18 tons/ha/year fresh and 18.5 tons/ha/year dry) when irrigated with 10,000 ppm sea water whereas the lowest values (8.8 tonw/ha/year fresh and 3.76 tons/ha/year dry) when irrigated with 20,000 ppm sea water. Irrigation with 5000 ppm produced biomass of 32 tons/ha/year fresh and 14 tons/ha/year dry) but irrigation with 15,000 ppm produced biomass of 12.5 tons/ha/year fresh and 8.2 tons/ha/year dry. On the other hand, the increase of salinity levels in the irrigation water (5%, 10%, 15% and 20%) tended to increase ash (10%, 12%, 10% and 18%), soluble carbohydrates (33%, 32%, 38% and 40%), crude protein (13%, 17%, 14% and 16%) and crude fat (2.3%, 2.3%, 2.7% and 2.4%) and decrease in crude fiber (41%, 37%, 33% and 30%) respectively.

(ii) The second trial was concducted in an inland salt affected land near lake Qaruon of Fayium Depression, Western Desert. The irrigatioin water was taken from the agricultural drainge with the following properties: pH = 7.4, EC = 2.8 mmohs/cm. $Na^+ = 22.7$ meq/L, $K^+ = 8.3$ meq/L, $Ca^{++} = 4.5$ meq/L, and $Mg^{++} = 3.0$ meq/L. Four plots (4 m^2 each) were transplanted with root stumps of *L. fusca*. Irrigation was weekly in autumn and winter and two times weekly in spring and summer. The first harvest was made after 2 months from transplantations.

The grass was repeatedly harvested at 2 months intervals. Total yield was 23.6 ton/ha/year (fresh weight) and 14.4 ton/ha/year (dry weight). The chemical composition of *L. fusca* of this experiment were: ash (14.7%), fat (1.83%), proteins (9.2%), carbohydrates (42.9%) and fibers (31.4%). Other elements determined were: potassium (1.02%), sodium (2.14%), calcium (0.54%), magnesium (0.4%) and phosphorus (0.16%). Thus, Kallar Grass (*L. fusca*) is a useful halophyte rich in its nutritive value and could be propagated in salt affected lands of the inland and coastal deserts of Egypt. Diluted sea water and brackish drainage water could be used for its irrigation. The grass tolerates salinity up to 20,000 ppm, however, high yield was obtained when it was irrigated with 10,000 ppm sea water. It is easily propagated through root stumps and produces green yield during summer where no traditional green fodder is available.

Experiment IV: *Panicum turgidum* is a common xerophytic fodder grass predominates in both arid and hyperarid zones of Egypt. It is a highly palatable grass and its seeds are used as human food. In addition, *P. turgidum* is an effective sand binding xerophyte and could be used to fix sand dunes. *P. turgidum* is also used in the nomadic medicine for removing white spots on the eye. The powder from its underground stems is used in healing wounds. Unfortunately, this multiuse drought resistant grass is usually overgrazed beyond its capacity to remain vigorous. However, it exhibits high growth rates in late spring and early summer month from its buds and rhizomes hidden underground soil surface and are thus protected from grazers (El-Kabalawy, 2004).

Batanouny et al. (2006) studied the reproductive capacity of this widespread grass in Egypt by conducting germination and growth experiments.

1. Germination Experiment

Germination of intact and dehusked seeds of *P. turgidum* has been tested under different temperature in light and dark. The results revealed that under constant temperature, the intact *P. turgidum* seeds did not germinate at 15 °C. By the rise of temperature, germination percentage increased up to 37% at 35 °C. Temperature ranging between 20 and 30 °C showed higher values of germination under dark conditions (31.7%) than those under light (21.7%). Also, alternation of temperature improved germination percentage reaching 51% under 10–20 °C increased to 84% under 20–30 °C.

On the contrary, germination of the dehusked seeds was favoured by light. Moreover, germinatioin attained its highest value up to 46% and 43% under constant temperature of 35 °C in light and dark conditions, respectively. Germination of the dehusked seeds recorded 63% at alternating temperature 10 °C and 25 °C, decreased to 51% when germination was tested under 20 °C and 30 °C.

2. Growth Experiment

(i) By transplants

This experiment was conducted in newly reclaimed area at El-Noubarya, Western Desert (148 km North of Cairo) along Cairo-Alexandria desert road. The transplants were obtained from a *Panicum turgidum* community in the nearby area. Clones were excavated by digging a furrow around each clone which was divided into units (transplants). The transplants were planted into holes (20–30 cm deep), rhizome and roots were completely buried in the soil. Irrigation using ground brackish water was carried out weekly. The experiment started on April 1994 and continued till January 1995. The results showed that the proper time for propagation of *P. turgidum* using transplants was April and October with sprouting percentage amounting 55.2% in April and 50.2% in October, climate during these 2 months is usually mild (mean temperature = 21.2 °C in April and 27.8 °C in October). Sprouting was reduced to half of its values during the hot months (July and August) to 27.7% and 25.8% respectively. (mean temperature ranged between 28 °C and 28.3 °C).

(ii). By seeds

Fully matured seeds collected from the natural stands of *P. turgidum* were sown in pairs, 10 cm apart within wooden frames filled with sandy-clay soil (3:1). The treatment was daily irrigated for 7 days and then irrigated every other day for a month. Irrigation water was artisian. The experiment continued for 4 months (May–September 1995). Five days after sowing, seedling emergence with a value of 10% was recorded, increased gradually by time till its highest value 85% after 25 days. Four month aged *Panicum turgidum* plants produced dense culms with a mean of 3450.6 culms/m^2, mean height = 107.3 cm and mean dry weight = 4395 gm/m^2. Though successful experiment yet Batanouny et al. (2006) did not recommend propagation of *P. turgidum* for dry-land agriculture by seeds. They proposed that the seed can be densely sown in nurseries and seedlings could be used as transplants for propagation. Transplants must be kept moist until plating in the field and the upper parts of their colons should be clipped before transplantation to reduce transpiration and to promote tillering.

C. Feeding Tests

The feeding values of three fodder halophytes, namely: *Kochia indica (Bassia indica), Juncus subulaus* and *Diplachne fusca* were tested by Zahran et al. (1999). Their tests were conduced on rams and rabbits using fresh materials of the three plants. Establishment of *K. indica* on salt affected land was also experimented. The results are summarized below:

1. *K. indica, J. subulatus* and *D. fusca* are rich in their nutritive values being: 18.63%, 16.10%, 8.62% crude protein, 35.9%, 24.6% and 28.29% crude fiber, 20.11%, 10.4% and 11.61% ash, 24.17%, 47.19% and 49.49% nitrogen free extract, 1.18%, 1.74% and 2.12% ether extract and 45.44%, 30.0% and 35.0% dry matter, respectively. The shoot and root systems of *K. indica* contain: 0.706%, and 0.373% soluble sugar, 4.738% and 6.295% total carbohydrates and 294 mg% and 85 mg% alkaloids, respectively.
2. The three studied plants are halophytes with higher fodder potentialities. Their nutritive values are comparable to those of clover and alfalfa.
3. *K. indica* could be used alone to feed livestock with no side effects. Rams fed on *K. indica* gained weight without side effects.
4. *J. subulatus* and *D. fusca* were not easily accepted by rabbits and, thus, it is preferable to use their materials in the commercial ration up to 40% instead of clover's hay which became expensive and not easily available in Egypt due to the limited areas usually cultivated by clover and alfalfa.
5. The total digestible nitrogen (TDN) of the three halophtes were high being 46.95%, 57.21% and 53.73% the TDN of alfalfa is 50.3%.
6. The digestible crude protein of *K. indica* was higher (14.12%) than those of *J. subulatus* (10.72%) and *D. fusca* (5.64%).
7. Propagation of *K. indica* conducted in salt affected soil in the deltaic Mediterranean Coast of Egypt was encouraging. Two vegetative yields could be obtained every year with productivity up to 5.3 tons/ha^{-1}/cut.

The above mentioned findings show the important role that could be played by the perennial xerophytes and halophytes (as well as the annual vegetation) in supplying green fodder for livestock all year around. Sustainability could be achieved through correct management and public awareness.

III. Sustainability of Range Lands

a. General Remarks

In arid countries, the pressure of land-use, overgrazing, over cutting, cleaning for agriculture etc., coupled with extreme climatic aridity and uncertainty of rainfall, has resulted in an advanced stage of desertification. Natural vegetation is currently regressing at a rate of 1%–2% annually (Le Hourerou, 1973). El-Kady (1987) stated that the intelligent use of rangeland in the deserts for optimum yields requires a detailed knowledge of what and how much the land can produce under given circumstances, i.e. knowledge of their biological potentialities. It requires furthermore that the land-use be managed so as to insure continuity (sustainability) of optimal yields. Not only it is necessary to conserve existing resources of soil, water, vegetation and wild animals, but also to monitor the condition of each with the eye on amelioration of change to allow for prediction of crucial long term effects of man-made manipulation. Thus, land protection usually results to an increase in density, cover and vigour of the gegetation and consequently better soil stabilization (Halwagy, 1962).

According to Ellison (1960), when left to rest, dryland ecosystems disturbed by overgrazing or stressed by drought, may recover. Recovery tends to advance at slow pace because of the low productivity of these ecosystems, enhanced productivity may occur in years of above average rainfall. However, monitoring of changes to allow for prediction of long term effects of man-made manipulations is therefore of crucial significances in the formulations of management plans of an ecosystem (Ayyad, 1983). Such monitoring may be viewed at three levels. The first level is concerned with the ecosystem components and human impacts on these components, in which changes are recorded, including soil characteristics, abundance of biodiversity (plants, animals, and microorganisms) above and below soil and the roles these biota play in energy flow and nutrient cycling. These records may then be used to construct simulation models for predicting future changes at the ecosystem levels (El-Kady, 1987). The second level includes monitoring of changes in the pattern of vegetation composition and of the physiographic features of land in a limited sector. This involves vegetation mapping (remote sensing, aerial photographs, etc...) and ground truth data. Successive year's photographs will provide means to assess the changes in vegetation patterns. The third level includes composition of the salient features of large areas in successive years for adequate evaluation of land transformation.

b. Grazing Experiments

The effect of protection and controlled grazing on the vegetation composition and productivity as well as the rate of consumption of phytomass by domestic animals had been assessed by El-Kady (1980), during 3 successive years 1977–1979 in the non-saline depression of El-Omayed area, 80 km west of Alexandria, Western Mediterranean Coastal desert. The climate of the study area belongs to the arid climate with mild winter and warm summer (world distribution of arid regions, UNESCO, 1977).

Five grazing treatments were studied, four of them were in fenced plots and the fifth represent the area outside the fences with free practised grazing pressure. The first plot was fenced on July 1974 and subjected to 50% of the free practiced grazing pressure (about 6 heads/10 ha). In May 1977, the second plot was fenced and subjected to 50% of the freely practiced grazing pressure since May 1977, the third plot was fenced and subjected to 25% of free grazing pressure and the fourth plot was fenced on July 1974 and kept without grazing by domestic animals till the July 1977. Changes in the density, cover, frequency, phytomass and the phenological sequence of species were recorded and compared to those of the same species outside the fenced plots. Grazing animals in each of the two plots with 50% grazing pressure (fenced on 1974 and 1977) were observed for 24 h every month. The species and part of the grazed plant parts were identified and the number of bites, size and weight of bites and grazing and resting times were recorded. A consumption index was then calculated to assess the relative preference of different plant species by the domestic animals. The following results had been obtained:

1. Seven perennial species constituted the major part of the animal diet in the controlled grazing plots: *Asphodelus microcarpus, Echiochelon fruticosum, Thymelaea hirsuta, Plantago albicans, Crucicnella eriocephalus, Helianthemum lippii* and *Gymnocarpos decander*, beside one annual species (*Rumex pictus*) during growing season. Outside the fenced plots, *A. microcarpus, E. fruticosum and T. hirsuta* were the only species contributed the main diet of animals. Consumption of animals, which indicated their relative preference, changed from season to season.
2. Remarkable increases were recorded in total density and cover of perennials, in frequency and presence of animals, and in phytomass as a result of protection and controlling grazing. However, some species exhibited negative responses and productivity of most species was more pronounced as the controlled grazing plots especially that with an initial period of full protection. Thus, partial protection and controlled grazing could be better consequences than full protection. Light nibbling and removal of standing dead biomass by domestic animals may promote vigour and growth of defoliated plants. Also, the availability of nutrients (especially nitrogen) may be enhanced by the passage of herbage through animal guts and out as faeces.
3. The consumption by domestic animals was about 20% and 40% of the net primary production of shoots in the controlled grazing plots and the uprooted area, respectively.
4. The ration of animals from the phytomass and necromass provided amounts of total digestable nutrients ranging from 520 gm/animal/day to 960 gm/animal/day in the controlled grazing plots, and almost half of these amounts in the free grazing area for sheep and goats during the growing season. The protein content was 10% less than the proper level. These amounts were estimated to be far in short of the requirements of animals in the free grazing area, and scarcely adequate for those under the rotational system with 50% grazing pressure.
5. To maximize the productivity of animals in the study area, adequate amounts of supplementary feed rich in protein should be supplied.
6. Similar studies need to be conducted on the plant life of the other rangelands of the coastal and inland deserts of Egypt to provide information necessary for the protection, recovery, management and consequently sustainability of these rangelands.

Fiber Producing Plants

A. General Remarks

The natural flora of the Egyptian deserts comprise considerable number of xerophytes, halophytes and psamophytes that could be considered as fiber producing plants. Among these we may mention: *Ammophila arenaria, Calotropis procera, Caralluma* spp., *Dracaena ombet, Desmostachya bipinnata, Euphorbia thi, Gossybium arboraem, Halopyrum mucronatum, Hibiscus micranthum, Juncus acutus, J. rigidus, Leptadenia pyrotechnica, Lygeum spartum, Phragmites asutralis, Thymelaea hirsuta* etc. All of these plants are growing under salinity and/or aridity stresses of the Egyptian deserts and some of them are dominant species. A plan aiming at the sustainable utilization of these plants should be encouraged to meat at least part of the needs of Egypt of raw materials for fiber industry e.g. paper production (Zahran, 1993b).

B. Representative Species

In Egypt, a country of arid region, paper mills import wood pulp to produce paper of good quality as well as to improve the strength properties and grade index of paper produced from locally available raw materials mainly rice straw and bagass. The amount of wood pulp imported is sharply increased annually and, hence the need to search for sufficient local raw materials for paper industry.

Juncus halophytes in paper Industry

Being salt tolerant and resistant to extreme desert temperatures rushes, *Juncus acutus* and *J. rigidus* had been selected for a study on likely fiber plants (El-Bagouri et al., 1976; Zahran and Abdel Wahid, 1982). These two rushes are characterized by their subsurface rhizomes that develop several leafy shoots (the culms) every year. These unbranched, smooth nodless culms represent the vegetative yields of *Juncus* plants used in fiber industry.

Juncus acutus and *J. rigidus* are cumulative halophytes accumulating excess salts they absorb from soil in the upper parts of their culms, an advantage characterizes *Juncus* species that may lead to the biological desalination of the saline soil.

In Egypt, *J. rigidus* is a common rush that flourishes in saline soils with total soluble salts up to 5% in the rhizosphere layer, mostly chlorides (>80%) and partly (<20%) sulphates. The distribution of *J. rigidus* in Egypt indicates that the plant is not only salt tolerant but also tolerates a wide range of climatic conditions. It is recorded in coastal as well as in inland salt marshes, i.e. in both arid and hyperarid provinces of the country. *J. acutus* and its community, however, are widespread in the salt marshes of the littoral and inland deserts of the northern arid territories of Egypt but rare or absent from those of the southern hyperarid province. Its habitat is usually saturated or inundated soils of the downstream zones of the sedge meadow habitat close to the reed swamp vegetation. The total soluble salts of the rhizosphere layers ranges between 1.2% and 1.7% mostly chlorides.

Economically, *J. rigidus* and *J. acutus* have manifold uses. Traditionally, both are used for making mats and their inflorescence and seeds are grazed by livestock. Their seeds are rich in fatty acids and their culms contain low ash (6.5%), low lignin (13.3%), high a-cellulose (39.8%) contents as well as high yield of unbleached pulp (36.8%). The strength properties of the depithed unbleached *Juncus* pulp are much higher than those of rice straw (grade index (GI) = 24%), and bagass (GI = 42%), it gave a grade index of 73% compared to softwood long-fiber unbleached imported pulp (GI = 100%). Consequently, if a sufficient amount of *Juncus* is available to paper mills in Egypt, the size of paper pulp imports may be changed. However, large scale economic production of paper entails large volume production of *Juncus* plants from homogenous vegetation. This may not be available from the heterogoneous natural growth of *Juncus* communities in Egypt. Hence, it was important to study the establishment of these rushes under salinity stress and to recognize the factors affecting their productivity in saline agricultural practice. Also, the effects of these factors on the paper-pulp production and qualities were considered (Zahran, 1993b).

a. Germination Experiment

This experiment was conducted under different salinity levels, temperatures and water regimes. It was found that germination of seed of both *Juncus* species generally decreased with increased salinitly level, but *J. rigidus* seeds treated with more saline solution (2–3% NaCl) showed relatively better germination than those of *J. acutus*. Increased temperature enhanced germination but high temperature (30 °C) during the day and >20 °C at night retarded it. Saturated or water logged soil proved more favourable for the seeds of *J. acutus* but seeds of *J. rigidus* germinated much better in moderately moistened soils (Zahran, 1993b).

b. Growth Experiment

Field growth experiments of *Juncus* plants had been conducted on two types of soils: saline and calcareous (El-Bagouri et al., 1976).

(1) The saline soils experiment (4032 m^2) was established in poorly drained soil under the influence of Manzala Lake. The water table (barackish) was shallow (10–15 cm in summer and 50–70 cm in winter below soil surface). The soil of the site was usually saturated with water, clayey in texture, black in colour, alkaline in reaction (pH = 7.8–8.4) with organic carbon (0.9–1.2%), calcium carbonate content of 3.8 –4.4% and total soluble salts at 0.9–1.8% in the surface and 0.3–0.65% in the subsurface layers. Rhizomes of both *Juncus* species were collected from its natural vegetation associated with Mariut Lake, Mediterranean coast and transplanted at equal distances about 5–10 cm deep. Irrigation was applied every 5 days in summer and 7 days in winter. After 1 year of growth, aerial green culms, were harvested and three parameters were measured. The results indicated that both *Juncus* species may be cultivated on poorly drained saline soils. The growth of *J. rigidus* was better than *J. acutus*. Mean height of *J. rigidus* culms (162 cm), mean fresh weight (5 kg/plot) and

mean dry weight (2 kg/plot) were higher than those of *J. acutus* being 85 cm, 2.8 kg/plot and 1.1 kg/plot, respectively.

Fertilization of *Juncus* experiment using six different concentrations of nitrogenous and phosphorus fertilizer and micro-nutrients was also tested. Generally, the total vegetative yields of *Juncus* culms (fresh and dry weights) were relatively higher in the plots treated with excess N fertilizer.

Measurements of fiber length in samples of the culms of *Juncus* species treated with various fertilizers showed that the addition of N-P fertilizer had positive effect on the fiber length. A reduction in the number of very short (<500 μm) and short (500–1000 μm) fibers of both species from 11% to 0.5% and from 12 to 0% in *J. rigidus* and *J. accutus*, respectively were determined. On the other hand, the percentage of the fibers with lengths ranging between 100 and 1500 μm increased by the addition of fertilizers from 23.5% to 36.5% and from 15% to 36.5% for *J. rigdius* and *J. acutus*, respectively. The same was true for the fibers with length between 1500 μm and >2000 μm.

(2) The second experiment was conducted in the Mariut Experimental Station of the Desert Research Center in America, located about 35 km south of Alexandria. The soil was calcareous ($CaCO_3$=32%), alkaline in reaction (pH=8.1–8.3), loamy in texture, low in productivity (cation exchange capacity=18 mEq/100 gm soil) and saline (EC=8.5 mmohs/cm). The area of the experiment (4536 m^2) was divided into 108 plots (6×7 m each), 54 plots for each *J. rigidus and J. acutus*. The transplantation was carried out with rhizomes of both species taken from the nearby Mariut salt marshes. The experiment was irrigated every 7, 14 and 28 days and continued for 6 months. Three nitrogen manuring treatments were applied after 1 month. The results confirmed that *J. acutus and J. rigiducs* succeeded to grow on the highly calcareous soil with all treatments. The biomass yields of *J. rigidus* was superior to those of *J. acutus*. Decreasing the frequency of irrigation reduced the yields of both species which is an indication of the high moisture requirements of *Juncus* plants. Nitrogen manuring increased the yields of both species.

Drug Producing Plants

Plants are the sources of chemical compounds and have always played a major role in the treatment of several diseases. Plants are, thus considered the nature's Green Pharmacy which provide drugs to maintain health of human being and animals. Prehistoric man used plants to battle diseases, induce hallucinogenic experiences and to fend off evil spirits. However, according to Boulos (1983), the adverse of western medicine diminished the importance of herbal remedies.

In Egypt, the use of herbs for therapeutic purposes dates back to immemorial times. In Egypt's deserts where people are living in tribes far away from each other, they are almost deprived of proper medicinal care. Naturally, a group of local prescribes have to fill the gap using folk medicine formula based on crude materials from their local environment. "Plants are, actually, used as drug with slight or almost no change leading in most cases to satisfactory results" (Boulos, 1983).

The Egyptian flora comprises more than 2000 species many of them proved to have considerable values in folk medicine. Unfortunately, little is known about their present status, most of them are seriously threatened due to the over use particularly in the deserts where there is no alternative for drugs. Thus, "there is a great need to provide a framework for the conservation and sustainable use of these valuable renewable natural resources" Hammouda (2005) stated. The following pages throw light on different aspects of 10 representative medicinal plants naturally growing in the Egyptian deserts.

1. Acacia nilotica

A. nilotica (Sant, Leguminosae) is a single-stemmed evergreen tree of 7–10 m high with glabrous or tomentose branches. The bark of the mature trees is rough, dark with longitudinal fissures. The paired spines at the base of each leaf are very noticeable. Leaves are bipinnate. The rounded flowering heads are bright yellow and sweet-scented. Fruits are long pods with marked constriction between the seeds, green when young changing to dark brown when ripe (Springuel, 2006).

A. nilotica is widely distributed in Egypt in the Nile Valley and Oases of the Western Desert and Sinai Peninsula.

Apart from being wood and charcoal producing plant, *A. nilotica* has many other uses. The bark and pods contain tannins used in the tanning of leather, pods are used for making dyes (yellow, red and black). Different parts of the tree are used in traditional medicine. The pods with enclosed seeds are popular for treating cough. The gum of *A. nilotica* is a demulcent and serves by the viscidity of its solution to cover and sheath inflamed surfaces. It is usually administered in the form of mucilage (*mucilago acaciae*). Infusion of fruits for diarrhea, powdered fruits for fever and diabetes (Boulos, 1983). Folk medicine uses include: anticancer, antitumors, antiscorbutic, astringent, diuretic, intestinal pain and diarrhea as well as nerve stimulants. The plant is used also for colds, congestion, cough, dysentery, fever, gallbladder problems, hemorrhage, leucorrhea, ophthalmic, sclerosis, small pox and tuberculosis (Ismail, 2006).

A. nilotica is easily regenerated from seeds. A mature tree can produce 2000–3000 pods/fruiting season, each with 8–16 seeds. Seed pretreatment is needed by mechanical or acid scarification or by pouring boiling water over the seeds and allowing them to cool. In addition, in some rural areas, the pods are fed to goats and the scarified seeds are removed from their droppings. *A. nilotica* seedlings are ready for planting in the permanent land after 5–6 months (Springuel, 2006).

2. Balanites aegyptiaca

B. aegyptiaca (Hegleeg, Balanitaceae) is a xerophytic thorny semi-deciduous tree or shrub dropping some but not all of its leaves during the dry season. Height of stem ranges between 6 and 12 m and its tap root is very deep (about 15 m depth). It has small green flowers and drup fruit about plum-size, green when young turning yellow at maturity. *B. aegyptiaca* begins fruiting at about 4–7 years and reaches maturity at 25 years old. It is a long living tree, some trees may reach more than 200 years old. According to Springuel (2006), fruiting of *B. aegyptiaca* can be heavy with yield of 45–150 kg/individual tree, a large tree in Kharga Oasis produced over 200 kg fruit throughout the year.

In Egypt, *B. aegyptiaca* has limited distribution and grows in the southern section of Egypt including the Nile Valley, the Eastern Desert, Red Sea mountain, Gebel Elba as well as in the southern Oases of the Western Desert, e.g. Kharga, Dakhla and Baris Oases. Relatively large undisturbed open woodland dominated by *B. aegyptiaca* occupies the upstream part of Wadi Allaqi in the Nubian Desert east of the Nile close to the Sudanese border.

B. aegyptiaca was one of the most widely used tree in ancient Egypt. The hard stones of its fruits had been found in Prehistoric Pharaonic and Greco-Roman sites. The most important Pharaonic use of *Balanites* was the extraction of Balanos Oil from the kernel of its fruits. This oil could be used as an unguent or for message.

Recently, it was found that the different parts of *B. aegyptiaca* tree including fruits, kernel, leaves, barks, seeds etc. are used to treat a wide range of illness and complains. For example, the fruit is used to kill the fresh-water snails that carry schistosomiasis (belharzia) flukes (Springuel, 2006). Boulos (1983) stated the following uses of *B. aegyptiaca*: anthelmintic, purgative, boils, leucoderma, herpes, vermifuge, malaria, emetic, wounds, syphilis, colds, liver and spleen problems.

Regeneration of *B. aegyptiaca* may occur either by seeds or by root sucker. For germination the mesocarp of the seeds should be removed by soaking the fruits in water of 24 h before sowing. Seeds germinate well at 30 °C–35 °C and seedlings could be planted in the permanent land when they are 5–8 months old. The best periods for planting are early spring (February–March) or early fall (October–November). However, growth of seedlings seems to be very slow during the first 4 years after which the stem reaches about 1 m high. The root develops intensively. The rate of growth increases considerably during the following years (50 cm or more every year).

3. Calotropis procera

C. procera (Oshar, Asclepiadaceae) is a shrub or small tree 3–5 m high, stem is softwood covered with thick cork-like bark which is light-brown in colour. Leaves are light green, simple large and broad up to 25 cm long. Flowers outside green, inside pink. Fruits are large (15 cm across), smooth, apple-like, green in colour, spongy. When mature the fruit open and reveals the seeds which are packed into a compact core and covered by long silky hairs, facilitating their dispersal by wind (Täckholm, 1974; Springuel, 2006).

C. procera is a common xerophytic plant throughout Egypt particularly in the southern extreme arid sections of the deserts and Nile regions.

C. procera has various uses including ropes from its inner bark. However, its most valuable composition is the milky latex which is used to treat many human and animal disorders. In the same time, the latex cause serious inflammation to eye that may lead to blindness. Decoction of bark and latex used in veterinary medicine, anti-leprosy for scabies. Powdered dried leaves vermifuge in small doses; dry leaves smoked as cigarettes for asthma, cataplasm of fresh leaves for sunstroke. Latex applied to teeth to loosen them, also for toothache. Leaf extract cardiotonic. Root emetic, expectorant; root bark for dysentery, elephantiasis, syphilitic ulcers, stomachic pains, and diaphoretic (Boulos, 1983).

C. procera is regenerated by seeds which are easily germinated without treatment and can be planted directly into moist ground. Seedlings grow rapidly when sufficient watering is applied.

4. Cassia senna

C. senna (*Senna alexandrina*, Senameki, Leguminosae) is a xerophytic glabrous undershrub, multi-stemmed reaching a height of 1–1.5 m and width of 2 m. The pale green stems are erect and densely branched at 20–30 cm above the ground. Leaves are compound with 3–7 pairs of elongating grayish-green mucilaginous leaflets having a peculiar odour and sweetish taste. Yellow racemed flowers, fruits are flat pods green when young and changed to yellow-brown when mature, each fruit contains 6 seeds. Root system is extensive widely spread in the ground.

In Egypt, *C. senna* is common in the southern extreme arid section of the Eastern Desert, Red Sea mountains, Gebel Elba, Sinai and the marginal zone between the Nile Valley and the desert (Springuel, 2006).

C. senna is used in traditional medicine and has a commercial values. It is well known as a purgative. Also, it is used in the treatment of influenza, asthma and nausea. "Infusion of powdered leaflets and pods a popular laxative and purgative" Boulos (1983).

C. senna is usually propagated by seeds and germination occurs without any treatments. However, to enhance germination scarification is recommended to break the outer cover of the seeds either by scratching or by putting the seeds in hot water for 1 day.

5. Cymbopogon schoenanthus

C. schoenanthus (Halfa Barr, Gramineae) is a perennial aromatic, stout, densely tufted grass (1 m high) with filiform narrow leaves and shallow fibrous roots. In Egypt, *C. schoenanthus* has a very narrow distribution. It grows and predominates in the far south of the Eastern Desert in the border with Sudan. For its medicinal value, the grass is highly threatened by over exploitation. It grows in the alluvial deposits of the desert wadis, in rocky habitat as well as in sandy soil. High temperature is very suitable for its growth and it tolerates a prolonged dry period, it is a true xerophyte.

Springuel (2006) stated that *C. schoenanthus* (halfa barr) was known in ancient Egypt as one of the ingredients to make the famous *Kyphi*, a scent free from oil and fat. It is used intensively in indigenous medicine as diuretic, painkiller for colic and antipyretic. Pharmaceutically, this grass is used in preparing the drug Proximol. Halfa barr is a healthy and refreshing hot drink especially popular in Upper Egypt. Boulos (1983) stated "Infusion of plant diuretic, emmenagogue, astringent, carminative, sudorific, antirheumatic, cataplasm for wounds of camels. Infusion of flowers febrifuge".

Seeds of *C. schoenanthus* are very small and difficult to germinate due to their long dormancy. They need to be in the soil for at least 8 months before germination. Thus, it is preferable to germinate this grass by root cuttings. Once established, halfa barr will be self propagated when moisture is available.

6. *Dracaena ombet*

D. ombet (the ombet or the Nubian Dragon tree, Dracaenaceae) is a small stout tree with forked stem up to 5 m high bearing on the top a rosette of large sword shaped leaves (60 cm long and 2 cm broad). Flowers pink forming panicle. Fruits arc one seeded small berries, spherical.

In Egypt, *D. ombet* is a threatened species occurs only in Gebel Elba and Gebel Shindeeb of the Red Sea coastal mountains. It thrives in rocky substratum and can survive prolonged drought period (Springuel, 2006). The mature fruit of the ombet trees are edible and the stem yields red resin which is used in traditional medicine.

D. ombet is difficult to germinate, however, recent trials succeeded to germinate its seeds collected from Gebel Elba Mountain. It is a slow growing plant, 4 years old plants are still growing in pots of a private nursery in Cairo. Trials to propagate this endangered xerophyte in the Egyptian desert is essential to protect it against extinction.

7. *Origanum syriacum subsp. sinaicum*

O. syriacum subsp. *sinaicum* (margoram, zaatar, bardagoosh, Labiatae) is an endemic tomentose herb or undershrub. Its stem (40–90 cm) is erect much branched, leaves broadly ovate, entire palmate-veined, verticillasters 2–8 flowered in dense spik-like inflorescences, often in panicles. Corolla lilac. It grows in the rocky habitat of the Sinai mountain. It is thought to be the tree Hyssop of the Bible (Hammouda, 2006). This medicinal plant is highly threatened being heavily collected for various medicinal uses and to prepare herbal tea. There is, thus a great need to conserve its natural growth and to propagate it in the Egyptian Deserts.

Zaatar has a long history as a medicinal and flavoring herb. It has been used in folk medicine to treat cold, coughs and gastrointestinal problems. Also, it is used as antibacterial, antifungal, anti-rheumatic, antiseptic, antispasmodic, carminative, cicatrizant, digestive, diuretic, emmenagogue, expectorant, nervine, sedative and stimulant. The essential oil produced from *Oroganum* (oregano Oreganol oil) is extracted from its leaves and flowering branches. The oil is rich in a long list of minerals including: calcium, magnesium, zinc, iron, coppeer, boron, and manganese in addition to vitamins.

Origanum is propagated by seed, division and basal cuttings (Hammouda, 2006). Seed germination (in a green house) usually takes place within two weeks. Seedlings planted into pots when they are large enough to handle and later plant them out into their permanent positions in early summer. Division propagation usually takes place either in March or October. Very easy, larger division, can be planted out direct into their permanent position on late spring or early summer.

Basal cuttings is also easy way for propagation of this medicinal plant. Collect the shoots with plenty of underground stems when they are about 8–10 cm above the ground during June. Plant them in individual pots and keep them in light shade in a cold farm or green house until they are rooted well. Plant them out in the summer.

8. *Salvadora persica*

S. persica, (Arak, Messwak, Salvadoraceae), the tooth brush tree, is a xerophytic tree or shrub. Its vernacular names are: Arak, Meswaak. It has opposite leaves and small greenish-white flowers in rich terminal panicle. Fruits are small white or pale purplish, globose drupe of pungent taste. Stem is heavily branched, both stems and branches are whitish bearing numerous coriaceous (leathery) leaves having a strange smell similar to mustard hence the English name: mustard tree. Roots are extensively branched from the base of the trunk and are very long spreading both horizontally and vertically, deeply penetrating the soil.

In Egypt, the tooth-brush tree grows, in certain sites abundantly in wadis of the extremely arid section of the Eastern Desert, Red Sea Coastal Desert and Sinai Peninsula. For its dense growth, one of the Wadis of the Red Sea Coastal Desert is called Wadi Arak. *Salvadora* is an evergreen plant that keeps its leaves even during the prolonged dry season.

S. persica is very well known in Egypt and throughout the middle east countries in traditional medicine. It is used to treat gonorrhea, spleen, boils, sores, gum diseases, headache, stomach pain, respiratory disorders. Leaves, roots, bark and flowers contain oil that is diuretic. Powdered leaves mixed with millet flour and honey are made into small balls and taken every morning for 40 days as antisyphilitic. Fruits edible, stomachic, carminative, febrifuge, fortify the stomach and bring good appetite (Boulos, 1983). The powdered bark of arak is used in the treatment of snake and scorpion bites. However, the most important use of arak and since the ancient times is to clean teeth. Its young stems and lower parts of the stems close to the roots are used as tooth-brush.

Propagation of *S. persica* could be by seeds that germinate easily without treatment. However, due to its low seed production, it is preferable to propagate it by cuttings that are obtained from naturally growing plants. Cuttings are taken from the base of the stem close to the root. Propagation by tissue culture is also recommended (Springuel, 2006).

9. *Solenostmma arghel*

S. arghel (argel, hargal, Asclepiadaceae) is an evergreen erect perennial xerophytic undershrub with 0.6–1.0 m in height. It is a blue-green finely velvety-pubsecent plant. Leaves elliptical lanceolate. Flowers white in axillary umbels, fruit ovate, smooth, very hard of dark purple colour. The plant flowers between September-April and fruiting between April-June. Seeds are turgid, ovoid channelled down one face. In Egypt, its occurrence is recorded to the Sinai Peninsula, southern Eastern and Red Sea Desert including Wadi Allaqi, Wadi Gimal, Wadi Umm Hargal, Wadi Umm Sider (Hamed, 2005). It usually thrives in the pebbly and gravelly habitats. Actually, *S. arghel* is a typical evergreen xerophytic undershrub that extends its vegetative activity for at least 3 years of a rainless periods. In dry periods some individuals may shed their leaves and even few branches but others remain evergreen throughout the vegetative periods. Reproductive activity is high in the third year from September to June. Flowers start to appear in September and fruits in April-June. Ripe fruits open and seeds are released.

S. arghel is a threatened plant due to its heavy collection from its natural stands to be sold as a folk medicinal plant in the Attarin shops all over Egypt. Propagation of this plant is successful by seeds which germinate under a wide range of temperature: 20–40 °C in Petri dishes and 25–40 °C in soil. The most suitable temperature for its germination is 35 °C. This may indicate its specific thermal adaptation to high temperature. Pre-sowing treatment of seeds with growth stimulators e.g. Thiourea 5 increased the percentage of seed germination up to 92.5%. Survival of seedlings is better when seeds are germinated in pots or plastic bags rather than sown directly into the open ground. The seedlings can be planted out in the permanent land at the age of 3–4 months in any season excep winter. Seeds can also be sown directly into the ground, preferably sandy soil at the beginning of summer (May–June). Daily watering is required until the first seedlings appear, hence watering can be reduced 3 times/week and 2 times/week when the seedlings are 2 months old. Germination percentage, is generally, low (about 30%) when seeds are sown directly in the ground (Springuel, 2006).

10. Tamarix aphylla

T. aphylla (Atl, Tamaricaceae) is an evergreen tree or tall shrub up to 12–15 m high. The stem of old trees may attain a diameter of more than 1 m. The bark of young trees is smooth, becoming thick and deeply furrowed with age. The leaves are minute and sheath-like. Branches have tiny salt glands that give a whitish appearance to the crown which could be more than 10 m across. Bunohes of small rose or white flowers are abundantly distributed all over the crown. It has long tap root that reaches a depth of 15–20 m with extensive side roots.

In Egypt, *T. aphylla* is widely distributed. It occurs along the Mediterranean and Red Sea Coastal Deserts as well as in the inland deserts (in the wadis of the Eastern Desert and Sinai Peninsula as well as in the Oases of the Western Desert). As mentioned before, the main channel of Wadi Allaqi is characterized by several terraces and numerous fossil hillocks with remains of extensive thickets of *T. aphylla*.

According to Springuel (2006), *T. aphylla* (tamarisk) has a broad ecological amplitude growing in the riverain, desert and seashore saline habitats. It may be considered a xerophyte having a halophytic nature being salt excretive plant. It was a sacred tree to the God Osiris and the biblical "manna" (Exodu 16) believed to be the sweet exudation produced by small scaly insects feeding on tamarisk branches.

T. aphylla can be planted throughout Egypt in different soil types: clayey, silty, sandy and saline. Also, it withstands long inundation (3–4 years) and tolerates high temperature as well as frost. The recycling of salts is another interesting ability of this plant.

Apart from being wood producing plant, *T. aphylla* has several medicinal values being: diuretic, depurative and sudorific. Boulos (1983) stated that the decoction of tamarisk leaves and young branches is used for oedema of spleen, same decoction mixed with ginger for uterus affection. Bark of large branches boiled in water with vinegar is used as lotion against lice. Infusion of galls astringent, used for enteritis and gastralgia.

T. aphylla naturally produces an enormous number of tiny seeds which are difficult to germinate because they remain viable for only a few days (Springuel, 2006). Tamarisk is easily propagated from woody cuttings of young branches (2 years old) during February-March period. Cuttings should be grown in a nursery for 7–8 months before planting out. Cuttings are advised to be placed with 1–2 buds above ground and 3–5 buds in underground. October-November is the most suitable period for transplanting the cuttings from the nursery to the permanent land.

Stabilization of Sand Dunes

A. General Remarks

Shifting sands in deserts are commonly considered to have resulted from aeolian processes due to high velocity winds, insufficient amount of precipitation and sparse vegetation. Sand drift depends upon the velocity and direction of the prevailing winds and its granulometric composition (Babyev, 1981).

Desertification involves the problem of shifting sands, that cause sand bodies to overwhelm irrigated farmlands, human settlements, villages, canals, railways, motor roads and various civil structures. These impacts incur enormous damage to the economy.

In Egypt's deserts sand formations occupy about 165,000 km^2, i.e. 16% of its total area distributed as follow: 135,000 km^2 in the Great Sand Sea of the Western Desert, 10,000 km^2 in both Qattara Depression and Siwa Oasis, 9000 km^2 in the Northern Mediterranean Coast, 4500 km^2 in the middle and South Oases of the Western Desert, 3000 km^2 in Wadi El-Natrun and Western section of the Nile Delta, 500 km^2 in the eastern section of the Nile Delta and 3000 km^2 in El-Fayium and Wadi El-Rayan Depressions. (Draz, 1993). Generally, the rate of sand dune movement differs in the Mediterranean coastal and inland deserts of Egypt. In the Mediterranean coastal desert, sand movement ranges between 1–13 m/year whereas in the inland desert it ranges between 20 and 100 m/year. This is mainly due to the extreme climatic aridity and higher wind velocity (mean = 28 m/s) in the inland deserts than in the Mediterranean coastal one (mean wind velocity = 16–21 m/s).

In Egypt's deserts, sand dune fixation takes place by two methods: temporary and permanent (Draz et al., 1992). Temporary fixation is implemented mechanically and chemically. In the mechanical method, artificial barriers are established at the windward side of the dunes. Local raw materials such as leaves of date palm, stems of olive trees and reeds are used for fencing. In the chemical fixation method, the following materials are sprayed over the leeward side of the wet sand dunes (i.e. mulching the sand dunes): bituminous emulsion, lubricating oil, Portland cement and polyethylene plastic sheets and nets. Evaluation of the success of the temporary means depends on its duration, toxicity, penetration rate of the liquid materials, easiness of handlings, pollution impacts and costs of labour and materials.

Permanent or biological method of sand dune fixation (phyto-reclamation) has the following advantages: 1. halts the expansion of areas under shifting sands, 2. excludes the formation of new tracts of shifting sands, 3. protects economic units from sand drifts and 4. reclamation of shifting sands to turn them into productive lands as e.g. pasture, forests, etc... Native psammophytes including xerophytic and halophytic trees, shrubs, grasses etc... may be used as sand fixtors. These include species of *Acacia, Ammophila, Atriplex, Desmestachya, Euphorbia, Elymus, Halopyrum, Imperata, Lotus, Nitraria, Panicum, Populus, Prosopis, Retama, Stipagrostis, Tamarix* and *Zygophyllum* etc.

B. Representative Experiments

In Egypt's deserts, the two methods (temporary and permanent) of sand dunes fixation have been experimented, for example: Hanna (1982), El-Hady and Hanna (1983), Maergner (1990), Draz et al. (1992, 1996), Draz (1993), and Draz and Zaghloul (2006, 2007).

In the following a brief account of representative trials.

a. Siwa Oasis Experiment

In Siwa Oasis, located at the northern edge of the Great Sand Sea, Western Desert, Egypt, movement of sand dunes represents one of the major hazards for its socioeconomic development. Draz et al. (1992), through a long term sand fixation program, selected an area in the oasis severely invaded by shifting sand to conduct temporary and permanent fixation experiments. In the temporary experiments (1988–1990) they tried the following:

(i) Establishment of physical barrier formed from leaves of date palm, stems of olive trees and reeds at the windward side of the experimental plot area (12,600 m^2),
(ii) Spraying of bituminous emulsion over an area of 1200 m^2 and lubricating oil on a stretch of 200 m^2,
(iii) Distributing Portland cement on previously wetted sand dunes over an area of 600 m^2,
(iv) Covering two areas (120 m^2 each) with polyethylene plastics and nets.

In the permanent experiment, Draz et al. (1992), tested 14 plant species, namely: *Acacia saligna, Agave sisalan, Atriplex nummularia, Ceratonia siliqua, Euphorbia mauritanica, Ficus carica* (fig), *Olea europaea* (olive), *Phoenix dactyliferea, Prosopis julifolia, P. pallida, Punica granatum* (pomegranate), *Simmondsia chinensis* (Jojoba), *Tamarix aphylla* and *Zizyphus jujubo*. Seedlings of these plants were transplanted during November 1988 in an area of 42,000 m^2 (10 feddans) at the windward sides of the dunes. The experiment continued till March 1989. Irrigation water was taken from the affected drains (3000–7000 ppm): drip irrigation system was used.

The results obtained revealed that the mechanical mean of sand fixation was effective being of low cost and high duration period. For the chemical ways, both lubricating oil and bituminous emulsion were the most effective treatments. This may be due to their high duration period, low costs and high penetration rate. On the other hand, from the environmental point of view, chemical treatments have bad side effects and limitations. For example, the lubricating oil shows negative effects on the natural vegetation inhabiting the treated dunes; application of bituminous emulsion needs personal skills and was negatively affected by the saline water used in the experiment (4000–5000 ppm). All chemical treatments are causing environmental pollutions.

Regarding the permanent fixation experiment, Draz et al. (1992) found that the 14 cultivated plants showed significant variations concerning the growth parameters recommended for the efficiency to fix sand dunes. The survival rate was as high as 94% in *Atriplex nummularia* shrubs and as low as 0.0% in *Zizyphus jujuba* and *Ceratonia siliqua*. 92%, 90%, 88%, 86%, 85%, 80% and 75% were those of: *A. sisalan, T. aphylla, P. juliflora, P. pallida, O. europaea* and *A. saligna*, respectively. The survival rates of other plants ranged between 56% (*E. mauritanica*), 30% (*P. dactylifera*), 28% (*P. granatum*) and 20% (*F. carica*).

Plant height was the highest (233 cm) in *A. saligna* followed by *P. pallida* (193 cm) and *A. nummularia* (145.4 cm). The lowest being that of *S. chinensis* (41.7 cm). the values of stem thickness ranged between 7.2 cm in *A. saligna* and 1.4 cm in *P. granatum*. The highest crown cover and crown volume were those of *A. saligna* (17.5 cm^2 and 15.5 cm^3) followed by those of *P. pallida* (9 cm^2 and 5.5 cm^2) respectively. The lowest were those of *E. mauritanica* being 0.13 cm^2 and 0.01 cm^3. The values of the lateral and vertical root distribution were the highest in *A. saligna* (5.74 cm and 148.7 cm) followed by those of *P. pallida* (2.27 cm and 168 cm) and *P. juliflora* (1.58 cm and 139 cm), respectively. The calculated scores pertaining the above mentioned properties show that *A. saligna* and *P. pallida* are of superior quality followed in a descending order by *A. nummularia* (scores = 9.4, 6.2 and 5.0, respectively). Such species, as in many other arid countries, can be recommended for large scale applications. However, although non of the fruit trees tested namely: *O. europaea* (olive), *P. granatum* (pomegranate), *P. dactylifera* (date palm), *F. carica* (fig), showed superiority on their growth parameters, yet Draz et al. (1992), recommended the use of olive with score of 2.9 as an intercropping species to encourage the participation of the local inhabitants in the sand dune fixation efforts.

1. Toshka Project Experiments

Toshka Project is the Southern Egypt Development project in the Western Desert. It aims at (Anonymous, 1998):

1. Addition of new areas of agricultural land which may reach two million feddans (feddan = 4200 m^2) at a distance of 220 km southwest of Aswan High Dam and

extending to the NE direction for about 50 km, in the west for 70 km and in the NW for 100 km to reach the cultivable land around Toshka Depression,
2. Establishment of agro-industrial complexes based on the agricultural raw materials available in the area,
3. Establishment of new communities expected to attract work force to solve the problem of overpopulation in the Nile Valley,
4. Promotion of touristic activities.

Toshka, being part of the Western Desert, is facing the problem of sand dune movement which is causing environmental problems. To control shifting of sands, Draz and Zaghloul (2006, 2007) carried out two field experiments. The first was during 2003–2004 to study the efficiency of bio- and mineral nitrogen fertilization on growth performance of *Acacia saligna* and *Prosopis juliflora* cultivated as shelterbelt for sand encroachment control. Obtained results showed that the growth characters of *A. saligna* and *P. juliflora* were significantly better with biofertilizered and nitrogen fertilization treatments compared to control treatment. Also, biofertilzation treatments in the presence of nitrogen application gave lower records of sand accumulation than with non-fertilization ones. Shelterbelt efficiency was higher in the treatment of biofertilization (*Rhizobia*) combined with nitrogen application at a rate of 100 and 150 kg N/feddan being 36.4% and 35% for *A. saligna* and *P. juliflora*, respectively.

The second experiment was conducted during 2001–2003 to study the possibility of controlling shifting sand along El-Sheikh Zayed irrigation canal which convey fresh water from the Lake to the newly reclaimed land of Toshka Project. One kilometer of green shelterbelt, included two strips was established perpendicular to the prevailing effective winds. Four plant species were cultivated on November 15th, 2001, namely: *Acacia saligna, Prosopis juliflora, Tamarix aphylla* and *Casuarina equesitifolia*. In each strip, 1 year old seedlings of each species were transplanted in ten plots (100 m length and 12 m width each) and distributed at variable distances: 3, 4, 6 and 12 m. The cultivated plants were irrigated by the brackish ground water (1980 ppm) using drip irrigation system. Experiment continued for 24 months and measurements of growth properties of the study plants, porosity of the cultivated shelterbelt and efficiency of various arrangements of plant species on the control of shifting sand were carried out four times: 6 months, 12 months, 18 months and 24 months after cultivation. The efficiency of each plot of the two strips was expressed by the reduction percentage of periodical cumulative amounts of the collected sands at the leeward side compared with that of the windward side. After 6 months from cultivation, the efficiency of the plots of the first and second strips increased from 43.3% to 75.3% and from 37.0% to 76.1%, respectively. Values of efficiency after 24 months increased from 57.8% to 85.6% and from 70.9% to 90.8% for the plots of the first and second strips, respectively.

From these results we may conclude that shelterbelt with native plants is efficient in the control of shifting sand in the Egyptian deserts. These phytobelts which enhance the deposition of the aeolian sand at a reasonable distance from the target areas (e., irrigation canal) are recommended for large scale application.

Palm Trees in Egypt

The climate of Egypt favours the growth of palm trees which may be considered the most ancient native trees in the country where they grow naturally and/or by cultivation. Thirteen palm species are recognized in Egypt belonging to 8 genera, namely: *Hyphaene thebaica, Medemia argun* and *Phoenix dactylifera* (naturally growing), *Arecastrum romanzoffianum, Cocos nucifera, Elaeis guineensis, Oreodoxa regia, Phoenix canariensis, P. loweirii, P. roebelenii, P. reclinata, Washingtonia filifera and W. robusta* (introduced, cultivated).

The three palm species naturally growing in Egypt have morphological and ecological interests. *H. thebaica* (doum palm) and *M. argun* (argun palm) are distinguished from date palm by their palmate leaves. On the other hand, argun palm has unbranched trunk (similar to date palm *P. dactylifera*) whereas the trunk of *H. thebaica* is biforked. The following are short notes on these three palm trees (Amer and Zahran, 1999; Springuel 2006; Boulos, 2007).

a. H. thebaica (doum palm)

The genus *Hyppaene* includes 40 species distributed in the tropical and subtropical areas of Africa, Madagascar, Arabian Peninsula, Southern Asia and Western India. *H. thebaica* is the only representative species of this genus in the flora of Egypt.

Doum-palm is a dioecious fan tree with a trunk up to 20 m high and about 30 cm across, repeatedly forked, rarely simple. Leaves are palmate 20–30 in terminal crown of each branch. Inflorescences on a spadix over 1 m long, surrounded by several cylindrical spathes. Flowers small, yellow, fruits bumpy (having an uneven surface) brown-glossy, punctuate, typically 7–8 cm long and almost as broad, edible, with spongy mesocarp enclosing the woody endocarp, seed 4.0×2.5 cm with white albumen. This palm produces fruits without artificial pollination. One palm tree will suffice for a great numbers of females even situated at a far distances (Täckholm, 1974).

Doum-palm has been mentioned in ancient history under the name Mara, i.e. divided into two and it is also depicted on old Egyptian monuments. It was described from Thebes, hence the epithet *thebaica*. Nowdays, doum-palm is naturally growing in the southern Egypt including Oases of the Western Desert, upstream parts of wadis in the Eastern Desert and Sinai Peninsula, as well as in Nubia, Aswan and Qena areas of the Nile Region. However, it is totally absent from the northern sections of the Egyptian Nile Region (Täckholm, 1974; Boulos, 1995).

The trunk of doum-palm is used for posts, beams, doors, water pipes and furniture. The leaves and stalks are used for roofing, mats, baskets, bags and rope making while its edible fruit could be considered as a gradient of folk medicine (Mannike, 1989).

b. Medemia agrun (argoun-palm)

Argoun-palm is a naturally growing fan palm with a dioecious unbranched trunk up to 20 m high and 30–40 cm diameter. Crown rounded, of 25–50 leaves. Leaf blade is coarse leathery, light green with 80–90 cm petiole. Inflorescences are inferior and arching; female ones 6–20 on tree, 120 cm long, branched to one order. Branches carry a single catkin-like rachilla carrying a densely hairy flowers. Male inflorescences 200–250 cm long, similar to female ones but first order branches bearing 1–4 digitally displayed rachillae. Argoun palm produces fruits without artificial pollination. Fruit-stone (edible), small, ellipsoid, about 4 cm long and 3 cm broad, deep purple with dry-yellow flesh with ruminated albumen, in cross section looking as white mess, penetrated by black needles. Mesocarp spongy, endocarp thin enclosed a large seed with red lines radiating to the center.

The genus *Medemia* has only one species, namely *M. argun* endemic to the Nubian desert in the southern section of Egypt and northern section of the Sudan, (Boulos, 2007). According to Zahran (1966), Boulos (1966a, 2005) and Springuel (2006), in Egypt, argoun palm grows only in Dungul Oasis in the Nubian Desert West of the Nile (220 km SW Aswan).

Täckholm (1974) stated that it may grow also in Nakhila Oasis (200 km South of Aswan). "This has been confirmed by Dr. Issawy (geologist) who recorded a few argoun trees that had been cut down in this Oasis", Springuel (2006) stated.

Historically, Täckholm and Drar (1950) reported that subsoil fruits were frequently found in ancient Egyptian tombs dating back as far as the 5th Dynasty. Recently, according to Boulos (2005), Rafik Khalil and Dina Ali visited Dungul Oasis in April 2005 and reported several argoun palms, of which two were bearing fruits. Amer and Zahran (1999) stated that *M. argun* was well known in ancient Egypt but recently feared to be endangered.

c. Phoenix dactylifera (date-palm)

Phoenix dactylifera is a dioecious tree, with unbranced trunk about 16–20 m high. 45 m high in certain areas is not unusual. The tree has a terminal crown with 30–150 leaves. Leaf is compound pinnate, about 6 m long and can persist from 3 to 7 years, bearing 120–240 leaflets, the leaf terminated with one or two leaflets. Leaves grow from the terminal bud in successive clusters each cluster has 3 or 5 leaves spirally arranged on the trunk. Staminate and pistillate flowers are morphologically indistinguishable until late in development. Fruit is a one seeded berry with smooth epicarp, variously fleshy mesocarp and silvery membranous endocarp. The seed is cigar-shaped consists mainly of hemicellulose with conspicuous longitudinal groove in one side and a small round protuberance on the other side. The embryo is enclosed in embryo sheath protruded from this protuberance. The fruit differs in colour according the type of date palm cultivar as well as the stage of fruit ripening. The plant is artificially pollinated to secure the fruit production. The fruit weight, diameter, moisture content, colour, and taste vary in the same female cultivar with the variation of the male cultivar used for pollination and/or the growth habitat.

Date palm is distributed in the northern subtropical zone between 10° and 30° north, the area between the Indus Valley in the east and the Canary Islands in the west. Date cultivation was expanded to the south Europe Lat: 45°24′N (in Italy) but not fruiting. This species flourishes in areas with direct sun-light and temperature between 9 °C and 45 °C; however, the optimum growth temperature ranges from 30 °C to 35 °C (Amer and Zahran, 1999).

The origin of date palm is not known, however, no doubt that its domestication was in the Old World. Date palm trunk was excavated from Kharga Oasis (Western Desert) dating back to the Old Stony Period. Mummy robes were excavated from Ruzikate (Sharkia Province) dating back to predynastic period (c. 3500 BC); also fruits as a gradient of beer industries were excavated from the same period (Bircher, 1990). Seeds recovered from Douch Necropolis (Kharga oasis) dating back to Graeco-Roman period excavated a large number of seeds, fruits and leaves from Abu Shaar site, Red Sea coast, Roman site. Nowadays, date-palm is naturally growing in the oases of the Western Desert, Sinai Peninsula, wadies of the Eastern Desert and Red Sea Coast (Zahran, 1965).

Date palm has a wide ecological range, it grows in the desert wadis, oases, mangrove margins, steep limestone cliffs and cultivated lands. It can thrive in a wide range of soil types from light sandy to heavy clay soils. The plant can also tolerate up to 1.5% of soil salinity. Individual trees occur in the delta of Wadi Gimal, along the Red Sea Coast at Mersa Alam, 700 km South Suez (Zahran, 1965). Groves of date palm trees grow in the salt marshes south of Gebel El-Nargis, Nile Delta and El-Arish coastal region. Regular irrigation and good drainage are required for proper growth. However, if the water table is high the roots are liable to be choked from the lack of aeration in a water-logged soils. Lack of aeration may be overcomed by the ability of the root to produced lateral superficial respiratory roots. On the other hand, in extremely dry soil the date palm roots attain 30% lateral horizontal roots (feeding root zone) not exceed 30 cm in depth and the absorption root-zone penetrate the soil to 10 m depth. Absorption zone is will prolonged and extend in different directions in search for water (Bircher, 1990).

Date palm trunk is used for roofing in rural areas. Leaves are used for animal feeding, chair manufacture, explosives, baskets and table cloth making. Fibers are used for bags and brushes making. The fruits (dates) are highly nutritive and are rich in Vitamins A, B_1, and B_2. The extract of the dates is diuretic, expectorant, anti-constipation, antipyretic, relief the bronchial asthma, hypertension, some gynecological diseases and is used also in the preparation of ointments for some skin diseases, pollen grains are aphrodisiac particularly when mixed with honey. The volatile oils of the spathe has anti-microbial effect. Even the kernels of the dates have medicinal importance, its paste is used for some ophthalmic diseases; the powder is a gradient of remedy used for kidney stone, whereas the smoke of the burned kernels relief the haemorroidal pains. Fruit has many uses in ancient religion and traditions e.g. remedies for cough, to kill worms, to accelerate the child birth, and used in many food industries.

Phoenix dactylifera comprises several native cultivars (14 types) of date palm in Egypt which are classified into three kinds based on the percentage of moisture in their fruits:

1. Fresh date: vernacular name "*Rutab*" with 50% moisture.
2. Semi-dry date: vernacular name "*Agua*". In some cultivars the ripened fruits get dry while they are on the tree till the moisture become about 20%.

3. Dry date: vernacular name "*Tamer*": This type gets dry after passing through fresh and semi-dry forms. The moisture content of this type is about 10% water.

Propagation of Palm Trees

Propagation of date palm is wide spread in Egypt in the desert and River Nile regions. On the other hand, doum and argoun palms have been propagated successfully for the first time in Aswan Desert Garden located in a desert area east of the River Nile. Springuel (2006) stated "Because the doum palms were planted few years earlier than argoun palms, some individuals already have a well-defined trunk and have begun fruiting, while the argoun palms, which are only 2–3 years old, are still very small. It seems that the growth rate of both doum and argoun palms is very slow".

Vegetative propagation of date and other palms by offshoots is the common method, whereas sexual propagation by seeds is not widely practiced due to heterozygosity Ahmed (1979). Most of seed-derived female progenies are not true to type (Wasel, 1999). Nevertheless, asexual propagation is not efficient because only a limited number of offshoots per tree is produced which remain attached to the parents for a long period (2–3 years) until they reach an appropriate size and adequate root system develops. On the other hand, in Vitro micropropagation provides a practical means to clone desired palm trees and obtain a large number of high quality and disease-free propagules. Wasel (1999) concluded that "tissue culture derived date palm trees had better growth habit and resulted in a more uniform trees".

Appendix: Photographs Covering Western Desert, Eastern Desert, Sinai Peninsula, Nile Region

A. Western Desert

Photo A.1 A community dominated by the psammophyte *Ammophila arenaria* inhabiting the coastal sand dunes of the Western Mediterranean Coast, Egypt

Photo A.2 Salt marsh vegetation with abundant growth of *Kochia indica* (*Bassia indica*) in the foreground. Mixed halophytes of *Juncus rigidus* and *Arthrocnemum macrostachyum* in the background, Western Mediterranean Coast, Egypt

Photo A.3 Dense growth of *Juncus rigidus* in the salt marshes of Siwa Oasis, Western Desert, Egypt

Photo A.4 Reed swamp vegetation dominated by *Typha domingensis*, Siwa Oasis, Western Desert, Egypt

Photo A.5 A *Populus euphratica* tree inhabiting a sand dune in Siwa Oasis, Western Desert, Egypt. A clump of *Stipagrostis scoparia* is seen in the foreground

Photo A.6 Dense stand dominated by *Typha elephantina*, Um Rishe Lake, Wadi El-Natrun Depression, Western Desert, Egypt

Photo A.7 A close up view of the succulent xerophyte *Zygophyllum coccineum*, Cairo-Alexandria desert road, Western Desert, Egypt

Photo A.8 *Pancratium sickenbergeri* bulbous herb, Mariut Plateau, northern section of the Western Desert, Egypt

Photo A.9 Close-up view of the annual herb *Asphodelus tenuifolius* growing in the Western Mediterranean Coast, northern section of the Western Desert, Egypt

B. Eastern Desert

Photo A.10 Mangal vegetation dominated by *Avicennia marina*, Red Sea Coast, Egypt

Photo A.11 Dense mangrove forest dominated by *Rhizophora mucronata*, Southern section of the Red Sea Coast, Egypt

Photo A.12 A close up view of *Rhizophora mucronata* mangrove tree, Shalateen swamps, southern section of the Red Sea Coast, Egypt

Photo A.13 Mangrove swamp of *Rhizophora mucronata* with a seedling in the forgroung, Mersa Abu Fissi, Red Sea Coast, Egypt

Photo A.14 A general view of the mangrove forest lining the shore-line of Mersa Abu Fissi, Red Sea coast, Egypt. The layer of long trees is that of *Rhizophora mucronata, Avicennia marina* trees occupies the short layer

Photo A.15 *Nitraria retusa* halophyte growing in the salt marsh habitat, Red Sea Coast, Egypt

Photo A.16 An almost pure stand of *Imperata cylindrica* in the delta of Wadi Hommath (36 km south of Suez, Western coast, Gulf of Suez), Egypt

Photo A.17 Delta of Wadi Di-ib with pure community of *Suaeda monoica*, Red Sea Coast, Egypt

Photo A.18 *Salsola baryosma* community, Red Sea coastal desert, Egypt. Camels are grazing the ephemeral plants covering the spaces between the dominant shrubs

Photo A.19 A close-up view of *Hammada elegans (Haloxylon salicornicum)* occupying a sand hillock, Wadi Di-ib, southern section of the Red Sea coastal desert, Egypt

Photo A.20 Mid-stream part of Wadi Laseitait with a stand of *Panicum turgidum* grassland with bushes of *Acacia tortilis*

Photo A.21 Wadi Lahmi, 110 km south of Mersa Alam, Red Sea Coast, Egypt with shrubs of *Acacia tortilis* reaching a height up to 5 m

Photo A.22 Midstream part of Wadi Aideib, Red Sea Coastal desert, Egypt with rich growth of *Acacia* scrub. Notice the woody climber *Cocculus pendulus* on an *A. raddiana* tree

Photo A.23 Tree of *Tamarix aphylla* growing in Wadi El-Ghoweibba, Red Sea coastal desert, Egypt

Photo A.24 Downstream part of Wadi Laseitait, Red Sea coastal desert, Egypt with dense thicket of *Capparis decidua*. Individuals of *Panicum turgidum* are seen in the foreground

Photo A.25 A Pot-hole with water on the slopes of Gebel Samiuki, Red Sea Coastal desert, Egypt. A small tree of *Moringa peregrina* is seen on the upper side of the hole

Photo A.26 General view of Wadi Aideib with *Acacia* scrubland. The slopes of the mountains are dominated by *Euphorbia cuneata*, Red Sea Coastal desert, Egypt

Photo A.27 General view of a rich growth of *Dracaena ombet* trees near the top of Gebel Elba, Red Sea coastal desert, Egypt

Photo A.28 Upstream part of a wadi in Gebel Elba area dominated by *Moringa peregrina* trees. Bushes of *Zilla spinosa* are growing on the Wadi, Red Sea coastal desert, Egypt

Photo A.29 A close-up view of *Maerua crassifolia* shrub, Gebel Nugrus area, Red Sea coastal desert, Egypt

Photo A.30 A close-up view of *Acacia ehrenbergiana* shrub in Wadi Umm Rilan, Eastern Desert, Egypt

Photo A.31 Big tree of *Balanites aegyptiaca* growing in Wadi Allaqi, Eastern Desert, Egypt. *Acacia* scrub is seen in the background of the photo

Photo A.32 A community dominated by *Hammada elegans* (*Haloxylon salicornicum*), Cairo-Suez desert road, Eastern Desert, Egypt

Photo A.33 Rich growth of *Cassia senna* in Wadi Marahel (an affluent of Wadi Allaqi), Eastern Desert, Egypt

Photo A.34 A fossil hillock with remains of *Tamarix aphylla* leaves, Wadi Allaqi, Eastern Desert, Egypt

C. Sinai Peninsula

Photo A.35 Stunted growth of *Tamarix nilotica* shrub, salt marsh habitat, Gulf of Aqaba Coast, Sinai Peninsula, Egypt

Photo A.36 A salt marsh community dominated by *Halocnemum strobilaceum*, Eastern Section of the Mediterranean Coast, Sinai Peninsula, Egypt

Photo A.37 A close-up view of *Zygophyllum dumosum* succulent xerophyte, Gebel Halal area, Sinai Peninsula, Egypt

Photo A.38 *Lavandula pubescens* growing in a rocky habitat in Wadi Hebra, Sinai Peninsula, Egypt

Photo A.39 A close-up view of a flowering head of the thistle-like spiny herb: *Blepharis edulis*, Gulf of Aqaba coastal desert, Sinai Peninsula, Egypt

Photo A.40 A stand dominated by *Iphiona mucronata* in Gebel Musa area, Sinai Peninsula, Egypt

Photo A.41 A community dominated by *Lygos raetam* (*Retama raetam*), Wadi Retama, Sinai Peninsula, Egypt

Photo A.42 A close-up view of the medicinal herb *Hyoscyamus muticus*, Wadi El-Raha, Sinai Peninsula, Egypt

Photo A.43 *Ochradenus baccatus* shrub growing in Wadi Retama, Sinai Peninsula, Egypt

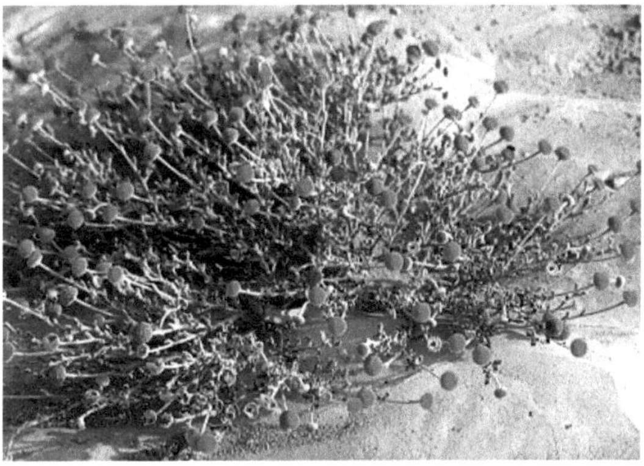

Photo A.44 The composite medicinal perennial herb *Achillea fragrantissima* growing in Wadi Gharandal, Sinai Peninsula, Egypt

Photo A.45 Trees and shrubs of *Juniperus phoenicea* inhabiting the high altitude of Gebel Halal, Sinai Peninsula, Egypt (by Prof. Dr. Kamal Shaltout)

Photo A.46 A close-up view of a *Juniperus phoenicea* tree growing in the rocky habitat of Gebel Halal, Sinai Peninsula, Egypt (by Prof. Dr. A. Kheder)

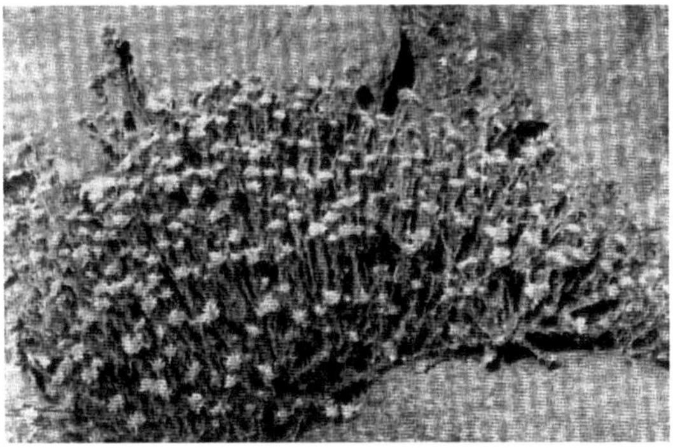

Photo A.47 A close-up view of the endemic mint-smell perennial herb *Origanum syriacum* subsp. *sinaicum* (labiatae), Gebel Musa, Sinai Peninsula, Egypt

D. Nile Region

Photo A.48 *Calligonum comosum* shrub growing on a partially stabilized sand dune of Abu Madi (Qalabshu) area, Deltaic Mediterranean coast, Nile Delta, Egypt

Photo A.49 Dense growth of *Desmostchya bipinnata* grass in a fallow land, Nile Delta, Egypt

Photo A.50 A salt marsh community dominated by *Zygophyllum aegyptium*, Deltaic Mediterranean coast, Egypt. In the background there are date palm trees (*Phoenix dactylifera*) inhabiting the coastal sand dunes

Photo A.51 Bushes of *Kochia indica* (*Bassia indica*) bordering the cultivated land, Nile Delta, Egypt

Photo A.52 *Calotropis procera* shrub growing in the Nile bank, Aswan area, Upper Egypt. Groves of date palm appear in the background of the photo

Photo A.53 A beautiful landscape showing a line of date palms (*Phoenix dactylifera*) bordering a River Nile Island, Nile Delta, Egypt

Photo A.54 Groves of *Hyphaene thebaicea* (doum palm) growing along the bank of the River Nile, Aswan area, Upper Egypt

Photo A.55 A close-up view a doum palm (*Hyphaene thebaicea*), River Nile bank, Aswan area, Upper Egypt

Photo A.56 Thick stand dominated by the floating hydrophyte *Eichhornia crassipes* covering an irrigation canal, Damietta area, Nile Delta, Egypt

Photo A.57 The submerged hydrophyte *Ceratophyllum demersum* growing in an irrigation canal, Fayium area, Nile Region, Egypt. Branches of *Eichhornia crassipes* appear in parts of the photo

Photo A.58 Pure stand of *Ceratophyllum demersum* submerged hydrophyte, Fayium area, Nile Delta, Egypt

Photo A.59 Dense growth of *Potamogeton nodosus* submerged hydrophyte, Nile Delta, Egypt

Photo A.60 Close-up view of two individuals, of the rare floating hydrophyte *Pistia stratiotes*, Nile Delta, Egypt

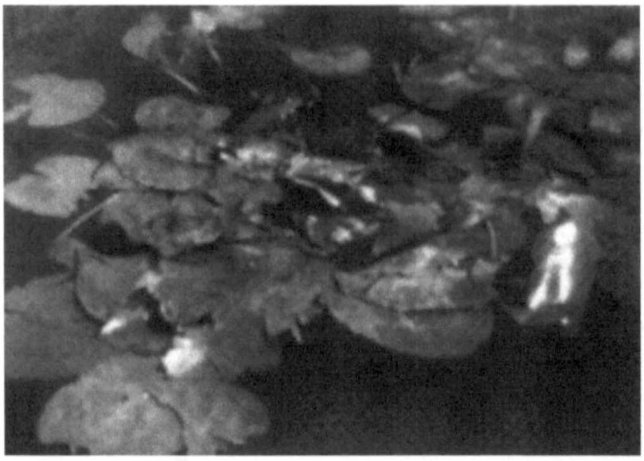

Photo A.61 General view of a stand dominated by *Nymphaea caerulea*, Nile Delta, Egypt

Photo A.62 Mixed hydrophytes formed of *Pharagmites australis, Ludwigia stolonifera* (*Jussiaea repens*) and *Ceratophyllum demersum*, Nile Delta, Egypt

Photo A.63 A close-up view of *Cyperus papyrus*, Damietta area, Nile Delta, Egypt

Photo A.64 Mixed aquatic vegetation: *Ceratophyllum demersum* (submerged), *Eichhornia crassipes* (floating), and *Cyperus papyrus* and *Saccharum spontaneum* (reeds, background), River Nile Island, Cairo area, Egypt

Photo A.65 Dense growth of *Eichhornia crassipes* and *Pharagmites australis*, Lake Manzala, Nile Delta, Egypt

References

Abdel Rahman, A.A., Ayyad, M.A. and El-Monayari, M. (1965a). Hydrobiology of the sand dune habitat at Burg El-Arab. *Bull. Fac. Sci. Cairo Univ.*, 40, 29–54.

Abdel Rahman, A.A., Shalaby, A.F., Balegh, M.S. and El-Monayari, M. (1965b). Hydroecology of date palm under desert conditions. *Bull. Fac. Sci. Cairo Univ.*, 40, 55–71.

Abdel Razik, M.S. (1994). *Avicennia marina "Al-Qurm" General Study and Propagation Experiment in Qatar*. Center for Scientific and applied Research, Doha, Qatar, 142pp. (in Arabic) + 18pp. (in English).

Abdel-Razik, M., Abdel-Aziz, M. and Ayyad, M. (1984). Multivariate analysis of vegetational variation in different habitats at Omayed (Egypt). *Vegetatio*, 57, 167–175.

Abduallah, M., Sa'ad, F., Eweida, A. and Mahmoud, M. (1984). Materials from CAIM Herbarium II: Flora of the Sinai Peninsula. *Notes Agric. Res. Centre Herb. Egypt*, 6, 15–214.

Abu Al-lzz, M.S. (1971). *Landforms of Egypt*. Translated by Dr Yusuf A. Fayid. The American University in Cairo Press, Cairo, Egypt, 281pp.

Abu Ziada, M.E.A. (1980). *Ecological Studies on the Flora of Kharga and Dakhla Oases of the Western Desert of Egypt*. PhD Thesis, Fac. Sci., Mansoura University, Egypt.

Abu Ziada, M.E.A. (1986). Autecological studies on *Echinochloa crus-galli* (L.). P. Beauv. *Bull. Fac. Sci. Mansoura Univ.*, 13(1), 295–305.

Ahmed, M.M. (1979). *Study on the Distribution of Indigenous Species of Palmae in Africa and Adaptation of Some African Species Growing Under Cairo Environment*. PhD Thesis, Institute of African Research and Studies, Cairo University, Egypt.

Ahmed, A.M. (1983).On the ecology and phytosociology of El-Qaa Plain, South Sinai, Egypt. *Bull. Inst. Desert Egypte*, 33(1–2), 281–314.

Ahmed, A.M. (2002). On the ecology and phyosociology of the Qattara Depression. *Bull. Desert Inst., Egypt* (Special Edn.), 1, pp. 1–42.

Ahmed, A.M. and Mounir, M.M. (1982). *Regional Studies on the Natural Resources of the NW Coastal Zone, Egypt*. US National Science Foundation, Oklahoma State University and the UNEP and Remote Sensing Centre, Academy of Scientific Research and Technology, Cairo, Egypt, 195pp.

Ahmed, A.M. & Nassar, Z.M. (1999). Chemical composition of some natural range plants from the north western coast of Egypt. *Proceed. 6th Natl. Conf. Eniron. Stud. Res, Ain Shams Univ., Cairo*, 551–568.

Ahmed, A.M., Kilany, S.S. & Khalifa, S.A. (2002). On the germination of seeds of range plants from the Egyptian desert. *J. Environ. Sci. Mansoura Univ., Egypt*, 23, 179–208.

Alaily, F., Bornkamm, R., Blume, H.R., Kehl, H. and Zielinsti, H. (1987). Ecological investigation on the Gilf Kebir (SW Egypt), *Phytocoenologia*, 15(1), 1–20.

Ali, M.M. (1998). Aquatic macrophytes in Egyptian Nubia pre- and post-formation of lake Nasser in relation to environmental factors: *ICLARM Symposium on Ecological Basis and Management*

Policy for Sustainable Fish Production in Lake Nasser, High Dam Lake Development Authority, June 20–23/1998, Aswan, Egypt, pp. 18–23.

Ali, M.A. (2004). *On the Ecology of Sinai Peninsula.* MSc Thesis, Faculty of Science, Mansoura University, Mansoura, Egypt.

Al-Kholy, A.A. (ed.). (1972). *Aquatic Resources of the Arab Countries.* Science Monograph Series, Arab League Education Cultural and Scientific Organization (ALECSO), 452pp. (In Arabic).

Amer, W.M. and Zahran, M.A. (1999). Palm trees in Egypt. *Proceedings of the International Conference on Date Palms*, Assiut University, Assiut, Egypt, 171–189.

Amin, S.A. (1998). *Ecological Study on the Plant Life in Wadi El-Rayan Area, El-Fayium, Egypt.* PhD Thesis, Faculty of Science, Cairo University, Egypt.

Andersen, G.L. (2006). How to detect desert trees using CORONA images: discovering historical ecological data. *J. Arid Environ.*, 65, 491–511.

Andersen, G.L. (2007). *Long-Term Dynamics of Wadi Trees in Hyper-Arid Cultural Landscap*: Introduction. Dissertation for the Degree of Doctor Scientiarum, University of Bergen, Bergen, Norway, pp. 1–42.

Andreossy, G. (1799). Notes on Lake Manzala (non-published).

Andrews, F.W. (1950–1956). *The Flowering Plants of the Sudan*, Vols. I–III. Sudan Gov., Khartoum, 237pp. (1950), 485pp. (1952), 597pp. (1956).

Anonymous (1960). *Climatic Normals of Egypt.* Ministry of Military Production, Meteorological Dept. Cairo, 237pp.

Anonymous (1980). *Firewood Crops: Shrub and Tree Species for Energy Production.* National Academy of Science, Washington, D.C., 237pp.

Anonymous (1981). *Agricultural and Water Investigations of Sinai.* Part IV. Plant Ecology, Desert Inst. Cairo and Dames and Moore, 91pp.

Anonymous (1982). *Sinai Development Study.* Phase I. Draft Final Report, Vol. VII. Environment. Dames and Moore, Chapter 4, Climatology and Meteorology (16pp.). and Chapter 7, Terrestrial Ecology (48pp.).

Anonymous (1985). *Sinai Development Study.* Phase I. Final Report, Vol. V. Water Supplies and Costs. Dames and Moore, Chapter 2, Water Resources Assessment, 90pp.

Anonymous (1987a). *Our Common Future: World Commission on Environment and Development*, Oxford University Press, Oxford, UK, 400pp.

Anonymous (1987b). *The Mangrove Ecosystem: Scientific Aspects and Human Impacts.* UNESCO, 41pp.

Anonymous (1998). *Southern Egypt Development Project.* Publication of the Ministry of Public Works and Water Rresources, Cairo, 68pp.

Anonymous (2004). *Status of Mangroves in the Red Sea and Gulf of Aden.* Technical series No. 11, PERSGA, UNDP, UNEP, GEF, 61pp.

Anonymous (2006). *Assessment and Management of Mangrove Forestes in Egypt for Sustainable Utilization Development.* ITTO, Japan in association with MALR, MSEA, EEAA, Cairo, Egypt. Progress Report No. 6, 176pp.

Anonymous (2007). *A Concise Report on the Expedition to the Gilf Kebir National Park, Nature Conservation Sector.* Egyptian Environmental Affairs (EEAA), ILICN, Italian Cooperation, 45pp.

Anyamba, A. and Tucker, C.J. (2005). Analysis of Sahelian vegetation dynamics using NOAA-AVHRR NDVI data from 1981–2003. *J. Arid Environ.*, 63(3), 596–614.

Arkell, A.J. (1949). *Early Khartoum.* Oxford University Press, London, 145pp.

Ascherson, P. and Schweinfurth, G. (1889a). Illustration de la Flora d' Egypte. *M6m. Inst. Egypte*, 3(1), 25–260.

Ascherson, P. and Schweinfurth, G. (1889b). Supplement a l'illustration de la Flora d' Egypte. *Mem. Inst. Egypte*, 2, 745–810.

Ashour, N.I., Serag, M.S., Adel Haleem, A.K., Mandoura, S., Mekki, B.B. and Arafat, S.M. (2002). Use of Kallar Grass (*Leptochloa fusca* L.). Kunth. in saline agriculture in Arid Lands of Egypt. *Egypt. J. Agron.*, 24, 63–87.

Attiah, M.I. (1954). *Deposits in the Nile Valley and the Delta, Egypt*. Geol. Survey, Cairo, 356pp.
Ayensu, E.S. (1979). Plants for medicinal uses with reference to arid zone. *Proc. Arid Land Plant Resour. Conf., Texas Tech. Univ.*, Lubbock, Texas, USA, pp. 177–178.
Ayyad, M.A. (1957). An *Ecological Study of Ras El-Hikma District*. MSc Thesis, Fac. Sci., Cairo University, Egypt.
Ayyad, M.A. (1969). An edaphic study of habitats at Ras El-Hikma. *Bull. Inst. Desert Egypte*, 19(2), 245–259.
Ayyad, M.A. (1973). Vegetation and environment of the Western Mediterranean coastal land of Egypt. I. The habitat of sand dunes. *J. Ecol.*, 61, 509–523.
Ayyad, M.A. (1976). Vegetation and environment of the Western Mediterranean coastal land of Egypt. IV. The habitat of non-saline depressions. *J. Ecol.*, 64, 713–722.
Ayyad, M.A. (1981). Soil vegetation-atmosphere interactions. In *Arid Land Ecosystems: Structure, Functioning and Management* (eds. O.W. Goodall and R.A. Perry), Vol. 2. Int. Biol. Programme, 17, 9–81, Cambridge University Press, Cambridge.
Ayyad, M.A. (1983). Some aspects of land transformation in the Western Mediterranean Desert of Egypt. *Adv. Space Res.*, 2(8), 19–29.
Ayyad, S.M. (1988). *Pollen Grain Ecology of the Mediterranean Sea Coast*, Egypt. PhD Thesis, University of Mansoura, Egypt.
Ayyad, M.A. and Ammar, M.Y. (1974). Vegetation and environment of the Western Mediterranean coastal land of Egypt. II. The habitat of inland ridges. *J. Ecol.*, 62, 439–456.
Ayyad, M.A. and Hilmy, S.H. (1974). The distribution of *Asphodelus micro-carpus* and associated species on the Western Mediterranean coast of Egypt. *Ecology*, 55, 511–524.
Ayyad, M.A. and El-Ghonemy, A.A. (1976). Phytosociological and environmental gradients in a sector of the Western Desert of Egypt. *Vegetatio*, 31(2), 93–102.
Ayyad, M.A. and El-Bayyoumi, M.A. (1979). On the phytosociology of sand dunes of the Western Desert of Egypt. *Glimpses of Ecology*, Jaipur, India (Professor R. Misra commemoration volume), pp. 219–237.
Ayyad, M.A. and El-Ghareeb, R.E.M. (1982). Salt marsh vegetation of the Western Mediterranean desert of Egypt. *Vegetatio*, 49, 3–19.
Ayyad, M.A. and El-Ghareeb, R.E.M. (1984). *Habitats and Plant Communities of the NE Desert of Egypt*. Communications in Agrisciences and Development Research, College of Agriculture, Univ. of Alexandria, 7(6), 1–34.
Ayyad, M.A. and Ghabbour, S.I. (1986). Hot deserts of Egypt and the Sudan. Chapter 5. In *Ecosystems of the World*, 12B, *Hot Deserts and Arid Shrublands* (eds. M. Evenari et al.). Elsevier, Amsterdam, pp. 149–202.
Ayyad, S.M. and Krzywinski, K. (1992). An archaeopalynological reflection upon vicia fabra type pollen from ancient Mendes (Tel El Roba area, Egypt). *Proc. Int. Conf. on Laminated Sediments and Archaeology*, Ravilla, Italy, 1991. *PACT*, pp. 663–675.
Ayyad, S.M., Krzywinski, K. and Pierce, R.I. (1992a). Mudbrick as a bearer of agricultural information: an archaeopalynological study. *Norwegian Archaeol. Rev.*, 24(2), 77–96.
Ayyad, S.K., Moore, P.D. and Zahran, M.A. (1992b). Modern Pollen rain studies of the Nile Delta, Egypt. *New Phyotolgist*, (2), 663–675.
Ba Kader, A.B.A., AL-Sabbagh, A.L.T., AL-Glend, M.A.S. and Izzidien, M.Y.S. (1983). *Islamic Priniples for the Conservation of the Natural Environment*. IUCN. Switzerland & MEPA Saudi Arabia, 25pp.
Babyev, A.G. (ed.). (1981). *Principles and Methods of Shifting Sands Fixation*, UNEP, USSR Commission, Mosco, 132pp.
Ball, J. (1902). *On the Topographical and Geological Results of a Reconnaissance Survey of Gebel Garra and the Oasis of Kurkur*. Egyptian Survey Dept., Public Works Ministry, Cairo, pp. 1–40.
Ball, J. (1912). *The Geography and Geology of South Western Desert*. Egyptian Survey Dept., Cairo, 394pp.
Ball, J. (1916). *The Geography and Geology of West Central Sinai*. Egyptian Survey Dept., Cairo, 219pp.

Ball, J. (1927). Problems of the Libyan Desert. *Geomorphol. J.*, 70, 21–38, 105–128, 259–264.

Ball, J. (1928). Remarks on lost oasis of the Libyan Desert. *Geogr. J.*, 72, 250-S.

Ball, J. (1939). *Contributions to the Geography of Egypt.* Survey and Mines Dept., Cairo, 308pp.

Batanouny, K.H. (1963). *Water economy of Desert Plants in Wadi Hof.* PhD Thesis, Fac. Sci., Cairo University, Egypt.

Batanouny, K.H. (1973). Habitat features and vegetation of deserts and semi-deserts in Egypt. *Vegetatio*, 27(4–6), 181–199.

Batanouny, K.H. (1985). Botanical exploration of Sinai. *Qatar Univ. Sci. Bull.*, 5, 187–211.

Batanouny, K.H. and Batanouny, M.H. (1969). Formation of phytogenic hillocks. I. Plants forming phytogenic hillocks. *Acta Bot. Acad. Sci. Hung.*, 14, 243–252.

Batanouny, K.H. and Abu El-Souod, S. (1972). Ecological and phytosociological study of a sector in the Libyan Desert. *Vegetatio*, 25(5–6), 335–356.

Batanouny, K.H. and Abdel Wahab, A.M. (1973). Ecophysiological studies on desert plants. VIII. Root penetration of *Leptadenia pyrotechnica* (Forssk.). Decne in relation to water balance. *Oecologia* (Berlin), 11, 151–161.

Batanouny, K.H. and Baeshin, N.A. (1982). Studies on the flora of Arabia. II The Medina-Badr Road, Saudi Arabia. *Bull. Fac. Sci. King Abdul Aziz Univ., Jeddah, Saudi Arabia*, 6, 1–26.

Batanouny, K.H., Zayed, K.M., Emad, H.M. and Kabeil, H.E. (2006). Reproductive ecology of *Panicum turgidum* Forssk. *Taeckhomia*, 26, 63–88.

Beadnell, H.J.C. (1909). *An Egyptian Oasis: An Account of the Oasis of Kharga in the Libyan Desert.* Murray, London, 248pp.

Beadnell, H.J.C. (1924). Reports on the geology of the Red Sea between Qusseir and Wadi Ranga. *Petroleum Res-Bull.*, 13, 37pp.

Beadnell, H.J.C. (1927). *The Wilderness of Sinai.* Arnold, London, 180pp.

Belgrave, C.D. (1923). *Siwa: The Oasis of Jupiter Ammon.* Bodley Head, London, 275pp.

Ben Haider, B.M. (1994). *Mangroves (Al-Qerm) Development in UAE*, 1st edn. Nadwat Al Thaqafa & Al-Ouloum, UAE, 53 (in Arabic).

Bioclimatic Map of the Mediterranean Zone (1963). UNESCO/FAO. Arid Zone Research, XXI(1963). NS 12 162 111, 22/A.

Bircher, W.H. (1990). *The Date Palm: A Boon For Mankind*, Cairo University Herbarium, Egypt, p. 100.

Bishai, H.M., Abdel-Malek, S.A. and Khalil, M.T. (2000). *Lake Nasser.* Publication of the National Biodiversity Unit, II Egyptian Environmental Affairs Agency (EEAA), Cairo, Egypt, 577pp. + 40 pages (in Arabic).

Blackenhorn, M.C.P. (1921). Aegypten: Handbuch d.Region. *Geol. Heidelberg*, VII, Heft 23, Abt. 9, 244pp.

Bock, J.H. (1969). Production of the water hyacinth: *Eichhornia crassipes* (Mart.). Solms. *Ecology*, 50, 460–464.

Bornkamm, R. (1986). Flora and vegetation of some oases in S-Egypt. *Phytocoenologia* 14(2), 275–284.

Bornkamm, R. and Kehl, H. (1990). The plant communities of the Western Desert of Egypt. *Phytocoenologia*, 19(2), 149–231.

Bornkamm, R., Springuel, I., Darius, F., Sheded, M.G. and Radi, M. (2000). Some observation on the plant communities of Dungul Oasis (Western Desert, Egypt). *Acta Bot. Croat.*, 59(1), 101–109.

Boulos, L. (1960). *Flora of Gebel El-Maghara, North Sinai.* General Organization for Gov. Printing Office, Ministry of Agr., Cairo, Egypt, 24pp.

Boulos, L. (1962). *Typha elephantina* Roxb. in Egypt. *Candollea*, 18, 129–135.

Boulos, L. (1966a). A natural history study of Kurkur Oasis, Libyan Desert, Egypt, rv The vegetation. *Postilla*, 100, 1–22.

Boulos, L. (1966b). Flora of the Nile Region in Egyptian Nubia. *Feddes Repert.*, 83(3), 183–215.

Boulos, L. (1980). Journey to the Gilf Kebir and Uweinat, southwest Egypt during 1978. Botanical results of the expedition. Chapter IV. *Geogr. J.*, 146, 68–71.

Boulos, L. (1982). Flora of Gebel Uweinat and some neighbouring regions of southwestern Egypt. *Conservatoire et Jardin Botaniques de Geneve, Candollea*, 37(1), 257–276.
Boulos, L. (1983). *Medicinal Plants of North Africa*. Library of Congress Cataloging in Publication Data, Reference Publications, Inc. Algonac, Michigan, U.S.A, 286pp.
Boulos, L. (1995). *Flora of Egypt Chicklist*. Al-Hadara Publishing, Cairo, Egypt, 283pp.
Boulos, L. (1999). *Flora of Egypt: Volume I (Azollaceae – Oxalidaceae)*. Al-Hadara Publishing, Cairo, Egypt, 419pp.
Boulos, L. (2000). *Flora of Egypt: Volume II (Geraminaceae – Boragniaceae)*. Al-Hadara Publishing, Cairo, Egypt, 352pp.
Boulos, L. (2002). *Flora of Egypt: Volume III (Verbenaceae – Compositae)*. Al-Hadara Publishing, Cairo, Egypt, 373pp.
Boulos, L. (2005). *Flora of Egypt: Volume IV Monocotyledons (Alismataceae – Orchidaceae)*. Al-Hadara Publishing, Cairo, Egypt, 617pp.
Boulos, L. (2007). *Biodiversity in Egypt*. Al-Hadara Publishing, Cairo, Egypt, 67pp.
Boulos, L. and El-Hadidi, M.N. (1984). *The Weed Flora of Egypt*. American University of Cairo Press, Cairo, 178pp.
Brown, L.R. (1981). *Building a Sustainable Society*, W. W. Norton Comp., New York, 433pp.
Butzer, K.W. (1959a). Contributions to the Pleistocene geology of the Nile Valley. *Erdkunde*, 11, 46–67.
Butzer, W.B. (1959b). Environment and human ecology in Egypt during Predynastic and early Dynastic times. *Bull. Soc. Geogr. Egypte*, 32, 36–88.
Butzer, K.W. (1964). Pleistocene palaeoclimates of the Kurkur Oasis, Egypt. *Can. Geogr.*, VIII(3), 125–140.
Butzer, K.W. (1976). *Early Hydraulic Civilization in Egypt: A Study in Cultural Ecology*. University of Chicago Press, Chicago.
Chapman, V.J. (1960). *Salt Marshes and Salt Deserts of the World*. Hill, London, 392pp.
Chapman, V.J. (1974). *Salt Marshes and Salt Deserts of the World*. 2nd edn, 392pp. (complemented with 102pp.), Cramer, Lehre.
Chapman, V.J. (1975). *Mangrove Vegetation*. Cramer, Lehre, 425pp.
Chapman, R.W. (1978). Geology. In *Quaternary Period in Saudi Arabia* (eds. S.S. Al-Sayari and J.G. Zotl). Springer-Verlag, Wien, pp. 3–19.
Chiras, D.D. (1991). *Science: Action for a Scientific Future*, 3rd edn. The Benjamin Cummings Publishing Company Inc., New York, 549pp.
Cope, T.A. and Hosni, H.A. (1991). *A Key to Egyptian Grasses*. Royal Botanical Gardens, Kew, p. 75.
Cufondontis, G. (1961–1966). Enumeratio plantarum Aethiopiae Spermatophyta, Sequentiae. *Bull. Jardin Bot. Natl. Belg.*, 31, 709–772 (1961); 36, 1059–1114 (1966).
Cunningham, W.P. and Saigo, B.W. (1992). *Environmental Sciences: A Global Concern*. Wm. C. Broan Publishers, U.S.A, 632pp.
Danin, A. (1969). An new *Origanum* from the Isthmic Desert (Sinai). *Origanum isthmicum* sp.n. *Israel J. Bot.*, 18, 191–193.
Danin, A. (1972). Mediterranean elements in rocks of the Negev and Sinai 'Deserts. *Notes Roy. Bot. Gard. Edinburgh*, 31, 437–440.
Danin, A. (1973). Contributions to the flora of Sinai. II. New Records. *Israel J. Bot.*, 22, 8–32.
Danin, A. (1981). Weeds of Eastern Sinai coastal area. *Willdenowia*, 11, 291–300.
Danin, A. (1983). *Desert Vegetation of Israel and Sinai*. Cana Pub. House, Jerusalem, Israel, 148pp.
Danin, A. and Hedge, I.C. (1973). Contributions to the flora of Sinai: I New and confused taxa. *Notes Roy. Bot. Gard. Edinburgh*, 32, 259–271.
Danin, A., Shimda, A. and Listen, A. (1985). Contributions to the flora of Sinai: III Checklist of the species collected and recorded by the Jerusalem team 1967–1982. *Willdenowia*, 15, 255–322.
Daoud, H.S. (1985). *Flora of Kuwait*, Vol. I *Dicotyledons*. KPI in association with Kuwait University, 224pp.

Dargie, T.C.D. and El Demerdash, M.A. (1991). A quantitative study of vegetation-environment relationships in two Egyptian deserts. *J. Veg. Sci.*, 2, 3–10.
Dasmann, R.E., Milton, J.P. and Freeman, P.H. (1960). *Ecological Principles for Economic Development*, John Wiley & Sons. Ltd., London, 235pp.
De Cosson, A. (1935). *Mareotis*. Country Life, London, 219pp.
Decaisne, J. (1834). Florula Sinaica. Enumeration des plantes recueilles par N. Bove dans les deux Arabies, la Palestine, la Syrie et l'Egypte. *Ann. Sci. Nat*, Ser. 2, 2, 5.18, 239–270.
Delile, A.R. (1809–1812). *Description de l'Egypte: Histoire Naturelle*, Vols. I and II. Imprimerie Imperiale, Paris, 49–82, 145–320.
Derby, W.J., Ghalioungui, P. and Grivetti, L. (1977). *Food: The Gift of Osiris*, Vol. I: 452pp. and Vol. II: 877pp. Academic Press, London.
Dixey, F. (1966). Water Supply, Use and Management. In: Chapter V. *Arid Lands: A Geographical Appraisal* (ed. E.S. Hills). 77–102, Methuen & Co., London and UNESCO, Paris.
Drar, M. (1936). Enumeration of the plants collected at Gebel Elba during two expeditions. *Fouad I Agricultural Museum Tech. and Sci. Ser.*, Ministry of Agriculture, Egypt, No. 149, 123pp., Cairo.
Draz, M.Y. (1993). Gaining experiences to stabilize sand dunes in Egypt. *Proceedings of the Symposium of Desertification and Land Reclamation of the Arab Gulf Countries*, Arab Gulf University, Baharain (in Arabic), pp. 1–15.
Draz, M.Y., Ahmed, A.M. and Afify, M.Y. (1992). Studies on sand encroachment in Siwas Oasis, Western Desert, Egypt. II feasibility of Sand dune fixation measures. *J. Eng. Appl. Sci.*, 39(4), 727–735.
Draz, M.Y., Wassif, M.M. and El-Maghraby, S.E. (1996). Improvement of Siwa dune sand for the growth of Acacia plants under saline water regime, Egypt. *J. Soil Sci.* 1(4), 153–164.
Draz, M.Y. and Zaghloul, A.K. (2006). Possibility of sand encroachment control using Prosopis juliflora and Tamarix articulata plants grown under different cultivation practices at Toshka Region. *Ann. Agr. Sci., Moshtoher, Egypt*, 44(3), 955–972.
Draz, M.Y. and Zaghloul, A.K. (2007). Impacts of different designs of shelterbelt on the control of shifiting sand at Toshka, South Egypt. *J. Environ. Sci., Mansoura Univ.*, 32(2), 601–602.
Dregne, H.E. (1976). *Soils of Arid Regions*. Elsevier, Amsterdam, 237pp.
Dürkop, E. (1903). Die Nutzpflanzen der Sahara, Beinh Z. Tropoenpflanzer, 4, 161–204.
Eig, A. (1931–1932). Les Elements et les groupes phytog6ographiques auxiliaries dans la Flora Palestinienne. *Feddes Repertorium, Spec. Nov., Beih.*, 63, 1–201.
Eisa, A.M.A. (2007). *Comparative Eco-Taxonomical Studies on Imprata cylindrica and Desmostachya bipinnata in Egypt*. MSc Thesis, Faculty of Science, Benha University.
El-Abyad, M.S.H. (1962). *Studies on the Ecology of Kutamiya Desert*. MSc Thesis, Fac. Sci., Univ. Cairo, Egypt.
El-Askary, M.A. (1968). *Geological Studies on Siwa Depression, Western Desert, Egypt*. MSc Thesis, Alexandria University, Egypt.
El-Bagouri, I.H., Zahran, M.A. and Abdel Wahid, A.A. (1976). Transplantation of Juncus spp. In calcarious soil, Egypt. *Bull. Fac. Sci., Mansoura Univ., Egypt*, 4, 59–61.
El-Bana, M.I. (2003). *Environmental and Biological Effects on Vegetation Composition and Plant Diversity of Threatened Mediterranean Coastal Desert of Sinai Peninsula*. PhD Thesis, Faculteit Westenschappear Univesteut Antwerpen, Antwerpern, 150pp.
El-Baz, F. (1980). Journey to the Gilf Kebir and Oweinat, South west Egypt. I – Narrative of the Journey. *Geo. J.*, 3, 146–151.
El-Beheiri, S.A. (1950). *The Desert SE of the Delta, a Geomorphological Study*. MSc Thesis, Fac. Sci., Cairo Univ. (In Arabic).
El-Demerdash, M.A., Zahran, M.A. and Serag, M.S. (1990). On the ecology of the deltaic Mediterranean coastal land, Egypt. III. The habitat of salt marshes of Damietta-Port Said coastal region. *Arab Gulf J. Sci. Res.*, 8(3), 103–119.
El-Fayoumi, I.F. (1964). *Geology and Ground Water Supplies in Wadi El-Natrun*. MSc Thesis, Fac. Sci., Cairo Univ., Cairo, Egypt.
El-Ghareeb, R.M. (1991). Suppression of annuals by *Tribulus terrestris* in an abandoned field in the sandy desert of Kuwait. *J. Veg. Sci.*, 2, 147–154.

El-Ghonemy, A.A. (1973). Phytosociological and ecological studies of the maritime sand dune communities in Egypt. I. Zonation of vegetation and soil along a dune side. *Bull. Inst. Desert Egypte*, 23(2), 463–473.

El-Ghonemy, A.A. and Tadros, T.M. (1970). Socio-ecological studies of the natural plant communities along a transect between Alexandria and Cairo. *Bull. Fac. Sci. Alexandria Univ. Egypt*, 10, 329–407.

El-Ghonemy, A.A., Shaltout, K, Valentine, W. and Wallace, A. (1977). Distributional pattern of *Thymelaea hirsuta* (L.). Endl. and associated species along the Mediterranean coast of Egypt. *Bot. Gaz.*, 138(4), 479–489.

El-Hadidi, M.N. (1965). *Potamogeton trichoides* Cham, and Schlecht. in Egypt. *Candollea*, 20, 159–165.

El-Hadidi, M.N. (1969). Observations on the flora of the Sinai mountain region. *Bull. Soc. Giogr. Egypte*, 40, 124–155.

El-Hadidi, M.N. (1971). Distribution of *Cyperus papyrus* and *Nymphaea lotus* in inland water of Egypt. *Mitt. Bot. Staatssamml.*, 10, 470–475.

El-Hadidi, M.N. (1976). The riverain flora in Nubia. In *The Biology of an Ancient River*, (Monograph Biol.). W. Junk, The Hague, Netherlands, 29, 87–91.

El-Hadidi, M.N. (1981). The vegetation of the Nubian Desert (Nabta), Region. I. *Prehistory of East Sahara. Appendix*, 5, 345–351.

El-Hadidi, M.N. (1985). Food plants of prehistoric and predynastic Egypt. In *Plants for Arid Lands* (eds. G.E. Wickens, J.R. Gooden and D.V. Field). George Allen and Unwin, London, pp. 87–92.

El-Hadidi, M.N. (1991). Annotated list of the flora of Sinai (Egypt). I. Introduction. The taxa of Peridophyta and Gymnospermae. *Taeckholmia*, 12, 1–6.

El-Hadidi, M.N., Kosinova, J. and Chartek, J. (1970). Weed flora of southern Sinai. *Acta Univ. Carol. Biol.*, 1969, 367–381.

El-Hadidi, M.N. and Kosinova, J. (1971). Studies on the weed flora of cultivated lands of Egypt. I Preliminary Survey. *Mitt. Bot. Staatssamml.*, 10, 354–367.

El-Hadidi, M.N. and Ayyad, M.A. (1975). Floristic and ecological features of Wadi Habis (Egypt). In *La Flore du Bassin Mediterrane'en: Essaide Systimatique Synthetique Colloques Internationaux du C.N.R.S.*, 235, 247–258.

El-Hadidi, M.N. and Springuel, I. (1978). Plant life in Nubia (Egypt): introduction: plant communities of the Nile Islands in Aswan. *Taeckholmia*, 9, 103–109.

El-Hadidi, M.N., Abdel Ghani, M.M. and Fahmy, A.G. (1991). *The Plant Red Data Book of Egypt. I-Woody Perennials*. Palm Press & Cairo University Herbarium, Cairo, Egypt, 155pp.

El-Hadidi, M.N. and Hosny, H. (2002). *Flora Egyptiaca*, Vol. 11, The Palm Press, Cairo 151pp.

El-Hady, O.A. and Hanna, A.H. (1983). Trials for sand dune stabilization using locally prepared bitumen emulsions. In *Final Report of a Research Project on Sand Soils Reclamation: Methods and Economic Aspects* (ed. A.E. El-Sherif). Academy of Scientific Research and Technology, Cairo, 154–167.

El-Henawy, M.T. (2008). *Ecology and Productivity of Wadi El-Rayan, Western Desert*, Egypt. PhD Thesis, Faculty of Science, Ain Shams University, Cairo.

El-Housini, A.A., Khalifa, S.A. and Nassar, Z.M. (2004). Growth and forage yield of *Atriplex canescens* as affected by different soil amendments under saline conditions of Wadi Sudr. *Ann. Agr. Sci. (Moshtohor)*, 42(2), 415–425.

El-Kabalawy, A. (2004). Salinity effect on seed germination of the common desert range grass *Panicum turgidum. Seed Sci. Technol.*, 32, 873–878.

El-Kady, H.F. (1980). *Effect of Grazing Pressures and Certain Ecological Parameters on Some Fodder Plants of the Mediterranean Coast of Egypt*. MSc Thesis, Faculty of Science, Tanta University, Tanta, Egypt.

El-Kady, H.F. (1987). *A Study of Range Ecosystems of the Western Mediterranean Coastal Desert of Egypt*. PhD Thesis, Faculty of Landscape Developments, Technical University, Berlin, Germany, 136pp.

El-Khatib, A.A. (1997). Former and present vegetation of Kraman Island, Upper Egypt. *Arab Gulf J. Sci. Res.*, 15(3), 661–682.

El-Kholy, A.A. (1989). *Biological and Ecological Studies of* Myriophyllum spicatum *L. as a Basis for a Better Control*. MSc Thesis, Institute of African Studies, University of Cairo, Egypt.
El-Kholy, A.A. (2001). Plant diversity in the dry land of Siwa Oasis, Western Desert, Egypt. *J. Environ. Sci. Mansoura Univ.*, Egypt, 22, 125–143.
Ellison, L. (1960). Influence of grazing on plant succession of rangelands. *Bot. Rev.*, 26, 1–78.
El-Qattamy, M.A. (1975). Transformation of sun energy into electric energy. *P. Int. Conf. Energy*, Sheraz, Iran, pp. 32–38.
El-Saiegh, A.M. (1976). Sun-energy of the desert. *Proceedings of the International Symposium on the Deserts: Dangers and Utilization*, Saudi Biological Society, Riyadh, Saudi Arabia, pp. 85–120 (in Arabic).
El-Sharkawi, H.M. (1961). Phytosociological studies on the vegetation of Bagoush area. *Bull. Inst. Desert Egypte*, 11(1), 1–18.
El-Sharkawi, H.M. and Fayed, A.A. (1975). Vegetation of inland desert wadis in Egypt. I. Wadi Bir El-Ain. *Feddes Repert.*, 86(9–10), 589–594.
El-Sharkawi, H.M., Fayed, A.A. and Salama, F.M. (1982a). Vegetation of inland desert wadis in Egypt. II. Wadi El-Matuli and Wadi El-Qarn. *Feddes Repert.*, 93(1–2), 125–133.
El-Sharkawi, H.M., Salama, F.M. and Fayed, A.A. (1982b). Vegetation of inland desert wadis in Egypt. III. Wadi Gimal and Wadi El-Muyah. *Feddes Repert.*, 93(1–2), 135–145.
El-Sharkawi, H.M. and Ramadan, A.A. (1983). Vegetation of inland desert wadis in Egypt, IV. Phytosociology of wadi system east of Minya Province. *Feddes Repert.*, 94(5), 335–346.
El-Sharkawi, H.M., Fayed, A.A. and Salama, F.M. (1984). Vegetation of desert wadis in Egypt. V. Wadi Qassab. *Feddes Repert.*, 95(7–8), 561–570.
El-Shazly, M.A. and Shata, A. (1960). Contributions to the study of the stratigraphy of El-Kharga Oasis. *Bull. Inst. Desert Egypte*, 10(1), 1–10.
El-Shenbary, S.H. (1985). A *Study of Recent Changes in Vegetation Composition in the North-Western Coastal Desert of Egypt in Burg El-Arab Area*. MSc Thesis, Tanta University, Egypt.
Emberger, L. (1951). *Rapport sur les regions arides et semi-arides de VAfrique du Nord*. Union Int. Soc. Biologiques, Serie B, Colloques, Paris, 9, 50–61.
Emberger, L. (1955). Afrique du Nord-Desert, ecologie vegetale. *Comptes rendus de Recherches. Plant Ecology, Rev. of Res., Paris*, UNESCO/219–249.
Entz, B. (1976). Lake Nasser and Lake Nubia. In *The Nile: Biology of an Ancient River* (ed. J. Rzdska), W. Junk, The Hague, Netherlands, pp. 271–298.
Evenari, M., Shanan, L. and Tadmor, N. (1971). *The Negev: The Challenge of a Desert*. Harvard University Press, Cambridge, MA, 345pp.
Ezzat, M.A., El-Badry, H.M. and Ibrahim, M.M. (1968). Hydrogeology of the Wadi el-Gidid Project, Western Desert, Egypt with special reference to Kharga Oasis. *Bull. Fac. Eng. Cairo Univ.*, 10, 479–500.
Fakhry, A. (1947). Wadi El-Rayan. *Annals du service des Antiquites de l'Egypte*, XLVI, Imprimerie de l'Institute Francais d'Archéologie Orientalis, Le Caire, pp. 1–19.
FAO (1980). *Range and Fodder Crop Development*. Syrian Arab Republic, National Range Management and Fodder Crop Production, FAO, Rome, 95pp.
FAO (1994). *Mangrove Forests: Management Guidelines*, FAO, Forestry: Paper No. 11.
Farag, I.A.M. and Ismail, M.M. (1956). Contribution to the stratigraphy of Wadi Hof area (NE of Helwan). *Bull. Fac. Sci. Cairo Univ.*, 34, 147–168.
Fayed, A.A. (1985). The distribution of *Myriophyllum spicatum* L. in the inland waters of Egypt. *Folio Geobotanice et Phytotaxonomice*, 20, 197–199.
Ferrar, H.T. (1914). Note on a mangrove swamp at the mouth of the Gulf of Suez. *Cairo Sci. J.*, CIH (88), 23–24.
Forsskal, P. (1775). *Flora Aegyptiaca-Arabica* (ed. C. Neibuhr). Hauniae Typ. Moller 32, CXXVI, 219 pp. + 1 map.
Fortin, M.J. and Dale, M.R.T. (2005). *Spatial Analysis: A guide for Ecology*. Cambridge University Press, Cambridge, UK, 365pp.
Fox, S. (1951). *The Geological Aspects of Wadi El-Rayan Project*. Government Press, Cairo, 92pp.

Fresenius, G. (1834). *Beitrage zur Flora von Aegypten und Arabien*. Museum Senckenbergianum, Frankfurt a.M., pp. 9–94 and pp. 165–188.
Friedman, J., Orshan, G. and Ziger-Cfir, Y. (1977). Suppression of annuals by *Artemisia herba-alba* in the Negev desert of Israel. *J. Ecol.*, 65, 413–426.
Garnsey, P. (1988). *Famine and Food Supply in the Graeco-Roman World*. Cambridge University Press, Cambridge.
Gibali, M.A. (1988). *Studies on the Flora of North Sinai*. MSc Thesis, Faculty of Science, Cairo University, Egypt.
Gilliland, H.B. (1952). The vegetation of Eastern British Somaliland. *J. Ecol.*, 40, 91–124.
Girgis, W.A. (1962). *Studies on the Ecology of the Helwan Desert*. MSc Thesis, Fac. Sci., Univ. Cairo, Egypt.
Girgis, W.A. (1965). *Studies on the Plant Ecology of the Eastern Desert, Egypt*. PhD Thesis, Fac. Sci., Univ. Cairo, Egypt.
Girgis, W.A. (1970). Phytosociological studies on the vegetation of Mariut area project. *Egypt. J. Bot.*, 13(2), 235–254.
Girgis, W.A. (1973). Phytosociological studies on the vegetation of Ras El-Hikma Mersa-Matruh coastal plain. *Egypt. J. Bot.*, 16, 393–409.
Girgis, W.A. (1977). *Ecological Study on Baharaya Oasis* (unpublished).
Girgis, W.A., Zahran, M.A., Reda, K.A. and Shams, H. (1971). Ecological notes on Moghra Oasis, Western Desert. *Egypt. J. Bot.*, 14, 145–155.
Girgis, W.A., El-Habibi, A.M. and Abu Ziada, M.E. (1981). Ecological studies on the New Valley. IV. Salt marsh ecosystem of Kharga and Dakhla Oases. *Delta J. Sci., Tanta, Egypt*, 5, 414– 440.
Girgis, W.A. and Ahmed, A.M. (1985). An ecological study of Wadis of SW Sinai, Egypt. *Bull. Inst. Desert Egypte*, 35(1), 265–308.
Godchild, M.F. (1994). Integrating GIS and Remote Sensing for vegetation analysis and modeling – methodological issues. *J. Veg. Sci.*, 5(5): 615–626.
Good, R. (1947). *The Geography of the Flowering Plants*. Longmans Green, London, 403pp.
Haines, R.W. (1951). Potential annuals of the Egyptian desert. *Bull. Inst. Fouad I du Desert, Egypte*, 1(2), 103–118.
Halwagy, R. (1962). The incidence of biotic factors in northern Sudan, *Oikos*, 13, 97–107.
Halwagy, M. (1973). *Ecological Studies on the Desert of Kuwait with Special Reference to the Salt Marshes*. MSc Thesis, University of Kuwait.
Hamed, A.I. (2005). *Solenostemma arghel* (Del.). Wayve. In *Encyclopedia of Wild Medicinal Plants in Egypt. Part I. Conservation and Sustainable Use of Medicinal Plants in Arid and Semi-Arid Ecosystems in Egypt* (ed. K.H. Batanouny). Egyptian Environmental Affairs Agency (EEAA), pp. 59–81.
Hammouda, F.M. (2005). *Conservation and Sustainable Use of Medicinal Plants: National Survey. Quarterly Report 3. The Western Desert and Oases*, EEAA, UNDP & GEF, 154pp.
Hammouda, F.M. (2006). *Origanum syriacum* L. subsp. *sinaicum*. In *Encyclopedia of Wild Medicinal Plants in Egypt* (ed. K.H. Batanouny), Vol. 2. EEAA, MPCP, GEF, UNDP, pp. 69–83.
Hanna, A.H. (1982). Studies on dune and their fixation with bituminous emulsion. *Bull. Natl. Res. Cent., Cairo*, V, 153–164.
Hass, H. (1977). Radiocarbon dating of charcoal and ostrich egg shells from Mushabi and Lagana sites. In *Prehistoric Inversitigations in Gebel Maghara Northern Sinai* (eds. O. Bar-Yousef and J.L. Phillips). The Hebrew University, Jerusalem, Qedom 7, pp. 261–263.
Hassan, L.M., Hegazy, A.K., Soliman, M.A. and El-Dawy, H.A. (2005). Phytosociological and floristic features of Bir Gendali hyper-arid desert, Egypt. *Egyptian J. Desert Res.*, 55(2), 1–17.
Hassanein Bey, A.M. (1924a). Crossing the untraversed Libyan Desert. *Nat. Geog. Mag.*, 46, 233–277.
Hassanein Bey, A.M. (1924b). Through Kufra to Darfur. *Geogr. J.*, 64, 273–291; 353–366.
Hassanein Bey, A.M. (1925). *The Lost Oasis*. Century Co., New York, London, 363pp.
Hassib, M. (1951). Distribution of plant communities in Egypt. *Bull. Fac. Sci. Univ. Fouad I, Cairo, Egypt*, 29, 59–261.

Hemming, C.F. (1961). The ecology of the coastal area of northern Eritrea. *J. Ecol.*, 49, 55–78.

Heneidy, S.Z. (1986). *A Study of the Nutrient Content and Nutritive Values of Range of Plants at Omayed, Egypt*. MSc Thesis, Fac. Sci., Alexandria Univ., Egypt.

Heneidy, S.Z. and Bidak, L.M. (1996). Halophytes as a forage source in the Western Mediterranean Coastal Region of Egypt. *Desert Inst. Bull., Cairo, Egypt*, 46(2), 283–304.

Hepper, F.N. (1990). *Pharaoh's Flowers: The Botanical Treasures of Tutank-hamun*. HMSO, London, 80pp.

Hills, E.S., Oilier, C.D. and Twidale, C.R. (1966). Geomorphology. In *Arid Lands: A Geographical Appraisal* (ed. E.S. Hills). Methuen, London, pp. 53–76.

Hobler, P.M. and Hester, J.J. (1969). Prehistory and environment in the Libyan desert. *S. Afr. Archaeol. Bull.*, 23, 120–130.

Hosny, A.I. (1977). New taxa of *Zygophyllum* in Egypt. *Bot. Notiser.*, 130, 467–468.

Hume, W.F. (1908). The Southwestern Desert of Egypt. *Cairo Sci. J.*, 2, 313–325.

Hume, W.F. (1925). Geology of Egypt, Vol. 1. The surface features of Egypt, their determining causes and relation to geological structure. *Egypt. Surv. Dept. Cairo*, 418pp.

Hurst, H.E. (1952). *The Nile*. Constable, London, 326pp.

Hussein, T.M.G. (2000). *Studies on the River Nile Vegetation in El-Khahera El-Kobra*, MSc Thesis, Faculty of Science, Helwan University, Cairo.

Ishac, Y.Z. and Moustafa, M.I. (1991). Prospects of VA-mycorrhizas in Egypt. *Mycorrhizas in Ecosystems-Structure and Function*. Abstracts, Third European Symposium on Mycorrhizas, Sheffield.

Ismail, S.I. (2006). Monograph on *Acacia nilotica*. In *Encyclopedia of Wild Medicinal Plants in Egypt*. (ed. K.H. Batanouny), Vol. 2. UNDP, MPCP, GEF, EEAA, The Palm Press, Cairo, 51–64.

Jewitt, Y.N. (1966). Soil of Arid lands. In *Chapter VI: 103–125. Arid Lands: A Geopraphical Appraisal* (ed. E.S. Hills), Methuen & Co. Lit., London & UNESCO, Paris.

Kamal El-Din, H. (1928). L' exploration du Desert Libyque. *La Olographic* 50, 171–183, 320– 336.

Kassas, M. (1952a). Habitat and plant communities in the Egyptian desert I. Introduction. *J. Ecol.*, 40, 342–351.

Kassas, M. (1952b). On the reproductive capacity and life cycle of *Alhagi maurorum. Egypt. Acad. Sci. Proc.*, VIII, 114–122.

Kassas, M. (1952c). On the distribution *of Alhagi maurorum* in Egypt. *Egypt. Acad. Sci. Proc.*, VIII, 140–151.

Kassas, M. (1953a). Landforms and plant cover in the Egyptian desert. *Bull. Soc. Geogr. Egypte*, 26, 193–205.

Kassas, M. (1953b). Habitat and plant communities in the Egyptian desert. II. The features of a desert community. *J. Ecol.*, 41, 248–256.

Kassas, M. (1955). Rainfall and vegetation belts in arid NE Africa. *Plant Ecology Proc. of the Montpellier Symp., UNESCO* (1955), 49–77.

Kassas, M. (1956). The mist oasis of Erkwit, Sudan. *J. Ecol.*, 44, 180–194.

Kassas, M. (1957). On the ecology of the Red Sea coastal land. *J. Ecol.*, 45, 187–203.

Kassas, M. (1960). Certain aspects of landform effects on plant-water resources. *Bull. Soc. Geogr. Egypte*, 33, 45–52.

Kassas, M. (1970). Desertification versus potential for recovery in circum-Saharan territories. In *Arid Lands in Transition*, (ed. H. Dregne). Amer. Assoc, for Adv. Sci., Washington, D.C., 13, 123–142.

Kassas, M. (1972a). Ecological consequences of water development project, Keynote paper. *The Environmental Future* 7. *Major Water ETC Development Projects* (ed. N. Polunin). Macmillan, London, pp. 215–246.

Kassas, M. (1972b). A brief history of land use in Mareotis region. *Minerva Biol.*, 1, 167–174.

Kassas, M. (1981). Egypt. In *Handbook of Contemporary Development in World Ecology* (eds. E.J. Kormondy and J.F. McCormick). Greenwood Press, London, pp. 447–454.

Kassas, M.A. (2004). *The Nile in Danger*, Eqraa Series, Dar El-Maaref, Cairo No. 705, 185pp.

Kassas, M. and Imam, M. (1954). Habitat and plant communities in the Egyptian desert. III. The wadi bed ecosystem. *J. Ecol.*, 42, 424–441.
Kassas, M. and Imam, M. (1959). Habitat and plant communities in the Egyptian desert. IV. The gravel desert. *J. Ecol.*, 47, 289–310.
Kassas, M. and El-Abyad, M.S. (1962). On the phytosociology of the desert vegetation of Egypt. *Ann. Arid Zone*, 1(1). 64–83.
Kassas, M. and Zahran, M.A. (1962). Studies on the ecology of the Red Sea coastal land, I. The district of Gebel Ataqa and El-Galala El-Bahariya. *Bull. Soc. Geogr. Egypte*, 35, 129–175.
Kassas, M. and Girgis, W.A. (1964). Habitat and plant communities in the Egyptian desert. V. The limestone plateau. *J. Ecol.*, 52, 107–119.
Kassas, M. and Girgis, W.A. (1965). Habitat and plant communities in the Egyptian desert. VI. The units of a desert ecosystem. *J. Ecol.*, 53, 715–728.
Kassas, M. and Zahran, M.A. (1965). Studies on the ecology of the Red Sea coastal land, II. The district from El-Galala El-Qibliya to Hurghada. *Bull. Soc. Geogr. Egypte*, 38, 155–193.
Kassas, M. and Zahran, M.A. (1967). On the ecology of the Red Sea littoral salt marsh, Egypt. *Ecol. Monogr.*, 37(4). 297–315.
Kassas, M. and Girgis, W.A. (1969–1970). Plant life in the Nubian desert east of the Nile, Egypt. *Bull. Inst. Egypte*, LI, 47–71.
Kassas, M. and Girgis, M.A. (1970). Habitat and plant communities in the Egyptian desert. VII. Geographical facies of plant communities. *J. Ecol.*, 58, 335–350.
Kassas, M. and Zahran, M.A. (1971). Plant life on the coastal mountains of the Red Sea, Egypt. *J. Ind. Bot. Soc.* Golden Jubilee Volume, 50A, 571–589.
Kassas, M. and Girgis, W.A. (1972). Studies on the ecology of the Eastern Desert of Egypt. I. The region between Lat. 27"30'N and Lat. 25'30'N. *Bull. Soc. Geogr. Egypte*, XLI–XLII, 43–72.
Keay, R.W.J. (1959). *Vegetation Map of Africa*. Oxford University Press, Oxford, 24pp.
Kehl, H. and Bornkamm, R. (1993). Landscape ecology and vegetation units of the Western Desert of Egypt. In *Geopotential and Ecology* (eds. B. Meiss ner and P. Wycisk). CATENA supplement 26, Cremlingen-Desteds, Germany, pp. 155–178.
Kerr, J.T. and Ostrovsky, M. (2003). From space to species: ecological application for remote sensing. *Trends Ecol. Evol.*, 18(6), 299–305.
Khalil, M.T. and Shaltout, K.H. (2006). *Lake Bardawil and Zaranik Protected Area*. Biodiversity Unit, EEAA, Cairo, 590pp.
Khedr, A. (1989). *Ecological Studies on Lake Manzala, Egypt*. MSc Thesis, Fac. Sci., Mansoura Univ., Egypt.
Khedr, A.A. and El-Demerdash, M.A. (1995). Distribution and environmental relations of the aquatic macrophytes in the north-east section, Nile Delta, Egypt. *Proc. 5th Intern. Conf. Environ. Protection is a must*. National Institute of oceanography, Alexandria, 197–223.
Khedr, A.A. and Zahran, M.A. (1999). Comparative study on the plant life of two Mediterranean deltaic lakes in Egypt. *Assiut University Bull. Env. Res.*, 2(1), 1–14.
Khedr, A.A. and Gazzar, A. (2006). *Phytoecology of Zaranik Lagoon, Lake Bardawil, North Sinai*, First Progress Report, Med-Wet Coast Project, EEAA, Cairo, 27pp.
Koppen, W. (1931). *Grundriss der Klimakunde*. W. de Gruyter, Berlin.
Krzywinski, K. and Pierce, R.H. (eds) (2001). *Deserting the desert: a threatened cultural landscape between the Nile and the sea*. Bergen, Alvheim og Eide Akademisk Forlag, 179pp.
Le Hourerou, H.N. (1973). Problèmes et Potentielles des terres and de L'Afrique du Nord. *Options Mediterranéen*, 26, 17–36.
Leonard, J. (1969). Expedition scientifique Beige dans le desert de Libye, Uweinat 1968–1969. *La Flore Africa-Tervuren*, 15(4), 110–116.
Leonard, J. (2001). *Flore et Végétation du Jebel Uweinat (Desert de Libye: Libya, Egypt, Sudan)*. Sixiem, et dernière, Partie. Etude de la Végétation Analyse Phyosociologique et Phytochrologique de Groupements Végétaux. Meise, Jardis Botanique Nationa de Belgique, 139pp.
Lewis, M.M., Jooste, V. and de Gasparis, A.A. (2001). Discrimnation of arid vegetation with Air Bore Multispectral Scanner Hyperspectral Imagery. *IEEE T. Geosci. Remote.*, 39, 1471–1497.

Lillisand, T.M. and Keifer, R.W. (1994). *Remote Sensing and Image Interpretation*, 1st edn. John Wiley & Sons Inc., New York, 750pp.
Lillisand, T.M. and Kiefar, R.W. (2000). *Remote Sensing and Image Interpretation*, 4th edn. John Wiley and Sons Inc., New York, 724pp.
Long, G.A. (1955). The study of natural vegetation as a basis for pasture improvement in the Western Desert of Egypt. *Bull. Inst. Desert Egypte*, 5(2), 18–45.
Lucas, A. (1912). Natural soda deposits in Egypt. *Surv. Dept. Cairo*, pp. 1–11.
Maergner, H. (1990). Sand dune and embankment profiling, an important drifts. *Egypt. J. Soil Sci.*, 30(1–2), 119–124.
Malcolm, C.V. (1972). *Establishing Shrubs in Saline Environment*. Technical Bulletin 14, Department of Agriculture Western Australia, 37pp.
Mannicke, Le (1989). *An Ancient Egyptian Herbal*. British Museum Publication, LTD, 133–134.
Mashaly, I.A. (1996). On the Phytosociology of Wadi Hagul, Red Sea Coast, Egypt. *J. Environ. Sci., Mansoura Univ*., Egypt, 12, 31–54.
Mehringer, J.R., Petersen, K.L. and Hassan, F.A. (1979). A pollen record from Birket Qarun and the recent history of the Fayum, Egypt. *Quaternary Res*., 11, 238–256.
Meigs, P. (1966). Geography of coastal deserts. *Arid Zone Res., UNESCO*, 28, 140pp.
Meikle, R.O. (1985). *Flora of Cyprus*, Vol. 2. Bentham-Moxon Trust, Royal Botanic Gardens, Kew, 1385pp.
Migahid, A.M. (1978). *Flora of Saudi Arabia*, 2nd edn. Riyadh Univ. Publ., 2 Vols., 939pp.
Migahid, A.M., Abdel Rahman, A-A., El-Shafei, A.M. and Hammouda, M.A. (1955). Types of habitat and vegetation at Ras El-Hikma. *Bull, Inst. Desert Egypte*, V(2), 107–190.
Migahid, A.M. and Ayyad, M.A. (1959). An ecological study of Ras El-Hikma district. IV. Structure of vegetation in the main habitats. *Bull. Inst. Desert Egypte*, 9(1), 99–120.
Migahid, A.M., El-Shafei, A.M. and Abdel Rahman, A.A. (1959). Ecological observations in the western and southern Sinai. *Bull. Soc. Geogr. Egypte*, 82, 165–205.
Migahid, A.M., El-Shafei, A.M., Abdel Rahman, A.A. and Hammouda, M.A. (1960). An ecological study of Kharga and Dakhla Oases. *Bull. Soc. Geogr. Egypte*, 33, 279–310.
Miller, T. JR. (1997). Environmental Science Sustaining the Earth, 4th edn. Wadsworth Publishing Company, California, U.S.A., 470pp. + Appendix.
Mitwally, M. (1952). History of the relation between the oases of the Libyan Desert and the Nile Valley. *Bull. Inst. Fouad I du Desert, Egypte*, 11(1), 114–131.
Mitwally, M. (1953). Physiographic features of the Libyan Desert. *Bull. Inst. Fouad I du Desert, Egypte*, III(l), 147–164.
Mohamed, M.K. and Hassan, L.M. (1998). Studies on the plant life of River Nile Islands in Minia Governorate. *Proceedings of the Sixth Egyptian Botanical Conference*, Cairo University, III, 481–492.
Montasir, A.H. (1937). On the ecology of Lake Manzala. *Bull. Fac. Sci. Egyptian Univ. Cairo*, 12, 50pp.
Montasir, A.H. (1938). Egyptian soil structure in relation to plants. *Bull. Fac. Sci., Egyptian Univ., Cairo*, 15, 47pp.
Moore, P.D. (1976). How far does pollen travel? *Nature*, 260, 388–389.
Moore, P.D. and Stevenson, A.C. (1982). Pollen studies in dry environments. In *Desertification and Development. Dryland Ecology in Social Perspective* (eds. B. Spooner and H.S. Mann), Academic Press, London, pp. 249–267.
Moreau, R.E. (1938). Climatic classification from the standpoint of East African biology. *J. Ecol*., 26, 467–496.
Mousa, E.E.A., Ata, E.M. and Zaghloul, A.K. (2006). Ecological and phytochmical studies on *Populus euphratica* growing on sand dunes at Siwa Oasis. *J. Eniron. Sci., Mansoura Univ., Egypt*, 32, 177–205.
Murray, G.W. (1947). A note on Sad El-Kafra: the ancient Egyptian dam in Wadi Garawi. *Bull. Inst. Egypte*, 28, 33–35.
Murray, G.W. (1951). The Egyptian climate. An historical outline. *Geogr. J*., 117(4), 422–434.
Murray, G.W. (1953). The land of Sinai. *Geogr. J*., 119, 140–153.

Muschler, R. (1912). *A Manual Flora of Egypt*, Vol. II. R. Friedlander, Berlin, 1312pp.
Newbold, D. (1928). Rock pictures and archaeology in the Libyan Desert. *Antiquity*, 2, 261–288.
Oliver, F.W. (1930–1931). Oasis impression, being a visit to the Egyptian Oasis of Kharga. *Trans. Norfolk and Norwich Naturalists' Soc.*, 13(2), 38–52.
Oliver, F.W. (1938). The flowers of Mareotis: an impression. Part I. *Trans. Norfolk and Norwich Naturalists' Soc.*, 14, 397–437.
Oliver, F.W. (1945). The flowers of Mareotis: an impression. Part II. *Trans. Norfolk and Norwich Naturalists' Soc.*, 16, 130–164.
Osborn, D.J. and Krombein, K.V. (1969). Habitat, flora, mammals and wasps of Gebel Uweinat, Libyan Desert. *Smithsonian Contributions to Zoology*, No. 11, Smithsonian Institute Press, City of Washington, pp. 1–18.
Pavlov, M.J. (1962). *Preliminary Report on the Geology, Hydrology and Groundwater of Wadi El-Natrun and Adjacent Area, Cairo*. UNESCO, 183pp.
Pearce, D., Barbier, E. and Markandya, A. (1990). *Sustainable Development: Economics and Environment in the Third World*. Edward Elgar, London, 217pp.
Peel, R.E. (1939). The Gilf Kebir. Part 4.1: R.A. Bagnold 'An expedition to the Gilf Kebir and Uweinat'. *Geogr. J.*, 93, 295–307.
Percheron, L. (1903). A jacinthe d'eau. *Bull, de l'Union Syndicate des Agr. d'Egypte*, 3eme annee, No. 23.
Pessarakli, M. (1993). *Handbook of Plant and Crop Stress*, Marcel Dekker Inc., New York, U.S.A., 692pp.
Pettorelli, N., Vik, J.O., Mysterud, A. Gillard, J.M., Tucker, C.J. and Stensenth, N.C. (2006). Using the Satellite – derived NDVI to assess ecological responses to environmental change. Trends Ecol. Evol., 20(9), 503–510.
Pickup, G., Chewings, V.H. and Nelson, D.J. (1993). Estimating change in vegetation cover over time in arid rangelands using Landsat MSS Data. *Remote Sens. Environ.*, 43(3), 243–263.
Raheja, P.C. (1973). *Man-Made Lakes. Their Problems and Environmental Effects*. America, Geophysical Union, Washington DC, Geophysical Monograph, 17, 234–240.
Rattray, J.M. (1960). The grass cover of Africa. *FAO Agr. Stud.*, 49, 37pp.
Reed, C.A. (1964). A natural history study of Kurkur Oasis, Libyan Desert, Western Governate, Egypt. I. Introduction. Peabody Museum of Natural History, Yale Univ. *Postilla*, 100, 1–22.
Ritchie, J.C. (1985). Modern pollen spectra from Dakhla Oasis, Western Egyptian desert. *Grana, Uppsala*, 1984, pp. 1–6.
Ritchie, J.C. (1987). A Holocene pollen record from Bir Atrun, Northwest Sudan. *Pollen Spores*, 29, 391–410.
Roubet, C. and El-Hadidi, M.N. (1981). 20 000 ans d'environment préhistorique dans la vallée du Nil et le Désert Egyptien. *Bull. Centenaire* (Suppl.). *Bull. Franc. Arch. Orient.*, 81, 70.
Ruprecht, F.J. (1849). Die Vegetation des Roten Meeres. *Mem. Soc. Sci. Nat. Petersburg*, 6, 71–84.
Saad, S.I. and Sami, S. (1967). Studies of pollen and spores content of the Nile Delta deposits (Berenbal Region). *Pollen Spores*, 9, 467–503.
Sadek, J. (1926). The geography and geology of the district between Gebel Ataqa and El-Galala El-Bahariya (Gulf of Suez). *Geol. Surv. Egypt, Cairo*, 120pp.
Sadek, J. (1959). The Miocene in the Gulf of Suez region (Egypt). *Egypt. Geol. Surv.*, 118pp.
Saenger, P. (2001). *Mangrove Ecology, Silviculture and Conservation*. Kluwer Academic Publishers, Netherlands, 372pp.
Saenger, P. (2003). *Rehabilitation, Conservation and Sustainable Utilization of Mangroves in Egypt*. EEAA, Cairo, 33pp.
Said, R. (1954). Remarks on the geomorphology of the area to the east of Helwan, Egypt. *Bull. Soc. Giogr. Egypte*, 27, 93–104.
Said, R. (1960). New light on the origin of the Qattara Depression. *Bull. Soc. Geogr. Egypte*, 33, 37–44.
Said, R. (1962). *The Geology of Egypt*. Elsevier, Amsterdam, 377pp.
Said, R. (1981). *The Geological Evolution of the River Nile*. Springer Verlag, New York, 151pp.

Said, R. and Issawy, B. (1964). Preliminary results of geological expedition to Lower Nubia and to Kurkur and Dungul Oases, Egypt. In *Contribution to Prehistory of Nubia* (ed. F. Wendorf). Southern University Press, Dallas, pp. 1–20.

Salama, F.M., Abdel Ghani, M.M., El-Naggar, S.M. and Baayo, K.A. (2005). Vegetation structure and environmental gradients in the Sallum area, Western Mediterranean Coast, Egypt. *Ecol. Mediterr.*, 31(1), 1–13.

Saleh, A.H. (1970). *Pedological Studies on Siwa Oasis*. MSc Thesis (Agr.). Fac. Agr., Cairo Univ.

Saleh, M.A. (1984). *Investigation of Inorganic Pollutants in El-Fayium Aquatic Environment.* Report 2: FRCU Grant No. 84202, Supreme Council of Universities, Cairo, 90pp.

Salem, B.B. (2003). Application of GIS to biodiversity monitoring. *J. Arid Environ.*, 54, 91–114.

Salem, B.B. (2007). *Remote Sensing Applied to Vegetation Classification* (non-published report). 6pp.

Salem, B.B. and Wassem, M. (2003). A study on Maghra Oasis by remote sensing. *Assiut Univ. J. Bot.*, 35(2), 337–387.

Sandford, K.S. (1929). The Pliocene and Pleistocene deposits of Wadi Qena and Nile Valley between Luxor and Assiut. *Quart. J. Geol. Soc. Lond.*, 75, 493–548.

Sandford, K.S. (1934). Paleolithic man and the Nile Valley in Upper and Middle Egypt. *Chicago Univ. Orient. Inst. Publ.*, 18, 1–131.

Sandford, K.S. and Arkell, W.J. (1939). Paleolithic man in the Nile Valley and in Lower Egypt with some notes upon a part of the Red Sea littoral. *Univ. Chicago Orient Inst. Publ.*, 46, 105pp.

Schweinfurth, G.A. (1865a). Flora der Soturba an der nubischen Kiiste. *Verh. Zool. Bot. Ges. Wien*, 15, 537–560.

Schweinfurth, G.A. (1865b). Reise an der Küste des Roten Meeres von Kosser bis Suakin. *Z. fur allgemeine Erdkunde, Berlin*, 18, 450–482.

Schweinfurth, G.A. (1883). Ueber die geologische Schichtengliederung des Mokattam bei Kairo. *Z. Deutschen geologischen Gesellschaft (Berl.)*. 35, 709–734.

Schweinfurth, G.A. (1896–1899). Sammlung Arabisch-Athiopischer Pflanzen. *Bull. Inst, de l'Herbier Boissier*, App. No. II, Vol. 4– 7., Geneve.

Schweinfurth, G.A. (1901). *The Flora of the Desert Surrounding Helwan and the Egyptian Desert.* George Allen, London, pp. 16–38.

Serag, M.M.S. (1991). *Studies on the Ecology and Control of Aquatic and Canal Bank Weeds of the Nile Delta, Egypt*. PhD Thesis, University of Mansoura, Egypt.

Serag, M.S. (2000). The rediscovery of the papyrus (*Cyperus papyrus* L.). on the bank of Damietta Branch, Nile Delta, Egypt. *Taedkholmia*, 20(2), 195–198.

Shabetai, J.K. (1940). Contribution to the flora of Egypt. Plants collected from southern Sinai in April 1937. *Tech. Sci. Service Fouad I Univ. Agr. Museum*, 234, 1–84.

Shalaby, A.F., Ghanem, S.S. and El-Habibi, A.M. (1975). Ecological study of *Prosopis stephaniana* (Willd.). Kunth. *Bull. Fac. Sci. Mansoura Univ.*, 3, 45–63.

Shaltout, K.H. (1983). An *Ecological Study* of *Thymelaea hirsuta (L.). Endl. In Egypt*. PhD Thesis, Fac. Sci., Tanta Univ., Egypt.

Shaltout, K.H., Al-Sodany, Y.M. and El-Sheikh, M.A. (2004). *Phragmites australis* (Cav.) Trin. ex Stud. In Lake Burullus, Egypt: is it an expanding or retreating population. *Proc. 3rd Conf. Biol. Society (ICBS).*, Faculty of Science, Tanta University, 3, 20–32.

Shaltout, K.H. and Khalil, M.T. (2005). *Lake Burullus (Burullus Protected Area).* National Biodiversity Unit, EEAA, Cairo. No. 13, 578pp.+25pp (in Arabic).

Sharaf El-Din, A. and Shaltout, K.H. (1985). On the phytosociology of Wadi Araba in the Eastern Desert of Egypt. *Proc. Egypt. Bot. Soc. IV Ismailia Conf.*, pp. 1311–1317.

Shata, A. (1955). Geomorphological aspects of the West Sinai foreshore Province. *Bull. Inst. Disert Egypte*, 5(2), 137–145.

Shata, A. (1956). Structural development of Sinai Peninsula, Egypt. *Bull. Inst. DeSert Egypte*, VI(2), 117–157.

Shata, A., Knetsch, G., Degens, E.T., Munnich, O. and El-Shazli, M. (1962a). The geology, origin and age of the ground water supplies in some desert areas of Egypt. *Bull. Inst. De'sert Egypte*, 12(2), 16–124.

Shata, A., Pavlov, M. and Saad, K. (1962b). *The geology, hydrology and ground water hydrology of Wadi El-Natrun*. The General Desert Development Organization, Cairo (Mimeographed).

Shaw, W.B.K. and Hutchinson, J. (1931). The flora of the Libyan Desert. *Bull. of Misc. Infor. of the R. Bot. Gard., Kew*, 4, 161–166.

Shaw, W.B.K. and Hutchinson, J. (1934). The flora of the Libyan Desert: botanical notes. *Bull, of Miscellaneous Information of the Royal Botanical Gardens, Kew*, 7, 271–289.

Sheded, M.G. and Shaltout, K. H. (1998). Weed flora in plantation of recently established tourist resorts along Red Sea coast, Egypt. *J. Union Arab Biologists*, 5(B), Botany, 109–119.

Shmida, A. and Orshan, G. (1977). The recent vegetation of Gebel Maghara. In *Prehistoric Investigations on Gebel Maghara, Northern Sinai-Qedem* (eds. O. Bar-Yosef and J.L. Phillips), 7, 32–36.

Shreve, F. (1942). The desert vegetation of North America. *Bot. Rev.*, 8, 195–246.

Simpson, N.D. (1932). *A Report on the Weed Flora of the Irrigation Channels in Egypt*. Ministry of Pub. Works, Gov. Press, Cairo, 124pp.

Springuel, I.V. (1981). *Studies on the Natural Vegetation of the Islands of the First Cataract at Aswan, Egypt*. PhD Thesis, Dept. Bot. Fac. Sci. at Aswan, Assiut Univ., Egypt.

Springuel, I.V. (1985a). The shore-line vegetation of the area between the two dams south of Aswan, Egypt. *Proc. Egypt. Bot. Soc. Ismailia*, 4, 1409–1421.

Springuel, I.V. (1985b). Study on shore-line vegetation of High Dam Lake at Aswan, Egypt. *Aswan Sci. Tech. Bull.*, 6, 297–310.

Springuel, I.V. (1987). Plant life in Nubia, V. Aquatic plants in Egyptian Nubia. *Aswan Sci. Tech. Bull.*, 8, 185–211.

Springuel, I. (2006). *The Desert Garden: A Practical Guide*. The American University in Cairo Press, Cairo, Egypt, 156pp.

Springuel, I. and Ali, M.M. (1990). Impact of Lake Nasser on desert vegetation. *Proc. Desert Development Conf. January 1987, Cairo, Egypt*. Part 1. *Desert Agriculture, Ecology and Biology* (eds. A. Bishay and H.E. Dregne). pp. 557–568.

Springuel, I., El-Hadidi, M.N. and Sheded, M. (1991). Plant communities in the southern part of the Eastern Desert (Arabian Desert). of Egypt. *J. Arid Environ.*, 31, 307–317.

Springuel, I., Sheded, M. and Murphy, K.J. (1997). The plant diversity of Wadi Allaqi Biosphere Reserve (Egypt). Impacts of Lake Nasser on the desert Wadi ecosystem. *Biodivers. Conserv.*, 6, 1259–1275.

Stocker, O. (1926–1927). Die aegyptisch-arabische Wuste. *Vegetationsbilder, Jena*, 17(5/6), 27pp.

Stocker, O. (1927). Das Wadi Natrun. *Vegetationsbilder, Jena*, 18, 6pp.

Stocker, O. (1928). Das Wasserhaushalt aegyptischer Wiisten und Salzpflanzen. *Botanische Abhandlungen*, 13(2), 200pp.

Sutton, L.J. (1947). *Rainfall in Egypt*. Phys. Dept., Gov. Press, Cairo, Paper No. 53, 129pp.

Täckholm, V. (1932). Some new plants from Sinai and Egypt. *Svensk Botanisk Tidskrift*, 26(1–2), 370–380.

Täckholm, V. (1951). *Faroas blomster*. Generalstabens Litografiska Anstalt, Stockholm, 295pp.

Täckholm, V. (1956). *Students'Flora of Egypt*. Anglo-Egyptian Bookshop, Cairo, 649pp.

Täckholm, V. (1974). *Students' Flora of Egypt*, 2nd edn. Cairo Univ. (Publ.), Cooperative Printing Company, Beirut, 888pp.

Täckholm, V., Tackholm, G. and Drar, M. (1941). *Flora of Egypt*. Fouad I Univ. Press, Cairo, Vol. 1, no. 17, 574pp.

Täckholm, V. and Drar, M. (1950). Flora of Egypt. II. *Bull. Fac. Sci. Fouad I Univ.*, 28, 99–145.

Täckholm, V. and Drar, M. (1954). *Flora of Egypt*, Vol. III. Cairo University Press, 638pp.

Tadros, T.M. (1949). Geobotany in Egypt: A historical review. *Vegetatio*, 2, 38–42.

Tadros, T.M. (1953). A phytosociological study of halophilous communities from Mareotis (Egypt). *Vegetatio*, 4, 102–124.

Tadros, T.M. (1956). An ecological survey of the semi-arid coastal strip of the western desert of Egypt. *Bull. Inst. Desert Egypte*, 6(2), 28–56.

Tadros, T.M. and Atta, B.A.M. (1958a). Further contribution to the study of the sociology and ecology of the halophilous plant communities of Mareotis (Egypt). *Vegetatio*, 8, 137–160.

Tadros, T.M. and Atta, B.A.M. (1958b). The plant communities of barley fields and uncultivated desert areas of Mareotis (Egypt). *Vegetatio*, 8, 161–175.

Tadros, T.M. and El-Sharkawi, H.M. (1960). Phytosociological and ecological studies on the vegetation of Ras El-Hikma area. II. Consistency and homogeneity of the open desert communities. *Bull. Inst. Desert Egypte*, 10(1), 16–63.

Tansley, A.G. (1939). *The British Islands and Their Vegetation*. Cambridge University Press, Cambridge, 930pp.

Tauber, H. (1965). Differential pollen dispersion and the interpretation of pollen diagrams. *Danmarks Geol. Undersogelse*, 11, 1–69.

Tawadrous, R.W. (1981). *Taxonomical and Ecological Studies on Water Plants in Egypt: Genus Potamogeton L.* MSc Thesis, Inst, of African Studies, Cairo University.

Tousson, O. (1932). Note sur les deserts d' Egypte. *Desert Inst. Bull. Cairo*, 14, 189–202.

Trewartha, G.T. (1954). *An Introduction to Climate*. McGraw Hill, New York, 377pp.

Troll, G. (1935). Wustensteppen und Nebeloasen in Siidnubischen Kustengebirge, Studien zur Vegetations- und Landschafts-Kunde der Tropen I. *Zeitschrift der Gesellschaft fur Erdkunde zu Berlin*, 7/8, 241–281.

UNESCO (1977). Map of the World Distribution of Arid Region. MAB Technical Notes: 7.

UNESCO/FAO (1963). Bioclimatic Map of the Mediterranean Zone. Explanatory Notes. *Arid Zone Res.*, 22, 17pp.

Verdcourt, B. (1968). French Somaliland Conservation of vegetation in Africa south of Sahara. *Symp. 6th Plenary meeting Assoc. Etud. Taxon Flora Afr. Trop.*, Uppsala, 12 Sept. 1966, 140–141.

Vesey-FitzGerald, D.F. (1955). Vegetation of the Red Sea coast south of Jeddah, Saudi Arabia. *J. Ecol.*, 43, 477–489.

Vesey-FitzGerald, D.F. (1957). The vegetation of the Red Sea coast north of Jeddah, Saudi Arabia. *J. Ecol*, 45, 547–562.

Volkens, G. (1887). *Die Flora der aegyptisch-arabischen Wuste aufGrundlage anatomisch-physiologischer Forschungen*. Gebrtider Borntraeger, Berlin, 156pp.

Waisel, Y. (1961). Ecological studies on *Tamarix aphylla* (L.). Karst. Ill The salt economy. *Plant Soil*, 4, 356–364.

Walter, H. (1961). The adaptation of plants to saline soils. In *Salinity Problems in the Arid Zones*. Proc. Teheran Symp., UNESCO, Paris, Arid Zone Res., 14, 129–134.

Wasel, A.S.A. (1999). Phenotypic composition of true culture-drived and conventionally propagated by offshoot date palm (*Phoenix dactylifera*) trees. I. Vegetation characteristics. *Proceedings International Conference of Date Palm*, Assiut University, Assiut, Egypt, 97–115.

Wendelbo, P. (1961). Studies in Primulaceae. II. An account of *Primula* subgenus *sphondylia. Arab. Univ. Bergen. Mat. Natur., Serie* 11, 5–49.

Wendorf, F. and Schild, R. (1984). Some implications of late Paleolithic cereal exploitation. In *Origin and Early Development of Food-Producing Cultures in North-Eastern Africa* (eds. L. Krzyaniak and M. Kobusiewicz). Polska Akademia Nauk., Poznan, pp. 117–127.

Wiens, J.A. (1989). Spatial scaling in ecology. Funct. Ecol., 3(4), 38–397.

Williams, M.A.J, and Hall, D.N. (1965). Recent expedition to Libya from Royal Military Academy Sandhurst. *Geogr. J.*, 131, 428–501.

Wright, H.E., McAndrews, J.H. and van Zeist, W. (1967). Modern pollen rain in western Iran, and its relation to plant geography and Quaternary vegetational history. *J. Ecol.*, 55, 415–443.

Younes, H.A., Zahran, M.A. and El-Qurashy, M.E. (1983). Vegetation-soil relationships of a sea landward transect, Red Sea coast, Saudi Arabia. *J. Arid Env.*, 6, 349–356.

Youssef, A.M., Hamed, A.F. and Salem, B.B. (2004). Spatial distribution of mangal-algal association in some sites along the Egyptian Red Sea Coast by Remote Sensing. *Egypt. J. Biol.*, 6, 39–51.

Youssef, A.M. and Ghallab, A. (2007). Using remotely sensed data, GIS and field investigation for preliminary considerations of sustainable development: West Qena area, Egypt. *Assiut Univ. Bull. Envir. Res.* 10(2), 31–46.

Zahran, M.A. (1962). *Studies on the Ecology of the Red Sea Coastal Land.* MSc Thesis, Fac. Sci., University of Cairo.
Zahran, M.A. (1964). *Contributions to the Study on the Ecology of the Red Sea Coast.* PhD Thesis, Fac. Sci., University of Cairo.
Zahran, M.A. (1965). Distribution of mangrove vegetation in Egypt. *Bull. Inst Desert Egypte*, 15(2), 7–12.
Zahran, M.A. (1966). Ecological study of Wadi Dungul. *Bull. Inst. Desert Egypte*, 16(1), 127–143.
Zahran, M.A. (1967). On the ecology of the eastern coast of the Gulf of Suez. I. Littoral salt marsh. *Bull. Inst Ddsert Egypte*, 17(2), 225–252.
Zahran, M.A. (1970–1971). Wadi El-Raiyan: A natural water reservoir, Western Desert, Egypt. *Bull. Soc. Geogr. Egypte*, XLIII–XLIV, 83–98.
Zahran, M.A. (1972). On the ecology of Siwa Oasis. *Egypt. J. Bot.*, 15, 223–242.
Zahran, M.A. (1976). The water hyacinth problem in Egypt. *Proc. Symp. On Nile Water and Lake Dam Project. Nat. Res. Centre, Cairo*, 6, 188–198.
Zahran, M.A. (1977). Africa. A. Wet formations of the African Red Sea coast. In *Wet Coastal Ecosystems* (ed. V.J. Chapman). Elsevier, Amsterdam, pp. 215–231.
Zahran, M.A. (1982a). Ecology of the halophytic vegetation of Egypt. Part I. In *Contributions to the Ecology of Halophytes, Tasks for Vegetation Science* (eds. D.N. Sen and K.S. Rajpurohit), Vol. 2. Dr W. Junk, The Hague, pp. 3–20.
Zahran, M.A. (1982b). *Vegetation Types of Saudi Arabia.* Publ. King Abdul Aziz Univ., Jeddah, Saudi Arabia, 61pp.
Zahran, M.A. (1984). A preliminary planning study for technology transfer of coastal vegetation of Egypt. *Supreme Council of Universities, FRCU, Final Report*, Cairo, 28pp.
Zahran, M.A. (1986). Forage potentialities *of Kochia indica* and *K. scoparia* in arid land, with particular reference to Saudi Arabia. *Arab Gulf J. Sci. Res.*, 4(1), 53–68.
Zahran, M.A. (1993a). Dry formations of the Asian Red Sea coast. In *Dry Coastal Ecosystems*, Part B (ed. E. van der Maarel). Elsevier, Amsterdam, pp. 17–30.
Zahran, M.A. (1993b). *Juncus* and *Kochia*: Fiber and Fodder producing halophytes under salinity and aridity stress. In *Handbook of Plant and Crop Stress* (ed. M. Pessarakli). Marcel Dekker Inc., New York, U.S.A., pp. 505–528.
Zahran, M.A. (2004). *The Natural Vegetation: A Renewable Resource for the Sustainable Development of the Deserts of the Arab World: (in Arabic). Zaied International Prize for Environment, Dubai, U.A.E*, 496pp.
Zahran, M.A. (2007a). *Mangrove Ecosystem of the Coastal Belts of the Red Sea and Arabian Peninsula (in Arabic).* Zaied International Prize for Environment, Dubai, U.A.E., 330pp.
Zahran, M.A. (ed.). (2007b). *Millennium Ecosystem Assessment, Sinai Sub Region Assessment, Assessment of the Biodiversity of Gebel Maghara Area, Sinai Peninsula, Egypt* (I Plant life), Suez Canal University, Ismailia, Egypt, 50pp.
Zahran, M.A. (2008). River Nile Hydrophytes in Egypt. In *The Nile* (ed. H. Dumont). 2nd edn, Springer Publishers, Netherland (in Press).
Zahran, M.A. and Girgis, W.A. (1970). On the ecology of Wadi El-Natrun. *Bull. Inst. Desert Egypte*, 20(1), 229–267.
Zahran, M.A., Kamal El-Din, H. and Boulos, S.T. (1972). Potentialities of the fibrous plants of the Egyptian flora in national economy. I. *Juncus rigidus* C.A. Mey and paper industry. *Bull. Inst. Desert Egypte*, 22(1), 193–203.
Zahran, M.A. and Boulos, S.T. (1973). Potentialities of the fiber plants of the Egyptian flora in national economy. II. *Thymelaea hirsuta* (L.). Endl. *Bull. Fac. Sci. Mansoura Univ., Egypt*, 1, 77–87.
Zahran, M.A., El-Demerdash, M.A. and Sharaf, A.A. (1987). *Mediterranean Coastal Vegetation*, North Sinai, Egypt. (Unpublished).
Zahran, M.A. and Abdel Wahid, A.A. (1982). Halophytes and human welfare. Part III. In *Contributions to the Ecology of Halophytes. Tasks for Vegetation Science* (eds. D.N. Sen and K.S. Rajpurohit), Vol. 2. Dr. W. Junk, The Hague, pp. 235–257.
Zahran, M-A., Younes, H.A. and Hajrah, H.H. (1983). On the ecology of mangal vegetation of the Saudi Arabian Red Sea Coast. *J. Univ. Kuwait (Sci.)*, 10(1), 87–99.

Zahran, M.A., El-Demerdash, M.A. and Mashali, I.A. (1985a). On the ecology of the deltaic coast of the Mediterranean Sea, Egypt. I. General Survey. *Proc. Egypt. Bot. Soc. Ismailia Conf.*, 4, 1392–1407.

Zahran, M.A., Younes, H.A. and El-Tawil, B.A. (1985b). Ecology of four community types, Red Sea coastal desert, Saudi Arabia. *J. Coastal Res.*, 1(3), 279–288.

Zahran, M.A., El-Demerdash, M.A., Abu Ziada, M.E. and Serag, M.S. (1988). On the ecology of the deltaic Mediterranean coastal land, Egypt. II. Sand formations of Damietta-Port Said Coast. *Bull. Fac. Sci. Mansoura Univ.*, 15(2), 581–606.

Zahran, M.A., Abu Ziada, M.E., El-Demerdash, M.A. and Khedr, A.A. (1989). A note on the vegetation on islands in Lake Manzala, Egypt. *Vegetatio*, 85, 83–88.

Zahran, M.A., El-Demerdash, M.A. and Mashali, I.A. (1990). Vegetation types of the deltaic Mediterranean coast of Egypt and their environment. *J. Veg. Sci.*, 1, 305–310.

Zahran, M.A. and Younes, H.A. (1990). Hema system: Traditional conservation of plant life in Saudi Arabia. *Bull. Fac. Sci. King Abdul Aziz Univ., Jeddah, Saudi Arabia*, 2, 19–41.

Zahran, M.A. and Mashaly, I.A. (1991). Ecological notes on the flora of the Red Sea coastal land of Egypt. *Bull. Fac. Sci. Mansoura Univ.*, 18(2), 251–292.

Zahran, M.A., Muhammed, B.K. and El-Dingawi, A.A. (1992). Establishment of *Kochia* forage halophytes in the salt affected lands of the Arab Countries. *J. Env. Sci., Mansoura Univ. Egypt*, 4, 93–119.

Zahran, M.A., Ayyad, S.M. and El-Khatib, A.A. (1995). Ecopalynological studies in the extreme arid part of Egypt. *Proceedings of the International Symposium on African Polynolgy*, Tervuren (Belgium). Publication OCCAS, CIFEG, Orlans, pp. 57–72.

Zahran, M.A., Mahmoud, B,K. and Mashaly, I.A. (1999). Introduction of non-conventional fodders under drought and salinity stresses of arid lands. *Proceedings Workshop on Livestock and Drought: Policies of Cooping with Changes*. Desert Research Center (DRC), Cairo, pp. 75– 79.

Zahran, M.A. and Willis, A.J. (2003). *Plant Life in the River Nile in Egypt*, Mars Publishers, Riyadh, Saudi Arabia, 531pp.

Zohary, M. (1935). Die phytogeographische Gliederung der Flora der Halbinsel Sinai. *Beih. Bot. Centralbl.*, 52, II, 549–621.

Zohary, M. (1944). Vegetation transects through the desert of Sinai. *Palestine J. Bot.*, Ill(2), 57– 78.

Zohary, M. (1962). *Plant Life of Palestine, Israel and Jordan*. The Ronald Press, New York, 262pp.

Zohary, M. (1966). *Flora Palaestina*. Part I, The Israel Academy of Sciences and Humanities, Jerusalem, 364pp.

Zohary, M. (1972). *Flora Palaestina*. Part II, The Israel Academy of Sciences and Humanities, Jerusalem, 489pp.

Zohary, M. (1978). *Flora Palaestina*. Part III, The Israel Academy of Sciences and Humanities, Jerusalem, 481pp.

List of Species

For ease of reference, species are listed by names widely used in the literature, and which have usually been followed in this book, although some of these names are outdated. The currently accepted (valid). names are generally also given, as well as some other common synonyms. Where a change of name involves a different genus, the names of both genera are listed. (Danin 1973, Danin & Hedge, 1973, Tackholm, 1974, Boulos, 1995, 1999, 2000, 2002, 2005).

Abutilon fruticosum
A. pannosum
Acacia albida (= Faidherbia albida).
A. arabica (= A. Nilotica suhsp tomentosa).
A. ehrenbergiana (= A. flava).
A. etbaica
A gerrardii and ssp. *negevensis*
A. laeta
A. mellifera
A. nilotica
A nubica (= A. oerfota var. oerfota).
A. raddiana (= A. raddiana ssp.*raddiana).*
A. Saligna
A. seyal
A tortilis (= A. raddiana ssp. *tortilis).*
Achillea fragrantissima
A santolina
Actiniopteris australis (= A. radiata).
Adiantum capillus-veneris
Adonis cupaniana
A. dentata
Aegialophila pumilo (= Centaurea pumila).
Aegilops kotschyi
Aeluropus lagopoides (= A. brevifolius, = A. massauensis, = A repens).
Aerva javanica (= A. persica). and v. *bovei* and v. *forsskalii)*
Aetheorhiza bulbosa (= Crepis bulbosa).
Agropyron junceiforme, see Elymus farctus
Agrostis semiverticillata (= Polypogon semi-verticillatus, = P. viridis).
Aizoon canariense
A. hispanicum
Ajuga iva
Alhagi graecorum (= A. mannifera, = A. maurorum).
Alisma gramineum
A. plantago-aquatica
Alkanna orientalis
A. tinctoria (=A. lehmani).
Allium artemisietorum
A. aschersonianum
A. barthianum
A. desertorum
A. erdelii
A. roseum
Alternanthera repens (=A. sessilis).
A. sessilis (= A. achyranthoides).
Amaranthus albus
A. blitoides
A. graecizans (= A. angustifolius).
A. hybridus (= A. chlorostachys, = A. paniculatus).
A. lividus (= A. ascendens).
Amberboa ltppii
Ambrosia maritima
Ammi majus
Ammophila arenaria
Anabasis orticulata
A. setifera
A. syriaca
Anacyclus alexandrinus

Anagallis arvensis
Anastatica hierochuntica
Anchusa aegyptiaca
A. hispida (= Gastrocotyle hispida).
A. milleri
Andrachne aspersa
A. racemosa
A. telephioides
Andropogon annulatus (= Dichanthium annulatum).
Anemone coronaria
Anethum graveolens
Anthemis cotula
A melampodina
A. microsperma
A. pseudocotula,
Antirrhinum orontium (= Misopates orontium).
Apium leptophyllum
Arenaria deflexa and v. glabrata Argemone mexicana
Argyrolobium saharae
A. uniflorum
Arisarum vulgare and v. veslingii
Aristida adscensionis (= A. arabica, = A. coerulescens v. arabica).
A. ciliata, see Stipagrostis ciliata
A. funiculata
A. mutabilis (= A. meccano, = A. mutabilis v. aequilonga).
A. plumosa, see Stipagrostis plumosa
Arnebia decumbens
A. hispidissima
A. linearifolia
Artemisia herba-alba (= A. inculta).
A. judaica
A. monosperma
Arthrocnemum macrostachyum (= A. glaucum).
Arundo donax
Asparagus aphyllus
A. stipularis and v. *tenuispinus*
Asperugo procumbens
Asphodelus microcarpus
A. tenuifolius (= A. fistulosus v. *tenuifolius).*
Aster squamatus
Asteriscus graveolens (= Nauplius graveolens).
A. pygmaeus (= A. hierochunticus).
Asthenatherum forsskaolii (= Centropodia forsskalii).
Astoma seselifolium
Astragalus alexandrinus (= A. caprinus).
A. annularis
A. asterias
A. boeticus
A. bombycinus
A. corrugatus
A. eremophilus
A. fresenii
A. gyzensis (= A. hauarensis).
A. hamosus
A. mareoticus
A. sieberi
A. sinaicus
A. spinosus (= A. forsskaolii).
A. tomentosus (= A. fruticosus).
A. tribuloides
A. trigonus and v. leucacanthus
A. vogelii
Atractylis flava (= A carduus).
A. prolifera
Atraphaxis spinosa v. *sinaica*
Atriplex dimorphostegia
A. farinosa
A. glauca
A. halimus
A. leucoclada (= A. inamoena).
A. portulacoides, see Halimione portulacoides
A stylosa (= A. glauca).
Avena barbata ssp. *barbata (= A. alba).*
A. fatua
A. sterilis and ssp. *ludoviciana*
Avicennia marina
Bacopa monnieri
Balanites aegyptiaca
Ballota undulata
Balsamodendron opobalsamum (= Commiphora gileadensis, = C. opobalsamum).
Bassia indica, see Kochia indica
B. muricata
Berula erecta
Beta vulgaris and ssp. *maritima*
Bidens bipinnata
B. schimperi
Blepharis edulis (= B. ciliaris). Boerhavia diandra (= B. Repens subs. *diandra).*
B. diffusa
Boissiera squarrosa
Borassus sp.
Brachiaria leersioides
B. reptans (= Urochloa reptans).
Brachypodium distachyum (= Trachynia distachya).
Brassica arabica (= Erucastrum araloicum).
B. nigra
B. tournefortii
Bromus fasciculatus

B. rubens
B. tectorum
Bupleurum nodiflorum
B. semicompositum
B.subovatum
Cadaba farinosa
C. rotundifolia
Cajanus cajan
Cakile maritima
Calendula arvensis (= C. aegyptiaca, = C. micrantha).
Calligonum comosum (= C. polygonoides subs. *comosum).*
Callipeltis cucullaria (= C. aperta). Calotropis procera
Campanula dulcis
Canna indica
Capparis decidua
C. orientalis (C. rupestris, C. spinosa var. *inermis).*
C sinaica (= C. cartilaginea).
C. spinosa v. *aegyptia (= C. aegyptia).*
Caralluma retrospiciens
C. sinaica
Carduncellus eriocephalus
Carduus arabicus (C. australis).
C. getulus
C. pycnocephalus
Carex distans
C. divisa
C. extensa
Carrichtera annua
Carthamus glaucus
C. lanatus
C. mareoticus (Carduncellus mareoticus).
C. tenuis
C. tinctorius
Cassia italica (= Senna italica).
C. senna (= C. acutifolia, = Senna alexandrina).
Casuarina equisetifolia
C. stricta
Caylusea hexagyna
Celtis integrifolia
Cenchrus pennisetiformis
Centaurea aegyptiaca
C. alexandrina
C. eryngioides
C. glomerata
C. pallescens
C. pumila (= Aegialophila pumila).
C. scoparia (= Phaeopappus scoparius).
C. sinaica
Centaurium pulchellum
C. spicatum
Centropodia forsskaolii, see Asthenatherum forsskaolii
Ceratonia siliqua
Ceratophyllum demersum
C. muricatum
C. submersum
Ceruana pratensis
Chara spp.
Cheilanthes villea (= Notholaena villea).
Chenolea arabica (= Bassia arabica).
Chenopodium album
C. ambrosioides
C. murale and v. *microphyllum*
Chrozophora obliqua (= C. oblongifolia, = C. verbascifolia).
C. plicata
Chrysanthemum coronarium
Cichorium pumilum (C. endivia subsp. *pumilum).*
Cistanche phelypaea (= C. tinctoria).
Citrullus colocynthis (= Colocynthis vulgaris).
A. lanatus (= C. vulgaris = Colocynthis citrullus).
Cladium mariscus
Cleome africana (= C. amblyocarpa).
C. arabica
C. brachycarpa
C. chrysantha
C. droserifolia
Cocculus pendulus
Colocynthis vulgaris, see Citrullus colocynthis
Colutea istria (= C. haleppica).
Commelina benghalensis
C. forsskaolii
C. latifolia
Commiphora opobalsamum (= C. gileadensis, = Balsamodendron opobalsamum).
Convolvulus althaeoides
C. austro-aegyptiacus
C. cancerianus
C. hystrix
C. lanatus (= C. elarishensis).
C. oleifolius
C. pilosellifolius
C. prostratus
Conyza aurita (= Laggera aurita).
C. bonariensis (= C. Linifolia)
C. dioscoridis (= Pluchea dioscoridis).
Corchorus depressus
C. olitorius
C. tricfens
Cordia gharaf (= C. sinensis).
Coriandrum sativum

Coris monspeliensis
Cornulaca monacantha
Coronopus didymus
C. niloticus
Cotoneaster orbicularis
Cotula cinerea
Crataegus sinaica
Cressa cretica
Crotalaria aegyptiaca
C. microphylla
C. thebaica
Crucianella herbacea (= C. aegyptiaca).
C. maritima
Crypsis aculeata
C. alopecuroides
C. schoenoides
Cucumis prophetarum
Cupressus sempervirens
Cuscuta brevistyla
C.pedicellata
Cutandia dichotoma
C.memphitica
Cymbopogon schoenanthus
Cymodocea ciliate (= Thalassodendron ciliatum).
C. major (= C. nodosa).
C. rotundata
C. serrulata
Cynanchum acutum
Cynara sibthorpiana (= C. cornigera).
Cynodon dactylon
Cyperus alopecuroides (= C. dives).
C. articulatus
C. capitatus
C. conglomerates (= C. jeminicus).
C. difformis
C. esculentus
C. laevigatus
C. longus
C. michelianus ssp. *pygmaeus (= C. pygmaeus).*
C. mundtii(= Pycerus mundtii).
C. papyrus
C. polystachyos
C. rotundus
Dactylis glomerata and ssp. *hispanica (= D. hispanica).*
Dactyloctenium aegyptium
Dalbergia sisso
Damasonium alisma
Daucus bicolor
D. *syrticus*
Delonix elata
Desmostachya bipinnata
Diceratella sahariana
Dichanthium annulatum (= Andropogon annulatus).
Digitaria nodosa
D. sanguinalis
Dipcadi erythraeum
Diplachne fusca (= D. malabarica, = Leptochloa fusca).
Diplotaxis acris
D. harra
D. muralis (= D. simplex).
Dipterygium glaucum
Dodonaea viscosa
Dracaena ombet
Echinochloa colona
E. crus-galli
E. stagnina
Echinops galalensis
E. glaberrimus
E. hussonii
E. spinosissimus
Echiochilon fruticosum
Echium angustifolium ssp. *sericeum (= E. sericeum).*
E. horridum
E. rauwolfii
E. rubrum (= E. setosum).
Eclipta alba (= E. erecta).
Eichhornia azurea
E. crassipes
Eleocharis caribaea (= E. geniculata).
E. palustris
Elodea canadensis
Elymus farctus (= Agropyron junceiforme).
Emex spinosa
Enarthrocarpus strangulatus
Ephedra alata
E. foeminea (= E. campylopoda).
E. foliata (= E. ciliata).
Equisetum ramosissimum
Eragrostis aegyptiaca
E. ciliaris
E. pilosa
Eremobium aegyptiacum
E. diffusum (= E.aegyptiacum var. *lineare).*
Eremopoa altaica (= E. persica).
Eremopogon foveolatus
Erodium cicutarium
E. crassifolium (= E. hirtum).
E. glaucophyllum
E. gruinum
E. laciniatum (=E. Laciniatum subsp. *laciniatum).*
E. malacoides

E. oxyrrhynchum (= *E. bryoniaefolium*).
E. pulverulentum (= *E. Laciniatum subsp. pulverulentum*).
Eruca sativa
Erucaria microcarpa
E. pinnata (= *E. uncata*).
Eryngium creticum
E. glomeratum
Ethulia conyzoides
Eucalyptus citriodora
E. rostrata
Euclea schimperi
Euphorbia candelabra
E. cuneato
E. erinacea
E. forsskaolii (= *E. aegyptiaca*).
E. granulata
E. hirta
E. mauritanica
E.nubica (= *E. schimperi*).
E. paralias
E. parvula
E. peplus
E. prostrata
E. retusa (= *E. kahirensis*).
E. scordifolia
E. terracina
Fagonia arabica
F. boulosii
F. bruguieri
F. cretica
F. glutinosa
F. indica (= *F. parviflora*).
F. kassasii
F. latifolia
F. mollis
F. schimperi
F. sinaica (= *F. kahirina*).
F. thebaica and v. *violacea*
F. tristis v. *boveana*
Farsetia aegyptia and v. *ovalis* (= *F. ovalis*).
F. longisiliqua
F. stylosa (= *F. ramosissima*).
Ficus carica and v. *rupestris*
F. palmata (= *F. pseudosycomorus*).
F. salicifolia (= *F. cordata*).
F. sycomorus
Filago desertorum (= *F. spathulata*).
Fimbristylis bis-umbellata
Foeniculum vulgare
Forsskaolea tenacissima
Francoeuria crispa (= *Pulicaria crispa*).
Frankenia hirsuta
F. pulverulenta
F. revoluta
Fuirena ciliaris
F. pubescens
Fumaria bracteosa
Gagea fibrosa
G. reticulata
Galium setaceum
G. sinaicum
G. spurium v. *tenerum*
G. tricorne (= *G. tricornutum*).
Gastrocotyle hispida (= *Anchusa hispida*).
Glaucium corniculatum
Glinus lotoides
Globularia arabica
Glossonema boveanum
Gnaphalium luteo-album (= *Pseudognaphalium luteo-album*).
G. pulvinatum (= *Homognaphalium pulvinatum*).
Gomphocarpus sinaicus (= *Asclepias sinaica*).
Gossypium barbadense
G. arboreum
Grewia tenax
Gymnarrhena micrantha
Gymnocarpos decander
Gynandropsis gynandra
Gypsophila capillaris
G. viscosa
Halimione portulacoides (= *Atriplex portulacoides*).
Halocnemum strobilaceum
Halodule uninervis
Halogeton alopecuroides (= *Agathophora alepecuroides* var. *papillosa*).
H. poore
Halopeplis perfoliata
Halophila ovalis
H. stipulacea
Halopyrum mucronatum
Hammada elegans (= *H. salicornica*, *Haloxylon salicornicum*).
H. scoparia (= *Haloxylon scoparium*)
Haplophyllum tuberculatum (= *H. longifolium*, = *H. obovatum*).
Helianthemum ciliatum
H. kahiricum
H. lippii
H. stipulatum
H. ventosum
H-. vesicarium
Helichrysum conglobatum
Heliotropium arbainense
H. bacciferum (= *H. ramosissimum*, = *H. undulatum*).

H. digynum (= *H. luteum*).
H. ovalifolium
H. pterocarpum (= *H. kassasi*).
H. strigosum
H. supinum
Hemarthria altissima
Herniaria hemistemon
Hibiscus esculentus
H. micranthus
Hippocrepis areolata (= *H. bicontorta*).
H. constricta
H. cyclocarpa
H. unisiliquosa
Homognaphalium pulvinatum (= *Gnaphalium pulvinatum*).
Hordeum leporinum (= *H. murinum* ssp. *leporinum*).
H. marinum
H. murinum
H. vulgare
Hymenocarpus nummularius (= *H. circinnatus*).
Hyoscyamus boveanus
H. muticus
Hyoseris lucida (= *H. radiata* subsp. *graeca*).
Hyparrhenia hirta
Hypericum sinaicum
Hyphaene thebaica
Ifloga spicata
Imperata cylindrica
Indigofera arenaria (= *I. argentea*).
I. articulata
I. argentea
I. lotononoides
I. oblongifolia
I. sessiliflora
I. spinosa
Inula crithmoides
Iphiona mucronata
Ipomoea stolonifera (=*I. imperati*).
Iris sisyrinchium (= *Gynandriris monophylla*).
Isatis microcarpa
Jasminum floribundum
J. fluminense and v. *blandum*
Jasonia montana, see Varthemia Montana
Juncus acutus
J. bufonius
J. rigidus (= *J. arabicus*).
J. subulatus
Juniperus phoenicea
Jussiaea repens (= *Ludwigia stolonifera*).
Kickxia aegyptiaca, see Linaria aegyptiaca
K. elatine
K. floribunda (= *Linaria floribunda*).
Kochia indica (= *Bassia indica*).
Koeleria phleoides (= *Lophochloa cristata*, = *Rostraria ciliata*).
Kohautia caespitosa
Koniga arabica (= *Lobularia arabica*).
Krascheninnikovia ceratoides
Lactuca orientalis
Lagonychium farctum (= *Prosopis farcta*).
Lamarckia aurea
Lantana viburnoides
Lappula sinaica
L. spinocarpos
Lasiurus scindicus (= *L. hirsutus*).
Lathyrus aphaca
L. cicera
L. hirsutus
L. pseudocicera
Launaea angustifolia
L. capitata (= *L. glomerata*).
L. cassiniana
L. massauensis
L. mucronata
L. nudicaulis
L. resedifolia
L. spinosa
L. tenuiloba
Lavandula pubescens
L. stricta (= *L.coronopifolia*).
Lemna gibba
L. minor
L. perpusilla
Leontice leontopetalum
Leontodon hispidulus
L. bulbosum (= *Crepis bulbosa*).
Lepidium sativum
Leptadenia arborea (= *L. heterophylla*).
L. pyrotechnica
Leptochloa fusca, see Diplachne fusca, = *D. malabarica*
Leucas neufliseana
Limoniastrum monopetalum
Limonium axillare
L. delicatulum (= *L. raddianum*).
L. pruinosum
L. sinuatum
L. thouini (= *L. lobatum*).
L. tubiflorum
Limosella aquatica
Linaria aegyptiaca (= *Kickxia aegyptiaca*).
L. floribunda (= *Kickxia floribunda*).
L. haelava
L. tenuis
Lindenbergia abyssinica
L. sinaica

Lippia nodiflora (= Phyla nodiflora).
Lobularia arabica (= Koniga arabica).
L. libyca
Lolium perenne
L. rigidum
Lophochloa cristata (= Koeleria phleoides, = Rostraria ciliata).
Loranthus acaciae (= Plicosepalus acaciae).
L. curviflorus (= Plicosepalus curviflorus).
Lotononis platycarpa
Lotus arabicus
L. corinculatus (= L. glaber).
L. creticus
L. glinoides
L. halophilus (= L. pusillus, L. villosus).
L. polyphyllos
L. schimperi
Lupinus varius ssp. orientalis (= L. hirsutus, = L. pilosus).
Lycium europaeum
L. shawii (= L. arabicum).
Lycopersicum esculentum
Lygeam spartum
Lygos raetam (= Retama raetam).
Lythrum junceum
Maerua crassifolia
Malva aegyptia
M. parviflora
Marsilea aegyptiaca
M. capensis
M. minuta
Matricaria aurea
Matthiola elliptica (= Diceratella elliptica).
M. longipetala (= M. humilis, = M. livida).
Maytenus senegalensis
Medemia argun
Medicago aschersoniana (= M. laciniata v. aschersoniana).
M. hispida
M. laciniata
M. littoralis
M. minima
M. polymorpha
M. sativa
M. truncatula
Melanocenchris abyssinica
Melhania denhamii
Melia azedarach
Melilotus indicus
M. siculus (= M. messanensis).
Mentha longifolia ssp. *typhoides*
Mesembryanthemum crystallinum
M. forsskaolii
M. nodiflorum
Micromeria nervosa (= M. biflora).
M. serbaliana
M. sinaica
Mimosa pigra
Mimusops schimperi
Minuartia geniculata v. communis (= Rhodalsine geniculata).
Misopates orontium (= Antirrhinum orontium).
Moltkiopsis ciliata
Monsonia nivea
Morettia philaena
Moricandia nitens
M. sinaica
M. suffruticosa
Moringa peregrina
Morus alba
M. nigra
Muscari comosum
M. pulchellum (= M. racemosum).
Myriophyllum spicatum
Najas delilei (= N. armata).
N. graminea
N. minor
N. pectinata
Nepeta septemcrenata
Neurada procumbens
Nicotiana glauca
N. rustica
Nitella spp.
Nitraria retusa
Noaea mucronata
Nonea viviani
Notholaena vellea (= Cheilanthes villea).
Nymphaea caerulea
N. caerulea v. *aschersoniana (= N. aschersoniana).*
N. lotus and *v. aegyptiaca*
Ochradenus baccatus
Ocimum basilicum
O. menthaefolium (= O. forsskaolii).
Olea europaea (= O. chryssophylla).
Oligomeris linifolia
Onobrychis crista-galli
O. ptolemaica
Ononis reclinata
O. serrata
O. vaginalis
Onopordum alexandrinum
O.ambiguum
Ophioglossum polyphyllum
Opuntia ficus-indica
Origanum dayi
O. isthmicum

O. ramonense
O. syriacum and v. *aegyptiacum*
Orlaya maritima (= O. pumila, = Pseudorlaya pumila).
Ornithogalum trichophyllum
Orobanche cernua
O.muteli and v. *sinaica (= O. Ramosa* var. *brevisspicata).*
O. ramosa
Oryza sativa
Oryzopsis miliacea (= Piptatherum miliaceum).
Osteospermum vaillantii
Otostegia fruticosa ssp. *schimperi (= O. fruticosa* ssp. *kaiseri).*
Ottelia alismoides
Oralis anthelmintica
Oxystelma alpini (= O. esculentum).
Pallenis spinosa (= Astericus spinosa).
Pancratium arabicum
P. maritimum
P. sickenbergeri
P. tortuosus
Panicum maximum
P. repens
P. turgidum
Papaver hybridum
P. rhoeas
Paracaryum intermedium
Parapholis incurva
P. marginata
Parietaria alsinifolia
Parkinsonia aculeata
Paronychia arabica (= P. desertorum).
P. argentea
P. sinaica
Parthenium hysterophorus
Paspalidium geminatum
P. obtusifolium
Peganum harmala
Pennisetum divisum (= P. dichotomum, = P. elatum).
Pergularia tomentosa
Periploca angustifolia
P. aphylla
Phaeopappus scoparius (= Centaurea scoparia).
Phagnalon barbeyanum
P. rupestre
P. sinaicum
Phalaris minor
P. paradoxa
Phlomis aurea
P. floccosa
Phoenix dactylifera
P. sylvestris
Phragmites australis (= P. communis).
Phyla nodiflora, see Lippia nodiflora
Physalis angulata
Picris radicata (= P. coronopifolia).
P. sprengeriana
Pimpinella etbaica
Piptatherum miliaceum (= Oryzopsis miliacea).
Pistacia atlantica
P. khinjuk and v. *glaberrima*
P. vera
Pistia stratiotes
Pituranthos tortuosus (= Deverra tortusa).
P. triradiatus (= Deverra triradiata).
Plantago albicans
P. amplexicaulis
P. ciliata
P. coronopus
P. crassifolia
P. crypsoides
P. cylindrica
P. indica (= P. arenaria).
P. lagopus
P. notata
P. ovata
P. pumila (= P. exigua).
P. squarrosa
Plicosepalus acaciae (= Loranthus acaciae).
P. curviflorus (= Loranthus curviflorus).
Pluchea dioscoridis, see Conyza dioscoridis
Polycarpaea repens
P. robbairea, see Robbairea delileana
Polycarpon succulentum (= P. arabicum).
Polygonum bellardii
P. equisetiforme
P. lanigerum (= Persicaria lanigera).
P. salicifolium (= Persicaria salicifolia).
P. senegalense (= Persicaria sanegalensis).
Polypogon maritimus
P. monspeliensis
P. viridis (= Agrostis semiverticillata, = Polypogon semiverticillatus).
Populus euphratica (= P. illicitana).
Portulaca oleracea
Posidonia oceanica
Potamogeton crispus
P. nodosus (= P. natans).
P. panormitanus (= P. pusillus).
P. pectinatus
P. perfoliatus
P. schweinfurthii (= P. lucens).
P. trichoides

Potentilla supina
Primula boveana
Priva cordifolia v. *abyssinica*
Prosopis farcta, see Lagonychium farctum
P. juliflora
P. pallina
Prunus amygdalus (= P. dulcis).
Pseudognaphalium luteo-album (= Gnaphalium luteo-album).
Pseudorlaya pumila (= Orlaya pumila, = O. maritima).
Psoralea plicata (= Cullen plicatum).
Pteranthus brevis (= P. papposus).
P. dichotomus
Pterogaillonia calycoptera
Pulicaria arabica
P. crispa, see Francoeuria crispa
P. desertorum (= P. incisa subsp. *candolleana).*
P. undulata (= P. incisa subsp. *incisa).*
Punica granatum
Pupalia lappacea
Pyrethrum santolinoides (= Tanacetum santolinoides).
Ranunculus asiaticus
R. rionii
R. saniculifolius (= R. peltatus subsp. *fucoides).*
R. sphaerospermus (= R. peltatus subsp. *sphaerospermus).*
R. trichophyllus
Raphanus raphanistrum
R. sativus
Reaumuria hirtella
R. negevensis
R. vermiculata
Reichardia tingitana (= R. orientalis).
Reseda alba
R. arabica
R. decursiva
R. pruinosa
Retama raetam, see Lygos raetam
Rhamnus disperma
R. oleoides v. *libyca (= R. lycioides* ssp. *oleoides, = R. oleoides).*
Rhizophora mucronata
Rhus abyssinica and v. *etbaica*
R. tripartita (= R. oxyacantha).
Ricinus communis
Robbairea delileana (= Polycarpaea robbairea).
Roemeria dodecandra (= R. Hybrida subsp. dodecandra).
R. hybrida
Rorippa integrifolia
Rostraria cristata, see Koeleria phleoides, = Lophochloa cristata
Rubia tenuifolia
Ruellia patula
Rumex dentatus
R. pictus
R. simpliciflorus
R. vesicarius
Ruppia maritima and v. *rostrata* and v. *spiralis*
Saccharum officinarum
S. spontaneum v. *aegyptiacum*
Sageretia brandrethiana (= S. thea).
Salicornia europaea (= S. herbacea).
S. fruticosa (= Sarcocornia fruticosa).
Salix babylonica
S. subserrata (= S. safsaf).
Salsola cyclophylla
S. imbricata (= S. baryosma).
S. kali
S. longifolia
S. rigida (= S. orientails)
S. schweinfurthii
S. tetragona
S. tetrandra
S. villosa (= S. delileana, = S. rigida, = S. vermiculata and v. *villosa).*
S. volkensii
Salvadora persica
Salvia aegyptiaca
S. lanigera
S. verbenaca
Samolus valerandi
Sarcocornia fruticosa, see Salicornia fruticosa
Savignya parviflora
Scabiosa arenaria (= S. rhizantha).
Scandix stellata
Schanginia aegyptiaca (= S. baccata, = S. hortensis, = Suaeda aegyptiaca).
Schismus barbatus
Schoenefeldia gracilis
Schoenus nigricans
Schouwia purpurea (= S. thebaica).
Scirpus fistulosus (= S. articulatus).
S.holoschoenus
S.litoralis and v. subulatus
S. supinus
S. tuberosus (= S. maritimus).
Scolymus hispanicus
Scorpiurus muricatus v. *subvillosus*
Scorzonera alexandrina (= S. undulata).
Scrophularia arguta
S. deserti
S. libanotica

S. xanthoglossa
Seddera latifolia
Sedum viguieri
Seidlitzia rosmarinus
Senecio aegyptius
S. desfontainei (= S. glaucus ssp. *coronopifolius, = S. coronopifolius).*
S. flavus
Senna alexandrina, see Cassia
Senna
S. italica, see Cassia italica
Sesbania sesban
Setaria verticillata
Sevada schimperi
Sida alba
Silene apetala
S. leucophylla
S. linearis
S. longipetala
S. nocturna
S. oliveriana (= S. colorata var. *oliveriana).*
S. rubella
S. schimperiana
S. setacea (= S. vivianii subsp. *vivianii).*
S. succulenta
S. villosa
Silybum marianum
Simmondesia chinensis
Sinapis alba
S. arvensis
Sisymbrium erysimoides
S. irio
Solanum dubium (= S. cougulans).
S. nigrum and v. *suffruticosum*
Solenostemma oleifolium (= S. argel).
Sonchus oleraceus
Sorghum bicolor (= S. durra).
S. halepense
S. virgatum
Spergula fallax
Spergularia diandra
S. media (= S. maritima).
S. salina (= S. marina).
Sphaeranthus suaveolens
Sphaerocoma hookeri and v. *intermedia (= S. aucheri).*
Sphenopus divaricatus
Spirodela polyrrhiza
S. punctata
Sporobolus pungens (= S. virginicus).
S. spicatus
Stachys aegyptiaca
Sternbergia clusiana
Stipa capensis
S. lagascae
S. parviflora
Stipagrostis acutiflora (= Aristida acutifolia) (= S. zittelii).
S. ciliata (= Aristida ciliata).
S. hirtigluma
S. plumosa (= Aristida plumosa).
S. scoparia
S. vulnerans
Striga asiatica
S. hermonthica
Suaeda aegyptiaca, see Schanginia aegyptiaca
S. calcarata
S. fruticosa (= S. vera).
S. maritima
S. monoica
S. palaestina
S. pruinosa
S. salsa (= S. maritima).
S. vera
S. vermiculata
S. volkensii
Tagetes minuta
Tamarix amplexicaulis
T. aphylla (= T. articulata).
T. arborea (= T. nilotica).
T. macrocarpa (= T. passerinoides v. macrocarpa).
T. mannifera (= T. nilotica).
T. nilotica (= T. arabica).
T. passerinoides
T. tetragyna
Tanacetum santolinoides (= Pyrethrum santolinoides).
Taverniera aegyptiaca
Telephium sphaerospermum
Tephrosia apollinea (= T. purpure subsp. *apollina).*
T. purpurea
Tetrapogon villosus
Teucrium leucocladum
T. polium
Thesium humile and v. *maritima*
Thymelaea hirsuta
Thymus capitatus
T. decussatus
Trachynia distachya (see Brachypodium distachyum)
Traganum nudatum
Tragus berteronianus
Trianthema crystallina (= T. salsoloides, = T. triquetra).
Tribulus kaiseri
T. mollis (= T. ochroleucus).

T. pentandrus (= *T. alatus*, = *T. longipetalus*).
T. terrestris (= *T. orientalis*).
Trichodesma africanum and v. *abyssinicum*
T. ehrenbergii
Tricholaena teneriffae
Trifolium alexandrinum
T. formosum (= *T. dasyurum*).
T. resupinatum
T. scabrum
T. stellatum
T. tomentosum
Trigonella hamosa (= *T. glabra*).
T. laciniata
T. maritima
T. stellata
Triticum aestivum (= *T. vulgare*).
Tulipa polychroma
Typha domingensis (= *T. australis*).
T. elephantina
Umbilicus botryoides
U. horizontalis
Urginea maritima and v. *pancratium*
U. undulata
Urochloa reptans (= *Brachiaria reptans*).
Urospermum picroides
Utricularia gibba ssp. *exoleta*
U. inflexa
Vaccaria pyramidata (= *V. hispanica*).
Vahlia dichotoma
Valantia hispida
Vallisneria spiralis
Varthemia candicans
V. iphionoides
V. montana (= *Jasonia montana*).
Verbascum fruticulosum
V. letourneuxii
V. schimperianum
Verbena supina
Verbesina encelioides
Veronica anagallis-aquatica
V. beccabunga
Vicia faba
V. monantha (= *V. calcarata*, = *V. cinerea*).
Vigna sinensis (= *V. unguiculata*).
V. vinifera
Withania obtusifolia
W. somnifera
Wolffia hyalina (= *Pseudowolffia hyalina*).
Xanthium brasilicum (= *X. strumarium*).
X pungens
X. spinosum
Zannichellia palustris v. *genuina* and v. major and v. *pedicellata*
Zea mays
Zilla *spinosa* and ssp. *macroptera* (= v. *microcarpa*).
Ziziphus spina-christi
Zosima absinthifolia
Zostera noltii
Zygophyllum aegyptium
Z. album
Z. coccineum
Z. decumbens
Z. sumosum
Z. simplex

MIX
Papier aus verantwortungsvollen Quellen
Paper from responsible sources
FSC® C105338

If you have any concerns about our products,
you can contact us on
ProductSafety@springernature.com

In case Publisher is established outside the EU,
the EU authorized representative is:
Springer Nature Customer Service Center GmbH
Europaplatz 3, 69115 Heidelberg, Germany

Printed by Libri Plureos GmbH
in Hamburg, Germany